PRINCIPLES OF TRIBOLOGY

PRINCIPLES OF TRIBOLOGY

Wen Shizhu
Tsinghua University, China

Huang Ping
South China University of Technology, China

John Wiley & Sons (Asia) Pte Ltd

TSINGHUA UNIVERSITY PRESS

This edition first published 2012

Published by John Wiley & Sons (Asia) Pte Ltd, 1 Fusionopolis Walk, #07-01 Solaris South Tower, Singapore 138628, under exclusive license by Tsinghua University Press in all media throughout the world excluding Mainland China and excluding Simplified and Traditional Chinese languages.

For details of our global editorial offices, for customer services and for information about how to apply for permission to reuse the copyright material in this book please see our website at www.wiley.com.

Library of Congress Cataloging-in-Publication Data

Wen, Shizhu, 1932-
Principles of tribology / Wen Shizhu, Huang Ping.
p. cm.
Includes bibliographical references and index.
ISBN 978-1-118-06289-0 (cloth)
1. Tribology. 2. Tribology–Research. I. Huang, Ping, 1957- II. Title.
TJ1075.W43 2012
621.8′9–dc23

2011025991

Print ISBN: 978-1-118-06289-0
ePDF ISBN: 978-1-118-06290-6
oBook ISBN: 978-1-118-06291-3
ePub ISBN: 978-1-118-06292-0
Mobi ISBN: 978-1-118-08302-4

Set in 9/11pt Times New Roman by Thomson Digital, Noida, India.
Printed and bound in Singapore by Markono Print Media Pte Ltd.

Contents

About the Authors xv

Preface xvii

Introduction xix

PART I LUBRICATION THEORY 1

1 Properties of Lubricants 3
- 1.1 Lubrication States 3
- 1.2 Density of Lubricant 6
- 1.3 Viscosity of Lubricant 7
  - 1.3.1 Dynamic Viscosity and Kinematic Viscosity 7
  - 1.3.2 Relationship between Viscosity and Temperature 9
  - 1.3.3 Relationship between Viscosity and Pressure 11
  - 1.3.4 Relationships between Viscosity, Temperature and Pressure 12
- 1.4 Non-Newtonian Behaviors 12
  - 1.4.1 Ree-Eyring Constitutive Equation 13
  - 1.4.2 Visco-Plastic Constitutive Equation 13
  - 1.4.3 Circular Constitutive Equation 14
  - 1.4.4 Temperature-Dependent Constitutive Equation 14
  - 1.4.5 Visco-Elastic Constitutive Equation 14
  - 1.4.6 Nonlinear Visco-Elastic Constitutive Equation 15
  - 1.4.7 A Simple Visco-Elastic Constitutive Equation 16
- 1.5 Wettability of Lubricants 17
  - 1.5.1 Wetting and Contact Angle 17
  - 1.5.2 Surface Tension 18
- 1.6 Measurement and Conversion of Viscosity 19
  - 1.6.1 Rotary Viscometer 19
  - 1.6.2 Off-Body Viscometer 20
  - 1.6.3 Capillary Viscometer 20
- References 22

2 Basic Theories of Hydrodynamic Lubrication 23
- 2.1 Reynolds Equation 24
  - 2.1.1 Basic Assumptions 24
  - 2.1.2 Derivation of Reynolds Equation 24

2.2 Hydrodynamic Lubrication 27
2.2.1 Mechanism of Hydrodynamic Lubrication 27
2.2.2 Boundary Conditions and Initial Conditions of Reynolds Equation 28
2.2.3 Calculation of Hydrodynamic Lubrication 29
2.3 Elastic Contact Problems 30
2.3.1 Line Contact 30
2.3.2 Point Contact 33
2.4 Entrance Analysis of EHL 36
2.4.1 Elastic Deformation of Line Contacts 36
2.4.2 Reynolds Equation Considering the Effect of Pressure-Viscocity 37
2.4.3 Discussion 37
2.4.4 Grubin Film Thickness Formula 39
2.5 Grease Lubrication 40
References 42

3 Numerical Methods of Lubrication Calculation 43
3.1 Numerical Methods of Lubrication 44
3.1.1 Finite Difference Method 44
3.1.2 Finite Element Method and Boundary Element Method 50
3.1.3 Numerical Techniques 53
3.2 Numerical Solution of the Energy Equation 57
3.2.1 Conduction and Convection of Heat 57
3.2.2 Energy Equation 59
3.2.3 Numerical Solution of Energy Equation 61
3.3 The Numerical Solution of Elastohydrodynamic Lubrication 62
3.3.1 EHL Numerical Solution of Line Contacts 62
3.3.2 EHL Numerical Solution of Point Contacts 67
3.4 Multi-Grid Method for Solving EHL Problems 70
3.4.1 Basic Principles of Multi-Grid Method 70
3.4.2 Nonlinear Full Approximation Scheme of Multi-Grid Method 72
3.4.3 V and W Iterations 74
3.4.4 Multi-Grid Solution of EHL Problems 74
3.4.5 Multi-Grid Integration Method 76
References 79

4 Lubrication Design of Typical Mechanical Elements 81
4.1 Slider and Thrust Bearing 81
4.1.1 Basic Equations 81
4.1.2 Solutions of Slider Lubrication 82
4.2 Journal Bearing 85
4.2.1 Axis Position and Clearance Shape 85
4.2.2 Infinitely Narrow Bearing 86
4.2.3 Infinitely Wide Bearing 88
4.3 Hydrostatic Bearing 92
4.3.1 Hydrostatic Thrust Plate 93
4.3.2 Hydrostatic Journal Bearings 94
4.3.3 Bearing Stiffness and Throttle 94
4.4 Squeeze Bearing 96
4.4.1 Rectangular Plate Squeeze 97

4.4.2 Disc Squeeze 98
4.4.3 Journal Bearing Squeeze 99
4.5 Dynamic Bearing 100
4.5.1 Reynolds Equation of Dynamic Journal Bearings 101
4.5.2 Simple Dynamic Bearing Calculation 103
4.5.3 General Dynamic Bearings 104
4.6 Gas Lubrication Bearings 107
4.6.1 Basic Equations of Gas Lubrication 107
4.6.2 Types of Gas Lubrication Bearings 108
4.7 Rolling Contact Bearing 111
4.7.1 Equivalent Radius R 111
4.7.2 Average Velocity U 112
4.7.3 Carrying Load Per Width W/b 112
4.8 Gear Lubrication 113
4.8.1 Involute Gear Transmission 113
4.8.2 Arc Gear Transmission EHL 117
4.9 Cam Lubrication 119
References 121

5 Special Fluid Medium Lubrication 123
5.1 Magnetic Hydrodynamic Lubrication 123
5.1.1 Composition and Classification of Magnetic Fluids 123
5.1.2 Properties of Magnetic Fluids 124
5.1.3 Basic Equations of Magnetic Hydrodynamic Lubrication 126
5.1.4 Influence Factors on the Magnetic EHL 129
5.2 Micro-polar Hydrodynamic Lubrication 129
5.2.1 Basic Equations of Micro-polar Fluid Lubrication 130
5.2.2 Influence Factors on Micro-polar Fluid Lubrication 133
5.3 Liquid Crystal Lubrication 135
5.3.1 Types of Liquid Crystal 136
5.3.2 Deformation Analysis of Liquid Crystal Lubrication 138
5.3.3 Friction Mechanism of Liquid Crystal as a Lubricant Additive 142
5.4 Electric Double Layer Effect in Water Lubrication 143
5.4.1 Electric Double Layer Hydrodynamic Lubrication Theory 144
5.4.2 Influence of Electric Double Layer on Lubrication Properties 148
References 151

6 Lubrication Transformation and Nanoscale Thin Film Lubrication 153
6.1 Transformations of Lubrication States 153
6.1.1 Thickness-Roughness Ratio λ 153
6.1.2 Transformation from Hydrodynamic Lubrication to EHL 154
6.1.3 Transformation from EHL to Thin Film Lubrication 155
6.2 Thin Film Lubrication 159
6.2.1 Phenomena of Thin Film Lubrication 159
6.2.2 Time Effect of Thin Film Lubrication 161
6.2.3 Shear Strain Rate Effect on Thin Film Lubrication 163
6.3 Analysis of Thin Film Lubrication 165
6.3.1 Difficulties in Numerical Analysis of Thin Film Lubrication 165
6.3.2 Tichy's Thin Film Lubrication Models 166

6.4 Nano-Gas Film Lubrication 168
6.4.1 Rarefied Gas Effect 168
6.4.2 Bounardy Slip 169
6.4.3 Reynolds Equation Considering the Rarefied Gas Effect 172
6.4.4 Calculation of Magnetic Head/Disk of Ultra Thin Gas Lubrication 173
References 176

7 Boundary Lubrication and Additives 177
7.1 Types of Boundary Lubrication 177
7.1.1 Stribeck Curve 177
7.1.2 Adsorption Films and their Lubrication Mechanisms 178
7.1.3 Chemical Reaction Film and its Lubrication Mechanism 183
7.1.4 Other Boundary Films and their Lubrication Mechanisms 185
7.2 Theory of Boundary Lubrication 186
7.2.1 Boundary Lubrication Model 186
7.2.2 Factors Influencing Performances of Boundary Film 187
7.2.3 Strength of Boundary Film 190
7.3 Lubricant Additives 191
7.3.1 Oily Additives 191
7.3.2 Tackifier 192
7.3.3 Extreme Pressure Additives (EP Additives) 193
7.3.4 Anti-Wear Additives (AW Additives) 193
7.3.5 Other Additives 194
References 195

8 Lubrication Failure and Mixed Lubrication 197
8.1 Roughness and Viscoelastic Material Effects on Lubrication 197
8.1.1 Modifications of micro-EHL 197
8.1.2 Viscoelastic Model 198
8.1.3 Lubricated Wear 199
8.2 Influence of Limit Shear Stress on Lubrication Failure 202
8.2.1 Visco-Plastic Constitutive Equation 203
8.2.2 Slip of Fluid-Solid Interface 203
8.2.3 Influence of Slip on Lubrication Properties 204
8.3 Influences of Temperature on Lubrication Failure 207
8.3.1 Mechanism of Lubrication Failure Caused by Temperature 208
8.3.2 Thermal Fluid Constitutive Equation 208
8.3.3 Analysis of Lubrication Failure 209
8.4 Mixed Lubrication 210
References 215

PART II FRICTION AND WEAR 217

9 Surface Topography and Contact 219
9.1 Parameters of Surface Topography 219
9.1.1 Arithmetic Mean Deviation R_a 219
9.1.2 Root-Mean-Square Deviation (RMS) σ or R_q 219
9.1.3 Maximum Height R_{max} 220
9.1.4 Load-carrying Area Curve 220
9.1.5 Arithmetic Mean Interception Length of Centerline S_{ma} 220

9.2 Statistical Parameters of Surface Topography 222
9.2.1 Height Distribution Function 222
9.2.2 Deviation of Distribution 224
9.2.3 Autocorrelation Function of Surface Profile 225
9.3 Structures and Properties of Surface 226
9.4 Rough Surface Contact 227
9.4.1 Single Peak Contact 228
9.4.2 Ideal Roughness Contact 229
9.4.3 Random Roughness Contact 230
9.4.4 Plasticity Index 232
References 232

10 Solid Friction and Control 233
10.1 Basic Characteristics of Friction 233
10.1.1 Influence of Stationary Contact Time 234
10.1.2 Jerking Motion 234
10.1.3 Pre-Displacement 235
10.2 Macro-friction Theory 236
10.2.1 Mechanical Engagement Theory 236
10.2.2 Molecular Action Theory 237
10.2.3 Adhesive Friction Theory 238
10.2.4 Plowing Effect 241
10.2.5 Deformation Energy Friction Theory 243
10.2.6 Binomial Friction Theory 245
10.3 Micro-friction Theory 246
10.3.1 "Cobblestone" Model 246
10.3.2 Oscillator Models 248
10.3.3 Phonon Friction Model 251
10.4 Sliding Friction 251
10.4.1 Influence of Load 252
10.4.2 Influence of Sliding Velocity 252
10.4.3 Influence of Temperature 253
10.4.4 Influence of Surface Film 253
10.5 Rolling Friction 254
10.5.1 Rolling Friction Mechanism 256
10.5.2 Resistances of Rolling Friction 256
10.6 Special Friction and Friction Control 257
10.6.1 Special Friction 257
10.6.2 Friction Control 258
References 261

11 Characteristics and Mechanisms of Wear 263
11.1 Classification of Wear 263
11.1.1 Wear Categories 264
11.1.2 Wear Process 264
11.1.3 Conversion of Wear 266
11.2 Abrasive Wear 266
11.2.1 Types of Abrasive Wear 267
11.2.2 Influence Factors on Abrasive Wear 267
11.2.3 Mechanism of Abrasive Wear 271

11.3 Adhesive Wear 272
11.3.1 Types of Adhesive Wear 272
11.3.2 Influence Factors on Adhesive Wear 273
11.3.3 Adhesive Wear Mechanism 275
11.3.4 Criteria of Scuffing 277
11.4 Fatigue Wear 280
11.4.1 Types of Fatigue Wear 280
11.4.2 Influence Factors on Fatigue Wear 281
11.4.3 Criteria of Fatigue Strength and Fatigue Life 285
11.5 Corrosive Wear 289
11.5.1 Oxidation Wear 289
11.5.2 Special Corrosive Wear 291
11.5.3 Fretting 291
11.5.4 Cavitation Erosion 292
References 295

12 Macro-wear Theory 297
12.1 Friction Material 298
12.1.1 Friction Material Properties 298
12.1.2 Wear-Resistant Mechanism 299
12.2 Wear Process Curve 300
12.2.1 Types of Wear Process Curves 300
12.2.2 Running-In 301
12.3 Surface Quality and Wear 304
12.3.1 Influence of Geometric Quality 304
12.3.2 Physical Quality 307
12.4 Theory of Adhesion Wear 308
12.5 Theory of Energy Wear 309
12.6 Delamination Wear Theory and Fatigue Wear Theory 311
12.6.1 Delamination Wear Theory 311
12.6.2 Fatigue Wear Theory 312
12.7 Wear Calculation 313
12.7.1 IBM Wear Calculation Method 313
12.7.2 Calculation Method of Combined Wear 314
References 319

13 Anti-Wear Design and Surface Coating 321
13.1 Selection of Lubricant and Additive 321
13.1.1 Lubricant Selection 322
13.1.2 Grease Selection 324
13.1.3 Solid Lubricants 325
13.1.4 Seal and Filter 326
13.2 Matching Principles of Friction Materials 326
13.2.1 Material Mating for Abrasive Wear 327
13.2.2 Material Mating for Adhesive Wear 328
13.2.3 Material Mating for Contact Fatigue Wear 329
13.2.4 Material Mating for Fretting Wear 330
13.2.5 Material Mating for Corrosion Wear 330
13.2.6 Surface Hardening 330

13.3 Surface Coating 331
13.3.1 Commonly Plating Methods 331
13.3.2 Design of Surface Coating 338
13.4 Coating Performance Testing 339
13.4.1 Appearance and Structure 339
13.4.2 Bond Strength Test 340
13.4.3 Hardness Test 344
13.4.4 Wear Test 345
13.4.5 Tests of Other Performances 345
References 346

14 Tribological Experiments 347
14.1 Tribological Experimental Method and Device 347
14.1.1 Experimental Methods 347
14.1.2 Commonly Used Friction and Wear Testing Machines 349
14.1.3 EHL and Thin Film Lubrication Test 349
14.2 Measurement of Wear Capacity 352
14.2.1 Weighing Method 352
14.2.2 Length Measurement Method 352
14.2.3 Profile Method 352
14.2.4 Indentation Method 353
14.2.5 Grooving Method 356
14.2.6 Precipitation Method and Chemical Analysis Method 357
14.2.7 Radioactive Method 357
14.3 Analysis of Friction Surface Morphology 358
14.3.1 Analysis of Surface Topography 358
14.3.2 Atomic Force Microscope (AFM) 358
14.3.3 Surface Structure Analysis 360
14.3.4 Surface Chemical Composition Analysis 362
14.4 Wear State Detection 363
14.4.1 Ferrography Analysis 363
14.4.2 Spectral Analysis 365
14.4.3 Lubricant Composition Analysis 365
14.4.4 Mechanical Vibration and Noise Analysis 365
14.4.5 Lubrication State Analysis 365
14.5 Wear Failure Analysis 365
14.5.1 Site Investigation 365
14.5.2 Lubricant and its Supply System 366
14.5.3 Worn Part Analysis 366
14.5.4 Design and Operation 366
References 367

PART III APPLIED TRIBOLOGY 369

15 Micro-tribology 371
15.1 Micro-friction 371
15.1.1 Macro-friction and Micro-friction 371
15.1.2 Micro-friction and Surface Topography 372
15.1.3 Plowing Effect and Adhesion Effect 375

15.2 Micro-contact and Micro-adhesion 377
 15.2.1 Solid Micro-contact 377
 15.2.2 Solid Adhesion and Surface Force 378
15.3 Micro-Wear 380
 15.3.1 Micro-Wear Experiment 380
 15.3.2 Micro-Wear of Magnetic Head and Disk 382
15.4 Molecular Film and Boundary Lubrication 385
 15.4.1 Static Shear Property of Molecular Layer 386
 15.4.2 Dynamic Shear Property of Monolayer and Stick-Slip Phenomenon 387
 15.4.3 Physical State and Phase Change 389
 15.4.4 Temperature Effect and Friction Mechanism 390
 15.4.5 Rheological Property of Molecular Film 390
 15.4.6 Ordered Molecular Film 393
References 395

16 Metal Forming Tribology 397
16.1 Mechanics Basis of Metal Forming 397
 16.1.1 Yield Criterion 397
 16.1.2 Friction Coefficient and Shear Factor 398
 16.1.3 Influence of Friction on Metal Forming 400
16.2 Forging Tribology 401
 16.2.1 Upsetting Friction 401
 16.2.2 Friction of Open Die Forging 403
 16.2.3 Friction of Closed-Die Forging 403
 16.2.4 Lubrication and Wear 404
16.3 Drawing Tribology 406
 16.3.1 Friction and Temperature 406
 16.3.2 Lubrication 407
 16.3.3 Wear of Drawing Die 409
 16.3.4 Anti-Friction of Ultrasound in Drawing 412
16.4 Rolling Tribology 415
 16.4.1 Friction in Rolling 415
 16.4.2 Lubrication in Rolling 417
 16.4.3 Roller Wear 419
 16.4.4 Emulsion Lubricity in Rolling 421
References 421

17 Bio-Tribology 423
17.1 Mechanics Basis for Soft Biological Tissue 423
 17.1.1 Rheological Property of Soft Tissue 423
 17.1.2 Stress-Strain Curve Analysis 423
 17.1.3 Anisotropy Relationship 425
17.2 Characteristics of Joint Lubricating Fluid 426
 17.2.1 Joint Lubricating Fluid 426
 17.2.2 Lubrication Characteristics of Joint Fluid 427
17.3 Lubrication of Human and Animal Joints 430
 17.3.1 Performance of Human Joint 430
 17.3.2 Joint Lubricating Fluid 431
 17.3.3 Lubrication Mechanism of Joint 432

17.4 Friction and Wear of Artificial Joint 434
17.4.1 Friction and Wear Test 434
17.4.2 Wear of Artificial Joint 435
17.5 Other Bio-Tribological Studies 438
References 438

18 Space Tribology 439
18.1 Features of Space Agency and Space Tribology 439
18.1.1 Working Conditions in Space 439
18.1.2 Features of Space Tribology Problems 441
18.2 Analysis of Performances of Space Tribology 442
18.2.1 Starved Lubrication 442
18.2.2 Parched Lubrication 443
18.2.3 Volatility Analysis 444
18.2.4 Creeping 446
18.3 Space Lubricating Properties 448
18.3.1 EHL Characteristics of Space Lubricant 448
18.3.2 Space Lubrication of Rolling Contact Bearing 449
References 450

Index 453

About the Authors

Wen Shizhu is a member of the Chinese Academy of Sciences and Professor of the Department of Precision Instruments and Technology, Tsinghua University. He is the honorary director of the State Key Laboratory of Tribology. His research interests include: elastohydrodynamic lubrication, thin film lubrication, mechanism of control of friction and wear, nano tribology and micro machine design. He was born in Feng County of Jiangxi Province in 1932, and graduated in 1955 in Tsinghua University. He obtains 19 national or ministerial prizes for his distinguished research achievements, including second prize of National Natural Science Award, third prize of National Technology Invention Award; the award of 2004 Teaching & Research of Tsinghua University; and the award of Science and Technology Achievement of Ho Leung and Ho Lee Foundation in 2002.

Huang Ping is Professor of School of Mechanical and Automotive Engineering, South China University of Technology. He was born in Qiqihar City, Heilongjiang Province in 1957. He graduated from the Department of Engineering Mechanics, Tsinghua University to obtain his PhD degree in 1989 and worked in the State Key Laboratory of Tribology of Tsinghua University for seven years. He now serves as the Director of the Design and Equipment Institute of South China University of Technology. He has published seven books and more than 160 articles. He won second prize of the National Natural Science Award, third prize of the National Invention Award and more than seven other provincial and ministerial scientific and technological progress awards.

Preface

The formation and development of tribology as a practical subject are closely related to the requirements of social production and the progress of science and technology so that its research styles and research area have been continuously evolving.

In the early eighteenth century, Amontons and Coulomb proposed the classic formulas of sliding friction after carefully studying a large number of friction tests and experiments. This was the early research style of tribology based on experiences.

At the end of the nineteenth century, Reynolds revealed the mechanism of the viscous fluid according to bearing lubrication to derive the famous equation of the hydrodynamic lubrication: the Reynolds equation, which laid the theoretical foundation of lubrication. Therefore, it created a new research style based on continuum mechanics.

In the twentieth century, due to production development, tribology research fields were further expanded. During the period, Hardy proposed the boundary lubrication theory, which was based on physical and chemical adsorption films of polar molecules of lubricant on the surface. This promoted studies of the lubricant and additive. Tomlinson explained the cause of solid sliding friction from the viewpoint of energy conversion in molecular motion. Furthermore, Bowdon and Tabor established the adhesion friction theory based on the plow effect. These achievements not only expanded the ranges of tribology, but prompted it to be a discipline involving mechanics, materials science, thermal physics and physical-chemistry so as to create a multidisciplinary research style.

In 1965, the British Ministry of Education and Science published the report "Tribology and Research." This was the first time that tribology was defined as the science on the friction process. Since then, tribology as a separate discipline has been paid wide attention by industry and academia wordwide and the tribology research has entered a new period of development.

With in-depth theoretical and applied research, it is recognized that in order to effectively realize the potential of tribology in the economy, research evolves from the macro to micro scale, from quality to quantity, from the static to dynamic and from single discipline to multi disciplines. At the same time, tribological research has gradually extended from the analysis of tribological phenomena to the analysis and control of them, or even to the control of tribological properties on a target. In addition, tribology research in the past mainly focused on equipment maintenance, but has now changed to innovative design of mechanical products.

Modern science and technology, especially information science, material science and nano technology, play significant roles in pushing the development of tribology. For example, because of the rapid development of computer science and numerical analysis, many complex tribological phenomena have been solved quite accurately with quantitative analysis. Therefore, the numerical methods used in lubrication simulation have pushed lubrication theory to consider a number of practical factors influencing the design of modern machinery. For another example, the electron microscope and micro-analytical instruments are widely used for the analysis of worn surfaces to provide useful tools in studying the wear mechanism.

At the same time, the development of material science has developed many new materials and surface treatment technologies so as to greatly promote research of the wear mechanism. The fields of modern wear have extended from metal material to nonmetallic material, including ceramics, polymers, composites and so on. Surface treatment technologies using physical, chemical and mechanical methods to modify the material properties of the surface are the most rapidly developing area of tribology in recent years.

The development of nanotechnology derives a series of new disciplines, including micro or nano tribology. It occurs because tribological phenomena are closely related to the micro-structural changes and the dynamic behaviors of the surface and interface. Nano tribology provides a new style from the macro to the atomic and molecular scales to reveal the mechanisms of friction, wear and lubrication so as to establish the relationship between the macroscopic properties and micro structures of the material. These are the basic tribology mechanisms. The emergence of nano tribology shows that tribology study has entered a new stage.

Furthermore, tribology is an interdisciplinary subject closely connected with other disciplines to form a new research field, which has distinctive features. Chemical tribology, biological tribology and ecological tribology appearing in the recent years may become hot fields in future tribological research.

This book is based on the Chinese version previously published by Tsinghua University Press, which achieved recognition of its excellence as a scientific work by gaining the National Book Award.

In the book, we try as much as possible to reflect the whole picture of modern tribology and introduce new areas of the tribological research and development tendencies. Obviously, the new areas currently are not well-known yet so we will give a brief exposition for the reader to promote development of these areas. For the classical contents of tribology, we try to clearly state the basis of knowledge.

Because the scope of tribology is wide and the nature of a book is essentially limited, some defects or deficiencies may exist and we therefore welcome criticisms and corrections from readers.

During the writing of the book, we have cited many researches of scholars both domestic and international. We present the most sincere thanks to them as well as the colleagues and graduate students at Tsinghua University for their enthusiastic support, help and contribution to the development of tribology research and this book.

Wen Shizhu
Huang Ping
2011 Lunar New Year

Introduction

This book is a compilation of the current developments from the tribology research of the authors and their co-workers over a long period. It is a systematic presentation of tribology fundamentals and their applications. It also presents the current state and development trend in tribology research.

There are 18 chapters in the book, consisting of three parts including lubrication theory and lubrication design, friction and wear mechanism and control, and applied tribology. Beside the classical tribology contents, it also covers the interdisciplinary areas of tribology. The book mainly focuses on the regularities and characteristics of tribological phenomena in engineering. Furthermore, it presents basic theories, numerical analysis methods and experimental measuring techniques as well as the applications of tribology.

The book is intended to be used as a textbook for senior-level or graduate-level students majoring in mechanical engineering or in related subjects in universities and colleges. It can also serve as a valuable reference for engineers and technicians in machine design and tribology research.

Part I

Lubrication Theory

1

Properties of Lubricants

Many fluids serve as lubricant in industry. Among them, oil and grease are the most commonly used. Air, water and liquid metals are also used as special lubricants. For example, liquid sodium is often used as a lubricant in nuclear reactors. In some situations, solid lubricants, such as graphite, molybdenum disulfide or polytetrafluoroethylene (PTFE) and so on can also be used. In this chapter we will discuss the viscosity and density of lubricants, which are the two important physical properties associated with lubrication.

In lubrication theory, the most important physical property of a lubricant is its viscosity, which is the most important factor to determine the lubrication film thickness. In hydrodynamic lubrication, the lubricant film thickness is proportional to the viscosity, while in elasto-hydrodynamic lubrication it is proportional to the viscosity of 0.7th powers. Although in boundary lubrication the viscosity does not directly influence the film thickness, the oil packages formed between peaks and valleys of roughness will carry part of the load. Therefore lubricant viscosity is closely related to its load carrying capacity.

Furthermore, viscosity is also an important factor influencing the frictional force. A high-viscosity lubricant not only causes a lot of friction loss, but produces a lot of heat that make cooling control difficult. Because temperature rise caused by friction may lead to failure of lubricant film, the surface will be worn increasingly. Therefore, a reasonable viscosity is required for practical lubrication.

The performance of elasto-hydrodynamic lubrication (EHL) also depend on the rheological characteristics of a lubricant. In point or line contacts, an EHL film is very thin, less than one micro-meter, but the pressure is very high, up to one GPa. And, because the contact area is often very small, the shear rate may be higher than $10^7\ s^{-1}$ such that the passing time is very short, less than 10^{-3} s. Therefore, a friction process is always accompanied by high temperature. For such conditions, the properties of a lubricant are quite different from those of a Newtonian fluid. In such cases, therefore, it is necessary to study the rheological properties of lubricants. Experiments show that although the film thickness formula derived from the Newtonian fluid model is usually applied to the elasto-hydrodynamic lubrication, the frictional force and temperature calculated by a Newtonian fluid model will cause a big error. Therefore, in thermo-elasto-hydrodynamic lubrication (TEHL), more realistic non-Newtonian fluid models should be used. These belong to a lubricant rheology study which will not only help us understand the lubrication mechanism more deeply but also has important significance in energy conservation and improvement in the life of mechanical elements.

1.1 Lubrication States

The purpose of lubrication is to form a lubricant film to separate the friction surfaces to carry a load with a low shear stress to reduce friction and wear of materials. A lubricant film can be a liquid, a gas or a solid.

Principles of Tribology, First Edition. Wen Shizhu and Huang Ping.

Table 1.1 Basic features of lubrication states

Lubrication state	Typical film thickness	Formation method of lubricant film	Applications
Hydrodynamic lubrication	1–100 μm	A relative movement between friction surfaces forms a dynamic lubricant film	For surface contacts in high speed situations such as journal bearings
Hydrostatic lubrication	1–100 μm	Through an external pressure fluid form a lubricant film between friction surfaces	For surface contacts in low speed situations such as journal bearings and guides
Elasto-hydrodynamic lubrication	0.1–1 μm	Same as hydrodynamic lubrication	For point or line contacts in high speed situations, such as gears and rolling bearing
Thin film lubrication	10–100 nm	Same as hydrodynamic lubrication	For point or line contacts in low speed and high precision situations, such as precision rolling contact bearing
Boundary lubrication	1–50 nm	Physical or chemical reaction such as adsorption between lubricant molecules and metal surfaces	For low speed situations, such as journal bearings
Dry friction	1–10 nm	Surface oxide film, gas adsorbed film, and so on.	For no lubrication or self-lubricating friction pairs

According to the mechanisms of lubricant film formation, lubrication states can be divided into the following six basic types: (1) hydrodynamic lubrication; (2) hydrostatic lubrication; (3) elasto-hydrodynamic lubrication; (4) thin film lubrication; (5) boundary lubrication; and (6) dry friction. The features of the lubrication states are listed in Table 1.1.

A lubrication state has its typical film thickness. However, we cannot determine the lubrication state simply and accurately based on the thickness alone because the surface roughness also needs to be considered. Figure 1.1 lists the thickness orders of different lubricant films and roughnesses. Only when a lubricant film thickness is high enough is it possible to form a full film that will completely lubricate to avoid the peaks of the two rough surfaces contacting each other. If several lubrication states exist at the same time, this is known as mixed lubrication, as shown in Figure 1.2.

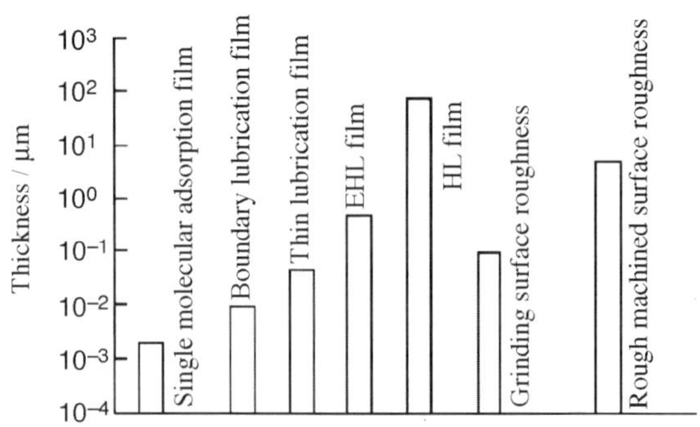

Figure 1.1 Lubricant film thickness and roughness height.

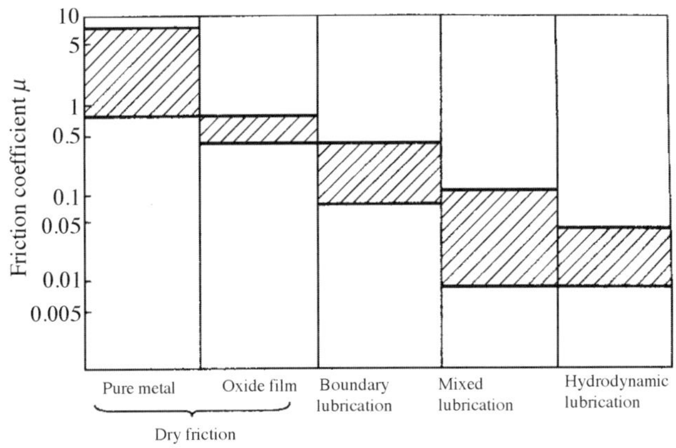

Figure 1.2 Typical friction coefficients of the lubrication states.

It is often inconvenient to determine a lubrication state based on lubricant film thickness because film thickness measurement is difficult. For convenience, the friction coefficient can also be used to determine a lubrication state. Figure 1.2 presents some typical friction coefficients corresponding to the lubrication states.

With varying working conditions, one lubrication state may transform into another. Figure 1.3 gives a typical Stribeck curve of a journal bearing. The curves indicate the transformation of lubrication states corresponding with the working conditions. Here, the dimensionless bearing parameter ($\eta U/p$) reflects the working conditions, where η is the lubricant viscosity, U is the sliding velocity, and p is the average pressure (carrying load per unit area).

It should be noted that methods of studying lubrication states may vary. For hydrodynamic lubrication and hydrostatic lubrication, theories of viscous fluid mechanics and heat transfer are necessarily used to analyze pressure and temperature distributions. As for elasto-hydrodynamic lubrication, elastic deformation of the contact surfaces and the rheological properties of lubricants must be added, while for boundary lubrication the perspectives of physical and chemical knowledge will help us understand the mechanisms of formation and failure of a boundary film. For dry friction, the main task is to avoid wear and tear. Therefore, its study involves material science, elastic and plastic mechanics, heat transfer, physical chemistry and so on.

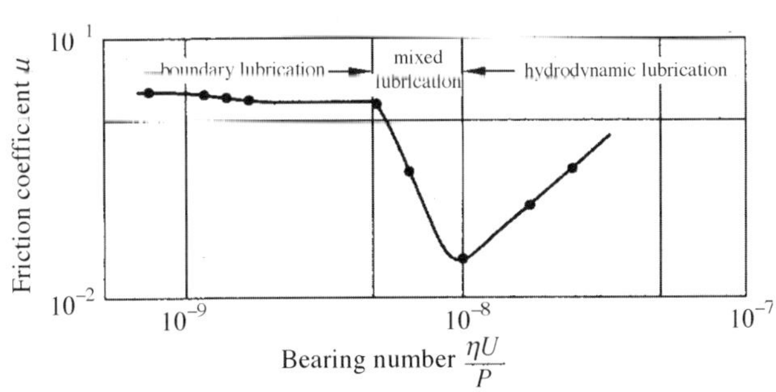

Figure 1.3 Stribeck curve of a journal bearing.

Table 1.2 Standard densities of some basic lubricants

Lubricant	Density g/cm^3	Lubricant	Density g/cm^3
Triguaiacyl phosphate	0.915–0.937	water-soluble polyalkylene glycol	1.03–1.06
Diphenyl phosphate	0.990	nonwater-soluble polyalkylene glycol	0.98–1.00
Hydroxymethyl-phenyl phosphate	1.161	dimethyl silicone oil	0.76–0.97
Hydroxymethyl-phenyl diphenyl phosphate	1.205	ethly-dimethyl silicone oil	0.95
Chlorinated diphenyl	1.226–1.538	phenyl-dimethyl silicone oil	0.99–1.10

1.2 Density of Lubricant

The density is one of the most common physical properties of a lubricant. A liquid lubricant is usually considered to be incompressible and its thermal expansion is ignored so that the density is considered as a constant. Generally, the density of 20 °C is considered the standard. In Table 1.2 the standard densities of some basic lubricants are given.

The density of a lubricant is actually the function of pressure and temperature. Under some conditions, such as in the elasto-hydrodynamic lubrication state, the density of a lubricant should be considered to be variable.

The volume of lubricant is reduced with increase of pressure so that its density increases. The relationship of density and pressure can be expressed as follows:

$$C = \frac{1}{\rho}\frac{d\rho}{dp} = \frac{V}{M}\frac{d(M/V)}{dp} = -\frac{1}{V}\frac{dV}{dp} \tag{1.1}$$

where C is the compression coefficient; V is the volume of lubricant; M is the mass of lubricant.

The following well-known density equation is available:

$$\rho_p = \rho_0[1 + C(p - p_0)] \tag{1.2}$$

where ρ_0 and ρ_p are the densities under the pressure p_0 and p respectively.

The desirable C can be obtained from the following expression:

$$C = (7.25 - \lg \eta) \times 10^{-10} \tag{1.3}$$

where η is the viscosity, mPa·s, and C is a constant, m^2/N.

Conveniently, the following density and pressure relationship is often used in lubrication analysis:

$$\rho_p = \rho_0\left(1 + \frac{0.6p}{1 + 1.7p}\right) \tag{1.4}$$

where p is the pressure, GPa.

The influence of temperature on density is due to thermal expansion, which increases the lubricant volume in order to decrease the density. If the thermal expansion coefficient of a lubricant is α_T, then

$$\rho_T = \rho_0[1 - \alpha_T(T - T_0)] \tag{1.5}$$

where ρ_T is the density under temperature T; ρ_0 is the density under temperature T_0; α_T is the constant, °C^{-1}.

Usually, α_T can be expressed in the following way. If the viscosity of a lubricant is less than 3000 mPa·s (i.e. $\lg\eta \leq 3.5$), then

$$\alpha_T = \left(10 - \frac{9}{5}\lg\eta\right) \times 10^{-4} \tag{1.6}$$

If the viscosity of a lubricant is larger than 3000 mPa·s (i.e. $\lg\eta > 3.5$), then

$$\alpha_T = \left(5 - \frac{3}{8}\lg\eta\right) \times 10^{-4} \tag{1.7}$$

1.3 Viscosity of Lubricant

Viscosity varies significantly with temperature and pressure. The properties of viscosity have a great influence on lubrication. In elasto-hydrodynamic lubrication, both the viscosity and density of a lubricant significantly vary with temperature and pressure.

1.3.1 Dynamic Viscosity and Kinematic Viscosity

Viscosity is the capability of a fluid to resist the shear deformation. When a fluid flows on a solid surface, due to adhesion to the solid surface and the interaction between the molecules of fluid, shear deformation of the fluid exists. Therefore, viscosity is the measurement of the resistance of the internal friction of a fluid.

1.3.1.1 Dynamic Viscosity

Newton first proposed the viscous fluid model. He considered that a fluid flow consists of many very thin layers. The adjacent layers slide relatively, as shown in Figure 1.4, where h is the thickness, U is the velocity of the moving surface; A is the area of the surface, and F is the drawing force. Due to viscous friction within layers of the fluid, movement is transferred from one layer to the other. Because of viscosity, relative sliding between the layers results in shear stress, that is, friction within the fluid. The movement is transferred to the adjacent layer such that the faster layer is decelerated, but the slower layer is accelerated. This forms a velocity difference. If the surfaces A and B are parallel to each other, the distribution of the velocity u is linear, as shown in Figure 1.4.

Newton assumed that the shear stress and shear rate are proportional to each other, which is known as Newton's viscosity law:

$$\tau = \eta\dot{\gamma} \tag{1.8}$$

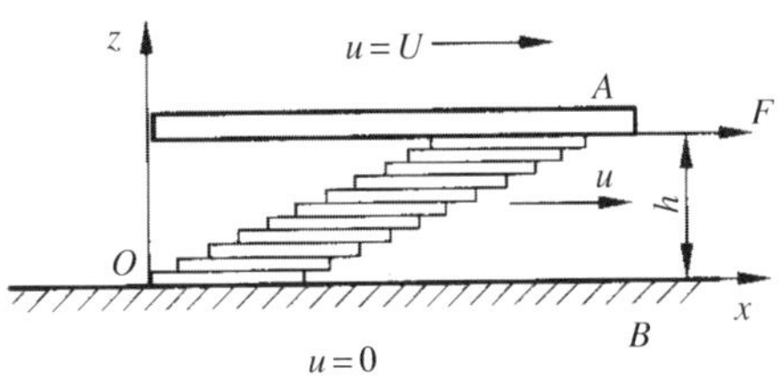

Figure 1.4 Newtonian fluid flow.

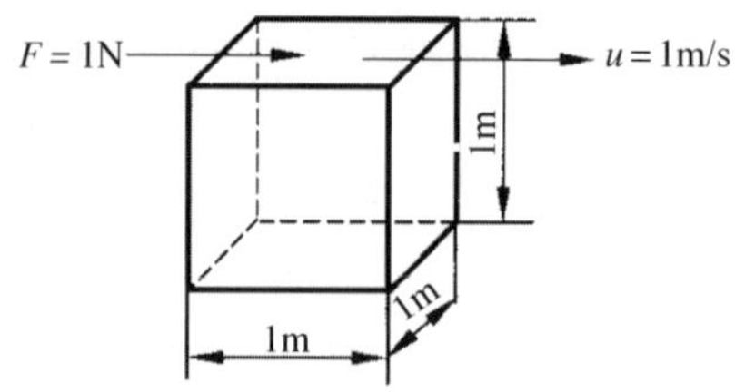

Figure 1.5 Viscosity definition.

where τ is the shear stress, $\tau = F/A$; $\dot{\gamma}$ is the shear rate, that is

$$\dot{\gamma} = \frac{dr}{dt} = \frac{d}{dt}\frac{dx}{dz} = \frac{d}{dz}\frac{dx}{dt} = \frac{du}{dz} \tag{1.9}$$

From the above equation we can see that the shear rate is equal to the gradient of the fluid flow velocity across the film thickness. Therefore, Newton's viscosity law can be written as

$$\tau = \eta \frac{du}{dz} \tag{1.10}$$

where η is the defined as the fluid dynamic viscosity.

Dynamic viscosity is the ratio of shear stress to shear rate. In the international system of units (SI), the unit of dynamic viscosity is N·s/m^2 or Pa·s, as shown in Figure 1.5.

In the CGS system often used in engineering, the dynamic viscosity unit is dyne·s/cm^2 or P (Poise).

$$1\ \mathrm{P} = 1\ \mathrm{dyne \cdot s/cm^2} = 0.1\ \mathrm{N \cdot s/m^2} = 0.1\ \mathrm{Pa \cdot s} \tag{1.11}$$

Because P is too large, 1% P or cP (centipoise) is often used.

If the imperial system is used, the unit of dynamic viscosity is Reyn.

$$1\ \mathrm{Reyn} = 1\ \mathrm{lbf \cdot s/in^2} = 6.89476 \times 10^4\ \mathrm{P} \tag{1.12}$$

Dynamic viscosities of fluids vary in a wide range. The viscosity of air is 0.02 mPa·s, the viscosity of water is 1 mPa·s, while the viscosity of molten asphalt is up to 700 mPa·s. The viscosities of engineering lubricants usually range from 2 to 400 mPa·s. Fluids obeying Newton's viscosity law are called Newtonian fluids; in contrast, those that do not are known as non-Newtonian fluids. Under optimal working conditions, most mineral lubricating oils are considered Newtonian fluids.

1.3.1.2 Kinematic Viscosity

In engineering, kinematic viscosity is often used rather than dynamic viscosity. Kinematic viscosity is equal to the ratio of the dynamic viscosity of a fluid to its density. If the density is ρ and the dynamic viscosity is η, the kinematic viscosity v is expressed as

$$v = \frac{\eta}{\rho} \tag{1.13}$$

The unit of kinematic viscosity in SI is m^2/s, and in the CGS system of units it is Stoke (St), 1 St = 10^2 mm^2/s = 10^{-4} m^3/s. Because St is too large, cSt (centi St) is more commonly used in practice; 1 cSt = 1 mm^2/s.

As the densities of common mineral oils are usually in the range of 0.7–1.2 g/cm^3, choosing the typical mineral oil density equal to 0.85 g/cm^3, the following approximation can be conveniently used in engineering.

$$\eta = 0.85\nu \tag{1.14}$$

1.3.2 Relationship between Viscosity and Temperature

Viscosity of lubricant varies with temperature significantly. Generally, the higher the viscosity, the more sensitive the lubricant is towards changes in temperature.

From a molecular viewpoint, fluid is composed of a large number of randomly moving molecules so that the viscosity of fluid is the result of gravitational forces and momentum of the molecules. The gravitational forces between the molecules significantly vary with the distance between molecules, while the momentum depend on velocity. When temperature rises, both the average molecular motion and average molecular distance of the fluid increase. This causes the momentum of molecules to increase, but the gravitational forces to decrease. Therefore, the viscosity of a liquid drops sharply with the increase of temperature and this significantly affects lubrication.

In order to accurately determine the lubrication performance, thermal analysis should be carried out to find out variation of viscosity. Temperature calculation therefore becomes an important part of lubrication analysis. The influence of temperature on gas viscosity is commonly neglected although viscosity of gas usually increases slightly with increase of temperature.

A lot of research on the relationships between viscosity and temperature has been carried out and as a result, a number of formulas are put forward. Some formulas are summaries of empirical data. To use these formulas, one must carefully consider their usage limitations.

1.3.2.1 Viscosity–Temperature Equations

Most lubricant viscosities drastically decline with increase of temperature. Their relationships are given in the following forms.

$$\text{Reynolds} \qquad \eta = \eta_0 e^{-\beta(T-T_0)} \tag{1.15}$$

$$\text{Andrado Erying} \qquad \eta = \eta_0 e^{\frac{a}{T}} \tag{1.16}$$

$$\text{Slotte} \qquad \eta = \frac{s}{(a+T)^m} \tag{1.17}$$

$$\text{Vogel} \qquad \eta = \eta_0 e^{b/(T+\theta)} \tag{1.18}$$

where η_0 is the viscosity under temperature T_0; η under temperature T; β is the viscosity–temperature coefficient, approximately equal to 0.03 1/°C; $m = 1, 2, \ldots$; θ is the temperature of "infinite viscosity," and for a standard mineral oil, θ is desirably equal to 95 °C; a, s and b are constants.

In the above equations, the Reynolds viscosity–temperature equation is more convenient to be used, but the Vogel viscosity-temperature equation is more accurate.

1.3.2.2 ASTM Viscosity-Temperature Diagram

ASTM (American Society for Testing and Materials) suggests using viscosity index (VI) to describe the viscosity–temperature relationship and giving their corresponding viscosity–temperature diagram.

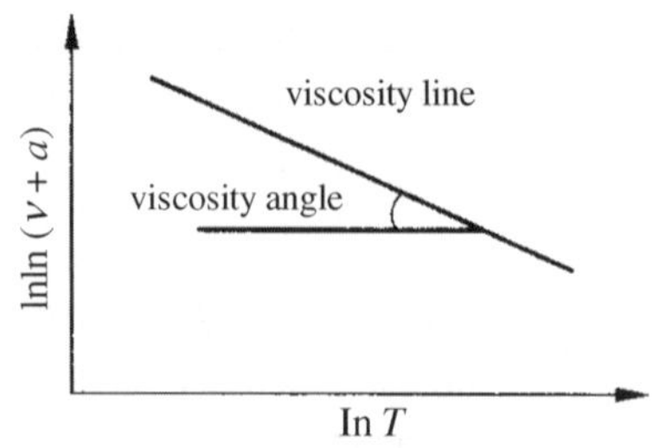

Figure 1.6 ASTM diagram.

The relationship is

$$(v + a) = bd^{1/T^c} \tag{1.19}$$

where v is the kinematic viscosity, mm^2/s; T is the absolute temperature; a, b and d are constants, $a = 0.6–0.75$, $b = 1$, $d = 10$.

For double logarithmic coordinates and single logarithmic abscissa, the formula is a straight line as shown in Figure 1.6.

Then, Equation 1.19 becomes

$$\ln \ln(v + a) = A - B \ln T \tag{1.20}$$

The advantage of Equation 1.20 is that only two viscosities under the corresponding temperatures need to be measured in order to determine the constants A and B. Then a straight line can be plotted to find other viscosities at any temperature.

For a typical mineral oil, an ASTM diagram is very effective. Furthermore, the viscosity angle in the diagram can be used as an index to evaluate the viscosity–temperature feature of a lubricant.

1.3.2.3 Viscosity Index

The viscosity index (*VI*) is used to represent variation of viscosity. Its expression is

$$VI = \frac{L - U}{L - H} \times 100 \tag{1.21}$$

In order to obtain *VI* experimentally, measure the kinematic viscosity $v_{210\,°F}$ of the oil to be tested at 210 °F (≈85 °C) first. Then select two standard oils having the same measured viscosity $v_{210\,°F}$ at 210 °F, but with *VI*s equal to 0 and 100 respectively. Then measure the kinematic viscosities v of the oil and the two standard oils at 100 °F (≈38 °C). If these kinematic viscosities are respectively represented by U, L and H, *VI* of the oil to be tested can be calculated by Equation 1.21.

VIs of some lubricating oils are given in Table 1.3.

Table 1.3 *VI* of some lubricating oils

Lubricant	VI	$v_{100\,°F}$ (mm^2/s)	$v_{210\,°F}$ (mm^2/s)
Mineral oil	100	132	14.5
Multi-grade oil 10 W/30	147	140	17.5
Silicon oil	400	130	53

As the larger the *VI*, the lesser the variation of viscosity with temperature, a lubricating oil with a large *VI* possesses a good viscosity–temperature property.

1.3.3 Relationship between Viscosity and Pressure

With increase of pressure, the distance between molecules of a fluid decreases such that its viscosity increases. Experiments show that when pressure is higher than 0.02 GPa, the viscosity of a mineral oil will obviously increase. Under the pressure of 1 GPa, the viscosity of a mineral oil is several orders larger than under atmospheric pressure. If pressure rises higher, a mineral oil may lose some of its liquid properties and become like a wax. Therefore, the viscosity–pressure relationship is essential to hydrodynamic lubrication under heavy loads, especially for elasto-hydrodynamic lubrication.

The following formulas are often used to describe the relationship of viscosity and pressure.

$$\text{Barus} \qquad \eta = \eta_0 e^{\alpha p} \tag{1.22}$$

$$\text{Roelands} \qquad \eta = \eta_0 \exp\left\{(\ln \eta_0 + 9.67)[-1 + (1 + p/p_0)^z]\right\} \tag{1.23}$$

$$\text{Cameron} \qquad \eta = \eta_0 (1 + cp)^{16} \tag{1.24}$$

where η is the viscosity under pressure p; η_0 is the viscosity under atmospheric pressure; α is the viscosity–pressure coefficient; p_0 is equal to 5.1×10^{-9} Pa; z is usually preferred to be equal to 0.68 for mineral oils; and c is approximately equal to $\alpha/15$.

Although the Barus equation is simple, the viscosity will be too large if pressure is higher than 1 GPa. Therefore, the Roelands equation is more reasonable for such a situation.

The viscosity–pressure coefficient α of mineral oils is around 2.2×10^{-8} m^2/N. Some are given in Tables 1.4 and 1.5.

Table 1.4 Viscosity–pressure coefficients α of mineral oils ($\times 10^{-8}$ m^2/N)

Temperature °C	Naphthenic base			Paraffin base		
	Spindle oil	Light machine oil	Heavy oil	Light machine oil	Heavy oil	Cylinder oil
30	2.1	2.6	2.8	2.2	2.4	3.4
60	1.6	2.0	2.3	1.9	2.1	2.8
90	1.3	1.6	1.8	1.4	1.6	2.2

Table 1.5 Viscosity–pressure coefficient α of base oils at 25 °C ($\times 10^{-8}$ m^2/N)

Lubricant type	α	Lubricant type	α
Paraffin base	1.5–2.4	Alkyl silicon oil	1.4–1.8
Naphthene base	2.5–3.6	Polyether	1.1–1.7
Aromatic base	4–8	Fragrant silicone oil	3–5
Polyolefin	1.5–2.0	Chloroalkane	0.7–5
Diester	1.5–2.5		

1.3.4 Relationships between Viscosity, Temperature and Pressure

When considering the influences of temperature and pressure on viscosity, the following viscosity-–temperature–pressure equations are used.

$$\text{Barus and Reynolds} \qquad \eta = \eta_0 \exp[\alpha p - \beta(T - T_0)] \tag{1.25}$$

$$\text{Roelands} \qquad \eta = \eta_0 \exp\left\{ (\ln \eta_0 + 9.67)\left[(1 + 5.1 \times 10^{-9} p)^{0.68} \times \left(\frac{T - 138}{T_0 - 138} \right)^{-1.1} - 1 \right] \right\} \tag{1.26}$$

Although Equation 1.25 is simpler and easier in calculation, Equation 1.26 is more accurate.

1.4 Non-Newtonian Behaviors

If a lubricant is considered as a Newtonian fluid, the relationship of the shear stress and shear rate is linear, as shown by Curve C in Figure 1.7. Viscosity of a Newtonian fluid changes with temperature and pressure, but without the shear rate.

Non-Newtonian fluids are different from Newtonian fluids, as shown by Curves A, B and D in Figure 1.7. A non-Newtonian fluid may present as plastic, pseudoplastic or expansive. For a pseudoplastic or expansive fluid, an index n is used to approximately describe its nonlinear nature

$$\tau = \eta \dot{\gamma}^n \tag{1.27}$$

where η and n are the constants, for a Newtonian fluid $n = 1$.

In Figure 1.7, Curve A representing plastic is known as the Bingham fluid. It has a yield stress τ_s. When the shear stress τ is less than τ_s, the shear rate is equal to zero. While τ is larger than τ_s, their relationship is

$$\tau = \tau_s + \eta \dot{\gamma} \tag{1.28}$$

Grease is similar to a Bingham fluid. However, the relationship of its shear stress and shear rate is nonlinear. The formula of rheological property for lubricating greases can be expressed approximately as

$$\tau = \tau_s + \eta \dot{\gamma}^n \tag{1.29}$$

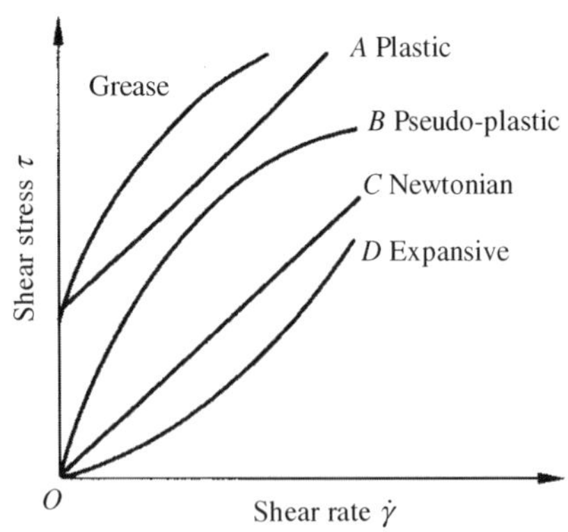

Figure 1.7 τ-$\dot{\gamma}$ curves of different types of fluids.

In order to improve lubrication performance, a modern lubricant is usually a combination of polymer materials, composed of additives and combined with extensive synthetic lubricants. Therefore, it often presents significantly non-Newtonian features. The rheological behaviors of lubricants cannot be ignored in lubrication design.

In lubrication analysis, the commonly used non-Newtonian fluid constitutive equations are as follows.

1.4.1 Ree-Eyring Constitutive Equation

The Ree-Eyring constitutive equation is the most commonly used non-Newtonian formula, as shown in Equation 1.30. Its shear rate slowly varies to infinite with the shear stress.

$$\dot{\gamma} = \frac{\tau_0}{\eta_0}\sinh\left(\frac{\tau}{\tau_0}\right) \tag{1.30}$$

where τ_0 is the characteristics stress; η_0 is the initial viscosity.

The Ree-Eyring model gives a fairly accurate description of the rheological property of some lubricants, especially for simple liquids. The relationship of the shear stress τ and the shear rate $\dot{\gamma}$ is similar to Curve 1 in Figure 1.8. τ_0 and η_0 are the two rheological parameters depending on the molecular structures of a lubricant.

1.4.2 Visco-Plastic Constitutive Equation

Curve 2 in Figure 1.8 is the visco-plastic fluid model. It has a limit shear stress τ_L. The variation of the shear stress with the shear rate is described by two straight lines.

$$\begin{aligned} \dot{\gamma} &= \frac{\tau}{\eta_0} \quad \eta_0|\dot{\gamma}| \leq \tau_L \\ \tau &= \tau_L \quad \tau_L \leq \eta_0|\dot{\gamma}| \end{aligned} \tag{1.31}$$

The oblique line of Curve 2 is Newtonian. When the shear stress reaches the limit τ_L, it does not change any more. Because the constitutive equation consists of two straight lines, its derivative is not continuous at the intersection point.

Experimental results show that the limit shear stress τ_L changes with pressure and temperature. The limit shear stresses of common lubricants are between 4×10^5 and 2×10^7 Pa.

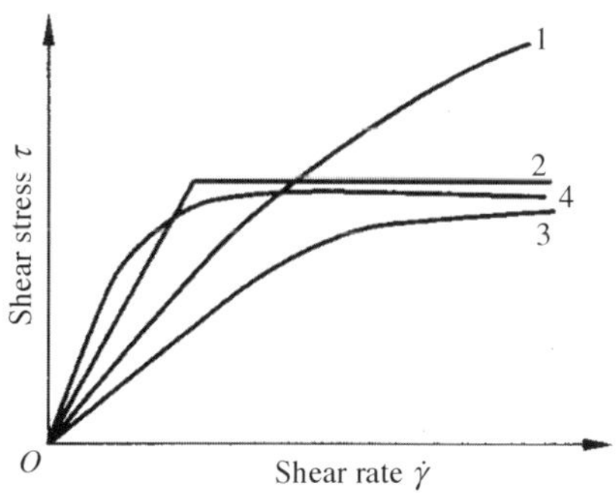

Figure 1.8 Constitutive curves of some lubricants. (1) Ree-Eyring fluid; (2) visco-plastic fluid; (3) circular fluid; (4) temperature-dependent fluid.

1.4.3 Circular Constitutive Equation

The circular constitutive model is asymptotic. It is used for the non-Newtonian fluid effect as shown by Curve 3 in Figure 1.8. It has a continuous derivative and the shear stress varying with the shear rate converges to the limit τ_L. The constitutive equation is as follows.

$$\dot{\gamma} = \frac{\tau_L \tau}{\eta_0 \sqrt{\tau_L^2 - \tau^2}} \tag{1.32}$$

1.4.4 Temperature-Dependent Constitutive Equation

The temperature-dependent constitutive model is shown by Curve 4 in Figure 1.8, considering the influence of temperature on viscosity [1]. The most important feature of the model is that after reaching the maximum, the shear stress begins to decline slightly with increase of the shear rate. The constitutive equation is as follows:

$$\tau = \frac{\eta_0 \dot{\gamma}}{\alpha \dot{\gamma}^2 + 1} \tag{1.33}$$

where $\alpha = 2\beta\eta_0 x/\rho c u_0$; β is the viscosity–temperature constant; η_0 is the initial viscosity; x is the distance away from the inlet; ρ is the density; c is the specific heat capacity of the lubricant; u_0 is the velocity of the moving surface.

1.4.5 Visco-Elastic Constitutive Equation

Experiments show that when a lubricant flows through contact region with dramatically changed stresses, it presents some elasticity, that is, it becomes a visco-elastic fluid. In EHL theory, the most commonly used visco-elastic model is the Maxwell model or the linear visco-elastic model, as shown in Figure 1.9. For a purely elastic material, it obeys Hooke's law.

$$\gamma_e = \frac{\tau}{G} = \frac{de_1}{dz} \tag{1.34}$$

where γ_e is the elastic shear strain; G is the shear elastic modulus.

Derivate du to time t we have $du = d\dot{e}$. Therefore, Equation 1.34 becomes

$$\dot{\gamma}_e = \frac{du_1}{dz} = \frac{d\dot{e}_1}{dz} = \frac{1}{G}\frac{d\tau}{dz} \tag{1.35}$$

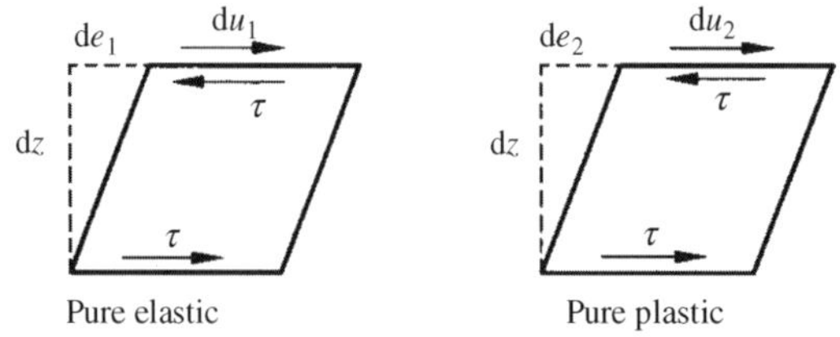

Figure 1.9 Visco-elastic body.

For a pure viscous material, it obeys Newton's viscosity law. Because $du = d\dot{e}$, we have

$$\dot{\gamma}_v = \frac{du_2}{dz} = \frac{d\dot{e}_2}{dz} = \frac{\tau}{\eta} \tag{1.36}$$

Thus, for the linear visco-elastic material, its constitutive equation is as follows:

$$\dot{\gamma} = \dot{\gamma}_v + \dot{\gamma}_e = \frac{\tau}{\eta} + \frac{1}{G}\frac{d\tau}{dt} \tag{1.37}$$

Equation 1.37 shows that the shear rate of a visco-elastic material is related to both time and shear stress. The two parameters, η and G, are used to describe a visco-elastic material. As the Maxwell model is obtained under conditions of little shear strain, it cannot be used to calculate large shear strain problems, such as EHL problems. The coefficient of friction of a linear visco-elastic model will be more reasonable for EHL problems.

1.4.6 Nonlinear Visco-Elastic Constitutive Equation

The friction coefficient obtained from the Maxwell model for an EHL problem is usually too large because of the Newtonian fluid viscosity of Equation 1.37. Therefore, a non-Newtonian constitutive equation is given as

$$\dot{\gamma} = F(\tau) + \frac{1}{G}\frac{d\tau}{dt} \tag{1.38}$$

where $F(\tau)$ is a nonlinear function of τ.

Johnson and Tevaarwerk [2] combined the Maxwell model with Ree-Eyring model to propose the following nonlinear visco-elastic constitutive equation.

$$\dot{\gamma} = \frac{\tau_0}{\eta_0}\sinh\frac{\tau}{\tau_0} + \frac{1}{G}\frac{d\tau}{dt} \tag{1.39}$$

If $\tau \ll \tau_0$, $\sinh \tau/\tau_0 \approx \tau/\tau_0$. Then, $F(\tau) \approx \tau/\eta_0$ such that $F(\tau)$ becomes the Newton's viscosity constitutive equation. Therefore, Equation 1.39 becomes the linear visco-elastic constitutive equation. Johnson and Tevaarwerk summarized that the proposed model is suitable for linear and nonlinear viscous materials, linear and nonlinear elastic materials, as well as for elastic and plastic materials as shown in Figure 1.10.

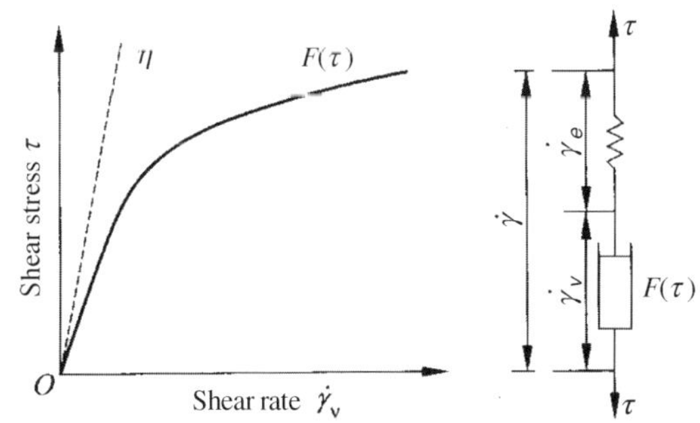

Figure 1.10 Nonlinear visco-elastic model.

1.4.7 A Simple Visco-Elastic Constitutive Equation

Bair and Winer [3] proposed a simple visco-elastic model. The relationship between the shear stress and the shear rate is

$$\dot{\gamma} = \frac{1}{G_\infty}\frac{d\tau}{dt} - \frac{\tau_L}{\eta_0}\ln\left(1 - \frac{\tau}{\tau_L}\right) \tag{1.40}$$

where G_∞ is the infinite shear elastic modulus derived experimentally under a variety of vibration frequencies; τ_L is the limit shear stress; η_0 is the initial viscosity. The three parameters are the functions of pressure p and temperature T, and can be determined by experiments.

In order to obtain the dimensionless form of Equation 1.40, set the dimensionless shear stress $\tau^* = \tau/\tau_L$, the dimensionless shear rate of $\dot{\gamma}^* = \dot{\gamma}\eta/\tau_L$, and we have $\dot{\tau}^* = (\eta_0/G_\infty\tau_L)/(d\tau/dt)$. Then, the dimensionless form of Equation 1.40 is as follows:

$$\dot{\gamma}^* = \dot{\tau}^* - \ln(1 - \tau^*) \tag{1.41}$$

According to Equation 1.41, the calculated friction curves of line contact EHL are fitted with the obtained experimental results.

In the relationship between τ^* and $\dot{\gamma}^*$ in the simple visco-elastic model, there is more than one rheological parameter. Usually following the Newton's viscosity law, the "apparent viscosity" represents the ratio of the shear stress to the shear rate. Apparently, the apparent viscosity of a non-Newtonian fluid varies with shear rate.

For non-Newtonian fluids, there are two important characteristics that have a significant effect on lubrication, namely pseudoplasticity and thixotropy.

1.4.7.1 Pseudoplasticity

Most liquids behave as non-Newtonian (e.g. 10^6–10^8 s^{-1}) at a high shear rate while viscosity decreases. However, two-phase liquids (such as emulsions or grease), high viscosity oils or oils containing polymers may present as non-Newtonian at a low shear rate (e.g. 10^2–10^6 s^{-1}). Such a phenomenon is called shear thinning or pseudoplasticity, as shown in Figure 1.11. A pseudoplastic fluid usually has long chain molecules but irregular arrangements. As the chains will be directionally arranged under shearing, the actions between adjacent layers are weakened so as to decrease its apparent viscosity.

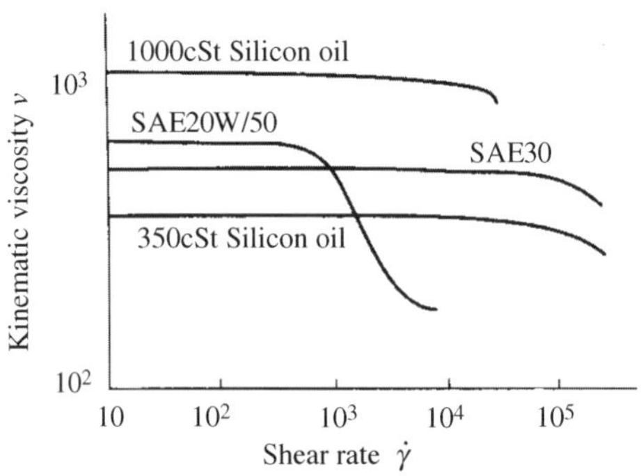

Figure 1.11 Pseudoplasticity.

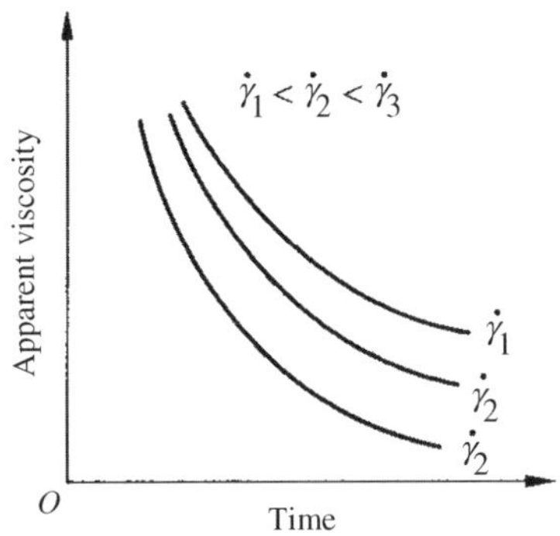

Figure 1.12 Thixotropy.

1.4.7.2 Thixotropy

The phenomenon that the apparent viscosity of a fluid diminishes gradually under shearing is known as thixotropy, as shown in Figure 1.12. Thixotropy is usually reversible. That is, if shearing has stopped, the viscosity recovers back to or close to its original value after sufficient time. For greases or thick emulsions, the effect of thixotropicity is that their structures continue to be disrupted under shearing, and then self-rebuild. When the structure damage develops, the apparent viscosity continues to decrease. When a new balance between destruction and reconstruction is established, the apparent viscosity becomes stable again.

1.5 Wettability of Lubricants

Wettability of a liquid is its capability to spread or gather on a solid surface. Usually, liquid surfaces tend to shrink, that is, when the external forces are very small, a small droplet tends to be spherical, such as a mercury drop on a table or a water drop on a lotus leaf. It is believed that the mechanism of boundary lubrication is connected with the wettability of the lubricant. In addition, the adhesion between two solid surfaces filled with a lubricant is also closely related to the surface tension of the lubricant because the adsorption of lubricant molecules on the surfaces relies on the adhesive energy closely related to wettability.

1.5.1 Wetting and Contact Angle

When a small amount of liquid contacts a solid and completely covers the solid surface, this is called wetting. If a liquid forms a spherical droplet, it is called nonwetting. Usually, partial wetting phenomena exist.

The phenomenon that a liquid surface automatically shrinks can be analyzed from energy. Usually, wetting can be measured by the contact angle of a liquid on a solid surface. As shown in Figure 1.13, the

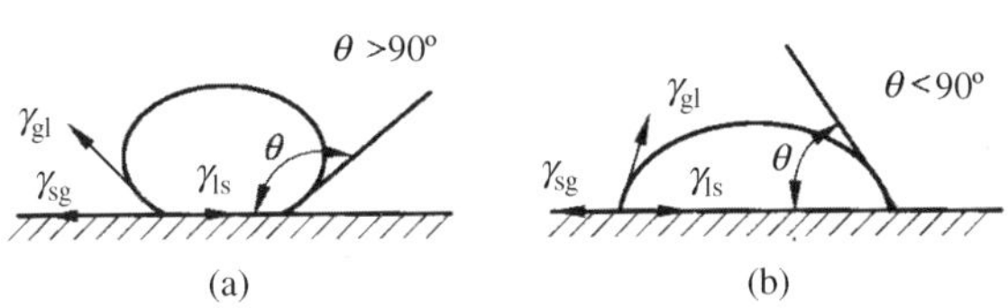

Figure 1.13 Wettability and contact angle.

contact angle θ is defined as the tangent angle between the solid–liquid interface and the liquid–gas interface at the junction point of solid, liquid and gas phases. The contact angle θ is from 0 to 180° for wetting to completely nonwetting. Liquid with a large contact angle θ is lipophobic, while a small contact angle θ is lipophilic, that is, the adhesion energy of a liquid is greater than its cohesive energy. The magnitude of the contact angle is determined by the solid and liquid surface tensions or the surface free energies. The surface tension presents the work done to increase each unit area of the surface. It is one of the basic physical and chemical properties, usually presented in the unit of mN/m.

Figure 1.13 shows the relationship of the contact angle and the surface tensions. If γ_{gl}, γ_{ls} and γ_{sg} are the surface tension of liquid–gas, solid–liquid and solid–gas respectively. Then,

$$\gamma_{sg} = \gamma_{ls} + \gamma_{gl} \cos\theta \tag{1.42}$$

The contact angle θ can be measured by experimental methods, such as the projection method. The gas–liquid surface tension γ_{gl} can be measured by a surface tension instrument. Then, γ_{sg}-γ_{ls} can be obtained by calculating the wetting energy (in general, γ_{sg} and γ_{sl} are difficult to be measured). In addition, the contact angle θ is related to the solid surface roughness, temperature and so on.

1.5.2 Surface Tension

A surface tension is actually the interface energy difference of the interactions of liquid and gas phases. The distances between molecules in liquid are not the same, although the summary force surrounding all directions of each molecule is equal to zero the average attraction force will prevent them (the liquid molecules) from thermally volatilizing. However, the molecules on the liquid surface are quite different, because the force of gas is much smaller than that of the liquid. Furthermore, because the gas density is smaller and the distance between molecules is larger, the summary force acting on the surface molecules points to the inside of the liquid, resulting in an increase in its energy. The energy is called surface free energy. As the distance between molecules on the surface is larger than that of the inner molecules, there is a lateral force acting on the surface molecules, known as the surface tension.

Wetting of a lubricant on solid surfaces and adhesion between two solid surfaces are all related to the surface tension.

If the width of a liquid film is w, and the length increment is Δl, the free energy increment is equal to

$$\Delta G = 2\gamma w \Delta l = \gamma \Delta A \tag{1.43}$$

where ΔA is the surface area increment; γ is the surface free energy, mJ/m^2.

For a liquid, the surface free energy is equivalent to the surface tension and has the same dimension.

There are various methods to measure the liquid surface tension, such as the capillary method, the maximum bubble pressure method, the stop dripping method, the hanging drop method, the drop weight method and so on. The most common method is the ring method. This method involves lifting a ring (usually a platinum ring) away from the surface of a liquid. While the ring which is placed in a horizontal plane parallel to the surface of the liquid (to ensure zero contact angle) is pulled upwards, it brings up some liquid to form a column. The forces applied on the measuring sensor include the weight of ring and the gravity of liquid, P. P increases with increase in the pulling height, but there is a limit. If the pulling height is larger than the limit, the ring and the liquid surface will be separated. The limit is related to the liquid surface tension and the size of the ring if the rise of the liquid column brought up by the external force is due to the liquid surface tension. Therefore, we have

$$P = 4\pi\gamma(R + r) \tag{1.44}$$

where R is the inner radius of the ring; r is the radius of the ring wire; γ is the surface tension to be tested.

Table 1.6 Surface tensions of some liquids (20 °C)

Liquid	Surface tension (mN/m)	Liquid	Surface tension (mN/m)
Water	72	Poly α olefin	28.5
Machine oil	29	Dioctyl sebacate	31
Pentaerythritol	30	Pentaerythritol ester	24
PFPE	20	Methyl silicone oil	21

As the liquid column is not a cylinder, a revised formula for the surface tension measurement is as follows:

$$\gamma = \frac{CP}{4\pi(R+r)} \tag{1.45}$$

where C is the correction factor, which is the function of R/r and R^3/V; V is the volume of the liquid brought up by the ring.

The liquid surface tension generally decreases linearly with increase of temperature. The surface tension is also affected by pressure, but the relationship is more complex. Some additives (such as surface-active agents) will significantly alter the surface tension of liquid. For a ferromagnetic fluid, its surface tension is affected by external magnetic field. Table 1.6 lists the surface tensions of some fluids at 20 °C.

1.6 Measurement and Conversion of Viscosity

Viscometers are used to measure viscosity. There are three types of viscometers according to their working principles. They are rotary, off-body and capillary viscometers.

1.6.1 Rotary Viscometer

A rotary viscometer consists of two parts filled with a liquid to be tested. One part is fixed and the other rotates. By measuring the shearing moment caused by the resistance of a liquid, the dynamic viscosity can be obtained. A rotary viscometer is shown in Figure 1.14a, and a cone-plate rheometer is shown in Figure 1.14b. The former is composed of two concentric cylinders, while the latter is composed of a plane and a conical surface. If the moving part rotates at different speeds, the relationship of the shear stress and the shear rate can be obtained, which is called the rhoelogical property. This is very useful, especially for non-Newtonian fluid.

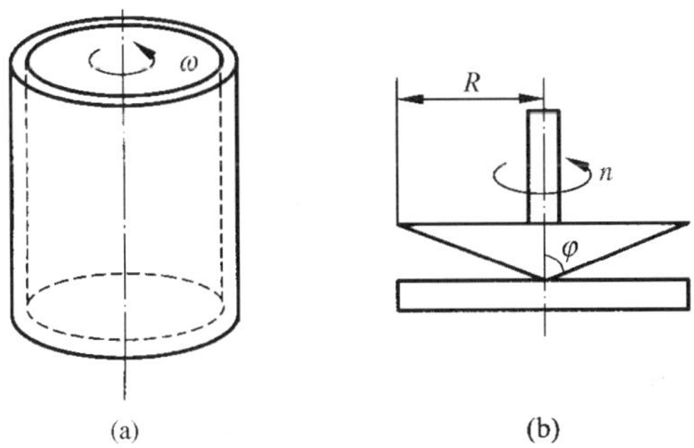

Figure 1.14 Rotary viscometers: (a) rotational viscometer (b) cone-plate rheometer.

1.6.2 Off-Body Viscometer

The most commonly used off-body viscometer is composed of a ball and a test tube filled with the fluid to be tested. In order to determine the viscosity, measure the final velocity of the falling ball. As the clearance between the ball and the tube is very small, the falling ball viscometer can be used to measure the viscosity of gas, or the viscosity of fluid under a high pressure. Another type of off-body viscometer consists of two vertical concentric cylinders. The fluid to be tested is filled between them. The outer cylinder is fixed, while the inner tube falls so that the viscosity can be obtained by measuring the final falling velocity. An off-body viscometer is mainly used to measure high-viscosity fluids.

1.6.3 Capillary Viscometer

The principle of a capillary viscometer is that while a certain volume of liquid flows through a standard capillary, because there exists a pressure difference and liquid weight, the passing time of flowing can be used to determine the viscosity of the liquid. There are two kinds of capillary viscometers, the absolute and the relative. An absolute capillary viscometer measures viscosity based on the viscous fluid mechanics formula, while the relative viscometer must be calibrated by a known viscosity liquid to obtain the viscometer constant first, only then can it be used to measure the viscosity of the liquid to be tested. As the scaling errors do not affect the measurement results of a relative viscometer, it is more reliable.

Figure 1.15 shows a relative capillary viscometer with a known constant c. Measure the time for the liquid surface to drop from A to B, the kinematic viscosity of the liquid being equal to

$$v = ct \tag{1.46}$$

If the density of the measured liquid is ρ, its dynamic viscosity is equal to

$$\eta = \rho v \tag{1.47}$$

Commonly commercial relative capillary viscometers include three kinds: Redwood, Saybolt and Engler types. Although their structures are similar, the volumes of the liquid and the capillary sizes are different. An Engler viscometer is shown in Figure 1.16 and its viscosity calculation formula is as follows.

$$\text{Viscosity (E}^\circ\text{)} = \frac{\text{Time used for 200 liters liquid flowing out}}{\text{Time for the same volume of water flowing out}} \tag{1.48}$$

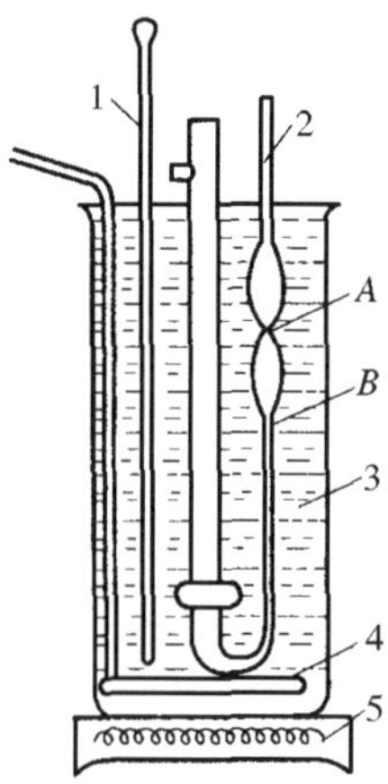

Figure 1.15 Common capillary viscometer: (1) thermometer; (2) capillary viscometer; (3) water or oil bath; (4) blender; (5) heater.

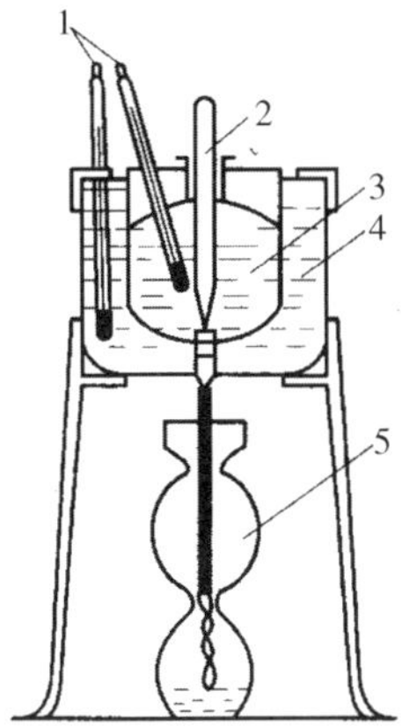

Figure 1.16 Engler capillary viscometer: (1) thermometer; (2) wood plug; (3) lubricating oil sample; (4) heat bath; (5) receiving bottle.

As different viscometers obtain different relative viscosities, some empirical formulas or charts are needed to convert them to kinematic viscosity. The conversion relationships of the three common viscometers are given in Figure 1.17.

It should be pointed out that usually a commercial viscometer under normal conditions measures only the body viscosity of a liquid, which does not fully reflect the rheological properties of a lubricant film.

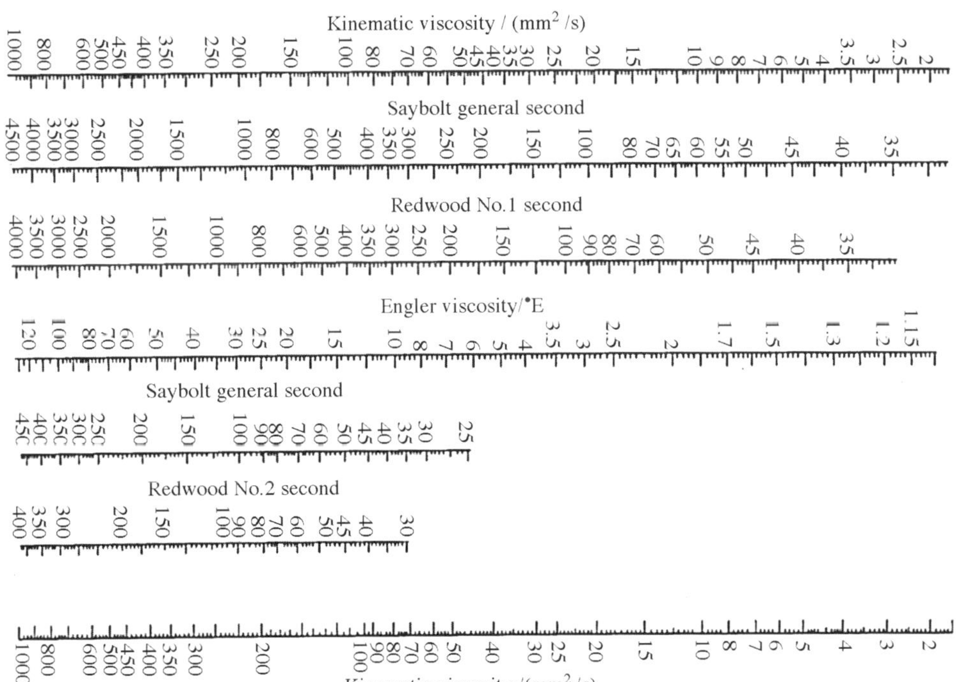

Figure 1.17 Viscosity conversion table.

Therefore, a number of special measuring devices are designed to measure some very high or very low viscosities. For example, a micro-viscometer is used to measure the viscosity under high pressure and high shear rate or it can be used to obtain the visco-elastic property of a liquid. The author used optical interference techniques to measure EHL lubricant film thicknesses, and took the standard liquid as benchmark to calibrate the viscosity–pressure coefficient of some oils [4]. Wang Renyou studied the influence of oil viscosity on high pressure and rheological properties [5].

References

1. Huang, P. and Wen, S. Z. (1996) Non-Newtonian effects of temperature and lubrication failure mechanism analysis. *Lubrication and Sealing*, **2**, 14–16.
2. Johnson, K. L. and Tevaarwerk, J. L. (1977) Shear behavior of elastohydrodynamic oil films. *Proceedings of the Royal Society of London*, **A356**, 215–236.
3. Bair, S. and Winer, W. O. (1979) A rheological model for elastohydrodynamic contacts based on primary laboratory data. *Transactions of the ASME Journal, Series F*, **101** (3), P 258-265.
4. Yu, X. G. and Wen, S. Z. (1984) Determination of optical interference pressure viscosity coefficient of lubricant. *Lubrication and Sealing*, **3**, 10–14.
5. Wang, R. Y. (1997) High Impact Technology and lubricating oil viscosity testing the rheological properties of, PhD thesis, Tsinghua University.

2

Basic Theories of Hydrodynamic Lubrication

Hydrodynamic lubrications include hydrodynamic elastohydrodynamic lubrications. The Reynolds equation is the basic equation of hydrodynamic lubrication.

In 1883, Tower first observed the hydrodynamic phenomenon on a train shaft bearing. In 1886, Reynolds derived the Reynolds equation to set up the foundation of hydrodynamic lubrication theory. The Reynolds equation successfully reveals the mechanism of hydrodynamic lubrication, that is, a flowing fluid generates dynamic pressure.

The Reynolds equation is a second-order partial differential equation. In the past, it was solved analytically. However, a lot of approximates must be made in order to get an analytical solution, resulting in some deviations. Today, with the rapid development of computer technology, many complex lubrication problems can be numerically and accurately solved.

In addition, advanced experimental technologies can investigate in depth the detailed nature of hydrodynamic lubrication to establish a more complete model. Thus, many engineering problems have come much closer to a solution using lubrication theoretical analysis. At present, lubrication calculations play a very important role in mechanical design.

The hydrodynamic lubrication of rigid surfaces is called hydrodynamic lubrication, which is based on the following basic equations:

1. **The momentum equations:** on the conservation principle of momentum, also known as the Navier-Stokes equations;
2. **The continuity equation:** on the conservation principle of mass;
3. **The energy equation:** on the conservation principle of energy;
4. **Density equation:** on the relationship between density, pressure and temperature;
5. **Viscosity equation:** on the relationship between viscosity, pressure and temperature.

For elastic surface lubrication problems, the elastic deformation of the surface should be considered. These lubrication problems are called elasto-hydrodynamic lubrication. The Reynolds equation derived from the momentum equation and the continuity equation are the basic equations of hydrodynamic lubrication theory.

Principles of Tribology, First Edition. Wen Shizhu and Huang Ping.

2.1 Reynolds Equation

2.1.1 Basic Assumptions

In order to derive the Reynolds equation, the following eight assumptions are made.

1. Ignore the body forces such as gravity force, magnetic force, and so on.
2. There is nonslip on the interface between solid and liquid. That is, the molecules of liquid are firmly affixed to the solid surface moving with the same velocity.
3. Across the film thickness, ignore the variations of pressure. Although this assumption is not accurate, however, as the film thickness is only a few micro-meters or less, pressure cannot significantly vary.
4. The film thickness compared to the radius of the bearing surface can be ignored. Therefore, to neglect the influence of film curvature, linear velocity is used to calculate the rotational speed instead.
5. Lubricants are Newtonian. This assumption is reasonable for most mineral oils in general working conditions.
6. The flow is laminar and it is not a boundary layer or turbulence. Only for high velocity and large-scale bearings may it be a turbulence.
7. Compared with the viscous force, the inertia force can be ignored, including the acceleration force and centrifugal force. However, the effects of the inertial force should be considered for high velocity and large-scale bearings.
8. Across the lubricant film thickness, viscosity remains unchanged. This assumption is just for mathematical convenience.

Assumptions (1)–(4) are basically correct for most hydrodynamic lubrication problems. Assumptions (5)–(8) are introduced in order to simplify calculations; therefore, as they are only conditionally applicable, they may need to be modified in certain operating conditions.

2.1.2 Derivation of Reynolds Equation

By using the above assumptions, the Reynolds equation can be derived from the Navier-Stoke equations and the continuity equation. However, to enable readers to understand the physical phenomena of hydrodynamic lubrication, the Reynolds equation is derived, based on the following method. The main steps are:

1. Balance the forces acting on an infinitesimal body to find the velocity distribution across the film thickness.
2. Obtain the flow rate across the film thickness.
3. Use the continuity condition to derive the general form of Reynolds equation.

2.1.2.1 Force Balance

The forces acting on a micro-element body in the x direction (along the flow) are shown in Figure 2.1. By using assumptions (1) and (7), only the fluid pressure p and the viscous shear stress τ are left. If u, v and w denote the fluid velocities along the directions of the coordinates x, y and z respectively, u is the primary component followed by v. As the z direction is across the thickness, w is much smaller than u and v. Compared with the gradient of u, the other velocity gradients can be neglected. So, no viscous shear stress is acting on the surfaces except the upper and the lower surfaces.

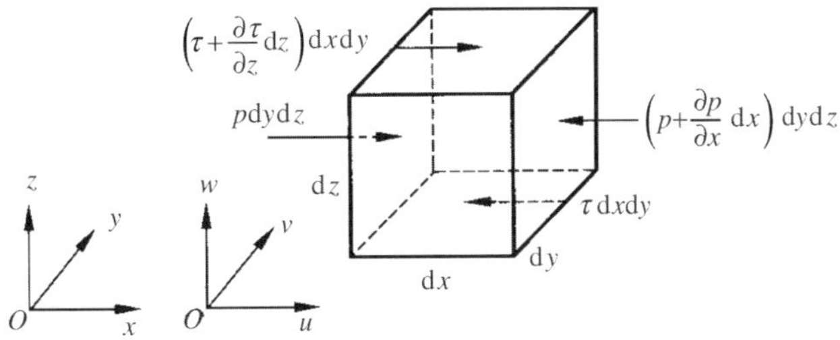

Figure 2.1 Forces on the micro-element.

Therefore in the x direction the following equations are available:

$$pdydz + \left(\tau + \frac{\partial \tau}{\partial z}dz\right)dxdy = \left(p + \frac{\partial p}{\partial x}dx\right)dydz + \tau dxdy \tag{2.1}$$

or

$$\frac{\partial p}{\partial x} = \frac{\partial \tau}{\partial z}. \tag{2.2}$$

According to Newton's viscosity law $\tau = \eta\, \partial u/\partial z$ (assumptions (5) and (6)), we have

$$\frac{\partial p}{\partial x} = \frac{\partial}{\partial z}\left(\eta \frac{\partial u}{\partial z}\right). \tag{2.3}$$

Similarly, in the y direction, there is

$$\frac{\partial p}{\partial y} = \frac{\partial}{\partial z}\left(\eta \frac{\partial v}{\partial z}\right). \tag{2.4}$$

Based on assumption (3), we have

$$\frac{\partial p}{\partial z} = 0. \tag{2.5}$$

From Equation 2.5, it is known that p is not a function of z. Furthermore, because η is not a function of z (on assumption (8)), integrate Equation 2.3 twice to z, we have

$$\begin{aligned} \eta \frac{\partial u}{\partial z} &= \int \frac{\partial p}{\partial x} dz = \frac{\partial p}{\partial x} z + C_1 \\ \eta u &= \int \left(\frac{\partial p}{\partial x} z + C_1\right) dz = \frac{\partial p}{\partial x}\frac{z^2}{2} + C_1 z + C_2. \end{aligned} \tag{2.6}$$

C_1 and C_2 can be determined because the fluid velocities are equal to the interface velocities of the surfaces (assumption (2)). If the two surface velocities are U_0 and U_h, that is, when $z=0$, $u=U_0$; when $z=h$, $u=U_h$, shown in Figure 2.2, we obtain

$$C_2 = \eta U_0; \quad C_1 = (U_h - U_0)\frac{\eta}{h} - \frac{\partial p}{\partial x}\cdot\frac{h}{2}. \tag{2.7}$$

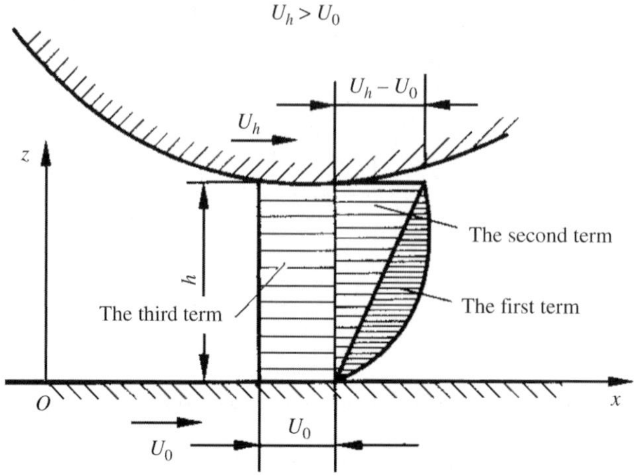

Figure 2.2 Velocity components of u.

Therefore, the flow velocity u across the lubricant film is equal to

$$u = \frac{1}{2\eta}\frac{\partial p}{\partial x}\left(z^2 - zh\right) + (U_h - U_0)\frac{z}{h} + U_0. \tag{2.8}$$

Similarly, the flow velocity v is equal to

$$v = \frac{1}{2\eta}\cdot\frac{\partial p}{\partial y}\left(z^2 - zh\right) + (V_h - V_0)\frac{z}{h} + V_0. \tag{2.9}$$

Figure 2.2 expresses the velocity u along the z direction. Equation 2.8 is composed of three parts: the first is a parabolic distribution, called "pressure flow"; the second is a linear (triangle) distribution, presenting the relative sliding velocity ($U_h - U_0$) of the two surfaces which is caused by the movement, known as "velocity flows"; and lastly the third is a constant.

Integrating Equation 2.9 to z, the flow q_x cross the lubricant film in the x direction is equal to

$$q_x = -\frac{1}{12\eta}\frac{\partial p}{\partial x}h^3 + (U_h + U_0)\frac{h}{2}. \tag{2.10}$$

Similarly, the flow q_y is equal to

$$q_y = -\frac{1}{12\eta}\frac{\partial p}{\partial y}h^3 + (V_h + V_0)\frac{h}{2}. \tag{2.11}$$

2.1.2.2 General Reynolds Equation

The fluid continuity equation is as follows.

$$\frac{\partial \rho}{\partial t} + \left[\frac{\partial(\rho u)}{\partial x} + \frac{\partial(\rho v)}{\partial y} + \frac{\partial(\rho w)}{\partial z}\right] = 0. \tag{2.12}$$

Integrate Equation 2.12 along the z direction:

$$\int_0^{h(x,y)} \frac{\partial \rho}{\partial t} dz + \int_0^{h(x,y)} \frac{\partial(\rho u)}{\partial x} dz + \int_0^{h(x,y)} \frac{\partial(\rho v)}{\partial y} dz + \int_0^{h(x,y)} \frac{\partial(\rho w)}{\partial z} dz = 0. \tag{2.13}$$

Exchange the orders of the integral and differential of Equation 2.13 and note that the upper limit h of Equation 2.13 is a function of x, y, and by substituting Equations 2.8 and 2.9 into Equation 2.13, it then becomes

$$\frac{\partial}{\partial x}\left(\frac{\rho h^3}{\eta}\frac{\partial p}{\partial x}\right) + \frac{\partial}{\partial y}\left(\frac{\rho h^3}{\eta}\frac{\partial p}{\partial y}\right) = 6(U_h - U_0)\frac{\partial(\rho h)}{\partial x} + 6(V_h - V_0)\frac{\partial(\rho h)}{\partial y}$$
$$+ 6\rho h \frac{\partial(U_h - U_0)}{\partial x} + 6\rho h \frac{\partial(V_h - V_0)}{\partial y} + 12\frac{\partial(\rho h)}{\partial t}. \tag{2.14}$$

Equation 2.14 is the general Reynolds Equation.

If set $U = U_h - U_0$, $V = V_h - V_0$, and suppose that the density ρ is a constant, Equation 2.14 can be written as follows.

$$\frac{\partial}{\partial x}\left(\frac{\rho h^3}{\eta}\cdot\frac{\partial p}{\partial x}\right) + \frac{\partial}{\partial y}\left(\frac{\rho h^3}{\eta}\cdot\frac{\partial p}{\partial y}\right) = 6\left[\frac{\partial}{\partial x}(U\rho h) + \frac{\partial}{\partial y}(V\rho h) + 2\rho(W_h - W_0)\right], \tag{2.15}$$

where $\partial h/\partial t = W_h - W_0$.

The Reynolds equation (2.15) can also be obtained by using the conservation principle of mass with the element-control volume method [1, 2].

2.2 Hydrodynamic Lubrication

2.2.1 Mechanism of Hydrodynamic Lubrication

The two terms on the left hand side of Reynolds equation (2.14) show the variation of pressure with the coordinates x and y; the terms on the right hand side show the effects of the lubrication parameters on pressure.

The physical meanings of the four effects of Equation 2.14 are as follows.

1. $U\rho\,(\partial h/\partial x), V\rho\,(\partial h/\partial y)$-------------- effect of hydrodynamics
2. $\rho h\,(\partial U/\partial x), \rho h\,(\partial V/\partial y)$-------------- effect of surface stretching
3. $Uh\,(\partial \rho/\partial x), Vh\,(\partial \rho/\partial y)$ ------------ effect of density variation
4. $\rho\,(\partial h/\partial t)$--------------------------------- effect of squeezing

Figure 2.3a shows the hydrodynamics effect. When the lower surface moves at the velocity U, the clearance along the motion direction gradually reduces to force the lubricant to flow from the large inlet to the small outlet. The flow along the motion direction gradually reduces (the triangle diagram in the figure). Based on the continuity condition, the pressure as shown in Figure 2.3a is inevitably produced. The flow induced by pressure will reduce the flow into the inlet, but increase the flow out of the outlet in order to maintain the flow balance of each section. Thus, the fluid flow along a convergence clearance will have a positive pressure, while the flow along a divergent clearance generally produces a negative pressure.

Figure 2.3b shows the stretching effect. When the solid surface changes its length due to elastic deformation or other reasons, the relative surface velocity will cause the surfaces to move relatively to generate pressure. In order to generate positive pressure, the relative velocity should be forward as shown in the figure.

Figure 2.3c is the effect of density variation. When the density of a lubricant gradually decreases along the direction of the motion, it will cause a different mass flow to generate pressure although the volume is

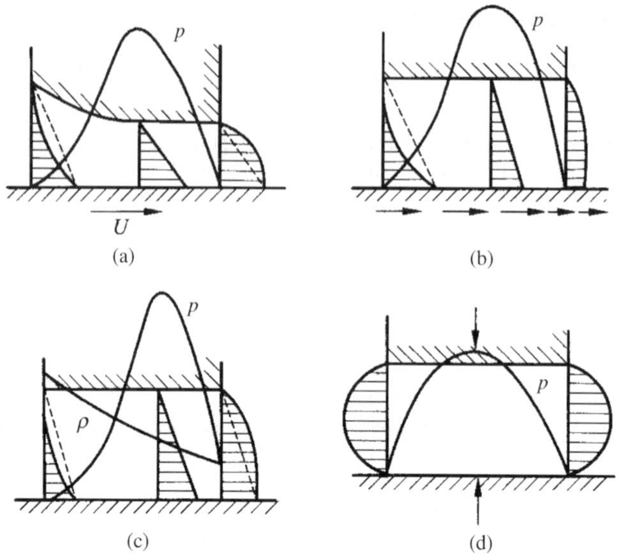

Figure 2.3 Four mechanisms of the pressure.

the same. The temperature variation may cause the variation of density. Although the pressure generated by the effect of density variation is not very high, it also allows a surface to possess a certain load-carrying capacity.

Figure 2.3d shows the effect of squeezing. When two parallel surfaces are under the action of a normal force, the thickness becomes thinner and pressure is produced. However, when the two surfaces are separated away, it will produce negative pressure.

Usually, the two main factors that form pressure are the dynamic pressure effect and the squeezing effect.

The Reynolds equation is a basic formula. Correctly understanding and using the Reynolds equation is the key to lubrication analysis. Furthermore, it is also important to correctly define the boundary conditions of the Reynolds equation that apply to each problem. In general, each of the borders should be set a boundary condition. However, when the location of the boundary is not given, more boundary conditions should be put forth, such as the Reynolds boundary conditions. For nonsteady lubrication problems, one should put forward the initial conditions. Finally, when one or more assumptions of the Reynolds equation do not hold, the fundamental equations should be modified in order to correctly solve the Reynolds equation.

EHL is another form of hydrodynamic lubrication based on Reynolds equation while considering elastic deformation of solid surfaces and the viscosity-pressure characteristics of a lubricant. An important element of this chapter is in helping readers to understand and master the basic characteristics of hydrodynamic lubrication.

2.2.2 *Boundary Conditions and Initial Conditions of Reynolds Equation*

2.2.2.1 Boundary Conditions

In order to solve the Reynolds equation, the integral constants must be determined, based on pressure boundary conditions. Pressure boundary conditions are generally in two forms:

Forced boundary condition $\quad p|_s = 0,$

Natural boundary condition $\quad \left.\dfrac{\partial p}{\partial n}\right|_s = 0,$

where s is the boundary of the solution domain; n is the normal direction of the boundary.

Usually, according to the geometric structure and the lubricant supply, it is not difficult to determine the boundaries of the oil film. However, if the sliding surface, such as a journal bearing, contains both convergent and divergent clearances, the outlet boundary position in the divergent region cannot be determined in advance. Its position is at the boundary where both the pressure and the pressure derivative are equal to zero. Such boundary conditions are called Reynolds boundary conditions, in the form below.

$$p|_s = 0 \quad \text{and} \quad \left.\frac{\partial p}{\partial n}\right|_s = 0.$$

Here are two examples of boundary conditions.

1. In $0 \le x \le L$, one-dimensional boundary conditions are:
   If the boundaries are known, the pressure boundaries are $p|_{x=0} = 0$ and $p|_{x=L} = 0$
   If the outlet boundary is unknown, the pressure boundaries are $p|_{x=0} = 0$; $p|_{x=x'} = 0$ and $\partial p/\partial x|_{x=x'} = 0$.
2. In $0 \le x \le L$ and $-B/2 \le y \le B/2$, two-dimensional boundary conditions are:
   If the boundaries are known, the pressure boundaries are $p|_{x=0} = 0$; $p|_{x=L} = 0$; and $p|_{y=\pm B/2} = 0$.
   If the outlet boundary is unknown, the pressure boundaries are $p|_{x=0} = 0$; $p|_{x=x'} = 0$; $\partial p/\partial x|_{x=x'} = 0$ and $p|_{y=\pm B/2} = 0$.
   Above, x' is the boundary to be determined.

2.2.2.2 Initial Conditions

For nonsteady lubrication problems, because velocity and load are time-dependent, the Reynolds equation contains a squeeze term, which is the last one on the right-hand side of Equation 2.14. The lubrication film thickness will change with time and therefore it needs to introduce the initial conditions. The general forms of the initial conditions are:

The initial film thickness is: $h|_{t=0} = h_0(x, y, 0)$.

The initial pressure is: $p|_{t=0} = p_0(x, y, 0)$.

If the lubricant viscosity and density also change with time, the respective initial conditions should also be given.

2.2.3 *Calculation of Hydrodynamic Lubrication*

After pressure is obtained from the Reynolds equation, other characteristics of the lubricant film can be calculated, including the load-carrying capacity, friction and flow.

2.2.3.1 Load-carrying Capacity *W*

Integrating pressure $p(x, y)$ through the whole area, the load-carrying capacity can be obtained as follows:

$$W = \iint p dx dy. \tag{2.16}$$

2.2.3.2 Friction Force F

The friction force can be obtained to integrate the shear stress on the surfaces along the entire lubricating region. To substitute the fluid velocity Equation 2.8 into Newton's law of viscosity, we have:

$$\tau = \eta \frac{\partial u}{\partial z} = \frac{1}{2} \cdot \frac{\partial p}{\partial x}(2z - h) + (U_h - U_0)\frac{\eta}{h}. \tag{2.17}$$

Substitute the down surface coordinate $z=0$ and the up surface coordinate $z=h$ into the above equation and integrate, then we have the friction forces on the two surfaces.

$$\begin{aligned} F_0 &= \iint \tau|_{z=0} dxdy \\ F_h &= \iint \tau|_{z=h} dxdy. \end{aligned} \tag{2.18}$$

Having obtained the friction forces, we can calculate the coefficient of friction $\mu = F/W$ as well as the viscous friction power loss and heat generated by friction.

2.2.3.3 Lubricant Flow Q

Flows through the boundaries in the x and y directions can be determined by integrating Equations 2.10 and 2.11.

$$\begin{aligned} Q_x &= \int q_x dy \\ Q_y &= \int q_y dx. \end{aligned} \tag{2.19}$$

The total flow will be the difference of the input and output flows of the borders. The calculated flow is used to estimate the oil supply, and it can also be used in calculating the temperature rise and the friction power loss by thermal equilibrium condition.

2.3 Elastic Contact Problems

Hertz contact theory, based on elastomer and derived from static contact conditions, is often used to calculate the elastic deformation and contact stress. In tribology, it is the basis of elastohydrodynamic lubrication and contact fatigue wear.

2.3.1 Line Contact

2.3.1.1 Geometry and Elasticity Simulations

A contact surface may be any shape. However, because the width of the contact area is usually much smaller than the radius of the contact point, the contact surface geometry can be properly simplified.

As EHL study only involves the area around the contact points, two arbitrary bodies in line contacts can be approximately equal to the contact of two cylinders with the same radius in the contact point as shown in Figure 2.4a. Furthermore, the two cylinders contact can be further transformed into the contact of an equivalent elastic cylinder with a rigid plane, as shown in Figure 2.4b. The clearance of the simulated problem is very similar to the shape of the actual situation.

In Figure 2.4a, the geometry of the clearance between the two cylinders can be written as follows:

$$h = h_0 + \left(R_1 - \sqrt{R_1^2 - x^2}\right) + \left(R_2 - \sqrt{R_2^2 - x^2}\right) \approx h_0 + \frac{x^2}{2R}, \tag{2.20}$$

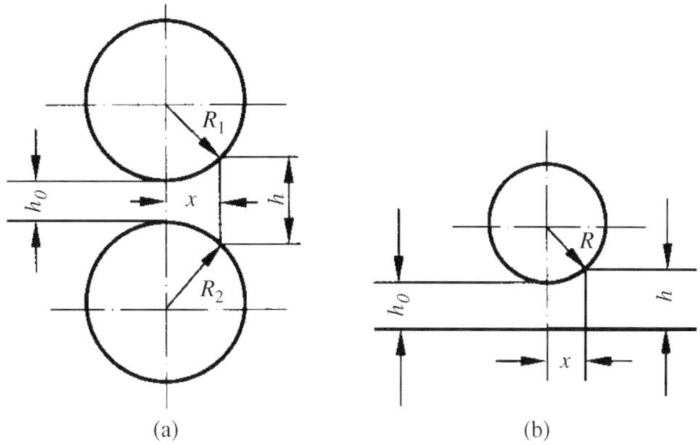

Figure 2.4 Clearance and the equivalent problem.

where R is the equivalent radius of curvature, as shown in Figure 2.4b. For the two conjugated surfaces, R is equal to

$$R = \frac{R_1 R_2}{R_1 + R_2}. \tag{2.21}$$

If the center points of the two columns are on the same side, and $R_1 > R_2$, R is equal to

$$R = \frac{R_1 R_2}{R_1 - R_2}. \tag{2.22}$$

The clearances in Figure 2.4a and b are the same. Therefore, their lubrication conditions are equivalent.

The principle of elasticity simulation is that the equivalent contact deformation of an elastic cylinder and a rigid flat surface is equal to the deformations of the two elastic cylinders. If the elastic moduluses of the two elastic surfaces are E_1 and E_2, and the Poisson ratios are μ_1 and μ_2, the equivalent elastic modulus E will be equal to

$$\frac{1}{E} = \frac{1}{2}\left(\frac{1-\mu_1^2}{E_1} + \frac{1-\mu_2^2}{E_2}\right). \tag{2.23}$$

Eventually, through geometry and elasticity simulations, the two elastic cylinders contact can be transformed into the contact of an elastic cylinder and a rigid plane with an equivalent radius of curvature R and the equivalent elastic modulus E. Therefore, in elasto-hydrodynamic lubrication, we only need to discuss the equivalent problem.

2.3.1.2 Contact Area and Stress

If two elastic cylinders are compressed by a load W, because of elastic deformation, the contact line will expand to a long, narrow plane as shown in Figure 2.5.

According to Hertz contact theory, the half width b of the contact area is equal to

$$b = \left(\frac{8WR}{\pi l E}\right)^{1/2}, \tag{2.24}$$

where l is the length of the cylinder.

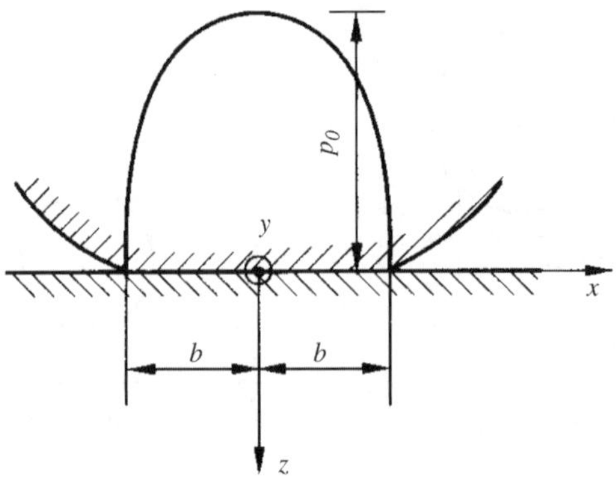

Figure 2.5 Line contact problem.

In the contact area, the contact stress distribution is a semi ellipse as follows.

$$p = p_0\left(1 - \frac{x^2}{b^2}\right)^{1/2}, \tag{2.25}$$

where p is the contact stress; p_0 is the maximum contact stress, which can be calculated as follows:

$$p_0 = \frac{2W}{\pi bl} = \left(\frac{WE}{2\pi Rl}\right)^{1/2} = \frac{E'b}{4R}. \tag{2.26}$$

In the contact body, the principal stresses σ_x, σ_y and σ_z acting on the centerline of the contact area (Figure 2.6 z-axis) are compressive stress. Figure 2.6 shows these major changes in stress along the z-axis.

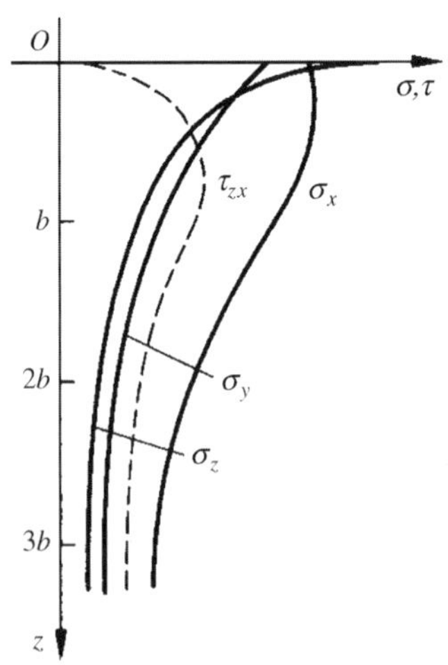

Figure 2.6 Contact stresses in the body varying with the centerline.

Although the maximum principal stress occurs on the contact surface, the maximum shear stress of 45° occurs inside the body. The maximum shear stress of 45° is composed of σ_x and σ_z, that is,

$$\tau_{zx} = \frac{1}{2}(\sigma_z - \sigma_x). \tag{2.27}$$

The maximum shear stress is equal to $0.301p_0$ and it acts away from the surface of 0.786b. It has an important effect in contact fatigue wear.

2.3.2 Point Contact

2.3.2.1 Geometric Relationship

The common area of a point contact is an ellipse. Two bodies of arbitrary shapes can be expressed by two oval bodies with the main radiuses in the contact point.

Figure 2.7 shows the geometric relationship of two bodies of arbitrary shapes near the contact point. The two bodies have two orthogonal principal planes in the contact point with the curvature radii of R_{1x}, R_{1y} and R_{2x}, R_{2y} respectively, with the corresponding axes x_1, y_1, x_2, and y_2. In the figure, γ is the angle between two axes. For engineering problems, usually $\gamma = 0$.

If we ignore the higher order terms, the surface near the contact point can be expressed by using the following equation.

$$\begin{aligned} z_1 &= A_1x^2 + A_2xy + A_3y^2 \\ z_2 &= B_1x^2 + B_2xy + B_3y^2, \end{aligned} \tag{2.28}$$

where A_1, A_2, A_3 and B_1, B_2, B_3 are constants.

The distance between the two surfaces along the z direction is equal to

$$s = z_2 - z_1 = (B_1 - A_1)x^2 + (B_2 - A_2)xy + (B_3 - A_3)y^2. \tag{2.29}$$

With appropriately selected axes x and y coordinates, Equation 2.30 will not contain the xy item, such that the distance between the two surfaces is expressed as

$$s = Ax^2 + By^2, \tag{2.30}$$

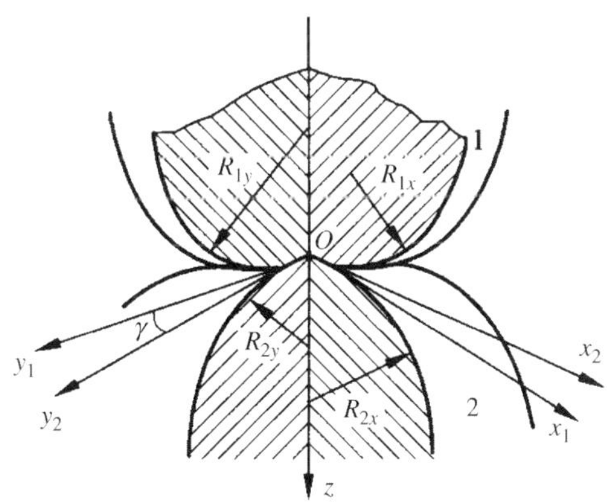

Figure 2.7 General Point contacts.

where A and B are constants, and connect with the geometry of the two bodies and can be obtained from the equations below.

$$A+B=\frac{1}{2}\left(\frac{1}{R_{1x}}+\frac{1}{R_{1y}}+\frac{1}{R_{2x}}+\frac{1}{R_{2y}}\right)$$

$$B-A=\frac{1}{2}\left[\left(\frac{1}{R_{1x}}-\frac{1}{R_{1y}}\right)^2+\left(\frac{1}{R_{2x}}-\frac{1}{R_{2y}}\right)^2+2\left(\frac{1}{R_{1x}}-\frac{1}{R_{1y}}\right)\left(\frac{1}{R_{1x}}-\frac{1}{R_{1y}}\right)\cos 2\gamma\right]^{1/2}. \tag{2.31}$$

From Equation 2.30, we can see that in the xOy plane, s is a family of elliptical contours. If the applied load compresses the two objects along the z axis, the elastic deformation of the contact area will have an oval border.

2.3.2.2 Contact Area and Stress

According to Hertz contact theory, the distribution of the contact stress is an ellipsoid. If a and b are set as the long and the short axes of the elliptical contact area respectively, and let the short axis coincide with the x-axis, the contact stress p will be

$$p=p_0\left(1-\frac{x^2}{b^2}-\frac{y^2}{a^2}\right). \tag{2.32}$$

The maximum Hertz contact stress p_0 is equal to

$$p_0=\frac{3W}{2\pi ab}. \tag{2.33}$$

a and b can be obtained by the following formula.

$$a=k_a\left[\frac{3W}{2E(A+B)}\right]^{1/3}, \tag{2.34}$$

$$b=k_b\left[\frac{3W}{2E(A+B)}\right]^{1/3}. \tag{2.35}$$

If set

$$\cos\theta=\frac{B-A}{B+A}. \tag{2.36}$$

k_a and k_b in Equation 2.36 can be obtained from Figure 2.8 with θ.

It can be seen from the above formula that the maximum contact stress and the load have no linear relationship. In line contacts, the maximum contact stress is proportional to the square root of the load, while in point contacts, the maximum contact stress is proportional to the cube root of the load. This is because as the load increases, the contact area also increases such that the increase of the maximum contact stress is less than that of the load. That the stress and the load are in a nonlinear relationship is an important feature of elastic contact problems. Another feature of contact problems is that the contact stress is related to the material elastic modulus and Poisson's ratio. This is because the contact area is relevant to the contact elastic deformation.

In engineering, the most common point contact problems of two objects belong to the main plane coinciding problems, as shown in Figure 2.9. As it is relatively simple and universal, so far the point contact EHL theory in this book is limited to such cases.

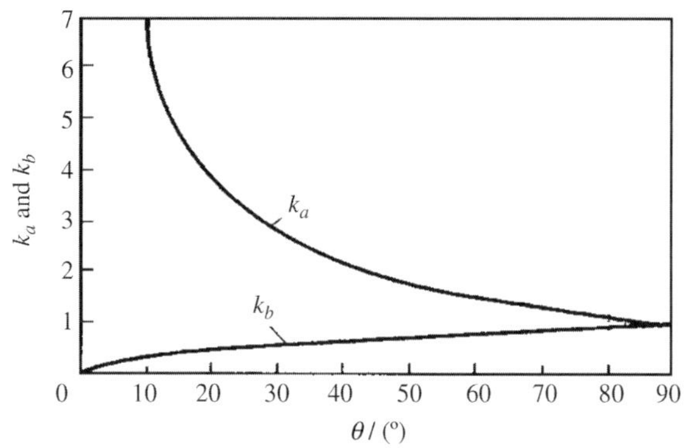

Figure 2.8 k_a and k_b via θ.

Figure 2.9 shows $\gamma = 0°$ or $90°$. If we select the two principal planes as the coordinate axes x and y, in the xOz plane the main curvature radii are R_{x1} and R_{x2}, and in the main yOz plane the main curvature radii are R_{1y} and R_{2y}, respectively. Then, from Equations 2.30 and 2.31 we have

$$s = \frac{x^2}{2R_x} + \frac{y^2}{2R_y}, \tag{2.37}$$

where

$$\frac{1}{R_x} = \frac{1}{R_{x1}} + \frac{1}{R_{x2}}, \tag{2.38}$$

$$\frac{1}{R_y} = \frac{1}{R_{y1}} + \frac{1}{R_{y2}}. \tag{2.39}$$

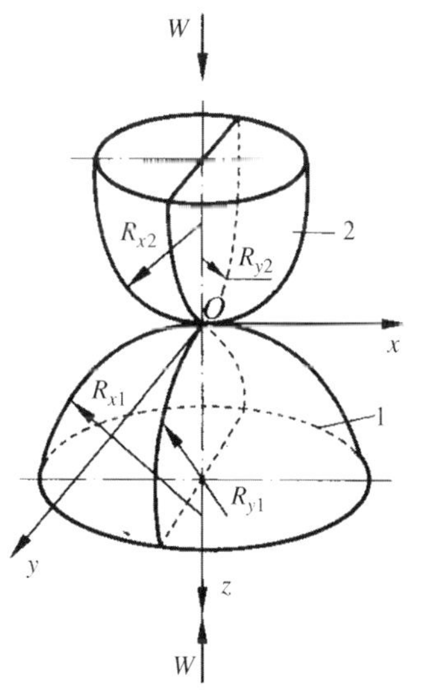

Figure 2.9 Coincidence of the main planes of the point contact.

Thus, a point contact problem of two elastic objects can be taken as the equivalent contact problem that a rigid plane contacts with an elastic object with an equivalent modulus E and two principal curvature radiuses of R_x and R_y.

As shown in Figure 2.9, if a lubricating oil film exists between the two surfaces, h_c is the thickness in the contact center and $\delta(x, y)$ is the elastic deformation, then the film thickness can be written as follows:

$$h(x,y) = h_c + s(x,y) + \delta(x,y) - \delta(0,0).$$

Substituting Equation 2.37 into the above equation, we have

$$h(x,y) = h_c + \frac{x^2}{2R_x} + \frac{y^2}{2R_y} + \delta(x,y) - \delta(0,0). \tag{2.40}$$

Set the center thickness $h_c = h_0 + \delta(0,0)$, where h_0 is for the rigid central film thickness, such that the film thickness can also be expressed as follows:

$$h(x,y) = h_0 + \frac{x^2}{2R_x} + \frac{y^2}{2R_y} + \delta(x,y). \tag{2.41}$$

In EHL calculation, either Equations 2.40 or 2.41 of the film thickness can be used. Note that h_0 is not a true film thickness, so it may be negative, but the film thickness h cannot be negative.

2.4 Entrance Analysis of EHL

This section describes the theory of Grubin [3], who presented the EHL entrance analysis in 1949. He proposed that the Reynolds hydrodynamic lubrication theory and the Hertz elastic contact theory [4] should be linked together to solve the EHL problem, and he first gave an approximate solution of the isothermal EHL.

2.4.1 Elastic Deformation of Line Contacts

The dotted line in Figure 2.10 shows an elastic cylinder with a radius R and a rigid flat surface. When the load W is applied, displacement is generated between the two surfaces. The deformed surface is shown

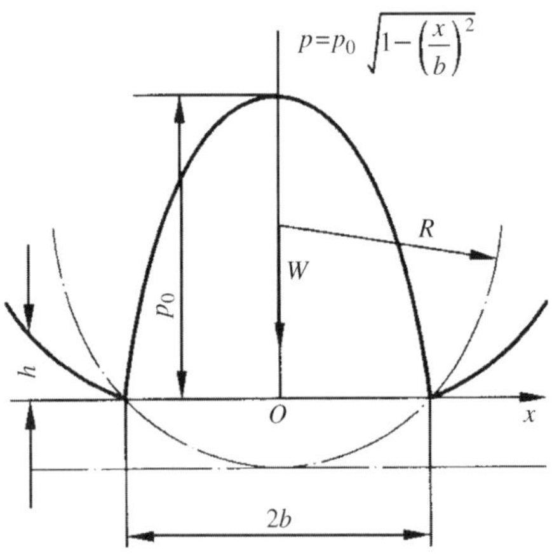

Figure 2.10 Hertz line contact deformation.

with the solid line in the figure. Obviously, the contact stress under the contact area causes the deformation of the elastic surface.

According to Hertz theory, the equation of the clearance outside the contact area is equal to

$$h = \frac{2bp_0}{E}\left[\frac{x}{b}\sqrt{\frac{x^2}{b^2}-1} - \ln\left(\frac{x}{b}+\sqrt{\frac{x^2}{b^2}-1}\right)\right]. \tag{2.42}$$

Set δ (called the Lame constant) is as follows.

$$\delta = 4\left[\frac{x}{b}\sqrt{\frac{x^2}{b^2}-1} - \ln\left(\frac{x}{b}+\sqrt{\frac{x^2}{b^2}-1}\right)\right]. \tag{2.43}$$

We can see that δ is a function of x, and only when $|x/b| \geq 1$, δ is meaningful.

By substituting Equation 2.43 into Equation 2.42, and using Equation 2.26, we have

$$h = \frac{W}{\pi El}\delta. \tag{2.44}$$

2.4.2 *Reynolds Equation Considering the Effect of Pressure-Viscocity*

If we substitute the Barus pressure-viscosity equation $\eta = \eta_0 e^{\alpha p}$ into the one-dimensional Reynolds equation, we have

$$\frac{dp}{dx} = 12U\eta_0 e^{\alpha p}\frac{h-\bar{h}}{h^3}. \tag{2.45}$$

If an induced pressure $q = (1 - e^{-\alpha p})/\alpha$ is introduced, then the above equation becomes

$$\frac{dq}{dx} = -\frac{1}{\alpha}\frac{d}{dx}(e^{-\alpha p}) = e^{-\alpha p}\frac{dp}{dx}. \tag{2.46}$$

If we substitute the above equation into Equation 2.45, the Reynolds equation considering pressure-viscosity effect can be obtained as follows:

$$\frac{dq}{dx} - 12U\eta_0\frac{h-\bar{h}}{h^3}. \tag{2.47}$$

Equation 2.47 shows that after the transformation in which the induced pressure q replaces the pressure p, the Reynolds equation considering viscosity-pressure and other relationships has the same form as the original one.

2.4.3 *Discussion*

Grubin very cleverly made the following inferences for EHL problems of line contacts.

1. The pressure on most contact area is so high that $e^{-\alpha p}$ tends to be 0. Therefore, the induced pressure $q = (1/\alpha)(1 - e^{-\alpha p})$ tends to be $1/\alpha$, such that q is nearly constant.

   If q is a constant, $dq/dx = 0$ in the contact region. From the Reynolds equation (2.47), we have $h = \bar{h} = h_0$. Therefore, the film thickness in the contact region is also a constant, which forms a parallel clearance. Furthermore, it can be inferred that in the contact area, the pressure distribution is the same as the Hertz pressure, regardless of whether the lubricant film exists.

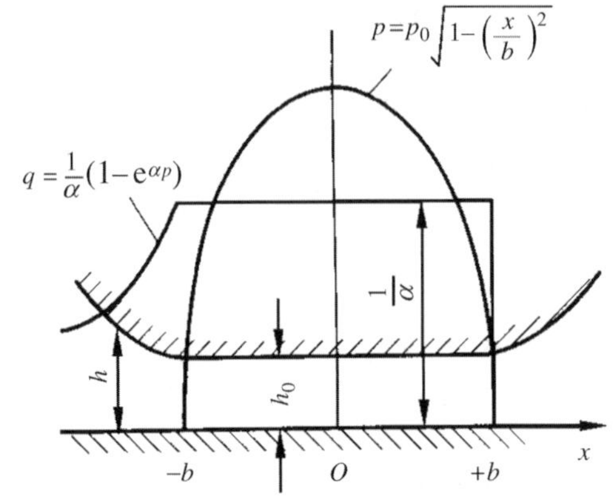

Figure 2.11 Pressure distribution and film shape of Grubin analysis.

2. Because the pressure in the contact region is much higher than outside the contact area ($x < -b$) of the entrance area, the elastic deformation of the cylinder depends only on Hertz contact pressure in the contact region, which means that outside the contact area the elastic deformation remains without an oil film. Therefore, the shape of the clearance can be calculated as follows:

$$h = h_0 + \frac{W}{\pi El}\delta. \tag{2.48}$$

As shown in Figure 2.11, the formation of hydrodynamic pressure p generated by the convergent clearance entrance should be equal to the pressure at $x = -b$, or $q = 1/\alpha$. This condition can be used to obtain the value of the oil film thickness h_0.

Figure 2.11 shows the pressure distribution and film shape of Grubin analysis. The above conclusions have been accurately proven by results of calculations and experiments.

It should be noted that Grubin theory is limited to the entrance area. As it is very complicated in the outlet zone, Hertz pressure distribution and deformation needs to be modified, otherwise it cannot meet the continuity condition. This is because in the contact center, $dp/dx = 0$, only the flow velocity exists and it is equal to Uh_0. However, at the outlet $x = +b$, $dp/dx = -2p_0/b$. Therefore, in addition to the velocity flow, there is also a considerable pressure flow. The total flow would be much greater than that of the contact center with no modification.

Figure 2.12 gives a numerical result of line contact EHL. It can be seen that in order to satisfy the flow continuity condition, in the outlet zone the elastic deformation of the surface tends to recover such that the

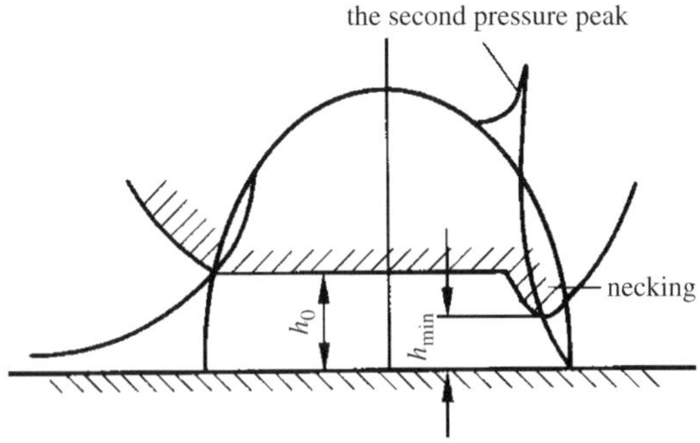

Figure 2.12 EHL pressure and film thickness of the numerical solution.

clearance forms a necking. The thickness necking is usually the minimum oil film thickness $h_{\min}$, which is smaller than h_0 at the center. Based on this formula, Grubin established that it is only h_0 of 75%. Due to the presence of necking, the corresponding pressure will appear as the second peak. The necking and the secondary pressure peak are important features of EHL.

2.4.4 Grubin Film Thickness Formula

Substituting the clearance Equation 2.48 of the entrance area into the Reynolds equation of Equation 2.45, Grubin obtained the following formula:

$$\frac{dq}{dx} = 12U\eta_0 \frac{W\delta}{\pi Elh^3}. \tag{2.49}$$

Set

$$Q = \left(\frac{W}{\pi El}\right)^2 \frac{q}{12U\eta_0 b}; \quad X = \frac{x}{b}; \quad H = \frac{\pi hEl}{W}; \quad H_0 = \frac{\pi h_0 El}{W},$$

the dimensionless clearance equation becomes

$$H = H_0 + \delta. \tag{2.50}$$

If we substitute the above dimensionless variables and Equation 2.50 into Equation 2.49, the dimensionless Reynolds equation becomes

$$\frac{dQ}{dX} = \frac{\delta}{H^3}. \tag{2.51}$$

According to boundary conditions: when $X \to -\infty$, $Q = 0$, the following definite integral can be used to obtain Q at $X = -1$.

$$Q|_{X=-1} = \int_{-\infty}^{-1} \frac{\delta}{H^3} dX = \int_{-\infty}^{-1} \frac{\delta}{(H_0+\delta)^3} dX. \tag{2.52}$$

In the integral equation, H_0 is not related to X, but δ is a function of X. By using the numerical method, an empirical formula is obtained as follows.

$$Q|_{X=-1} = 0.0986\, H_0^{11/8} \tag{2.53}$$

As mentioned earlier, in $x = -b$, the induced pressure q should be equal to $1/\alpha$ or $Q|_{X=-1} = (W/\pi El)^2\,(1/12U\eta_0 b\alpha)$. Substituting $b - (8WR/\pi lE)^{1/2}$ into Equation 2.53, finally, Grubin obtained

$$\frac{h_0}{R} = 1.95\left(\frac{U\eta_0\alpha}{R}\right)^{8/11}\left(\frac{ElR}{W}\right)^{1/11}. \tag{2.54}$$

This is the famous Grubin formula of EHL theory.

In order to facilitate analysis and comparison, Dowson and Higginson used the following dimensionless parameters [7]:

Film thickness parameter $H_0^* = h_0/R$;
Material parameter $G^* = \alpha E$;
Velocity parameter $U^* = \eta_0 U/ER$;
Load parameter $W^* = W/ERl$.

Then, Equation 2.54 becomes:

$$H_0^* = 1.95\frac{(G^* U^*)^{8/11}}{W^{*1/11}}. \tag{2.55}$$

Equation 2.55 gives a fairly accurate approximation of the average film thickness. It is about 20% larger than the measured value.

The approximation method of entrance area analysis of EHL proposed by Grubin has been widely used. For example, if a ball contacts with a plane, the Grubin EHL formula of the equivalent film thickness will be

$$\frac{h_0}{R} = 1.73\left(\frac{U\eta_0\alpha}{R}\right)^{5/7}\left(\frac{ER^2}{W}\right)^{1/21}. \tag{2.56}$$

2.5 Grease Lubrication

Grease is a kind of lubricant that results from adding some thickening agents into oil to form a semi-solid jelly-like substance. Thickening agents are commonly metallic soaps. The fibers of the soaps form mesh frameworks for storing oil. As the grease is of a three-dimensional frame structure composed of fiber, it cannot be a laminar flow. In the lubrication process it shows complex mechanical properties, that is, a time-dependent visco-plastic fluid. Figure 2.13 indicates the rheological behavior of grease. The main characteristics of grease can be summarized as follows.

1. The viscosity of grease usually increases with decrease in shear strain rate, and thus the relationship of the shear stress and shear rate is nonlinear.
2. As shown in Figure 2.13, grease is of a yield shear stress τ_s. Only when the shear stress τ is larger than τ_s, will grease present the properties of a fluid. When $\tau \leq \tau_s$, grease becomes solid in nature. It may have a certain amount of elastic deformation. As grease has a yield shear stress, this makes the grease lubricant film appear as a nonflow layer when $\tau \leq \tau_s$. In a flowing layer, the velocities perpendicular to the layer are the same.
3. Grease has a thixotropic property, which means that when grease flows under a certain shear rate, with increase of time the shear stress decreases gradually and its viscosity also decreases. And after the shear process ceases, its viscosity will partially recover. Thus, the grease lubrication state is a time-dependent process. While we call a grease lubrication problem "steady," this only refers to a relatively stable state.

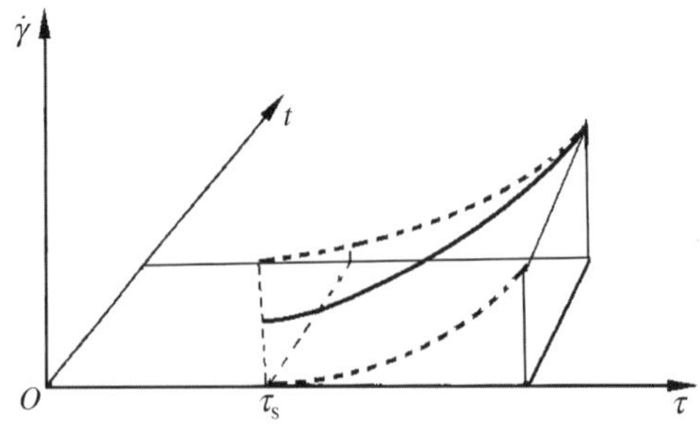

Figure 2.13 Rheological properties of grease.

Three kinds of constitutive equations currently used to describe the rheological properties of grease are as follows:

1. Ostwald model $\tau = \eta\dot{\gamma}^n$,
2. Bingham model $\tau = \tau_s + \eta\dot{\gamma}$,
3. Herschel-Bulkley $\tau = \tau_s + \eta\dot{\gamma}^n$,

where n is the rheological index; η is the plastic viscosity.

Practically, Herschel-Bulkley model is much closer to the experimental results, and is especially more accurate at low velocity. In addition, when $n = 1$, the Herschel-Bulkley model will turn into the Bingham model, and when $\tau_s = 0$, it turns into the Ostwald model. Therefore, the Herschel-Bulkley model is more universal. Strictly speaking, the three rheological parameters τ_s, η and n should be functions of temperature and pressure. For isothermal lubrication problems we do not need to consider the influence of temperature. For simplification, the rheological index n is usually assumed not to be related to pressure p, while the yield shear stress τ_s and plastic viscosity η vary with the pressure p according to the following relationships:

$$\begin{aligned} \tau_s &= \tau_{s0}e^{ap} \\ \eta &= \eta_0 e^{ap}, \end{aligned} \tag{2.57}$$

where τ_{s0} is the initial yield shear stress and η_0 is the initial plastic viscosity of grease at atmosphere pressure respectively; α is the viscosity-pressure coefficient of the base oil used for manufacturing grease.

The equations of grease lubrication are similar to those of oil lubrication. The Reynolds equation of grease lubrication is also derived based on the constitutive equation, the equilibrium equation and the continuity equation. However, because the constitutive equation of a grease model contains the yield shear stress τ_s, the lubricant film will be divided into two parts, known as the nonshear flow part and the shear flow part. The two parts must be dealt with separately, so the deriving process is complicated. For example, the one-dimensional Reynolds equation based on the Herschel-Bulkley model is as follows:

$$\frac{dp}{dx} = 2\eta\left[2\left(2+\frac{1}{n}\right)\right]^n \frac{U^n(h-\bar{h})^n}{h^{2n+1}}\left[1-\frac{2\tau_s}{\frac{dp}{dx}h}\right]^{-(n-1)}\left[1+\frac{n}{n+1}\frac{2\tau_s}{\frac{dp}{dx}h}\right]^{-n}. \tag{2.58}$$

where $\bar{h}$ is the film thickness at $dp/dx - 0$.

The derivation of Equation 2.58 can be found in references [5, 6]. If set $\tau_s = 0$, Equation 2.58 will become the Reynolds equation of Ostwald model as follows:

$$\frac{dp}{dx} - \frac{2\eta}{h^{2n+1}}U^n\left[2\left(2+\frac{1}{n}\right)\right]^n(h-\bar{h})^n. \tag{2.59}$$

If set $\tau_s = 0$ and $n = 1$, Equation 2.58 turns into the common Reynolds equation of Newtonian fluid.

$$\frac{dp}{dx} = 12\eta U\frac{h-\bar{h}}{h^3} \tag{2.60}$$

It should be noted that the Reynolds equation can be derived according to different rheological models, but its applications are not exactly the same. In order to obtain solutions of the practical problems to meet the requirements, the process must be carefully chosen according to their own circumstances and transformation.

References

1. Wen, S.Z. (1990) *Principles of Tribology*, Tsinghua University Press, Beijing.
2. Wen, S.Z. and Huang, P. (2002) *Principles of Tribology*, 2nd edn, Tsinghua University Press, Beijing.
3. Grubin, A.N. (1949) Fundamentals of the Hydrodynamic Theory of Lubrication of Heavily Loaded Cylindrical Surfaces, Central Scientific Research Institute for Technology and Mechanical Engineering Book no. 30 Moscow, Kh. F. Ketova (ed.), (DSIR Translation No. 33, 115–166).
4. Timoshenko, S. and Goodier, J.N. (1973) *Theory of Elasticity*, McGraw-Hill.
5. Ying, Z.N. (1985) The rheological properties of grease and Elastohydrodynamic Lubrication Mechanism. A master's degree thesis, Tsinghua University.
6. Wen, S.Z. and Yang, P.R. (1992) *Elastohydrodynamic Lubrication*, Tsinghua University Press, Beijing.
7. Dowson, D. and Higginson, G.R. (1997) *Elasto-hydrodynamic Lubrication*, Pergamon Press, London.

3

Numerical Methods of Lubrication Calculation

A variety of hydrodynamic lubrication problems are related to the viscous fluid flow in a narrow gap. The basic equation describing the physical phenomena is the Reynolds equation, and its general form is:

$$\frac{\partial}{\partial x}\left(\frac{\rho h^3}{\eta}\frac{\partial p}{\partial x}\right)+\frac{\partial}{\partial y}\left(\frac{\rho h^3}{\eta}\frac{\partial p}{\partial y}\right)=6\left(U\frac{\partial \rho h}{\partial x}+V\frac{\partial \rho h}{\partial y}+2\frac{\partial \rho h}{\partial t}\right). \tag{3.1}$$

The analytic solutions of this elliptic partial differential equation may only be obtained for some special clearances, while for the complex geometric shapes or working conditions, the exact analytical solutions cannot be obtained. With the rapid development of computing technology, numerical methods are now widely used as the effective way to solve the lubrication problem.

With numerical methods partial differential equations are transformed into the form of algebraic equations. The general principles of numerical methods are: first, divide the solution region into a mesh with a limited number of elements and make each element small enough so that the unknown variables (such as oil film pressure p) in each element will be considered to be uniform or a linear change without causing any significant error.

Then, with discrete methods, the partial differential equations are transformed into a set of linear algebraic equations. The algebraic equations express the relationship between unknown variables of the element and the unknown variables of the surrounding elements.

Finally, with the help of elimination or iteration methods to solve the algebraic equations of the whole region, all the unknown variables may be solved.

Many of the numerical methods can be used to solve the Reynolds equation. The most commonly used are the finite difference method, finite element method or boundary element method. Although all the methods will divide the region to be solved into many elements, their numerical solutions are different. For the finite difference method and finite element method, the solved results in the region are only approximate, but meet the boundary conditions given exactly. However, for the boundary element method, the solved results are satisfied by the basic equation in the region, but are approximate on the boundary.

The energy equation and the elastic deformation equation are used to consider the thermal effect and surface elastic deformation. In this chapter, their numerical methods will also be introduced. Furthermore, the multi-grid method has been widely used in lubrication calculation because of its effectiveness.

Principles of Tribology, First Edition. Wen Shizhu and Huang Ping.

Therefore, the multi-grid method used for solving differential equations and integral equations are also introduced at the end of the chapter.

3.1 Numerical Methods of Lubrication

3.1.1 Finite Difference Method

If the boundary conditions are given in solving a differential equation, this is known as the boundary value problem. In hydrodynamic lubrication calculations, the finite difference method is commonly used to solve the Reynolds equation. The major steps of finite difference method are as follows.

First, change the partial differential equations into the dimensionless forms. This is accomplished by expressing the variables in the universal form.

Then divide the solution region into a mesh with uniform or non-uniform grids. Figure 3.1 is a uniform mesh, in the x direction with nodes m and in the y direction with nodes n such that the total nodes are equal to $m \times n$. The division of a mesh is determined by calculation accuracy. For a common hydrodynamic lubrication problem, $m = 12$–25 and $n = 8$–10 will usually meet the requirement of accuracy. Sometimes, in order to improve the accuracy, the unknown variables have a rapid change in the region and the grid needs refining by using two or more different sub-grids.

Take the pressure p as an example. The distribution of p in the whole region can be expressed by each node p_{ij}. According to the differential regularities, the partial derivatives at the node $O(i, j)$ can be represented by the surrounding node variables.

As shown in Figure 3.2, the expression of the partial derivatives of the intermediate difference at the node $O(i, j)$ have the following forms:

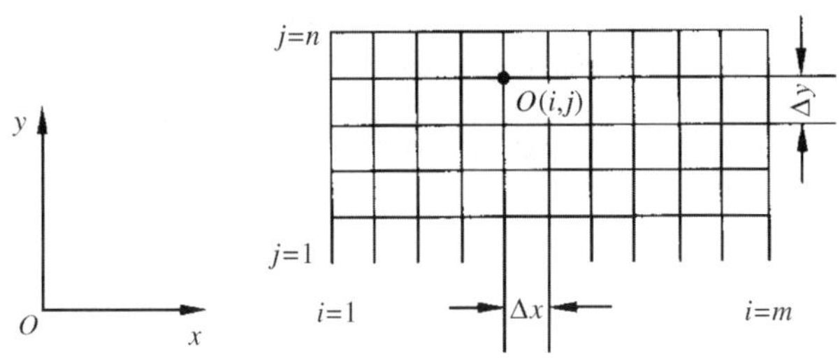

Figure 3.1 Uniform mesh.

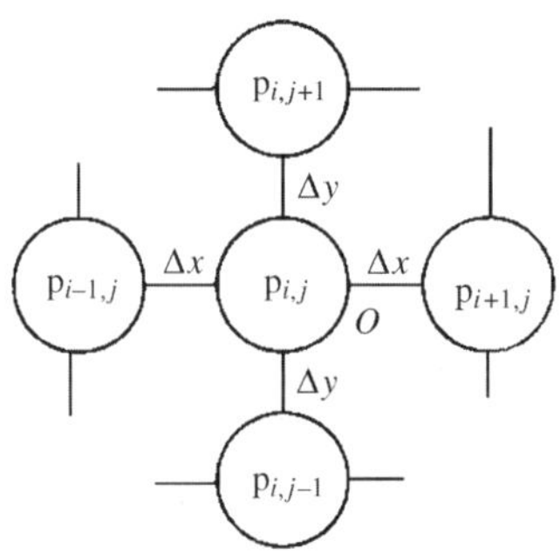

Figure 3.2 Relationship of difference.

$$\left(\frac{\partial p}{\partial x}\right)_{i,j} = \frac{p_{i+1,j} - p_{i-1,j}}{2\Delta x}$$

$$\left(\frac{\partial p}{\partial y}\right)_{i,j} = \frac{p_{i,j+1} - p_{i,j-1}}{2\Delta y}. \tag{3.2}$$

The second-order partial derivatives of the intermediate difference are as follows:

$$\left(\frac{\partial^2 p}{\partial x^2}\right)_{i,j} = \frac{p_{i+1,j} + p_{i-1,j} - 2p_{i,j}}{(\Delta x)^2}$$

$$\left(\frac{\partial^2 p}{\partial y^2}\right)_{i,j} = \frac{p_{i,j+1} + p_{i,j-1} - 2p_{i,j}}{(\Delta y)^2}. \tag{3.3}$$

In order to obtain the unknown variables near the border, the forward or the backward difference formulas are used as follows:

$$\left(\frac{\partial p}{\partial x}\right)_{i,j} = \frac{p_{i+1,j} - p_{i,j}}{\Delta x}$$

$$\left(\frac{\partial p}{\partial y}\right)_{i,j} = \frac{p_{i,j+1} - p_{i,j}}{\Delta y}, \tag{3.4}$$

$$\left(\frac{\partial p}{\partial x}\right)_{i,j} = \frac{p_{i,j} - p_{i-1,j}}{\Delta x}$$

$$\left(\frac{\partial p}{\partial y}\right)_{i,j} = \frac{p_{i,j} - p_{i,j-1}}{\Delta y}. \tag{3.5}$$

Usually, the accuracy of the intermediate difference is high. The following intermediate difference formulas can also be used in calculation:

$$\left(\frac{\partial p}{\partial x}\right)_{i,j} = \frac{p_{i+1/2,j} - p_{i-1/2,j}}{\Delta x}. \tag{3.6}$$

According to the above formulas, the two-dimensional Reynolds equation can be written in a standard form of the second-order partial differential equation:

$$A\frac{\partial^2 p}{\partial x^2} + B\frac{\partial^2 p}{\partial y^2} + C\frac{\partial p}{\partial x} + D\frac{\partial p}{\partial y} = E, \tag{3.7}$$

where A, B, C, D and E are the known parameters.

Equation 3.7 can be applied to each node. According to Equations 3.2 and 3.3, the relationship of pressure pi, j at the node O(i,j) with the adjacent pressures can be written as follows:

$$p_{i,j} = C_1 p_{i,j+1} + C_2 p_{i,j-1} + C_3 p_{i+1,j} + C_4 p_{i-1,j} + G \tag{3.8}$$

where

$$
\begin{aligned}
C_1 &= \left[\frac{B}{\Delta y^2}+\frac{D}{2\Delta y}\right] \\
C_2 &= \frac{1}{K}\left[\frac{B}{\Delta y^2}-\frac{D}{2\Delta y}\right] \\
C_3 &= \frac{1}{K}\left[\frac{A}{\Delta x^2}+\frac{C}{2\Delta x}\right] \\
C_4 &= \frac{1}{K}\left[\frac{A}{\Delta x^2}-\frac{C}{2\Delta x}\right] \\
G &= -\frac{E}{K} \qquad K = 2\left[\frac{A}{\Delta x^2}+\frac{B}{\Delta y^2}\right]
\end{aligned}
$$

By using Equation 3.8, we can write out the calculation equations of finite difference method at each node. And, in the border nodes, variables should satisfy the boundary conditions because their values are known. In this way, a set of linear algebraic equations can be obtained. The number of the equations obtained is equal to the unknown variables so it is a definite problem to be solved. By using an elimination method or iterative method we can solve algebraic equations. When the convergent result meets the given condition, the solution of each node has been found.

The following describes how to solve hydrodynamic lubrication problems with the finite difference method.

3.1.1.1 Hydrostatic Lubrication

In stable operating conditions, the oil film thickness h of hydrostatic lubrication is a constant. Not considering the relative sliding and the thermal effect, the viscosity η is also a constant. Then, its Reynolds equation can be simplified to the Laplace equation:

$$\nabla^2 p = \frac{\partial^2 p}{\partial x^2}+\frac{\partial^2 p}{\partial y^2} = 0. \tag{3.9}$$

Set the dimensionless parameters as $X = x/l$, $Y = y/b$, and $P = p/p_r$ where l is the length of the bearing, b is the width, and p_r is the pressure of oil chamfers, and $a = l^2/b^2$. Then, the dimensionless Reynolds equation will be:

$$\frac{\partial^2 P}{\partial X^2}+a\frac{\partial^2 P}{\partial Y^2} = 0. \tag{3.10}$$

The boundary conditions of Equation 3.10 are:

1. In the oil chambers $P = 1$.
2. At the borders of the bearing $P = 0$.

Substitute Equation 3.2 into Equation 3.10, we have:

$$\frac{P_{i+1,j}+P_{i-1,j}-2P_{i,j}}{\Delta X^2}+\alpha\frac{P_{i,j+1}+P_{i,j-1}-2P_{i,j}}{\Delta Y^2} = 0. \tag{3.11}$$

Substitute the boundary conditions into Equation 3.11, the numerical solution of oil film pressure distribution can be easily obtained with an elimination or iteration method.

3.1.1.2 Hydrodynamic Lubrication

For incompressible hydrodynamic lubrication, its Reynolds equation is

$$\frac{\partial}{\partial x}\left(\frac{h^3}{\eta}\frac{\partial p}{\partial x}\right)+\frac{\partial}{\partial y}\left(\frac{h^3}{\eta}\frac{\partial p}{\partial y}\right)=6U\frac{\partial h}{\partial x}. \tag{3.12}$$

If h is a known function of x and y, and for isoviscosity lubrication Equation 3.12 is linear. If the viscosity varies with temperature or pressure, Equation 3.12 will be nonlinear. In this case, the solution process will be complex.

1. Two-dimensional quasi problem
   Because a nonlinear equation of a two-dimensional problem is complicated, a simplified method is to change the equation to a quasi linear one. The key point is to set the pressure p as a known function along one coordinate (such as along the y direction). Then, substituting p into the equation, we can transfer the two-dimensional problem into a one-dimensional problem, which is much easier to solve.
   According to Ocvirk's analysis on the infinite narrow bearing, the distribution of pressure p along y direction (i.e. the axial direction) is close to a parabola. Therefore, we can assume

$$p=p_c(1-y^n), \tag{3.13}$$

where p_c is the central pressure along the x direction; n is an index, usually $n=2$.

Below is an example of isoviscosity hydrodynamic lubrication of a wedge-shaped slider (see Figure 3.3).

Set

$$\begin{aligned} x&=Xl\\ y&=Yb\\ p&=P\frac{6\eta Ul}{h_0^2}\\ a&=\frac{l^2}{b^2}\\ h&=h_0\left[1+X\frac{h_1-h_0}{h_0}\right]=H\,h_0. \end{aligned}$$

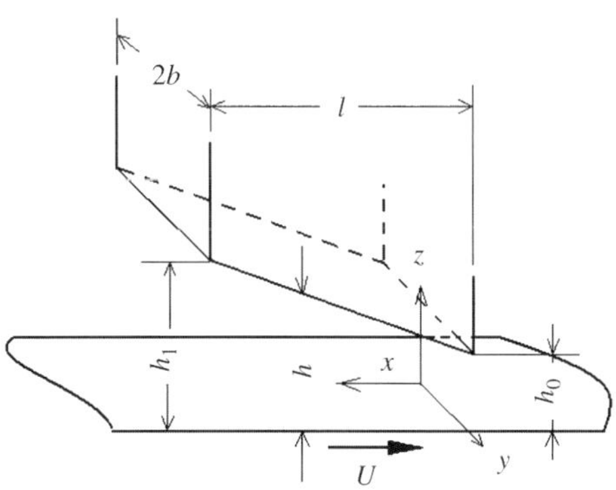

Figure 3.3 A wedge-shaped slider.

Substitute the above parameters into the Reynolds equation of Equation 3.12, we have the dimensionless equation as follows:

$$\frac{\partial^2 P}{\partial X^2}+a\frac{\partial^2 P}{\partial Y^2}+\frac{3}{H}\frac{dH}{dX}\frac{\partial P}{\partial X}=\frac{1}{H^3}\frac{dH}{dX}. \tag{3.14}$$

The above equation is known as the Poisson equation.

Then, substitute $P=P_c(1-Y^2)$ into Equation 3.14, we have

$$\frac{\partial^2 P_c}{\partial X^2}+\frac{3}{H}\frac{dH}{dX}\frac{\partial P_c}{\partial X}-2aP_c=\frac{1}{H^3}\frac{dH}{dX}. \tag{3.15}$$

Or

$$\frac{\partial^2 P_c}{\partial X^2}+\alpha\frac{\partial P_c}{\partial X}+\beta P_c=\gamma, \tag{3.16}$$

where $Y=y/b$; b is the half bearing width; and

$$\begin{aligned}\alpha &= \frac{3}{H}\frac{dH}{dx}\\ \beta &= -2a=-\frac{2b^2}{l^2}\\ \gamma &= \frac{1}{H^3}\frac{dH}{dX}.\end{aligned} \tag{3.17}$$

Therefore, the differential equation of Equation 3.16 can be written as

$$P_i+C_1P_{i-1}+C_2P_{i+1}=C_3, \tag{3.18}$$

where C_1, C_2 and C_3 are the functions of the known coefficients α, β and γ of Equation 3.17.

2. Two-dimensional problem

   The isoviscosity lubrication of a journal bearing is shown in Figure 3.4. The film thickness h is a function of x.

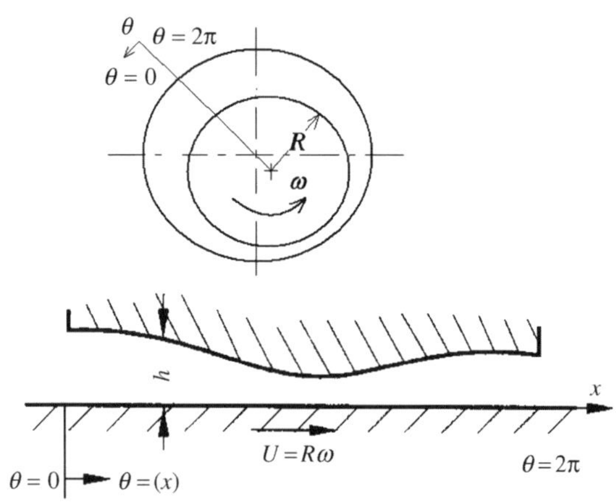

Figure 3.4 Journal bearings expand.

Substituting $x = R\theta$ and $U = R\omega$, the Reynolds equation becomes

$$\frac{\partial}{\partial \theta}\left(h^3 \frac{\partial p}{\partial \theta}\right) + h^3 R^2 \frac{\partial^2 p}{\partial y^2} = 6\eta R^2 \omega \frac{dh}{d\theta}. \tag{3.19}$$

Set

$$\begin{aligned} y &= Yb \\ a &= (R/b)^2 \\ h &= c(1 + \varepsilon \cos \theta) = Hc \\ p &= P\frac{6\eta R^2 \omega}{c^2}. \end{aligned}$$

where R is the shaft radius; b is the half bearing length; ε is the eccentricity, $\varepsilon = e/c$; e is the eccentricity; η is the viscosity; and c is the radius clearance of the shaft and bearing.

Then, the dimensionless Reynolds equation is as follows:

$$\frac{\partial}{\partial \theta}\left(H^3 \frac{\partial P}{\partial \theta}\right) + aH^3 \frac{\partial^2 P}{\partial Y^2} = \frac{dH}{d\theta}. \tag{3.20}$$

Or

$$\frac{\partial^2 P}{\partial \theta^2} + a\frac{\partial^2 P}{\partial Y^2} - \frac{3\varepsilon \sin \theta}{1 + \varepsilon \cos \theta}\frac{\partial P}{\partial \theta} = -\frac{\varepsilon \sin \theta}{(1 + \varepsilon \cos \theta)^2}. \tag{3.21}$$

It can be seen from Equation 3.21 that for a journal bearing, y varies from 0 to 1, and θ from 0 to 2π. The boundary conditions of this problem are:

1. In the axial direction: $P|_{Y=1} = 0; \partial P/\partial Y|_{Y=0} = 0$.
2. In the circumferential direction: $P|_{\theta=0} = 0$. And, because the outlet position of the film is unknown, a more boundary condition should be given to determine it. Therefore, $P|_{\theta_2} = 0$ and $\partial P/\partial \theta|_{\theta_2} = 0$ are used to determine both $\theta 2$ and another integration constant. In order to obtain the outlet of the film, an iterative method can be used. Set the pressure to zero at the point where $P < 0$. Then, the first zero pressure point along x direction is the outlet boundary of the oil film.

Another difficulty in calculation of isoviscosity lubrication is how to determine the viscosity value. In hydrodynamic lubrication, the viscous friction makes the temperature different at each point. Thus, their viscosities are also different. The accurate method to calculate viscosity of lubricant is based on the temperature field. Obviously, this is quite complicated.

Sometimes, an equivalent viscosity η_e is substituted into the Reynolds equation in order to consider the influence of temperature. The equivalent viscosity should be determined by the equivalent temperature T_e of the bearing.

If we assume that all the heat converted from friction work is taken away by oil flow, the heat balance equation can be written as

$$FU = Jc_v \rho Q \Delta T, \tag{3.22}$$

or

$$\Delta T = \frac{FU}{Jc_v \rho Q} = \frac{4\pi\eta U^2 Rb}{Jc_v \rho Qc}, \tag{3.23}$$

where ΔT is the temperature increment of lubricating oil; F is the friction force of the journal. By Petlov friction theory, $F = 4\pi U\eta Rb/c$; U is the sliding velocity; J is the mechanical equivalent of heat; c_v is the specific heat of oil; ρ is the density of oil; Q is the volumetric flow; η is the viscosity of oil; R is the radius of the bearing; $2b$ is the length of the bearing; and c is the radius clearance.

The equivalent temperature T_e is between the import and export oil temperatures. Therefore, it can be written as

$$T_e = T_i + k\Delta T, \tag{3.24}$$

where T_i is the import oil temperature; k is a constant between 0 and 1.

According to the calculation results of the tilting-pad bearing lubrication by Wang *et al.* [1], when the equivalent temperature is set as equal to 0.9 times the average temperature for each tilt, the calculated load-carrying capacity of isoviscosity lubrication is much closer to the result of a variviscosity lubrication calculation and experiments.

The equivalent viscosity is a basic parameter in solving the Reynolds equation. However, it depends on the temperature. And, the determination of temperature rise depends on pressure, which must be obtained from the Reynolds equation. Therefore, an iterative method must be used to solve these equations.

3.1.2 Finite Element Method and Boundary Element Method

The following gives a brief introduction to the use of the finite element method and the boundary element method in solving lubrication problems.

3.1.2.1 Finite Element Method (FEM)

The finite element method was first developed in terms of elasticity theory and was applied to hydrodynamic lubrication in the 1990s. Compared with the finite difference method the main advantages of the finite element method are its adaptability and convenience in complicated geometric shapes. Its element size and node number can be arbitrarily selected using accurate calculation. However, its calculating procedures are more complex.

The finite element method obeys the variation principles for solving functional equations. The general form of Reynolds equation used for incompressible hydrodynamic lubrication is as follows:

$$\frac{\partial}{\partial x}\left(\frac{h^3}{12\eta}\frac{\partial p}{\partial x}\right) + \frac{\partial}{\partial y}\left(\frac{h^3}{12\eta}\frac{\partial p}{\partial y}\right) = \frac{1}{2}\frac{\partial(hU)}{\partial x} + \frac{1}{2}\frac{\partial(hV)}{\partial y} + \frac{\partial h}{\partial t}. \tag{3.25}$$

The vector form of Equation 3.25 is

$$\nabla \cdot \left(\frac{h^3}{12\eta}\nabla p\right) = \frac{1}{2}\nabla \cdot (hU) + \dot{h}, \tag{3.26}$$

where $\nabla = \boldsymbol{i}\,\partial/\partial x + \boldsymbol{j}\,\partial/\partial y$; U is the velocity vector; $\dot{h} = \partial h/\partial t$.

As shown in Figure 3.5, the lubrication region is divided into a number of triangular elements. On the border there are two boundary conditions. The pressure on s_p is known as $p = p_0$; and the flow on s_q is known as $q = q_0$.

Suppose e is an element with pressure p_e, the functional equation of the element can be written as

$$J_e = -\iint_A \left[-\frac{h^3}{12\eta}\nabla p_e \cdot \nabla p_e + h\mathbf{U} \cdot \nabla p_e - 2\dot{h}p_e\right] dA + 2\int_{s_q} q_0 p_e ds, \tag{3.27}$$

where A is the solution zone; s is the border.

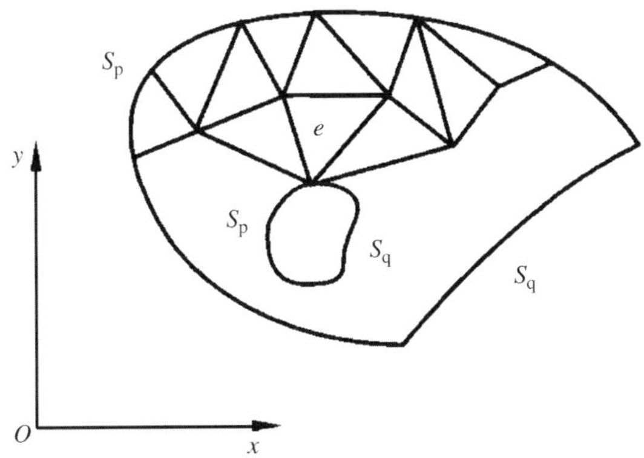

Figure 3.5 Finite element division of lubrication zone.

If the lubrication region is divided into a total of n elements, the total function shall be equal to the sum of functions of all elements.

$$J = \sum_{e=1}^{n} J_e. \tag{3.28}$$

According to the variation principle, the extreme of the total function exists while

$$\delta J = \sum_{e=1}^{n} \delta J_e = 0. \tag{3.29}$$

By using the Euler-Lagrange equation it can be proven that the solution $p(x, y)$ of Equation 3.26 satisfies Equation 3.29 and the given boundary conditions. Or, $p(x, y)$ obtained from Equation 3.29 must be the solution of the Reynolds equation (3.26) and satisfy the given boundary conditions. Therefore, the finite element method need not directly solve the two integral equations, but it transforms the Reynolds equation into a functional equation, and by solving Equation 3.29, we can also obtain the solution.

The solution process of the finite element method can be generally summarized as follows:

1. Divide the solution region into a number of triangular or quadrilateral elements;
2. Write out the functional equation according to the variation principles;
3. Establish interpolation functions to express the variables by the node values of each element;
4. Based on the boundary conditions algebraic equations are established in terms of the unknown variables of each node;
5. Use an iteration or elimination method to solve the algebraic equations.

3.1.2.2 Boundary Element Method

The basic feature of the boundary element method is to solve the unknown parameters in the region by the known borders. First, divide the border into a number of elements. Then, solve the other unknown border variables by the known and then the unknown variables in the solution region. Therefore, the main advantage of the boundary element method is that it has a very limited number of algebraic equations so as to significantly reduce the amount of data. In addition, the boundary element method has a higher accuracy than the other methods and can be easily used in a mixed problem. However, the establishment of equations of the boundary element method is not so easy.

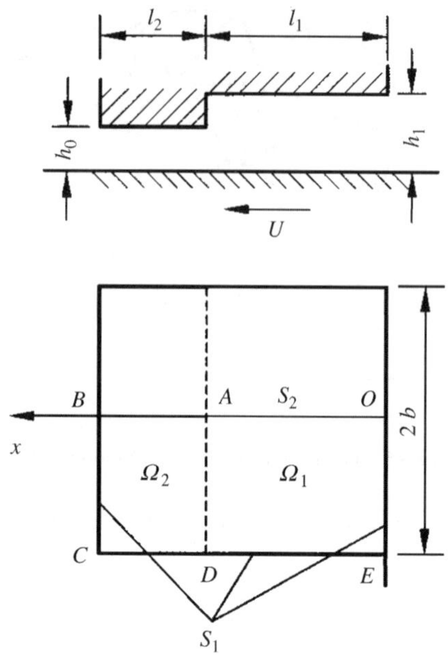

Figure 3.6 Rayleigh step slider.

At present, the boundary element method is mainly used in the analysis of theory of elasticity and heat transfer. The author took the Rayleigh step slider lubrication as an example to calculate its lubrication properties by the boundary element method [2].

The slider is as shown in Figure 3.6. It can be divided into two different parts Ω_1, and Ω_2. The pressure p in each part depends on the following Reynolds equation.

$$\nabla^2 p = \frac{\partial^2 p}{\partial x^2} + \frac{\partial^2 p}{\partial y^2} = 0. \tag{3.30}$$

Because the problem is symmetric to the x axis, only half of the slider needs to be considered in the analysis, for an example, *OBCE*. If the total boundary is s, it is divided into s_1 and s_2, or $s = s_1 + s_2$. The known boundary conditions are: $p|_{s1} = p_0 = 0$ and $q|_{s_2} = \partial p/\partial y|_{s_2} = q_0 = 0$. Now, let us introduce a weighting function f to meet the basic Equation 3.30. The equation of the boundary element method by the weighted residual method is as follows.

$$\int_{\Omega} (\nabla^2 p) f d\Omega = \int_{s_2} (q - q_0) f ds - \int_{s_1} (p - p_0) Q ds, \tag{3.31}$$

where $Q = \partial f / \partial y$.

The weight function f can be obtained through mathematical analysis as follows.

$$f = -\frac{\ln r}{2\pi}, \tag{3.32}$$

where r is the distance from point i to other points.

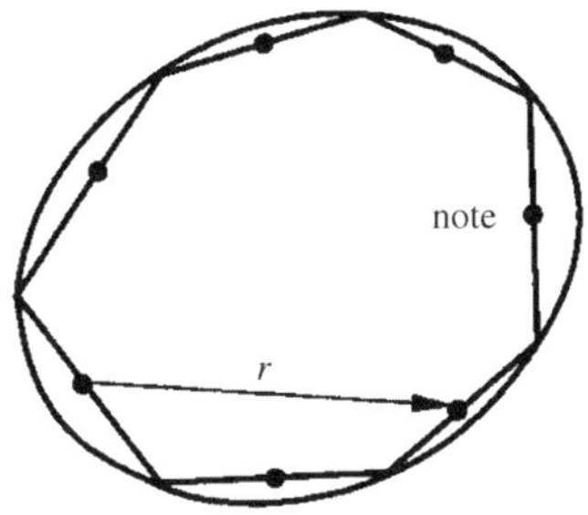

Figure 3.7 Boundary element division.

The unknown variables p_i in the region has a relationship with the boundary variables as follows:

$$p_i + \int_s pQds = \int_s fds. \tag{3.33}$$

Similarly, the unknown variables p_i on the border can be solved by the known border variables from the following integral equations:

$$\frac{1}{2}p_i + \int_s pQds = \int_s qfds. \tag{3.34}$$

With Equation 3.34 the unknown variables along the border can be obtained. Then, by using Equation 3.33, the unknown variables can be calculated.

Simply use straight lines to divide the boundary into n elements, as shown in Figure 3.7. Then, use the n straight line segments instead of the actual curve border. If the midpoint of each segment is taken as the node, the unknown variables on each element will linearly vary.

Applying Equation 3.34 to the equivalent element boundaries, we have

$$\frac{1}{2}p_i + \sum_{j=1}^{n} p_j \int_{s_j} Qds = \sum_{j=1}^{n} q_j \int_{s_j} fds. \tag{3.35}$$

Because each node has two variables p and q, the total variables are $2n$. Here, $n = n_1 + n_2$, while n_1 is the number of the known p_i and n_2 of the known q_j. Therefore, there are n unknown variables. With Equation 3.35, we have n algebraic equations, which are equal to the number of the unknown variables. Therefore, the total equations have definite solutions to obtain the unknown p and q of each node on the border. Then, use the border variables to calculate the inner unknown variables with Equation 3.33. The discrete form of Equation 3.33 is as follows:

$$p_i = \sum_{i=1}^{n} q_i \int_{s_j} fds - \sum_{j=1}^{n} p_j \int_{s_j} Qds. \tag{3.36}$$

3.1.3 Numerical Techniques

3.1.3.1 Parameter Transformation

When the eccentricity ε of a journal bearing is larger than 0.8 or the tilt angle of a wedge-shaped slider is large, the minimum film thickness h_{min} will be very small so that dh/dx in the vicinity of h_{min} will be very large too. This may cause pressure to change dramatically in a narrow range such that the solution is difficult to obtain. A common method in solving the problem is to use fine grids. However, the fine grids will increase the computational task. Furthermore, dramatic changes of pressure often lead to instability in the process of solving the Reynolds equation.

In order to overcome the instability, a parameter transformation can be used, commonly set as $M = ph^{3/2}$, known as the Vogenpohl transformation.

Substitute $p = M/h^{3/2}$ into the Reynolds equation to solve the variable M. Then carry out the inverse transformation and p can be obtained indirectly. Although p near h_{min} changes dramatically, M is much smooth because h is very small. Therefore, a high accurate solution can be guaranteed.

3.1.3.2 Numerical Integration

After the pressure distribution is obtained, it will be used to calculate the load-carrying capacity, friction force, flow, and so on. Therefore, a numerical integral method should be used. Here, take the slider of Figure 3.3 as an example:

Load-carrying capacity

or

$$W = \iint p dx dy$$

$$W^* = \frac{Wh_0^2}{12U\eta bl^2} = \iint PdXdY. \tag{3.37}$$

Friction force

or

$$F = \iint \tau dx dy = \iint \left(\frac{\eta U}{h} \mp \frac{h}{2} \frac{\partial p}{\partial x} \right) dx dy$$

$$F^* = \frac{Fh_0}{2U\eta bl} = \iint \left(\frac{1}{H} \mp 3H \frac{\partial P}{\partial X} \right) dXdY. \tag{3.38}$$

Flow

or

$$Q_x = \int \left(\frac{Uh}{2} - \frac{h^3}{12\eta} \frac{\partial p}{\partial x} \right) dy$$

$$Q_y = \int \left(- \frac{h^3}{12\eta} \frac{\partial p}{\partial y} \right) dx$$

$$Q_x^* = \frac{Q_x}{2Ubh_0} = \int \left(\frac{H}{2} - \frac{H^3}{2} \frac{\partial P}{\partial X} \right) dY = \int q_x^* dY$$

$$Q_y^* = \frac{Q_y}{2Ubh_0} = \int a \left(- \frac{H^3}{2} \frac{\partial P}{\partial Y} \right) dX = \int q_y^* dX, \tag{3.39}$$

where W^*, F^*, Q_x^*, Q_y^*, q_x^* and q_y^* are the dimensionless properties of lubrication.

One commonly used numerical integral method is the Simpson method. For an example, the flow Q_x^* along the x direction is equal to

$$Q_x^* = \frac{1}{3(m-1)} \Delta x (q_{x1}^* + 4q_{x2}^* + 2q_{x3}^* + 4q_{x4}^* + \cdots + q_{xm}^*), \tag{3.40}$$

where m is the number of nodes, and if the Simpson method is used, m should be odd; q_{x1}^*, q_{x2}^*,…q_{x1}^*, q_{x2}^*, …, and q_{xm}^* are the flows of each node; Δx is the distance between the two neighbor nodes.

In order to obtain W^* and F^* two-dimensional integration should be used. Furthermore, in order to obtain F^*, Q_x^* and Q_y^*, numerical derivatives of pressure should be pre-calculated. We can use Equation 3.2 to obtain the derivatives. For nodes that are on the border, a three-point parabolic formula can be used as follows.

Suppose the pressure distribution is in the form of $p = ax^2 + bx + c$. Then, we have

$$\frac{\partial p}{\partial x} = 2ax + b. \tag{3.41}$$

Therefore,

$$\left(\frac{\partial p}{\partial x}\right)_{1j} = b. \tag{3.42}$$

Accordingly, its numerical derivative is

$$\left(\frac{\partial p}{\partial x}\right)_{1j} = \frac{4p_{2j} - p_{3j} - 3p_{1j}}{2\Delta x}. \tag{3.43}$$

3.1.3.3 Empirical Formula

The advantage of a numerical method is to accurately solve a complex problem so that it is undoubtedly an effective way for the important design and theoretical research. However, a numerical method is an individual solution without a general versatility. To overcome this, we can collect massive solution data to fit some empirical formulas.

For example, if the influences on the unknown variable p are the relevant parameters A, B, C, D …, select an appropriate function, usually an exponential function, to show their relationship, namely,

$$p - KA^a B^b C^c D^d \ldots \tag{3.44}$$

where K, a, b, c, d… and so on arc constants to be determined.

Then, according to a group of a sufficient number of solutions (for example, 500) theoretically or experimentally determine the above constant. Obviously, the fitting formula may not be very accurate but it is fairly credible. It should be pointed out that it must be repeated and modified many times to get a satisfactory fitting formula.

3.1.3.4 Sudden Thickness Change

In Equation 3.1 of the Reynolds equation, set $\rho h^3/\eta$ as a flow coefficient k. The differential of Equation 3.6 is:

$$\left(\frac{\partial Q}{\partial x}\right)_{i,j} = \frac{Q_{i+1/2,j} - Q_{i-1/2,j}}{\Delta x}, \tag{3.45}$$

where $Q = \rho h^3/\eta.(\partial p/\partial x) = k\,\partial p/\partial x$.

Because k is a function of x, it is related to the coefficients at the point $i-1$, i and $i+1$, that is, k_{i-1}, k_i and k_{i+1}. In order to obtain the derivative of Equation 3.45, $k_{i-1/2}$ and $k_{i+1/2}$ must be obtained first. The following discussion is about the non-uniform coefficient.

Usually, the differential flow coefficient $k_{i-1/2}$ at the middle interface is obtained under the most simple and intuitive approach as follows.

$$k_{i-1/2} = (1-\alpha)k_{i-1} + \alpha k_i, \tag{3.46}$$

where α is the insertion factor, which can be determined by the following formula:

$$\alpha \equiv \frac{(\delta x)_{i-1/2+}}{(\delta x)_{i-1/2}}, \tag{3.47}$$

where $(\delta x)_{i-1/2}$ is the distance between nodes $i-1$ and i; $(\delta x)_{i-1/2+}$ is between $(\delta x)_{i-1/2}$ and $(\delta x)_{i-1/2}$. If the interface is in the middle of the nodes $i-1$ and i, $\alpha=0.5$ in Equation 3.47, that is, $k_{i-1/2}$ is the arithmetic mean of k_{i-1} and k_i.

Such a simplistic approach would lead to quite inaccurate results while calculating the flow coefficient with a sudden film change. There is a simple and much better approach which can be used for such situations. It should be pointed out that our main concern is not on the local value of the coefficient at the interface $i-1/2$, but to get a pressure drop to describe the correct expression of interface flow rate $Q_{i-1/2}$, and finally, to get the correct pressure. With analyzing Reynolds equation we know that the relationship between the flow and pressure difference is as follows:

$$Q_{i-1/2} = \frac{k_{i-1/2}(p_i - p_{i-1})}{(\delta x)_{i-1/2}}. \tag{3.48}$$

In order to get the correct $k_{i-1/2}$ of Equation 3.48 let us to discuss the following situation.

Around the grid point i, there is a uniform flow rate coefficient k through the control volume while around $i-1$ the flow rate coefficient is k_{i-1}. For a step film thickness between $i-1$ and i, we have

$$p_i - p_{i-1} = \frac{Q_{i-1/2}(\delta x)_{i-1/2-}}{k_{i-1}} + \frac{Q_{i-1/2}(\delta x)_{i-1/2+}}{k_i} \tag{3.49}$$

or

$$Q_{i-1/2} = \frac{p_i - p_{i-1}}{(\delta x)_{i-1/2-}/k_{i-1} + (\delta x)_{i-1/2+}/k_i}. \tag{3.50}$$

Compare Equation 3.48 with Equation 3.50, we know

$$k_{i-1/2} = \left(\frac{\alpha}{k_{i-1}} + \frac{1-\alpha}{k_i}\right)^{-1}. \tag{3.51}$$

Therefore, when the intermediate interface is located in the midpoint of $i-1$ and i, we have $\alpha=0.5$. Thus

$$k_{i-1/2}^{-1} = 0.5(k_i^{-1} + k_{i-1}^{-1}) \tag{3.52}$$

or

$$k_{i-1/2} = \frac{2k_i k_{i-1}}{k_i + k_{i-1}}. \tag{3.53}$$

Equations 3.52 and 3.53 indicate that $k_{i-1/2}$ is the harmonic mean of k_{i-1}, and k_i, but not the arithmetic mean as given in Equation 3.47.

3.2 Numerical Solution of the Energy Equation

In the above calculation, the influence of temperature is ignored. However, temperature is an important factor affecting the lubrication properties because temperature significantly changes the viscosity of lubricant, therefore affecting the pressure distribution and load-carrying capacity. In addition, temperature causes the thermal deformation of solid surfaces that results in a change in the gap shape, thus again affecting lubrication performances. Extremely high temperatures may cause lubrication failure or surface material failure. Therefore, the local temperature is usually limited to below 120–140 °C.

In order to obtain the temperature distribution of a lubricant film it is necessary to solve the energy equation.

3.2.1 Conduction and Convection of Heat

Heat dissipation in lubricant film occurs in the following two ways.

1. **Conduction:** across the film thickness (z direction), heat passes through the solid surface;
2. **Convection:** along the film length and width directions (x and y directions), heat will be brought away by the lubricant flow.

Because lubrication conditions are different, the dissipation of these two cooling ways of heat varies. The relationship of the film heating and cooling of two parallel plates can be analyzed in Figure 3.8.

Assume that the temperature on the stationary plate is linear, at both ends the temperatures are T_0 and T_1 respectively, and the temperature on the moving plate is T_0. Thus, at the outlet the temperature rise is $\Delta T = T_1 - T_0$. If the two plates are with a width b, and the lubricant film thickness is h, now let us analyze the heat dissipation.

3.2.1.1 Conduction Heat H_d

If the temperature gradient across the film thickness direction is linear, thus

$$\frac{dT}{dz} = \frac{x}{b}\frac{\Delta T}{h}. \tag{3.54}$$

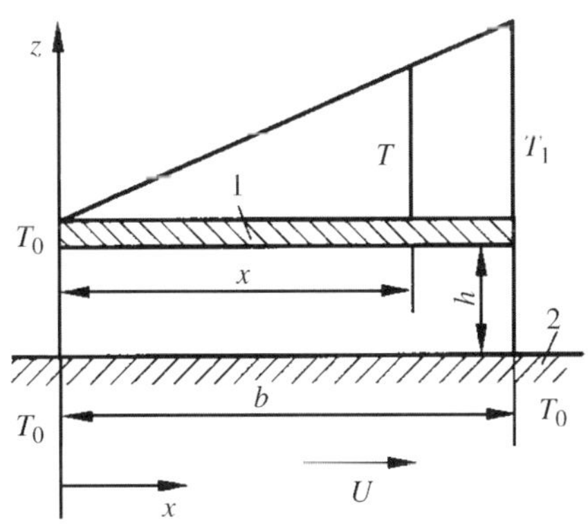

Figure 3.8 Thermal analysis of the two plates.

Therefore, the amount of conduction heat per unit length will be

$$H_d = \int_0^b K \frac{dT}{dz} dx = \int_0^b K \frac{x}{b} \frac{\Delta T}{h} dx = \frac{Kb\Delta T}{2h}, \tag{3.55}$$

where K is the thermal conductivity of the lubricant film.

3.2.1.2 Convection Heat H_v

If q_x is the flow per unit length along the x direction; ρ is the density of the lubricating oil; c is the heat capacity of the lubricating oil, and the average oil film temperature rise is $\Delta T/2$, the convection heat will be:

$$H_v = q_x \rho c \frac{\Delta T}{2} = \frac{1}{4} Uh \rho c \Delta T. \tag{3.56}$$

The ratio of the conduction heat and the convection heat, which is called the Peclet number, is equal to:

$$\text{Peclet number} = \frac{H_d}{H_v} = \frac{K}{\rho c} \frac{2b}{Uh^2}. \tag{3.57}$$

The Peclet number can be used to characterize the heat dissipation of a lubrication system. When the ratio is infinite, the lubrication system has no convection. Therefore, all heat dissipation relies on the conduction. So, because of the absence of the convection heat, and if the flow rate q_x is not zero, ΔT must be zero. This means that the temperature of oil film along the x direction is constant, or the flow of lubricating oil is isothermal. When the Peclet number is equal to zero, all the heat dissipation relies on convection. Without heat conduction, the temperature along the z direction is equal. No heat exchange occurs across the film through the solid surfaces. The flow of oil is adiabatic. However, the above two extreme situations cannot occur in the actual lubrication.

For mineral oils, usually $K/\rho c = 8 \times 10^{-8}\,\text{m}^2/\text{s}$. If we choose $b = 25$ mm, for different h and U, the corresponding Peclet numbers are presented in Table 3.1.

Table 3.1 shows that for an actual lubrication system the Peclet number is finite, and conduction and convection all exist. When the Peclet number is larger than 0.4, conduction dominates, while if it is less than 0.1, convection dominates.

From Equation 3.58 we can see that the film thickness h is a main factor influencing heat dissipation. The Peclet number will be sharply reduced with increase of h, leading to a strengthening in convection.

Clearly, hydrodynamic lubrication is based on convection. Usually, the conduction heat need not be considered. So, the flow can be thought as adiabatic. In the EHL state, however, heat dissipation mainly relies on conduction. In the temperature distribution calculation, convection heat is often ignored. However, for a very high-speed EHL, convection heat can no longer be ignored.

Table 3.1 Peclet number ($b = 25$ mm)

Sliding velocity U (m/s)	Film thickness h (μm)		
	100	30	10
10	0.04	0.4	4
30	0.01	0.1	1
100	0.004	0.04	0.4

3.2.2 *Energy Equation*

In this section the energy equation of hydrodynamic lubrication is derived by using a simple method.

For a hydrodynamic lubrication, variations of the kinetic energy and potential energy of a fluid flow can be ignored so that the fluid energy is only a function of temperature. For a steady flow, all the variables do not vary with time. In addition, because the convection can be ignored under hydrodynamic lubrication conditions, conduction heat across the thickness is equal to zero, that is, $\partial T/\partial z = 0$. Then, the temperature T of lubricant film is only a function of x and y.

Now, let us analyze the variations of heat and mechanical work of the fluid. As shown in Figure 3.9, take a micro-column with the width δx in the x direction, a unit length in the y direction, and the height h in the z direction. Set q_x and q_y as the flows of the micro-column in the x and y directions, the heat flow into the micro-columns are equal to

$$H_x = q_x T \rho c \tag{3.58}$$

and

$$H_y = q_y T \rho c, \tag{3.59}$$

where q_x and q_y are the flows given by Equations 2.11 and 2.12.

In the x and y directions, the heat flows out of micro-columns as follows:

$$H_x + \frac{\partial H_x}{\partial x} dx \tag{3.60}$$

and

$$H_y + \frac{\partial H_y}{\partial y} dy. \tag{3.61}$$

If $dx = 1$, the sum of the heat flows on the cross-sectional areas of the micro-column is

$$H_x - \left(H_x + \frac{\partial H_x}{\partial x}\right) + H_y - \left(H_y + \frac{\partial H_y}{\partial y} dy\right) = -\frac{\partial H_x}{\partial x} - \frac{\partial H_y}{\partial y}. \tag{3.62}$$

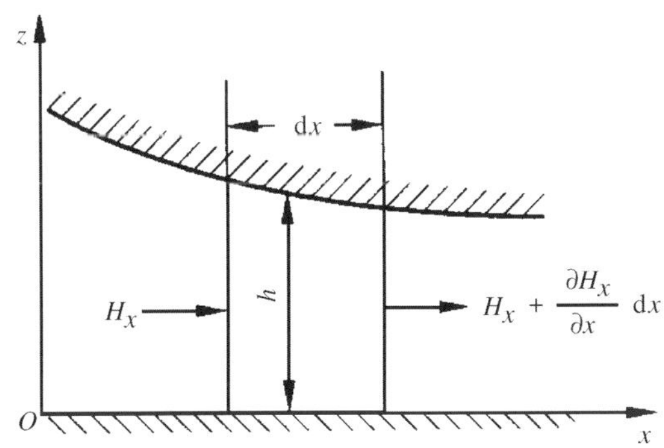

Figure 3.9 Heat flow.

If S is the mechanical work done on the unit cross-sectional area of the micro-column, according to the principle of the energy conservation, we have:

$$-\frac{\partial H_x}{\partial x}-\frac{\partial H_y}{\partial y}=\frac{S}{J}, \tag{3.63}$$

where J is the mechanical equivalent of heat.

Substitute H_x and H_y of Equations 3.58 and 3.59 into the above equation, we have

$$\frac{\partial(q_xT)}{\partial x}+\frac{\partial(q_yT)}{\partial y}=-\frac{S}{J\rho c}. \tag{3.64}$$

Because the flow continuity condition is

$$\frac{\partial q_x}{\partial x}+\frac{\partial q_y}{\partial y}=0. \tag{3.65}$$

Therefore, we have:

$$q_x\frac{\partial T}{\partial x}+q_y\frac{\partial T}{\partial y}=-\frac{S}{J\rho c}. \tag{3.66}$$

Now, let us discuss the work done by the cross-sectional area of the micro-column. A viscous flow contains two parts of the work: the flow work and the friction work. The former is a resistance to the pressure; the latter is the work consumed on the surface by the shear stress as shown in Figure 3.10.

From the figure we can see that the flow work in the x direction is equal to

$$\left(p+\frac{\partial p}{\partial x}dx\right)\left(q_x+\frac{\partial q_x}{\partial x}dx\right)-pq. \tag{3.67}$$

For the unit cross-sectional area along the x direction of the micro-column, that is, $\delta x=1$, if we ignore the high-order terms, the flow work will be

$$q_x\frac{\partial p}{\partial x}+p\frac{\partial q_x}{\partial x}. \tag{3.68}$$

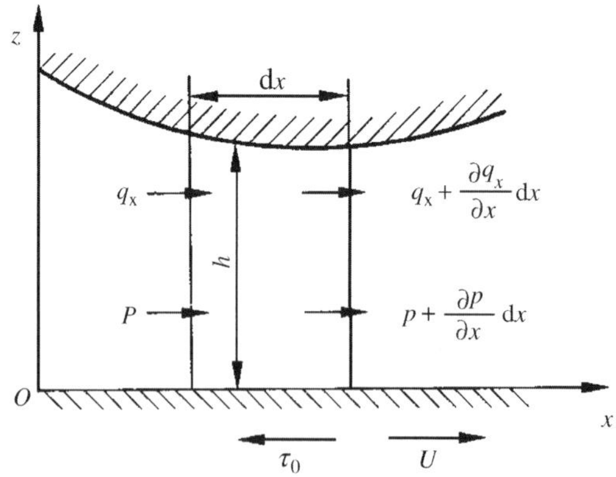

Figure 3.10 Fluid flow.

If we consider the flows in both the x and y directions, the total flow work W of the unit cross-sectional area of the micro-column is equal to:

$$W = q_x \frac{\partial p}{\partial x} + q_y \frac{\partial p}{\partial y} + p\left(\frac{\partial q_x}{\partial x} + \frac{\partial q_y}{\partial y}\right). \tag{3.69}$$

Because $\partial q_x/\partial x + \partial q_y/\partial y = 0$ (see Equation 3.66), the total flow work W can be written as

$$W = q_x \frac{\partial p}{\partial x} + q_y \frac{\partial p}{\partial y}. \tag{3.70}$$

In Chapter 2, the shear stress τ_0 on the moving plate is given. Then, the friction work consumed by the unit cross-sectional area of the micro-column is equal to

$$\tau_0 U = \left(-\frac{h}{2}\frac{\partial p}{\partial x} - \frac{\eta U}{h}\right)U. \tag{3.71}$$

Therefore, the total work S consumed by the cross-sectional area of the micro-column is

$$S = q_x \frac{\partial p}{\partial x} + q_y \frac{\partial p}{\partial y} - \left(\frac{h}{2}\frac{\partial p}{\partial x} + \frac{\eta U}{h}\right)U. \tag{3.72}$$

If we substitute q_x and q_y of Equations 2.11 and 2.12 into the above equation, we have

$$S = -\frac{\eta U^2}{h} - \frac{h^3}{12\eta}\left[\left(\frac{\partial p}{\partial x}\right)^2 + \left(\frac{\partial p}{\partial y}\right)^2\right]. \tag{3.73}$$

Then, if we substitute the above equation into Equation 3.66, we obtain the energy equation of hydrodynamic lubrication as follows:

$$q_x \frac{\partial T}{\partial x} + q_y \frac{\partial T}{\partial y} = \frac{\eta U^2}{J\rho c h} + \frac{h^3}{12\eta J\rho c}\left[\left(\frac{\partial p}{\partial x}\right)^2 + \left(\frac{\partial p}{\partial y}\right)^2\right]. \tag{3.74}$$

3.2.3 Numerical Solution of Energy Equation

If set

$$X = x/b;\ Y = y/l;\ H = h/h_0;\ a = b/l;\ P = \frac{h_0^2}{6U\eta_0 b}p;\ Q = \frac{q}{Uh_0};\ \eta^* = \frac{\eta}{\eta_0};\ T^* = \frac{2J\rho c h_0^2}{Ub\eta_0}T.$$

The dimensionless energy equation of Equation 3.74 is

$$\frac{\partial T^*}{\partial X} = \frac{1}{Q_x}\left\{-aQ_y\frac{\partial T^*}{\partial Y} + \frac{2\eta^*}{H} + \frac{6H}{\eta^*}\left[\left(\frac{\partial P}{\partial X}\right)^2 + \alpha^2\left(\frac{\partial P}{\partial Y}\right)^2\right]\right\}, \tag{3.75}$$

where η_0 is the initial viscosity; and

$$Q_x = \frac{H}{2} - \frac{H^3}{2}\frac{\partial P}{\partial X}$$

$$Q_y = -\frac{H^3}{2}\frac{\partial P}{\partial Y}. \tag{3.76}$$

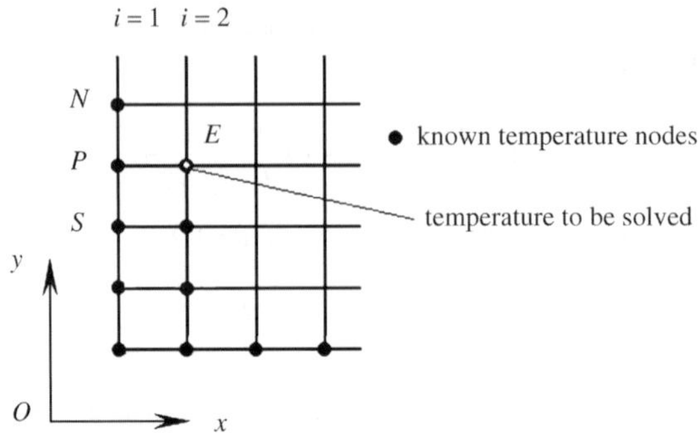

Figure 3.11 Temperature computational mesh.

From Equations 3.74 and 3.75, we can see that in order to obtain temperature we must first know pressure because in the equation there are $\partial P/\partial X$ and $\partial P/\partial Y$. However, pressure is also influenced by temperature inversely. Therefore, to solve a thermal lubrication problem, we must solve the Reynolds equation and the energy equation together.

In addition, the temperature analysis is two-dimensional, and in order to know $\partial T/\partial X$, $\partial T/\partial Y$ should be known in advance.

Calculation of a temperature field of a lubricant film is known as an initial value problem, because the temperature of the lubricant at the inlet must be known. To solve an initial value problem the marching method is generally used, that is, to solve it step by step. The basic steps are as follows:

As shown in Figure 3.11, divide the solution region into grids and select the ordinate system xOy. The given temperatures of the first row nodes at the boundary are used as the initial values. For an example, set the ambient temperature as the known temperature value at the boundary nodes so that T_{1j}^* is known.

Second, if T_{ij}^* is known, use the intermediate differential formulas to calculate $(\partial T^*/\partial Y)_{ij}$.

In addition, use the known pressure field to calculate $(\partial P/\partial X)_{ij}$ and $(\partial P/\partial Y)_{ij}$ as well as Q_{ij} and so on.

Third, substitute them into Equation 3.76 to determine $(\partial T^*/\partial X)_{ij}$.

Finally, the temperature T_{2j}^* of the second row nodes $i = 2$ can be obtained. Repeat the above steps to calculate T_{3j}^* and so on until the last row.

It should be pointed out that by using the above method to solve the temperature field the following situations should be paid attention to. First, the forward step in the direction must be the same as the flow direction of lubricant. If the direction is along the x axis, the $Q_x > 0$ condition should be met. However, when the pressure of supplied oil or at the inlet area is much higher, $Q_x < 0$ may occur, that is, the inverse flow area. Obviously, for the inverse flow area we cannot simply use the above-mentioned method. In addition, when $Q_x = 0$, $\partial T^*/\partial X$ will be infinite (see Equation 3.77). In this situation, the above method will not be able to be used either.

3.3 The Numerical Solution of Elastohydrodynamic Lubrication

For an EHL problem, in order to obtain an exact solution while fully considering the influences of the elastic deformation and pressure distribution, we must rely on numerical calculation.

3.3.1 EHL Numerical Solution of Line Contacts

Petrusevich gave an isothermal EHL numerical solution in line contacts first, and proposed a thickness formula. Although the formula is limited, the characteristics of the typical EHL pressure distribution and film shape are clear. Since then, Dowson and Higginson have given a series of systematic numerical calculations on the isothermal EHL in line contacts [3]. Based on the results, they proposed a more accurate formula of the thickness, which has been verified by experiments and is widely used.

3.3.1.1 Basic Equations

The equations to solve EHL problems in line contacts are as follows.

1. Reynolds equation

$$\frac{d}{dx}\left(\frac{\rho h^3}{12\eta}\frac{dp}{dx}\right) = U\frac{d(\rho h)}{dx} \tag{3.77}$$

where U is the average velocity, $U = (u_1 + u_2)/2$; h is the film thickness; η is the viscosity of lubricant, ρ is the density of lubricant. h, η and ρ are the function of x.

The boundary conditions of Reynolds equation are:

$$\text{At the inlet,}\quad p|_{x=x_1} = 0$$

$$\text{At the outlet,}\quad p|_{x=x_2} = 0;\quad \left.\frac{\partial p}{\partial x}\right|_{x=x_2} = 0.$$

where x_1 is the inlet position. The inlet position is based on the lubricant supply, usually $x_1 = (5\text{–}15)b$ is chosen; b is the half width of the contact region; x_2 is the outlet position, and it will be determined in the solution process.

2. Film thickness equation

   As shown in Figure 3.12, for the contact of an elastic cylinder and a rigid plane, the film thickness is expressed as follows:

$$h(x) = h_c + \frac{x^2}{2R} + v(x), \tag{3.78}$$

where h_c is the center thickness without elastic deformation; R is the equivalent radius. For two cylinders, $1/R = 1/R_1 + 1/R_2$; $v(x)$ is the elastic deformation generated by pressure.

3. The elastic deformation equation

   For the line contact problems, the length and radius of a contact body is always much larger than the width of contact region so that the problem can be considered as a plane strain state. Such an elastic

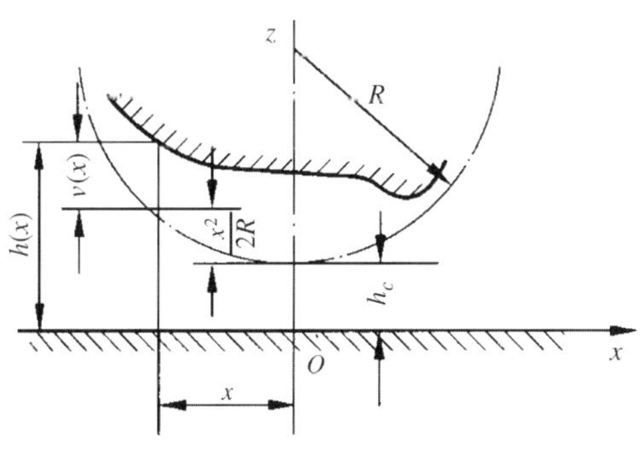

Figure 3.12 Film shape.

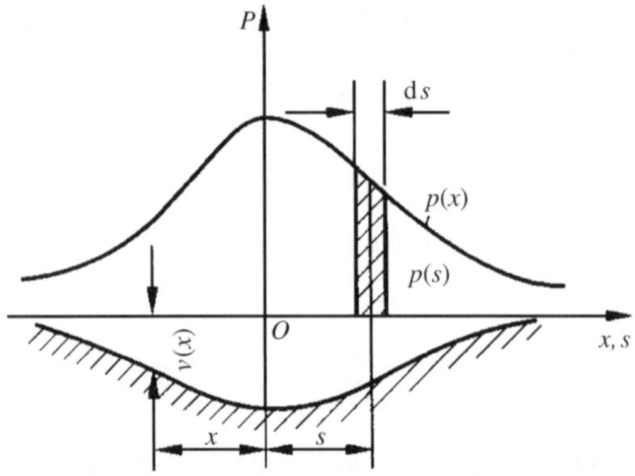

Figure 3.13 Elastic deformation.

deformation is as shown in Figure 3.13. According to the theory of elasticity, the elastic displacement along the vertical direction can be derived as follows:

$$v(x) = -\frac{2}{\pi E}\int_{s_1}^{s_2} p(s)\ln(s-x)^2 ds + c, \tag{3.79}$$

where $p(s)$ is the load distribution, or pressure; s_1 and s_2 are for the starting-point and the ending-point coordinates of $p(s)$; E is the equivalent modulus of elasticity, $1/E = 1/E_1 + 1/E_2$; c is a constant to be determined in calculation.

4. Viscosity-pressure relationship

   Barus viscosity-pressure formula is commonly used for convenience.

$$\eta = \eta_0 e^{\alpha p}, \tag{3.80}$$

where η_0 is the viscosity of the lubricant under $p = 0$.

5. Density–pressure relationship

   Fitting with the experimental data, a density–pressure relationship can be obtained as follows:

$$\rho = \rho_0\left(1 + \frac{0.6p}{1 + 1.7p}\right), \tag{3.81}$$

where ρ_0 is the density of the lubricant under $p = 0$.

3.3.1.2 Solution of the Reynolds Equation

From Equation 3.77, it can be seen that the pressure distributions are influenced by η, h_0 and ρ. Because the maximum increment of the density ρ with pressure p is about 33%, the density variation has little influence on solutions. Therefore, the lubricant is usually considered as an incompressible fluid, or a simple density–pressure relationship may be used for convenience. However, η having an exponential relation with p will be dramatically changed and the film thickness h has a cubic form in the Reynolds equation. Therefore, the visco-pressure effect and the elastic deformation have very significant influences so more attention must be paid to them in EHL.

In addition, from an EHL pressure distribution we can see that pressure p and its derivative dp/dx rapidly vary in a very narrow range. In order to solve the process stably, a parameter transformation normally needs to be used so that pressure varies slightly.

One commonly used transformation variable is the induced stress $q(x) = 1/\alpha(1 - e^{-\alpha p})$. If we consider the viscosity-pressure effect, the Reynolds equation of EHL becomes:

$$\frac{d}{dx}\left(\rho h^3 \frac{dq}{dx}\right) = 12\eta_0 U \frac{d(\rho h)}{dx}. \tag{3.82}$$

After having obtained $q(x)$, we can use an inverse parameter transformation to obtain $p(x)$, that is:

$$p(x) = -\frac{1}{\alpha}\ln[1 - \alpha q(x)].$$

In EHL calculation, the Vogelpohl transformation to set $M(x) = p(x)[h(x)]^{3/2}$ is also often used. If so, the Reynolds equation will be:

$$\frac{d}{dx}\left(\frac{\rho h^{3/2}}{\eta}\frac{dM}{dx}\right) - \frac{3}{2}\frac{d}{dx}\left(\frac{\rho h^{1/2} M}{\eta}\frac{dh}{dx}\right) = 12U\frac{d(\rho h)}{dx}. \tag{3.83}$$

3.3.1.3 Calculation of Elastic Deformation

If the pressure distribution $p(x)$ has been obtained, the deformation $v(x)$ can be obtained to integrate Equation 3.79. However, the deformation equation is singular at point $s = x$. This is one difficulty for calculating elastic deformation:

$$I = \int_{s_1}^{s_2} p(s)\ln(s-x)^2 ds. \tag{3.84}$$

To avoid the singularity, a simple way is to take sectional integrations. As the integral function is continuous except for $s = x$, it can be treated as follows:

$$I = \int_{s_1}^{x-\Delta x} p(s)\ln(s-x)^2 ds + \int_{x+\Delta x}^{s_2} p(s)\ln(s-x)^2 ds. \tag{3.85}$$

However, the difficulty of this approach is how to determine Δx properly. If incorrect, it may cause a considerable calculation error.

Another way to overcome singularity is to use a discrete integration method, referring to reference [4]. The main steps are as follows:

Divide the integral region $[x_1, x_2]$ into a number of sub-regions and express pressure distribution $p(x)$ approximately as a polynomial function of x:

$$p(x) = c_1 + c_2 x + c_3 x^2. \tag{3.86}$$

The coefficients c_1, c_2 and c_3 can be determined according to the known pressure at the nodes. For example, on the interval $[x_i, x_{i+1}]$, the pressure distribution is expressed as follows:

$$p_i(x) = c_{1i} + c_{2i}x + c_{3i}x^2. \tag{3.87}$$

Therefore, the deformation integration becomes as follows:

$$I_i = \int_{x_i}^{x_{i+1}} (c_{1i} + c_{2i}s + c_{3i}s^2)\ln(x-s)^2 ds$$
$$= 2\left[c_{1i}\int_{x_i}^{x_{i+1}} \ln|x-s|ds + c_{2i}\int_{x_i}^{x_{i+1}} s\ln|x-s|ds + c_{3i}\int_{x_i}^{x_{i+1}} s^2\ln|x-s|ds\right] \quad (3.88)$$

The analytical integral formula, such as $\int \ln s ds$ and $\int s \ln s ds$, can be used in the calculation of *Ii* of Equation 3.88.

Furthermore, in the above calculation, x is the coordinate and should be selected in the three intervals, $x \leq x_i$, $x_i < x < x_{i+1}$ and $x_{i+1} \leq x$. Except for $x_i < x < x_{i+1}$, the singularity will appear in the other two intervals, that is, $x = x_i$ or $x = x_{i+1}$. For example, when $x \leq x_i$, if set $DX = x_{i+1} - x_i$ and $X = x_i - x$, we have

$$\frac{I_i}{2} = (c_{1i} + c_{2i}x + c_{3i}x^2)[(X+DX)\ln(X+DX) - X\ln X - DX]$$
$$+ (c_{2i} + 2c_{3i}x)\left[(X+DX)^2\ln(X+DX) - X^2\ln X - \frac{2XDX + DX^2}{2}\right]\Big/2$$
$$+ c_{3i}\left[(X+DX)^3\ln(X+DX) - X^3\ln X - \frac{3XDX(X+DX) + DX^3}{3}\right]\Big/3 \quad (3.89)$$

As long as $X \neq 0$, I_i can be obtained. If $X = 0$, that is, $x = x_i$, I_i is a singular integral. In this case, we can use $\lim_{t\to 0^+} t\ln t = 0$ to obtain I_i. Therefore, Equation 3.89 becomes:

$$\frac{I_i}{2} = (c_{1i} + c_{2i}x + c_{3i}x^2)[DX\ln DX - DX] + (c_{2i} + 2c_{3i}x)\left[DX^2\ln DX - \frac{DX^2}{2}\right]\Big/2$$
$$+ c_{3i}\left[DX^3\ln DX - \frac{DX^3}{3}\right]\Big/3 \quad (3.90)$$

For $x_i < x_{i+1} \leq x$, using the same method above can overcome the singularity at $x = x_{i+1}$, and a similar formula may be obtained as well.

3.3.1.4 Dowson-Higginson Film Thickness Formula of Line Contact EHL

Based on a large number of systematically numerical calculations, Dowson and Higginson proposed twice the minimum film thickness formula of EHL in line contacts. Their experimental results showed that their formula results are very close to the most measured values of film thickness. The dimensionless formula of 1967 is

$$H^*_{\min} = 2.65\frac{G^{*0.54}\,U^{*0.7}}{W^{*0.13}}. \quad (3.91)$$

The dimensional form of the above formula is

$$h_{\min} = \frac{2.65\,\alpha^{0.54}(\eta_0 U)^{0.7}\,R^{0.43}\,l^{0.13}}{E^{0.03}\,W^{0.13}}, \quad (3.92)$$

where H^*, G^*, U^* and W^* are the dimensionless parameters given in Section 2.4.

As we can see from the above formula, the minimum film thickness $h_{\min}$ of line contact EHL increases significantly with the initial viscosity η_0 and the average speed U, but the load effect is very weak, that is, a substantial increase of load decreases the film thickness very little. This is one of the basic special features of EHL.

It should be pointed out that the Dowson-Higginson formula is used to calculate the minimum necking thickness $h_{\min}$, but the Grubin formula is used to calculate the thickness h_0 at the inlet of the contact zone, that is, $x = -b$. Dowson and Higginson showed that numerical calculated contact center thickness h_c is very close to the calculated results of Grubin formula. The ratio of the minimum film thickness and the central film thickness $h_{\min}/h_c = 3/4$.

It should be noted that both Dowson-Higginson formula and Grubin formula have their applications. When the material parameter $G^* < 1000$, that is, the low elastic modulus of the solid materials and a low viscosity pressure coefficient of a lubricant, or when the load parameter $W^* < 10^5$, corresponding to a light load condition, the calculation errors of Equation 3.91 are quite large. In addition, the above formulas are derived under the condition that the lubricant supplication is sufficient for an isothermal EHL. If the oil supply is short, the film thickness will reduce, while in high-speed conditions when heat causes the viscosity to decrease significantly, the film thickness will decrease.

3.3.2 EHL Numerical Solution of Point Contacts

Generally, point contact problems include two spherical bodies to form an elliptical contact area. This is more complex than line contact problems. In 1965, Archard and Cowking proposed the first Grubin approximate solution for the circular contact EHL [5]. In 1970, Cheng gave a solution for an elliptic contact EHL problem [6]. Later, Hamrock and Dowson proposed the formula for calculating the minimum film thickness according to their numerical results of the elliptical contact EHL problems [3]. The first author and Zhu Dong presented a full numerical solution for elliptical contact EHL problem [7]. Below, we will briefly introduce its main points.

3.3.2.1 The Reynolds Equation

If the surface speed is not along the contact zone axis, the Reynolds equation should be written as

$$\frac{\partial}{\partial x}\left(\frac{\rho h^3}{\eta}\frac{\partial p}{\partial x}\right)+\frac{\partial}{\partial y}\left(\frac{\rho h^3}{\eta}\frac{\partial p}{\partial y}\right)=12\left(U\frac{\partial \rho h}{\partial x}+V\frac{\partial \rho h}{\partial y}\right) \tag{3.93}$$

Figure 3.14 expresses the coordinates and the solution region. x is the short axis of the ellipse contact zone. If the velocity components of the two surfaces in the x and y directions are respectively u_1, u_2, v_1 and

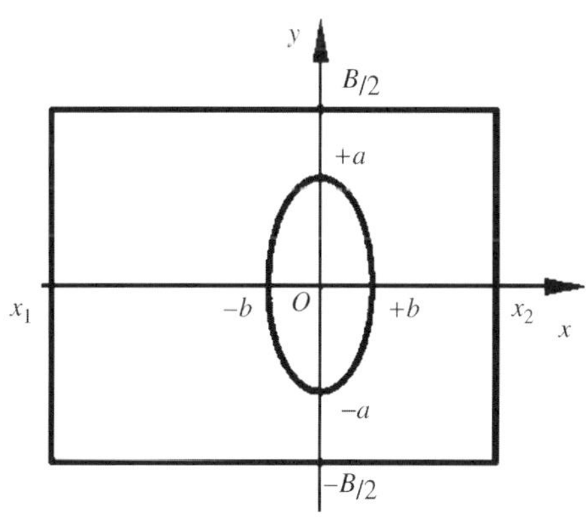

Figure 3.14 Point contact solution region.

v_2, the average velocities are noted as follows:

$$u = \frac{1}{2}(u_1 + u_2)$$
$$v = \frac{1}{2}(v_1 + v_2). \tag{3.94}$$

The boundary conditions of Equation 3.93 are: the inlet and the side pressures at the borders are equal to zero, that is, $p = 0$ where $x = x_1$ and $y = \pm B/2$. At the outlet, we use Reynolds boundaries conditions, that is, $p = 0$ and $\partial p/\partial x = 0$ where $x = x_2$.

As this is the same situation as the line contact EHL, the induced pressure $q(x, y)$ can be introduced as follows.

$$q(x, y) \equiv \frac{1}{\alpha}\left[1 - e^{-\alpha p(x,y)}\right], \tag{3.95}$$

because

$$\frac{\partial q}{\partial x} = e^{-\alpha p}\frac{\partial p}{\partial x}, \qquad \frac{\partial q}{\partial y} = e^{-\alpha p}\frac{\partial p}{\partial y}. \tag{3.96}$$

Substituting them into Equation 3.93, we have:

$$\frac{\partial}{\partial x}\left(\rho h^3\frac{\partial q}{\partial x}\right) + \frac{\partial}{\partial y}\left(\rho h^3\frac{\partial q}{\partial y}\right) = 12\eta_0\left[u\frac{\partial}{\partial x}(\rho h) + v\frac{\partial}{\partial y}(\rho h)\right] \tag{3.97}$$

Equation 3.97 is the two-dimensional Reynolds equation with the viscosity–pressure relationship of lubricant considered.

3.3.2.2 Elastic Deformation Equation

According to the theory of elasticity, if the surface pressure is $p(x, y)$, the surface deformation $\delta(x,y)$ can be described as follows:

$$\delta(x, y) = \frac{2}{\pi E}\iint_{\Omega}\frac{p(s, t)}{\sqrt{(x-s)^2 + (y-t)^2}}dsdt, \tag{3.98}$$

where s and t are the integral variables in the x and y directions; Ω is the solution region.

Obviously, when $s = x$, $t = y$, Equation 3.98 is singular. To overcome this, similar approaches as the line contact EHL are adopted. Moving the coordinate origin to $\xi = x - s$ and $\zeta = y - t$, Equation 3.98 becomes:

$$\delta(x, y) = \frac{2}{\pi E}\iint_{\Omega}\frac{p(\xi, \zeta)}{\sqrt{\xi^2 + \zeta^2}}d\xi d\zeta. \tag{3.99}$$

For the polar coordinates, set $x = r\cos\theta$, $y = r\sin\theta$, then we have

$$\delta(x, y) = \frac{2}{\pi E}\iint_{\Omega}p(r, \theta)drd\theta. \tag{3.100}$$

Usually, the calculation task of elastic deformation is excessive. A very effective way to overcome this difficulty is to use a deformation matrix. The steps are as follows:

First, divide the solution region into a mesh, for example, m nodes in the x direction and n in the y direction, that is, $i = 1, 2, \ldots, m$ and $j = 1, 2, \ldots, n$. Define D_{ij}^{kl} as the elastic deformation of Node k and l caused by pressure p_{ij}; the total deformation of Node k and l is equal to

$$\delta_{kl} = \frac{2}{\pi E}\sum_{i=1}^{n}\sum_{j=1}^{m} D_{ij}^{kl} p_{ij}. \tag{3.101}$$

Therefore, D_{ij}^{kl} only need to be calculated once and stored up to be used repeatedly in the iterative process. Thus it may reduce a large amount of computation work.

Because the total number of matrix D_{ij}^{kl} is $(m \times n)^2$, a uniform mesh will save much more storage. If the mesh is uniform in the y direction, we have $D_{ij}^{kl} = D_{is}^{kl}$, where $s = |j - l| + 1$. So the total number of D_{ij}^{kl} is reduced to $m^2 \times n$. If the uniform mesh is used in the x directions, the number will be further reduced to $m \times n$.

When all the deformations are obtained, the film thickness will be

$$h(x, y) = h_0 + \frac{x^2}{2R_x} + \frac{y^2}{2R_y} + \delta(x, y), \tag{3.102}$$

where R_x and R_y are the equivalent radius in the x, and y directions, respectively. Then, substituting Equation 3.102 into the Reynolds equation, pressure distribution can finally be obtained.

3.3.2.3 Hamrock-Dowson Film Thickness Formula of Point Contact EHL

Hamrock and Dowson proposed the following film thickness formula for isothermal point contact EHL after carrying out numerical analysis [3]:

$$H_{\min}^* = 3.63\frac{G^{*0.49}U^{*0.68}}{W^{*0.073}}\left(1 - e^{-0.68k}\right), \tag{3.103}$$

$$H_c^* = 2.69\frac{G^{*0.53}U^{*0.67}}{W^{*0.067}}\left(1 - 0.61e^{-0.73k}\right), \tag{3.104}$$

where $H_{\min}^* = h_{\min}/R_x$ is the dimensionless minimum film thickness, $H_c^* = h_c/R_x$ is the dimensionless central film thickness; $G^* = \alpha E$ is the dimensionless material elastic module; $U^* = \eta_0 u/ER_x$ is dimensionless speed; $W^* = w/ER_x^2$ is dimensionless load; $k = a/b$ is the ellipticity, which is approximately equal to $k = 1.03\left(R_x/R_y\right)^{0.64}$.

From the above formula, it can be known that if other parameters except the ellipticity are kept unchanged, the film thickness rapidly decreases with increase of the ellipticity. If $k > 5$ the film thickness changes slowly with k. With comparison, it is easy to know that when $k > 5$, the central film thickness of the point contact EHL is approximately equal to the film thickness of the line contact EHL.

In Figures 3.15 and 3.16, there are the pressure distribution and film shape of point contact EHL by Ranger [8]. They are much more complex than those of the line contact EHL.

Figure 3.15 shows that in the contact area of the point contact EHL, oil film gas appears as a horseshoe-shaped depression. The minimum film thickness appears at both sides of the neck.

We can see from Figure 3.16 that pressure distribution of point contact EHL has a crescent of the secondary pressure peak region, but the pressure peak in the center of this region is the highest, and is far from the contact center.

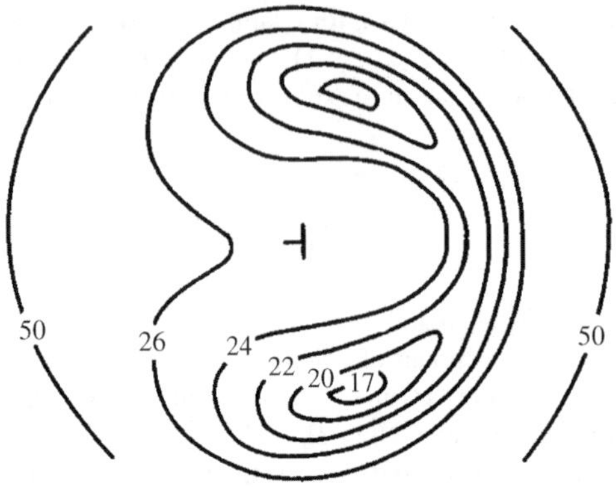

Figure 3.15 Film thickness contours.

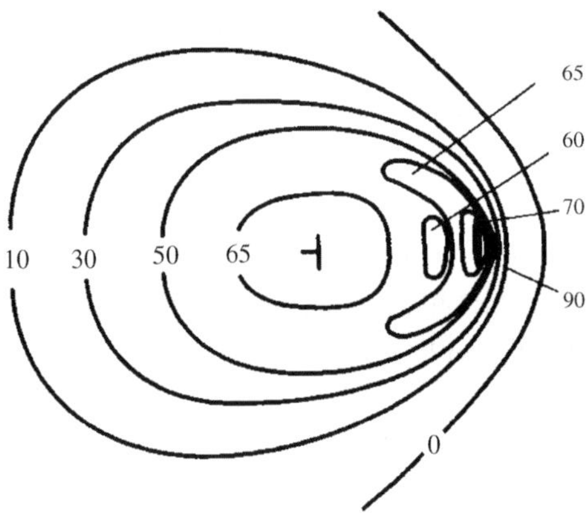

Figure 3.16 Pressure contours.

3.4 Multi-Grid Method for Solving EHL Problems

The multi-grid method is a very powerful tool in solving EHL problems. Furthermore, the multi-grid integration method has rapid speed characteristics. We will briefly introduce its calculation steps below.

3.4.1 Basic Principles of Multi-Grid Method

Multi-grid method is put forward for solving large algebraic equations. With the iterative method of solving algebraic equations, the deviations of an approximate solution to the exact solution can be decomposed into a variety of frequency deviation components. The higher frequency components can be quickly eliminated in the fine grids, while the lower frequency components can only be eliminated in the coarse grids. The basic idea of the multigrid method is to carry out iteration among the fine and coarse grids so as to eliminate all the deviative components [9].

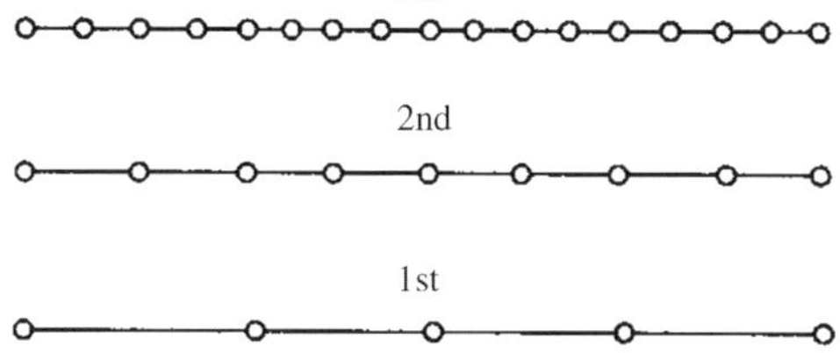

Figure 3.17 One-dimensional uniform multi-grid structure.

3.4.1.1 Grid Structure

Take a one-dimensional problem with three uniform grids as an example. The finest grid has 17 nodes, the middle 9 and the coarsest 5, as shown in Figure 3.17.

Note the coarsest grid as the first mesh, and the finest is the *m*th mesh, where m = 3. For convenience, it is usually better to choose $m = 2^n + 1$. In addition, the uniform mesh is appropriate although it should not be. For a two-dimensional problem, the increments of the coordinates in the two directions may not be equal, that is, $\Delta x \neq \Delta y$. However, the uniform mesh will be quite convenient.

3.4.1.2 Discrete Equation

If the solution region is Ω, the solving equation is generally written as

$$\mathrm{L}\boldsymbol{u} = f, \tag{3.105}$$

where *l* is an operator, which can be differential, integral, or another operator; $\boldsymbol{u}$ is the variable vector to be solved; f is the right-hand side item vector, which has been known.

When using a numerical method to solve Equation 3.105, first divide Ω into a mesh. Then, discretize Equation 3.105 into the algebraic equations. For multi-grid method, the grids must be given in each mesh. For the *k*th mesh, the discrete formula is recorded as

$$\mathrm{L}^k \boldsymbol{u}^k = \boldsymbol{f}^k, \tag{3.106}$$

where $\boldsymbol{u}^k = (u^k) = [u_1^k, u_2^k \ldots, u_{nk-1}^k]^T; f = (f^k) = [f_1^k, f_2^k, \ldots, f_{nk-1}^k]^T$.

3.4.1.3 Transformation

For application of multi-grid method, one should generally choose one iterative method, such as Gauss-Seidel iterative method, to obtain the approximate solution from the algebraic equations of Equation 3.106. The iterative process is generally carried out to iterate several times on one mesh, and then the results are transferred to another mesh. In the coarsest mesh a large number of iterations are usually carried out. Because the number of nodes of the coarsest grid is very few, there is little time spent on iterating. Between the two adjacent meshes, the process in which the results of a finer grid are transferred to a coarser mesh is called restriction. This is achieved by a restriction operator. The contrary process is called the extension. This is achieved by an extension operator. Some simple restriction and extension processes are as follows:

1. Mapping operator
   A mapping operator is a special operator which can be used as a restriction or extension operator. It transfers the results of nodes of one mesh directly to the corresponding node of the adjacent mesh, as shown in Figure 3.18.

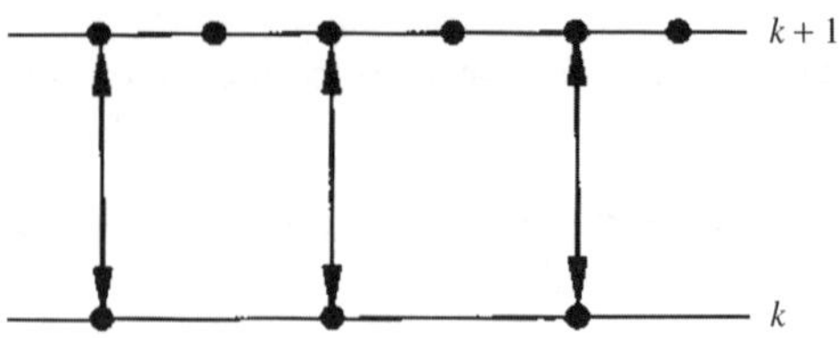

Figure 3.18 Mapping operator.

2. Weighted operator
   The weighted operator will transfer the present results to the adjacent mesh by weighting the corresponding and neighbor variables and then transferring them to the coaser mesh, as shown in Figure 3.19. The weighted operator is suitable for the linear problem. As for a strong nonlinear problem, a higher order weighted operator can be used.
3. Interpolation operator
   As shown in Figure 3.20, the operator will transfer the results to the finer mesh by mapping and weighing both.

3.4.2 *Nonlinear Full Approximation Scheme of Multi-Grid Method*

In multi-grid method, if we solve a linear problem, a coarse grid modification is usually selected, but the full approximation scheme (FAS) is used for a nonlinear problem. Because EHL problems are nonlinear, we will discuss FAS in detail although FAS is also suitable for a linear problem.

Applying FAS to a nonlinear problem, the algebraic equations of the kth mesh can be written in the form of Equation 3.107.

$$L^k \boldsymbol{u}^k = \boldsymbol{f}^k \tag{3.107}$$

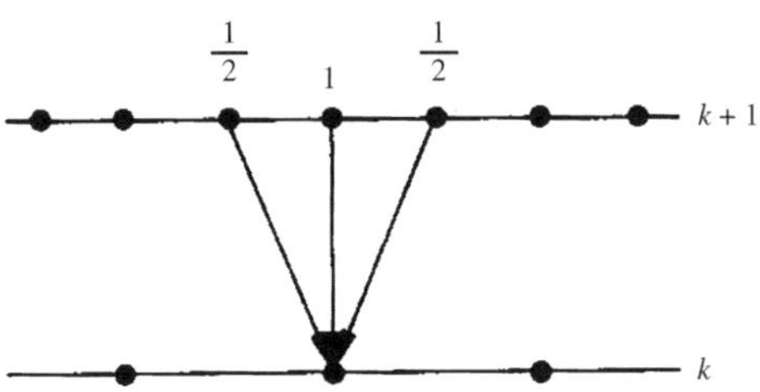

Figure 3.19 Weighted restriction operator.

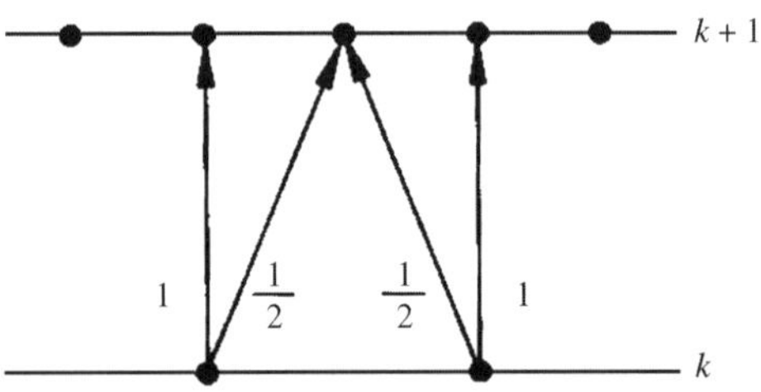

Figure 3.20 Interpolation operator.

If $k \neq 1$, take $\bar{\boldsymbol{u}}^k$ as the initial value of Equation 3.107 to do m_1 times relaxation iteration. Then use the transfer operator I_k^{k-1} to transfer the obtained approximate solution $\tilde{\boldsymbol{u}}^k$ to the next mesh. Then, take $\tilde{\boldsymbol{u}}^k$ as the initial value of the new mesh to iterate again, that is

$$\bar{\boldsymbol{u}}^{k-1} = I_k^{k-1}\tilde{\boldsymbol{u}}^k. \tag{3.108}$$

In the k-1th mesh, the algebraic equations are

$$L^{k-1}\boldsymbol{u}^{k-1} = \boldsymbol{f}^{k-1}. \tag{3.109}$$

The key factor of FAS is to determine $\boldsymbol{f}^{k-1}$ in Equation 3.109. Because the goal to solving Equation 3.108 is to modify the approximate solution $\tilde{\boldsymbol{u}}^k$ of Equation 3.106, in order to analyze $\boldsymbol{f}^{k-1}$, we have to analyze Equation 3.106 and $\tilde{\boldsymbol{u}}^k$ first.

From Equation 3.106, subtracting $\mathrm{L}^k\tilde{\boldsymbol{u}}^k$ from both right-hand and left-hand sides, we have:

$$L^k\boldsymbol{u}^k - L^k\tilde{\boldsymbol{u}}^k = \boldsymbol{f}^k - L^k\tilde{\boldsymbol{u}}^k. \tag{3.110}$$

We denote the right-hand side of the above equation as $\boldsymbol{r}^k$, which is the error of the equation:

$$\boldsymbol{r}^k = \boldsymbol{f}^k - L^k\tilde{\boldsymbol{u}}^k. \tag{3.111}$$

Because the error on the k-1th mesh is quite different from the error on the kth mesh, with transformation of $\tilde{\boldsymbol{u}}^k$, $\boldsymbol{r}^k$ must be transferred to the k-1th mesh, that is

$$L^{k-1}\boldsymbol{u}^{k-1} - L^{k-1}(I_k^{k-1}\tilde{\boldsymbol{u}}^k) = I_k^{k-1}\boldsymbol{r}^k. \tag{3.112}$$

The above transformation is accurate for a linear problem, but approximate for a nonlinear problem. Substituting Equation 3.111 into the above, we have

$$L^{k-1}\boldsymbol{u}^{k-1} = L^{k-1}(I_k^{k-1}\tilde{\boldsymbol{u}}^k) + I_k^{k-1}(\boldsymbol{f}^k - L^k\tilde{\boldsymbol{u}}^k). \tag{3.113}$$

Compare Equation 3.106 with Equation 3.113, we can see that the right-hand side of Equation 3.111 should be

$$\boldsymbol{f}^{k-1} = L^{k-1}(I_k^{k-1}\boldsymbol{u}^k) + I_k^{k-1}(\boldsymbol{f}^k - L^k\tilde{\boldsymbol{u}}^k). \tag{3.114}$$

From Equation 3.114 it is known that only on the finest mesh, that is, $k = m$, the numerical calculation of the right-hand side items of the equation can be directly obtained from the original equation. However, on the coarser mesh, all the right-hand side items of the equation contain the errors of the approximate solution on the coarser meshes.

Having obtained $\boldsymbol{f}^{k-1}$ of Equation 3.106, the algebraic equations of the k-1th mesh have been determined. Then, set $k = k - 1$, calculation can be carried out continuously on the finer mesh. If $k \neq 1$ carry out relaxation iteration m_1 times or m_0 times if $k = 1$.

If the kth mesh smooth process has been finished, the obtained $\tilde{\boldsymbol{u}}^k$ will be sent to the other mesh to obtain an approximate solution. We usually do not transfer $\tilde{\boldsymbol{u}}^k$ to the mesh to be smoothened directly, but modify $\tilde{\boldsymbol{u}}^k$ on the original mesh first, and then transfer to the finer mesh with interpolation. So, the result combining $\tilde{\boldsymbol{u}}^k$ with $\tilde{\boldsymbol{u}}^{k+1}$ is taken as the beginning of the $k+1$th mesh to carry out relaxation iteration m_2 times until the finest mesh. The process can be expressed as the following formula.

$$\bar{\boldsymbol{u}}^{k+1} = \tilde{\boldsymbol{u}}^{k+1} + I_k^{k+1}(\tilde{\boldsymbol{u}}^k - I_{k+1}^k\tilde{\boldsymbol{u}}^{k+1}). \tag{3.115}$$

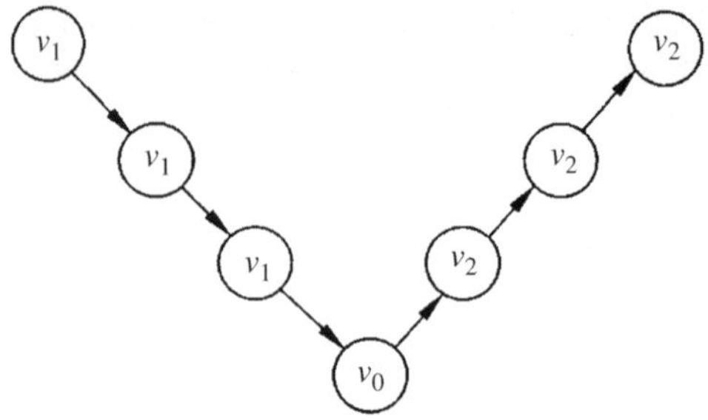

Figure 3.21 V iteration ($k_{max} = 4$).

3.4.3 *V and W Iterations*

Multi-grid method is an iterative process, by using restriction and extension operators alternatively on different meshes to smooth Equation 3.106. V iteration and W iteration are the typical processes. Figure 3.21 shows V iteration with $k_{max} = 4$, while Figure 3.22 shows a W iteration with $k_{max} = 4$. Here, k_{max} is the maximum number of the meshes; m_0, m_1 and m_2 are the numbers of restricting or extending relaxation iterations at the bottom, top and middle meshes respectively.

3.4.4 *Multi-Grid Solution of EHL Problems [10]*

3.4.4.1 Iteration Methods

The iterative process of the multi-grid method includes pressure correction and load balancing by adjusting the rigid displacement. All these calculations are carried out on the same mesh. For pressure correction, the Gauss-Seidel iteration method is commonly used while pressure is low. If pressure is high, the method may cause divergence. Therefore, Jacobi bipolar iteration will be chosen in a high pressure area. Both iterative methods can be written as

$$\bar{p}_i = \tilde{p}_i + c_1 \delta_i, \tag{3.116}$$

where c_1 is the relaxation factor; δ_i is the pressure correction quantity; $\bar{p}_i$ and $\tilde{p}_i$ are the pressures before and after iteration respectively.

For the kth mesh, the solving equation can be simply written as

$$L_i(p_i) = 0. \tag{3.117}$$

In Equation 3.116, $\delta_i = (\partial L_i / \partial p_i)^{-1} \bar{\gamma}_i$ is for the Gauss-Seidel method; $\delta_i = (\partial L_i / \partial p_i - \partial L_i / \partial p_{i-1})^{-1} \tilde{\gamma}_i$ is for the Jacobi bipolar method. $\bar{\gamma}_i = -\left[\varepsilon_{i-1/2}\bar{p}_{i-1} - (\varepsilon_{i-1/2} + \varepsilon_{i+1/2})\bar{p}_i + \varepsilon_{i+1/2}\bar{p}_{i+1}\right]/\delta^2 + (\bar{\rho}_i \tilde{h}_i - \bar{\rho}_{i-1}\tilde{h}_{i-1})/\delta$; $\tilde{\gamma}_i$ is obtained

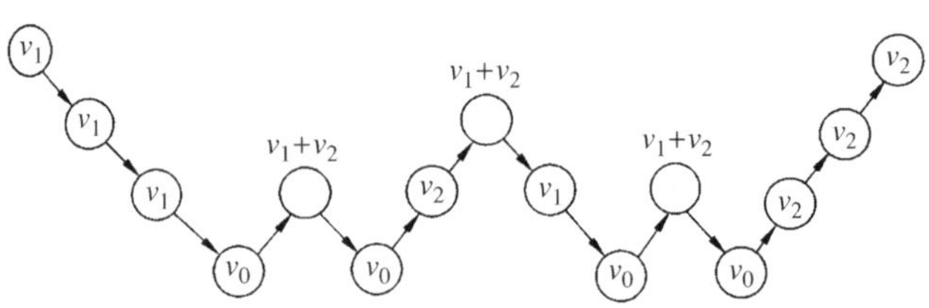

Figure 3.22 W iteration ($k_{max} = 4$).

Because film thickness h_i is the function of pressure, by replacing $\bar{p}_{i-1}$ by $\tilde{p}_{i-1}$ in the formula of $\tilde{\gamma}_i$ the derivative $\partial L_i/\partial p_i$ should be considered. For convenience, the derivative of ε need not be counted. Therefore,

$$\frac{\partial L_i}{\partial p_i} = -\left(\varepsilon_{i-1/2} + \varepsilon_{i+1/2}\right)/\delta^2 + \frac{1}{\pi}\left(\bar{\rho}_i K_{ii}^{\delta\delta} - \bar{\rho}_i K_{i-1,i}^{\delta\delta}\right)\Big/\delta. \tag{3.118}$$

The calculation of $\partial L_i/\partial p_{i-1}$ is similar to that of $\partial L_i/\partial p_i$.

By using Jacobi bipolar iteration to add δ_i to pressure of this point, one must remember to subtract δ_i from the pressure of the front point as well. That is,

$$\bar{p}_i = \tilde{p}_i + c_2\delta_i, \tag{3.119}$$

$$\bar{p}_{i-1} = \tilde{p}_{i-1} - c_2\delta_i, \tag{3.120}$$

The load balance condition can be realized by modifying the rigid displacement h_0 as follows:

$$\bar{h}_0 = \tilde{h}_0 + c_3\left[g^\Delta - \frac{\Delta}{\pi}\sum_{j=1}^{N-1}\left(p_j + p_{j+1}\right)\right], \tag{3.121}$$

where c_2 and c_3 are the relaxation factors; Δ is the increment distance of the coarsest mesh; g^Δ is the dimesionless load on the coarsest mesh.

In order to stabilize the iteration process, modification of the rigid body displacement is only carried out on the coarsest mesh. This is also of benefit in reducing the computational work.

3.4.4.2 Iterative Division

Besides modifying pressure, the two different iteration methods can be used in different zones for a single problem. Although we can separate the whole region into the high and low pressure zones, we should give a criterion to divide the region for the two iterative methods. There are two parts in Equation 3.118 to influence pressure. They are:

$$\text{Pressure part} \quad A_1 = \left(\varepsilon_{i-1/2} + \varepsilon_{i+1/2}\right)/\delta^2,$$

$$\text{Thickness part} \quad A_2 = \frac{1}{\pi}\left(\bar{p}_i K_{ii}^{\delta\delta} - \bar{p}_i K_{i-1,i}^{\delta\delta}\right)/\delta.$$

When A_1 is larger, the Gauss-Seidel method is more effective. When A_2 is larger, because γ_i does not allow modification of the film thickness, the Gauss-Seidel method is not effective as it is easy to diverge. Therefore, Jacobi bipolar iteration should be used. Our calculations show that when $A_1 \geq 0.1A_2$, the Gauss-Seidel method will be adopted. When $A_1 < A_2$, the Jacobi bipolar method is more effective. Between 0.1 A_2, $< A_1 < A_2$, both methods can be used.

3.4.4.3 Relaxation Factors

In a multi-grid method iterative process, there are three relaxation factors that need to be selected. They are the Gauss-Seidel iterative relaxation factor c_1, the Jacobi bipolar iterative relaxation factor c_2 and the rigid body displacement iterative relaxation factor c_3. The choice of these factors usually depends on experience. According to the author's experience, the ranges of the first two factors are: $c_1 = 0.3$–1.0 and $c_2 = 0.1$–0.6. Actually, c_2 has a greater influence on convergence, especially in heavy load conditions.

For such a situation, c_2 should be smaller. There is no rule to determine the region of c_3. In the following, we give a method to determine c_3 using the existing empirical formula. The film thickness usually has a relationship with the load as follows:

$$h = G^{\alpha} U^{\beta} W^{\gamma}, \tag{3.122}$$

where G is the shear elastic modulus; α, β, and γ are the indexes of the empirical formula. If unbalanced, the corresponding load increment and a film thickness increment have the following relation:

$$dh = \gamma G^{\alpha} U^{\beta} W^{\gamma-1} dW. \tag{3.123}$$

Because $-\mathrm{dW} = g^{\Delta} - \Delta/\pi \sum_{j=1}^{N-1} \left(p_j + p_{j+1}\right)$, therefore the relationship of dW and dh is

$$dh = \bar{h}_0 - \tilde{h}_0 = -c_3 dW. \tag{3.124}$$

Using Equations 3.123 and 3.124, it is easy to determine c_3 as follows:

$$c_3 = -\frac{dh}{dW} = -\gamma G^{\alpha} U^{\beta} W^{\gamma-1}. \tag{3.125}$$

3.4.4.4 Numbers of Iteration Times

Usually, in each mesh iteration will be carried out several times. In the coarsest mesh, iterative times are equal to m_0 for the downward process, m_1 and m_2 for the upward process. According to the author's experience: $m_1 = 2$, $m_0 = 5$–20, $m_2 = 1$; see Figure 3.22 [10–13].

3.4.5 *Multi-Grid Integration Method*

The computing time of multi-grid integration method is nearly proportional to the node number. Therefore, for an EHL problem with many nodes, the advantage of multi-grid integration method is obvious [9].

We introduce integration for a line contact problem between two meshes first. If we understand this, the multi-grid integration will be similar.

The multi-grid integration method is first use to transfer the variable from the finer mesh to the coarser mesh. Then, integrate the variable on the coarser mesh and transfer the integrated result back to the finer mesh. After modification, a result meeting the accuracy requirement is obtained.

Suppose there are two meshes. The finer one is indicated by a superscript h and the node number with subscripts i and j; for the coarser one with H, I, and J. If I or J is equal to 0, 1. . ., or N, and i or j is equal to 0, 1, . . ., n, we have $n = 2N - 1$ because we always assume that the node numbers of the finer mesh is twice as much as that of the coarser one.

For a line contact EHL problem, the integral formula of elastic deformation is

$$w(x) = \int_{x_a}^{x_b} \ln\left|x - x'\right| p(x') dx'. \tag{3.126}$$

The numerical integral calculation formula on the finer mesh is equal to

$$w_i^h = \sum_{j=0}^{n} K_{i,j}^{hh} p_j^h, \tag{3.127}$$

where $K_{i,j}^{hh}$ is the integration coefficient. It uses two superscript "h"s. The first h indicates the finer mesh and node number i, while the second h corresponds to the finer mesh and node number j; p_j^h is the dimensionless pressure and its superscript and subscript are similar to $K_{i,j}^{hh}$.

We know that the amount of calculation on the finer mesh is very heavy. If the numerical integration is carried out on the coarser mesh, it will be

$$w_I^H = \sum_{J=0}^{N} K_{I,J}^{HH} p_J^H. \tag{3.128}$$

Although the computational times of Equation 3.128 are much less than those of Equation 3.127, because the result of the coarser grid is not as accurate as that of the finer, it needs to be modified in order to get the same accuracy. The modification includes the following steps:

3.4.5.1 Transfer Pressure Downwards

Although we can directly transfer the pressure from the finer mesh to the coarser mesh node to node, in order to consider the variation of pressure, it is better to transfer pressure with an interpolation formula as follows.

$$p_I^H = \frac{1}{32}\left(-p_{2I-3}^h + 9p_{2I-2}^h + 16p_{2I-1}^h + 9p_{2I}^h - p_{2I+1}^h\right) \qquad I = 2, 3, \cdots, N-1. \tag{3.129}$$

In the above equation, the pressure on the left-hand side is on the coarser mesh, while the pressures on the right-hand side are on the finer mesh. For $I = 1$ and N, the formulas become

$$p_I^H = \frac{1}{32}\left(16p_1^h + 18p_3^h - 2p_5^h\right), \tag{3.130}$$

$$p_N^H = \frac{1}{32}\left(-p_{n-2}^h + 18p_{n-1}^h + 16p_n^h\right). \tag{3.131}$$

3.4.5.2 Transfer Integral Coefficients Downwards

A mapper operator is used to transfer the integral coefficients.

$$K_{I,J}^{HH} = K_{2I-1,2J-1}^{hh} \tag{3.132}$$

In the above equation, the superscripts and subscripts are the same as previously mentioned.

3.4.5.3 Integration on the Coarser Mesh

The integration on the coarser mesh is the same form of Equation 3.128. However, the integral coefficient and the pressure are transferred downwards from the finer mesh rather than generated in the mesh itself:

$$w_I^H = \sum_{J=0}^{N} K_{I,J}^{HH} p_J^H. \tag{3.133}$$

3.4.5.4 Transfer Back Integration Results

Because the integral value is not calculated on the finer mesh, the known value should be interpolated back. First, map the results on the coarser nodes to the finer nodes:

$$\tilde{w}_{2I-1}^h = w_I^H. \tag{3.134}$$

For the node for which the coarser mesh has no related node to the finer mesh, interpolation is as follows:

$$\tilde{w}_{2I}^{h} = \frac{1}{16}\left(-w_{I-1}^{H} + 8w_{I}^{H} + 8w_{I+1}^{H} - w_{I+2}^{H}\right). \tag{3.135}$$

For $i = 2$ or $i = n - 1$, the finer mesh nodes can be calculated as the average of two adjacent nodes.

3.4.5.5 Modification on the Finer Mesh

Modifications on the finer mesh include three parts: the integrated coefficient modification, the mapped value modification and the interpolated value modification.

1. Integral coefficient modification
   First, calculate the interpolated coefficients. Then, substract the interpolated values from the integrated coefficients to obtain a difference. As a mapped value of the nodes is usually not equal to an interpolated value, we must modify them. The interpolated value of the integral coefficient is determined by the following formula:

$$\tilde{K}_{2I-1,2J-1}^{hh} = \frac{1}{16}\left(9K_{I+1,J}^{HH} + 9K_{I-1,J}^{HH} - K_{I+3,J}^{HH} - K_{I-3,J}^{HH}\right). \tag{3.136}$$

Because the adjacent integral nodes are not suitable for high-order interpolation, the following interpolation formulas can be used instead:

$$\tilde{K}_{1,2J-1}^{hh} = \frac{1}{8}\left(9K_{2,J}^{HH} - K_{4,J}^{HH}\right), \tag{3.137}$$

$$\tilde{K}_{2,2J-1}^{hh} = \frac{1}{16}\left(9K_{1,J}^{HH} + 9K_{3,J}^{HH} - K_{3,J}^{HH} - K_{5,J}^{HH}\right), \tag{3.138}$$

$$\tilde{K}_{3,2J-1}^{hh} = \frac{1}{16}\left(9K_{2,J}^{HH} + 9K_{4,J}^{HH} - K_{2,J}^{HH} - K_{6,J}^{HH}\right). \tag{3.139}$$

For a mapped node, the difference between the calculated integral coefficient and the interpolated integral coefficient is equal to:

$$\Delta\tilde{K}_{i,j}^{hh} = K_{i,j}^{hh} - \tilde{K}_{i,j}^{hh}. \tag{3.140}$$

For an interpolated node, the difference between the calculated integral coefficient and the interpolated integral coefficient is equal to:

$$\Delta\tilde{K}_{i,j}^{hh} = \begin{cases} 0 & \text{mapped node} \\ K_{i,j}^{hh} - \tilde{K}_{i,j}^{hh} & \text{interpolated node.} \end{cases} \tag{3.141}$$

2. Mapped value modification
   Use the difference between the integral coefficient to modify the integral value of a mapped node:

$$w_{2I-1}^{h} = \tilde{w}_{I}^{H} + \sum_{j=1}^{M} \Delta\tilde{K}_{i,j}^{hh} p_j \Delta x. \tag{3.142}$$

3. Interpolated value modification

Use the difference between the integral coefficient to modify the integral value of an interpolated node:

$$w_{2I}^{h} = \tilde{w}_{2I}^{h} + \sum_{j=1}^{M} \Delta \hat{K}_{i,j}^{hh} p_j \Delta x. \tag{3.143}$$

In the above equations, $M \geq 3 + 2\ln(n)$, M and it should be rounded; n is the node number.

The above steps are only for the two meshes, for an entire multi-grid mesh, the multi-grid integration method follows the following steps.

1. According to $M \geq 3 + 2\ln\ (n)$ calculate M;
2. According to Equations 3.129–3.132 transfer node parameters (pressure, integral coefficient, etc.) downwards to the coarsest mesh;
3. According to Equation 3.133 numerically calculate integration on the coarsest mesh;
4. According to Equation 3.132 calculate the integral coefficient of the corresponding node on the upper mesh;
5. According to Equations 3.136–3.140 interpolate the integral coefficients of the upper mesh;
6. According to Equation 3.134 map the corresponding value to the upper mesh node;
7. According to Equation 3.142 modify the mapped integral values;
8. According to Equation 3.135 interpolate the noncorresponding values of the upper mesh nodes;
9. According to Equation 3.143 modify the interpolated integral values;
10. Return to the 4th step for other calculations until all the nodes integral values have been obtained and calculation is completed.

It should be pointed out that by using the multi-grid integration method, we must correctly understand the importance of all the modifications. The main purpose of the modifications is through a less coarse grid calculation and modifications to obtain the result with the same accuracy of the fine grid.

References

1. Wang, Y.L., Huang, T.T., and Wen, S.Z. (1987) Tilting transition temperature journal bearing static and dynamic performance calculation. *Journal of Tsinghua University*, **27** (1), 84–91.
2. Wen, S.Z. (1982) Boundary element method in the application of a lubrication problem Rayleigh Step bearing. *Lubrication and Sealing*, **3**, 10–16.
3. Dowson, D. and Higginson, G.R. (1997) *Elasto-Hydrodynamic Lubrication*, Pergamon Press, London.
4. Wen, S.Z. and Zhu, D. (1985) Isothermal Elastohydrodynamic Lubrication Direct iterative solution of the problem. *Lubrication and Sealing*, **4**, 20–25, 1986, 4, 9–15.
5. Archard, J.F. and Cowking, E.W. (1965) Elastohydrodynamic lubrication of point contacts. *Proceedings of the Institution of Mechanical Engineers*, **180** (3B), 47–56.
6. Cheng, H.S. (1970) A numerical solution of the elastohydrodynamic film thickness in an elliptical contact. *Journal of Lubrication Technology –Transactions of the ASME, Series F*, **92** (1), 155–162.
7. Zhu, D. and Wen, S.Z. (1984) A full numerical solution for the thermoelastohydrodynamic problem in elliptical contacts. *Journal of Tribology –Transactions of the ASME*, **106** (2), 246 254.
8. Ranger, A.P. (1974) PhD thesis, Imperial College, University of London.
9. Lubrecht, A.A., ten Narel, W.E., and Bosma, R. (1989) Multigrid, An alternative method for calculating film thickness and pressure profiles in elastohydrodynamic lubricated line contacts. *Journal of Tribology – Transactions of the ASME*, **108** (4), 551–556.
10. Huang, P. and Wen, S.Z. (1992) Multi-grid method for solving the problem EHL line contact. *Journal of Tsinghua University*, **32** (5), 26–34.
11. Wen, S.Z. (1990) *Tribology Principle*, Tsinghua University Press, Beijing.
12. Wen, S.Z. and Huang, P. (2002) *Tribology Principles*, 2nd edn, Tsinghua University Press, Beijing.
13. Wen, S.Z. and Yang, P.R. (1992) *Elastohydrodynamic Lubrication*, Tsinghua University Press, Beijing.

4

Lubrication Design of Typical Mechanical Elements

In this chapter lubrication designs of some typical mechanical elements are described. First, as an example a slider is given to illustrate the calculation of lubrication performances. Then, lubrication calculations of plain bearings, rolling contact bearings, gears and cam mechanisms are discussed [1].

4.1 Slider and Thrust Bearing

Design of a wedge-shaped slider is the simplest lubrication problem. If the geometric shape of a slider is not very complex, an analytical solution can often be obtained. Furthermore, the analysis of the problem will not only help in understanding the basic characteristics of lubrication, but also form the basis of the lubrication design of a thrust bearing.

4.1.1 Basic Equations

To solve an infinite-wide slider so as not to consider the side leakage, the Reynolds equation is simplified into a one-dimensional ordinary differential equation. When the film thickness is known, we can obtain a general pressure solution. Then, by substituting the pressure boundary conditions the pressure distribution can be obtained. The pressure distribution can be used to obtain the load, friction and flow.

4.1.1.1 Reynolds Equation

The Reynolds equation to solve a slider lubrication problem is

$$\frac{d}{dx}\left(h^3\frac{dp}{dx}\right) = 6U\eta\frac{dh}{dx}. \tag{4.1}$$

Integrate Equation 4.1 twice, we have:

$$p = \int \frac{6U\eta}{h^2}dx + C_1\int\frac{dx}{h^3} + C_2, \tag{4.2}$$

where C_1 and C_2 are the integral constants to be determined by the pressure boundary conditions.

Principles of Tribology, First Edition. Wen Shizhu and Huang Ping.

4.1.1.2 Boundary Conditions

Commonly, two kinds of pressure boundary conditions are:

1. $p|_{x=0} = 0; p|_{x=x'} = 0$ (x' is the outlet border, $x' = l$, l is the width of the slider);
2. $p|_{x=0} = 0; p|_{x=x'} = 0$ and $\mathrm{d}p/\mathrm{d}x|_{x=x'} = 0$ (x' to be determined by the outlet boundary, $x' \leq l$).

4.1.1.3 Continuous Conditions

If the pressure or its derivative is discontinuous, both sides of the pressure at the discontinuous boundary should be solved independently. Then, set the pressures or flows of the two sides equal to determine the integral constants.

1. Pressure continuous condition is

$$p|_{x=x^*-0} = p|_{x=x^*+0}. \tag{4.3}$$

2. Flow continuous condition is

$$\left[-\frac{h^3}{12}\frac{\mathrm{d}p}{\mathrm{d}x} + (U_1 + U_2)\frac{h}{2}\right]_{x=x^*-0} = \left[-\frac{h^3}{12}\frac{\mathrm{d}p}{\mathrm{d}x} + (U_1 + U_2)\frac{h}{2}\right]_{x=x^*+0}. \tag{4.4}$$

4.1.2 *Solutions of Slider Lubrication*

In addition to a linear slider, other types of sliders are the curved slider, the composite slider, the ladder slider and so on. An infinitely wide linear slider is shown in Figure 4.1.

1. Thickness equation
   If set $K = (h_1 - h_0)/h_0$ and $X = x/l$, the film thickness can be expressed as:

$$h = h_0(1 + KX). \tag{4.5}$$

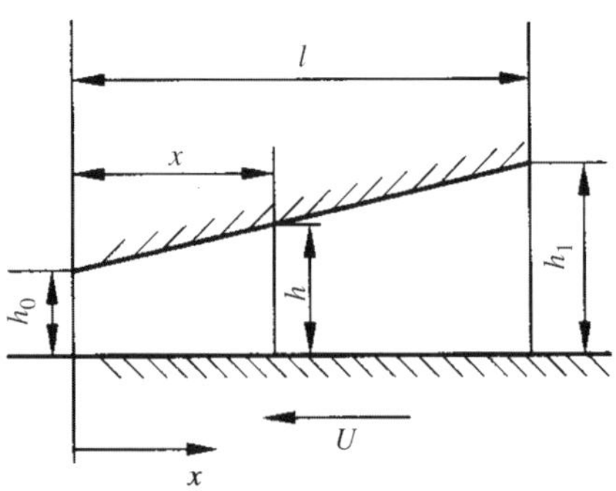

Figure 4.1 A simple slider.

2. Pressure solution

Because the film thickness h is linear to x, the differential of the film thickness to the x is equal to $dh = h_0 K dX$. If we substitute $dX = 1/Kh_0\, dh$ into Equation 4.1 and integrate it, we have:

$$p = \frac{6U\eta l}{Kh_0}\left(-\frac{\bar{h}}{2h^2} + \frac{1}{h} - C\right), \tag{4.6}$$

where $\bar{h}$ is the film thickness at $dp/dx = 0$ or the maximum pressure position.
By using boundary conditions $p|_{h=h_0} = 0$ and $p|_{h=h_1} = 0$, we have:

$$\bar{h} = \frac{2h_0 h_1}{h_0 + h_1}, \quad C = \frac{1}{h_0 + h_1}. \tag{4.7}$$

Therefore, the pressure distribution is equal to:

$$p = \frac{6U\eta l}{Kh_0}\left(-\frac{h_0 h_1}{h_0 + h_1}\frac{1}{h^2} + \frac{1}{h} - \frac{1}{h_0 + h_1}\right). \tag{4.8}$$

3. Load

The carrying load per unit width is equal to:

$$\frac{W}{b} = \int_0^l p dx = \frac{l}{h_0 K}\int_{h_0}^{h_1} p dh = \frac{6U\eta l^2}{K^2 h_0^2}\left[\ln\frac{h_1}{h_0} - \frac{2(h_1 - h_0)}{h_0 + h_1}\right] = \frac{6U\eta l^2}{K^2 h_0^2}\left[\ln(K+1) - \frac{2K}{K+2}\right] \tag{4.9}$$

where b is the width of the slider in the y direction.
Deriving W to K, and setting $dW/dK = 0$, we will find the maximum carrying load W_{max}. The corresponding $K = 1.2$, that is, $h_1/h_0 = 2.2$.

4. Pressure center

The pressure center is the point the sum of the load is acting on. We can obtain it by finding the origin point of the moment. As shown in Figure 4.1, let the pressure center or the origin be equal to x_0, and the load-carrying capacity per unit length be equal to W/l; we have:

$$\frac{x_0 W}{b} = \int_0^l p x dx. \tag{4.10}$$

Substituting Equations 4.8 and 4.9 into the above equation and integrating it, we have:

$$\frac{x_0}{l} = \frac{K(6+K) - 2(2K+3)\ln(1+K)}{2K[(2+K)\ln(1+K) - 2K]}. \tag{4.11}$$

5. Frictional force

The shear stress on the surface is equal to:

$$\tau = \eta\frac{du}{dz} = \frac{dp}{dx}\left(z - \frac{h}{2}\right) + \frac{\eta}{h}U. \tag{4.12}$$

The frictional forces per unit length are:

$$F_{h,0}/b = \int_0^b\int_0^l \tau_{h,0} dx dy = \int_0^b\int_0^l \left(\pm\frac{dp}{dx}\frac{h}{2} + \eta\frac{U}{h}\right) dx dy, \tag{4.13}$$

where F_h and F_0 are the frictional forces at the surfaces of $z=h$ and $z=0$, respectively; the positive sign $+$ is for the surface $z=h$; the negative $-$ for $z=0$.
The first integral of Equation 4.13 is equal to:

$$\int_0^l \pm\frac{\mathrm{d}p}{\mathrm{d}x}\frac{h}{2} = p\frac{h}{2}\bigg|_0^l \mp \int_0^l p\frac{dh}{2} = \frac{h_0 K}{2l}\int_0^l p dx = \mp\frac{h_0 K}{2l}\cdot\frac{W}{b}. \tag{4.14}$$

Therefore, the frictional forces are:

$$F_{h,0}/b = \frac{\eta U l}{h_0}\frac{\ln(K+1)}{K} \mp \frac{h_0 K}{2l}\cdot\frac{W}{b}. \tag{4.15}$$

6. Flow
   Because there is no side leak for an infinitely wide slider, that is, $q_y=0$, the flow is equal to:

$$Q_x = \int_0^b q_x dy = \int_0^b\left(-\frac{h^3}{12\eta}\cdot\frac{dp}{dx}+\frac{Uh}{2}\right)dy. \tag{4.16}$$

Because $dp/dx|_{h=\bar{h}} = 0$, the flow per unit length is equal to:

$$\frac{Q_x}{b} = \frac{1}{b}\int_0^b \frac{U\bar{h}}{2}dy = Uh_0\frac{K+1}{K+2}. \tag{4.17}$$

Some sliders formed by straight lines are shown in Figures 4.2 and 4.3. They are the composite slider and the Rayleigh slider. The thickness discontinuity conditions of Equations 4.3 and 4.4 must be used at the joint while solving the Reynolds equation [16].

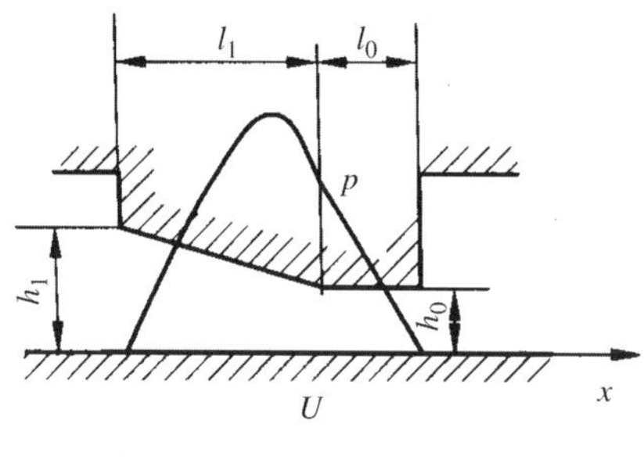

Figure 4.2 Composite slider.

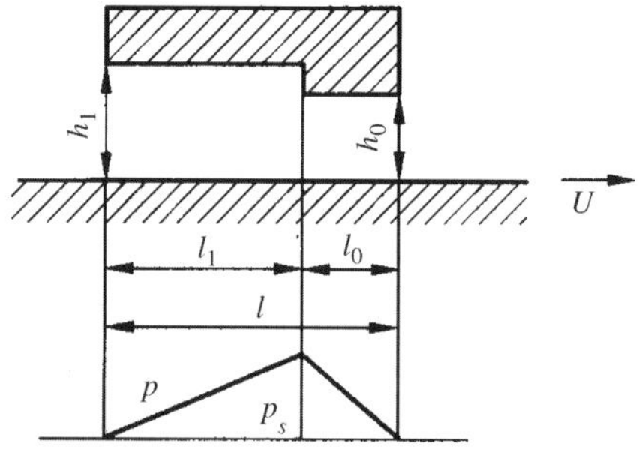

Figure 4.3 Rayleigh slider.

Calculation of lubrication of a finite slider is needed to solve the two-dimensional Reynolds equation. Because few analytical solutions can be obtained, numerical methods are usually used. The main points of numerical calculation have been mentioned in Chapter 3.

4.2 Journal Bearing

Journal bearings are the most useful mechanical elements in hydrodynamic lubrication. The diameter of a bearing is a little larger than that of the journal, about 0.002 less. When the journal is in an eccentric position, two wedge-shaped clearances are formed. While the journal rotates, the lubricant is brought into the convergent wedge to generate hydrodynamic pressure. Because the actual situations are very complicated, some simplifications will be made to solve the problems by the current lubrication theory.

4.2.1 Axis Position and Clearance Shape

As the journal rotates to bring lubricant into the convergent gap to form a hydrodynamic pressure, its resultant will balance the load. The equilibrium position of the journal inside the bearing is as shown in Figure 4.4.

The journal center O_2 at the equilibrium position can be determined by two parameters, the deviation angle ψ and the eccentricity e. The deviation angle ψ is the angle between the line of the load W and the connection line O_1O_2 of the bearing and the journal centers. The dimensionless eccentricity ε is often used in calculation, where $\varepsilon = e/c$. Here, c is the clearance of the radii of the bearing and the journal, that is, $c = R_1 - R_2$.

We can see from Figure 4.4 that the film thickness h is the function of θ, which is the angle counted from the deviation angle ψ. In ΔO_1O_2P, with the cosine law we have:

$$R_1^2 = e^2 + (R_2 + h)^2 - 2e(R_2 + h)\cos\theta, \tag{4.18}$$

or

$$h = e\cos\theta + R_1\sqrt{1 - \left(\frac{e}{R_1}\right)^2 \sin^2\theta} - R_2. \tag{4.19}$$

Because $e/R_1 << 1$, it can be omitted. Therefore, we have the geometric relationship

$$h = e\cos\theta + c = c(1 + \varepsilon\cos\theta). \tag{4.20}$$

Equation 4.20 shows that the shape of the bearing film is a cosine function; the error expression is only about 0.1%.

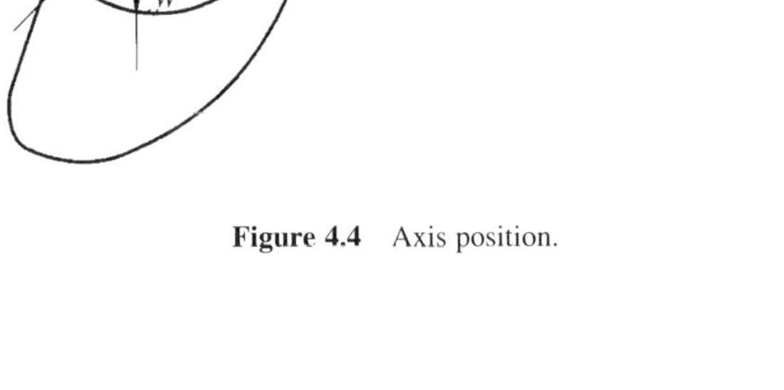

Figure 4.4 Axis position.

4.2.2 Infinitely Narrow Bearing

When the width of a bearing in the y direction is much smaller than the length in the x direction, $\partial p/\partial y$ is much larger than $\partial p/\partial x$. Therefore, $\partial p/\partial x$ can approximately be omitted. For example, if the ratio *b/D* is less than 0.25, the satisfactory approximation of the narrow bearing can be accepted. A smaller *b/D* bearing is often used, especially for high-speed bearing.

Because h is often only the function of x, the Reynolds equation of the narrow bearing theory becomes:

$$\frac{d}{dy}\left(h^3 \cdot \frac{dp}{dy}\right) = 6U\eta \frac{dh}{dx}. \tag{4.21}$$

The boundary conditions become $p|_{y=\pm b/2} = 0$. Because of the symmetry, $dp/dy|_{y=0} = 0$. Substitute the boundary conditions into Equation 4.21, the pressure distribution can be obtained as follows:

$$p = 3U\eta \frac{1}{h^3} \cdot \frac{dh}{dx}\left(y^2 - \frac{b^2}{4}\right), \tag{4.22}$$

or

$$p = \frac{3U\eta\varepsilon \sin\theta}{c^2 R(1+\varepsilon\cos\theta)^3}\left(\frac{b^2}{4} - y^2\right). \tag{4.23}$$

4.2.2.1 Load-carrying Capacity

Here, the semi-Sommerfeld boundary condition is adopted, that is, only consider the positive pressure in the convergent gap ($0 \le \theta \le \pi$). The load W can be divided into the components W_x in the horizontal direction and W_y in the vertical direction, that is, the weight. They are:

$$\begin{aligned} W_x &= \frac{U\eta b^3}{c^2} \frac{\varepsilon^2}{(1-\varepsilon^2)^2} \\ W_y &= \frac{U\eta b^3}{c^2} \frac{\pi\varepsilon}{4(1-\varepsilon^2)^{3/2}}. \end{aligned} \tag{4.24}$$

The load-carrying capacity is equal to

$$W = \sqrt{W_x^2 + W_y^2} = \frac{U\eta b^3 \varepsilon}{c^2(1-\varepsilon^2)}\sqrt{\frac{16\varepsilon^2 + \pi^2(1-\varepsilon^2)}{16(1-\varepsilon)^2}}. \tag{4.25}$$

The dimensionless carrying load will be presented by a comprehensive parameter known as the Sommerfeld number as follows:

$$\Delta = \frac{W/b}{\upsilon\eta} \cdot \frac{c^2}{R^2}.$$

Therefore, Equation 4.25 becomes

$$\Delta = \left(\frac{b}{D}\right)^2 \frac{\pi\varepsilon}{(1-\varepsilon^2)^2}\sqrt{\left(\frac{16}{\pi^2} - 1\right)\varepsilon^2 + 1}. \tag{4.26}$$

Equation 4.26 shows that Δ is related to *b/D* and ε. When *b/D* increases, Δ increases and the carrying load W also increases. A larger *b/D* will raise the bearing temperature so as to decrease the effective viscosity of lubricant. If ε increases, $h_{\min}$ decreases. Therefore, increase of ε is limited by preventing the two rough surfaces from contacting each other.

4.2.2.2 Deviation Angle and Axis Track

The deviation angle can be calculated as follows:

$$\tan\psi = \frac{W_y}{W_x} = \frac{\pi}{4}\frac{\sqrt{1-\varepsilon^2}}{\varepsilon}. \tag{4.27}$$

Because $\tan\psi = \sin\psi/\cos\psi = \sqrt{1-\cos^2\psi}/\cos\psi$, if we take $\pi/4 \approx 1$ and compare the above two equations, we have:

$$\cos\psi = \varepsilon = \frac{e}{c}. \tag{4.28}$$

The above result means that the axis track is a semi-circle as shown in Figure 4.5. Therefore, the two parameters ε and ψ are not independent while they determine an equilibrium position of the journal axis. With variation of the bearing operating parameters we can plot the track of the axis. The relationship of ε-ψ is a semi-circle, which is also called a track circle.

4.2.2.3 Flow

While a journal is rotating, the oil continues to leak from the sides of the bearing. In order to maintain effective lubrication and cooling, we must continuously supply lubricant. The side leakage can be calculated as follows.

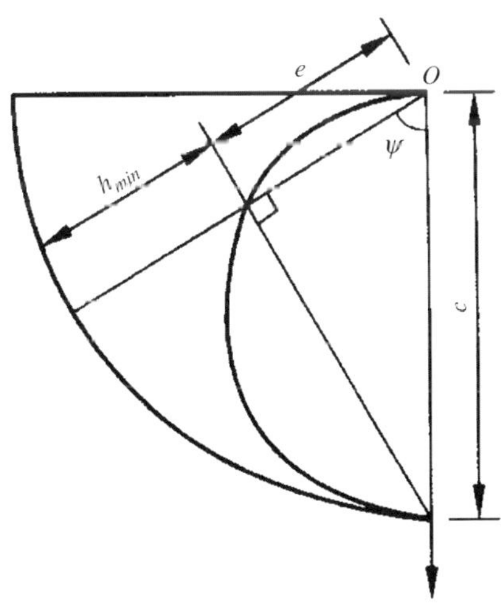

Figure 4.5 Infinitely narrow bearing track.

For a journal bearing, the flow of unit width in the circumferential direction is equal to:

$$q_\theta = -\frac{h^3}{12\eta R}\frac{\partial p}{\partial \theta} + \frac{1}{2}Uh. \tag{4.29}$$

For a narrow bearing, because $\partial p/\partial\theta = 0$, we have:

$$q_\theta = \frac{1}{2}Uh. \tag{4.30}$$

Therefore, the total flow is

$$Q_\theta = \frac{1}{2}Uhb. \tag{4.31}$$

As the maximum film thickness is equal to $h = c(1+\varepsilon)$ and the minimum film thickness $h = c(1-\varepsilon)$, the leaked flow is

$$Q_c = \frac{1}{2}Ubc(1+\varepsilon) - \frac{1}{2}Ubc(1-\varepsilon) = Ubc\varepsilon. \tag{4.32}$$

The structure of a lubricant inlet is usually of three kinds: circumferential oil slots, inlet holes, and axial slots or facilities. The magnitude of the axial flow depends on the oil pressure, viscosity and the supplement structure, but is not related to the shaft rotation speed. The axial structures for the oil supplement can be found in reference [1].

4.2.2.4 Frictional Force and Friction Coefficient

According to the infinitely narrow bearing theory, $\partial p/\partial\theta = 0$. This means that the shear stress caused by pressure variation does not exist along the circumferential direction. Thus, $\tau = \eta U/h$. Then, the frictional force of the journal is

$$F = \int_{-b/2}^{b/2}\int_0^{2\pi} \tau R d\theta dy = \frac{2\pi\eta URb}{c}\frac{1}{(1-\varepsilon^2)^{1/2}}. \tag{4.33}$$

The friction coefficient is equal to:

$$\mu = \frac{F}{W} = \frac{8Rc}{b^2}\frac{(1-\varepsilon^2)^{3/2}}{\varepsilon(0.62\varepsilon^2+1)^{1/2}}. \tag{4.34}$$

4.2.3 *Infinitely Wide Bearing*

As infinitely wide bearing means $dp/dy = 0$, that is, no side leakage. Therefore, the Reynolds equation becomes:

$$\frac{d}{dx}\left(h^3\frac{dp}{dx}\right) = 6U\eta\frac{dh}{dx}. \tag{4.35}$$

With integration, we have:

$$\frac{dp}{dx} = 6U\eta \frac{h - \bar{h}}{h^3}, \tag{4.36}$$

where $dp/dx|_{h=\bar{h}} = 0$.

For a journal bearing, Equation 4.36 can be rewritten as

$$\frac{dp}{d\theta} = 6U\eta R \frac{h - \bar{h}}{h^3}. \tag{4.37}$$

If we substitute $h = c(1 + \varepsilon \cos \theta)$ into Equation 4.37, and integrate it twice, the equation becomes:

$$p = \frac{6\eta UR}{c^2}\left[\int \frac{d\theta}{(1+\varepsilon \cos \theta)^2} - \frac{\bar{h}}{c}\int \frac{d\theta}{(1+\varepsilon \cos \theta)^3}\right] + C_1, \tag{4.38}$$

where $\bar{h}$ and C_1 are the integral constants.

In order to obtain the solution of Equation 4.38 analytically, the Sommerfeld integral transformation can be used as follows:

$$\cos \gamma = \frac{\varepsilon + \cos \theta}{1 + \varepsilon \cos \theta}, \tag{4.39}$$

or

$$\cos \theta = \frac{\cos \gamma - \varepsilon}{1 - \varepsilon \cos \gamma}.$$

Thus, we have:

$$d\theta = \frac{(1-\varepsilon^2)^{1/2}}{1 - \varepsilon \cos \gamma} d\gamma. \tag{4.40}$$

Integrate the transformed Equation 4.38 twice, it becomes

$$\begin{aligned} \int \frac{d\theta}{(1+\varepsilon \cos \theta)^2} &= \frac{1}{(1-\varepsilon^2)^{3/2}}(\gamma - \varepsilon \sin \gamma) \\ \int \frac{d\theta}{(1+\varepsilon \cos \theta)^3} &= \frac{1}{(1-\varepsilon^2)^{5/2}}\left(\gamma - 2\varepsilon \sin \gamma + \frac{\varepsilon^2 \gamma}{2} + \frac{\varepsilon^2}{4} \sin 2\gamma\right). \end{aligned} \tag{4.41}$$

Finally, we have

$$p(\gamma) = \frac{6U\eta R}{c^2}\left[\frac{\gamma - \varepsilon \sin \gamma}{(1-\varepsilon^2)^{3/2}} - \frac{\bar{h}}{c(1-\varepsilon^2)^{5/2}} \times \left(\gamma - 2\varepsilon \sin \gamma + \frac{\varepsilon^2 \gamma}{2} + \frac{\varepsilon^2}{4} \sin 2\gamma\right)\right] + C_1. \tag{4.42}$$

Usually, Reynolds boundary conditions are used as shown in Figure 4.6. That is, in the inlet $p|_{\theta=0} = 0$ and in the outlet $p|_{\theta=\theta 2} = 0$ and $dp/dx|_{\theta=\theta 2} = 0$. If the maximum pressure is at the point of $\theta_1 = \pi - \alpha$, the outlet is at the point of $\theta_2 = \pi + \alpha$, where the conditions of $p|_{\theta=\theta 2} = 0$ and $dp/dx|_{\theta=\theta 2} = 0$ are satisfied.

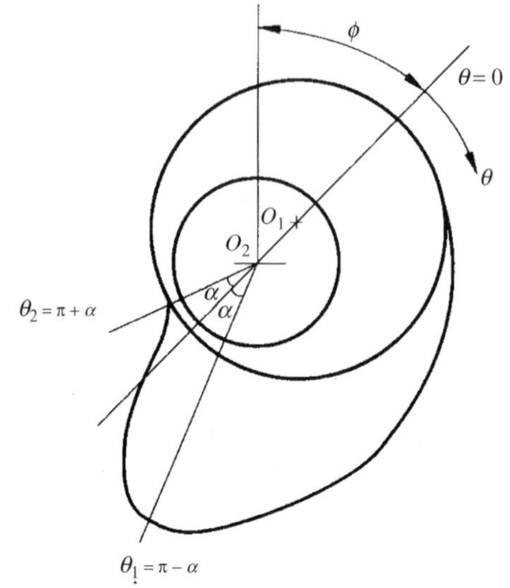

Figure 4.6 Reynolds boundary conditions.

At $\theta_1 = \pi - \alpha$, the pressure p reaches the maximum, that is, $dp/d\theta = 0$, so that the film thickness is equal to $\bar{h}$ and the corresponding transformed angle $\gamma_1 = \pi - \beta$. With Equation 4.39, the relationship of θ_1 and γ_1 can be obtained as follows:

$$\begin{aligned} 1-\varepsilon\cos\gamma_1 &= 1-\frac{\varepsilon(\varepsilon+\cos\theta_1)}{1+\varepsilon\cos\theta_1} = \frac{1-\varepsilon^2}{1+\varepsilon\cos\theta_1} \\ 1+\varepsilon\cos\theta_1 &= \frac{1-\varepsilon^2}{1-\varepsilon\cos\gamma_1} = \frac{1-\varepsilon^2}{1+\varepsilon\cos\beta}. \end{aligned} \tag{4.43}$$

Because $\bar{h} = c(1+\varepsilon\cos\theta_1) = c(1-\varepsilon^2)/(1+\varepsilon\cos\beta)$, substitute $\bar{h}$ into Equation 4.42 we have

$$p(\gamma) = \frac{6U\eta R}{c^2(1-\varepsilon^2)^{3/2}}\left[\gamma - \varepsilon\sin\gamma - \frac{\gamma(2+\varepsilon^2) - 4\varepsilon\sin\gamma + \varepsilon^2\sin\gamma\cos\gamma}{2(1+\varepsilon\cos\beta)}\right] + C_1. \tag{4.44}$$

According to the conditions: $p|_{\theta=0} = 0$, we have $\gamma = 0$, $C_1 = 0$. Because at the outlet, $p = 0$, we have the following equation to obtain β for $\gamma = \gamma_2 = \pi + \beta$:

$$\varepsilon\sin\beta\cos\beta + 2(\pi+\beta)\cos\beta - (\pi+\beta)\varepsilon - 2\sin\beta = 0. \tag{4.45}$$

After γ_1 and γ_2 have been obtained, the pressure distribution is known. Therefore, the carrying load and the deviation angle will be:

$$W\sin\psi = \int_0^{\pi+\alpha} bRp\sin\theta d\theta = \frac{6U\eta b(R/c)^2[(\pi+\beta)\cos\beta - \sin\beta]}{(1-\varepsilon^2)^{1/2}(1+\varepsilon\cos\beta)}, \tag{4.46}$$

$$W\cos\psi = \int_0^{\pi+\alpha} bRp\cos\theta d\theta = -\frac{3U\eta b(R/c)^2\varepsilon(1+\cos\beta)^2}{(1-\varepsilon^2)(1+\varepsilon\cos\beta)}. \tag{4.47}$$

Then, the carrying load is equal to:

$$W = \sqrt{(W\sin\psi)^2 + (W\cos\psi)^2}$$
$$= \frac{3U\eta b(R/c)^2}{(1-\varepsilon^2)^{1/2}(1+\varepsilon\cos\beta)}\left\{\frac{\varepsilon^2(1+\cos\beta)^4}{1-\varepsilon^2} + 4[(\pi+\beta)\cos\beta - \sin\beta]^2\right\}^{1/2} \tag{4.48}$$

And, the deviation angle is equal to:

$$\tan\psi = -\frac{2(1-\varepsilon^2)^{1/2}[\sin\beta - (\pi+\beta)\cos\beta]}{\varepsilon(1+\cos\beta)^2}. \tag{4.49}$$

The frictional forces of an infinitely wide bearing can be obtained by integrating the surface shear stresses:

$$\tau_{h,0} = \pm\frac{dp}{dx}\frac{h}{2} + \eta\frac{U}{h}, \tag{4.50}$$

where the positive sign + is for the surface $z = h$; the negative – for $z = 0$.

On both surfaces, the frictional forces are equal to:

$$F_{h,0} = b\int \tau_{h,0}dx = b\left(\pm\int_0^{\pi+\alpha}\frac{dp}{d\theta}\frac{h}{2}d\theta + \int_0^{2\pi}\eta\frac{UR}{h}d\theta\right), \tag{4.51}$$

where F_h and F_0 are the surface frictional forces for $z = h$ and $z = 0$ respectively.

It should be pointed out that the first item on the above equation is the pressure flow friction. Thus, the upper integral limit should be equal to the outlet determined by the Reynolds boundary conditions, that is, $(\pi + \alpha)$. The second term is the frictional force generated by the velocity flow. The upper and lower integral limits will be 0 and 2π, that is, the whole circumference of the bearing if we assume that the bearing is full of oil with no bubbles and strips.

With step integration, the frictional forces are finally obtained:

$$F_{h,0} = \pm\frac{c\varepsilon}{2R}W\sin\psi + \frac{2\pi U\eta Rb}{c(1-\varepsilon^2)^{1/2}}, \tag{4.52}$$

$$\mu\left(\frac{R}{c}\right) = \pm\frac{1}{2}\varepsilon\sin\psi + \frac{2\pi}{(1-\varepsilon^2)^{1/2}}\frac{1}{\Delta}, \tag{4.53}$$

where μ is the friction coefficient of the journal surface, $\mu = F/W$.

As shown in Figure 4.7, the difference between the journal frictional force F_h and the bearing frictional force F_0 is caused by the eccentricity. Suppose the radius of the journal (also the bearing) is R, the difference of friction torque is equal to

$$F_hR - F_0R = Wc\varepsilon\sin\psi. \tag{4.54}$$

Then the moment acting on the journal and the bearing caused by the eccentricity is

$$m = We\,\varepsilon\sin\psi = Wc\varepsilon\sin\psi. \tag{4.55}$$

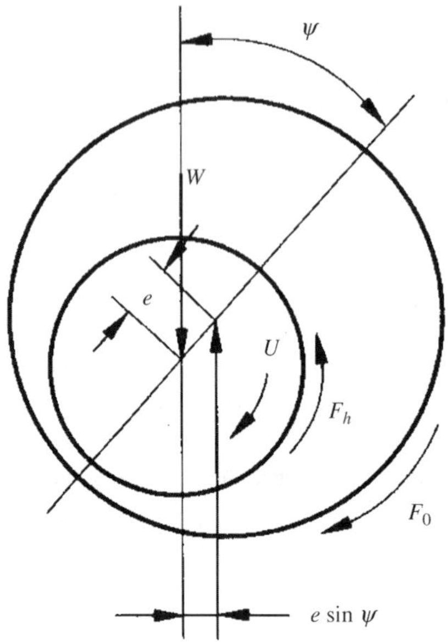

Figure 4.7 Friction.

That is, the moment balances the friction torque difference caused by the frictional forces of the journal and the bearing.

4.3 Hydrostatic Bearing

The oil film of hydrostatic lubrication depends on an external pressure to form a load-carrying capacity. Even with no relative sliding velocity, the lubrication film can also be achieved. Hydrostatic lubrication has the following advantages:

1. it does not need a sliding velocity;
2. the oil film possesses a strong stiffness so that it can realize a high supporting accuracy;
3. it has a low friction coefficient so as to eliminate the static friction.

The disadvantages of hydrostatic lubrication are: its structure is complicated; it needs an external pressure oil supply system, which may weaken the service life and the reliability of hydrostatic lubricated bearings.

If there is no relative sliding velocity, the Reynolds equation for hydrostatic lubrication is as follows:

$$\frac{\partial}{\partial x}\left(\frac{\rho h^3}{\eta}\frac{\partial p}{\partial x}\right)+\frac{\partial}{\partial y}\left(\frac{\rho h^3}{\eta}\frac{\partial p}{\partial y}\right)=0. \tag{4.56}$$

Furthermore, if we suppose lubrication is under an equal film thickness and the lubricant is incompressible and has equal viscosity, the above Reynolds equation can be further simplified as follows:

$$\frac{\partial^2 p}{\partial x^2}+\frac{\partial^2 p}{\partial y^2}=0. \tag{4.57}$$

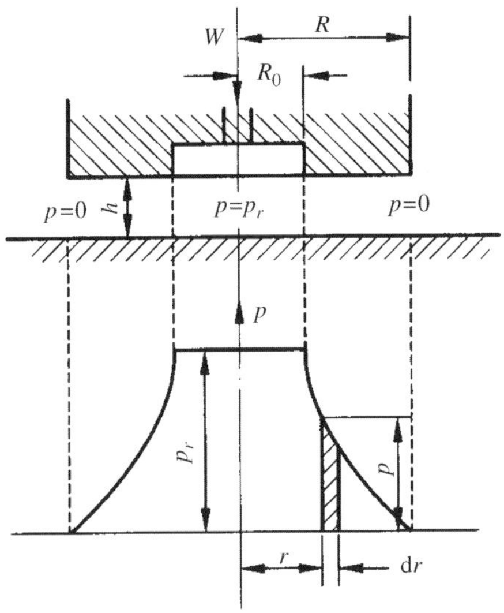

Figure 4.8 A circular thrust plate.

4.3.1 Hydrostatic Thrust Plate

As shown in Figure 4.8, there is a circular thrust plate with a single oil chamber. The following solutions will be obtained if the outside diameter of the circler is R, the chamber radius is R_0, the supplied pressure into the chamber is p_r, and the oil chamber is deep enough.

1. Reynolds equation
   The cylindrical coordinate Reynolds equation of Equation 4.57 is

$$\frac{\partial}{\partial r}\left(r\frac{\partial p}{\partial r}\right)+\frac{\partial}{\partial \theta}\left(\frac{\partial p}{r\partial \theta}\right)=0. \tag{4.58}$$

Because the problem is axially symmetric, that is, the pressure is independent to θ, the Reynolds equation can be written as:

$$\frac{\partial}{\partial r}\left(r\frac{\partial p}{\partial r}\right)=0. \tag{4.59}$$

2. Pressure distribution, carrying load and flow
   By integrating Equation 4.59 twice and substituting the pressure boundary conditions, we obtain the pressure distribution as follows:

$$p=\frac{p_r}{\ln R/R_0}\ln\frac{R}{r}, \tag{4.60}$$

where $R_0 \leq r \leq R$.

The flow Q and the carrying load W are respectively:

$$Q = \frac{\pi h^3 p_r}{6\eta} \frac{1}{\ln(R/R_0)}, \tag{4.61}$$

$$W = \frac{\pi p_r}{2}\left[\frac{R^2 - R_0^2}{\ln(R/R_0)}\right] = \frac{3\eta R^2 Q}{h^3}\left[1 - \left(\frac{R_0}{R}\right)^2\right]. \tag{4.62}$$

Equation 4.62 shows that the carrying load W is affected by the chamber pressure p_r and the structural sizes, such as R and R_0, but is independent of the film thickness h.

4.3.2 Hydrostatic Journal Bearings

In Figure 4.9, there is a four-chamber hydrostatic journal bearing with four throttles C_1, C_2, C_3 and C_4 operated by a high-pressure pump, which supplies the oil with the pressure p_s. The load W is balanced by the chamber pressure p_{r1}, p_{r2}, p_{r3} and p_{r4}. If the load W increases, the journal moves downwards. The lower clearance decreases to cause p_{r3} increase, but the opposite pressure p_{r1} decreases when its clearance increase. In this way, the resulting pressure allows the load to maintain its balance.

4.3.3 Bearing Stiffness and Throttle

The load is balanced by the pressure in the chambers. At the same time, the flow supplied to the chambers is equal to the leakage. Therefore, when the structure of a hydrostatic bearing and the input pressure p_r are given, the load-carrying capacity is determined. In order to balance the varying load, the flow control equipment should be used to adjust the pressure in the chambers.

In Figure 4.10, there are typical lubricant supply systems of hydrostatic bearings. The easiest way is to use a constant flow system shown in Figure 4.10a. In the system, the flow is controlled to be constant, which does not vary with the pressure. When the load increases, the film thickness decreases. Because the flow remains unchanged, the pressure will increase to balance the load and vice versa.

However, the most commonly used system of a hydrostatic bearing is the constant pressure system, shown in Figure 4.10b. With a pressure valve, the supplied pressure keeps constant regardless of variation in flow. Then, a throttle controls the chamber pressure by changing the flow into the oil chamber to balance the load.

The function of a throttle is to produce a flow resistance to increase the stiffness and stability of a lubrication system. In Figure 4.10b, when the load increases, the film thickness decreases and the outflow

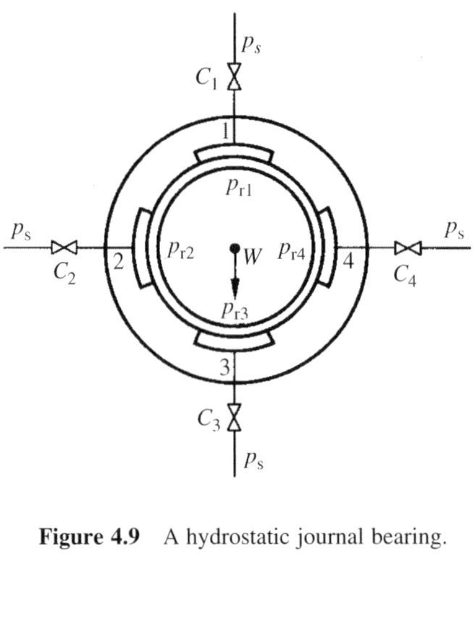

Figure 4.9 A hydrostatic journal bearing.

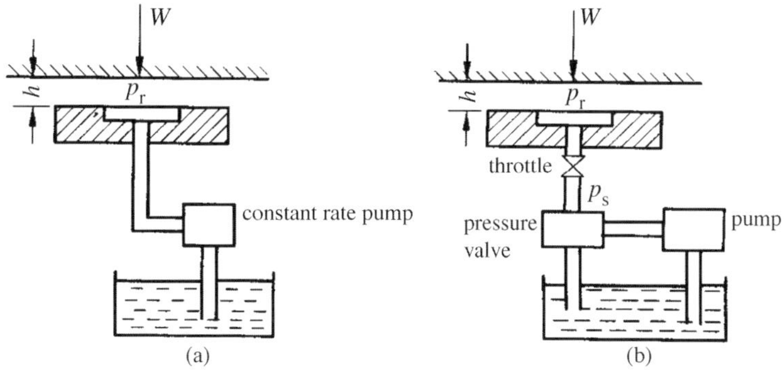

Figure 4.10 Static pressure lubrication system: (a) constant flow system, (b) constant pressure system.

from the oil chamber is reduced. Flow through the throttling device depends on the pressure difference at both ends. When the flow is reduced the pressure difference across the throttle is also reduced. Thus, if the supplied pressure p_s is constant, the oil chamber pressure p_r will rise, allowing the carrying load to increase.

Now let us take a circular thrust plate with a single oil chamber, for example, to discuss the film stiffness of a flow control device, as shown in Figure 4.8.

4.3.3.1 Constant Flow Pump

The film stiffness is an ability to resist a load change, that is, the variation of the unit film thickness is needed in order to produce the carrying load. Accordingly, we define the stiffness coefficient K as follows:

$$K \equiv -\frac{dW}{dh}. \tag{4.63}$$

With Equation 4.62, the load for Figure 4.8 is equal to

$$W = \frac{3\eta R^2 Q}{h^3}\left[1 - \left(\frac{R_0}{R}\right)^2\right]. \tag{4.64}$$

Because the flow Q is a constant, according to Equation 4.63 the stiffness coefficient of a constant flow system is equal to

$$K = \frac{3W}{h} = \frac{9\eta R^2 Q}{h^4}\left[1 - \left(\frac{R_0}{R}\right)^2\right]. \tag{4.65}$$

Equation 4.65 means that the stiffness of a hydrostatic bearing with a constant flow system is inversely proportional to the four powers of the film thickness. It has high stiffness.

4.3.3.2 Capillary Throttle

The length of a capillary throttle is much larger than its diameter. The capillary throttles have the forms of a needle tube, a circular tube, a spiral groove and so on. The lubricant flow passing through a capillary with a diameter d and length l into the oil chamber can be calculated by the Poiseuille formula.

$$Q = \frac{\pi d^4 \nabla p}{128\eta l} \tag{4.66}$$

where ∇p is the difference of the capillary pressure between the two ends, $\nabla p = p_s - p_r$.

Because Q calculated by Equation 4.66 is equal to that obtained by Equation 4.61, we have

$$\frac{\pi d^4 \nabla p}{128\eta l} = \frac{\pi h^3 p_r}{6\eta} \cdot \frac{1}{\ln(R/R_0)}, \tag{4.67}$$

or

$$\frac{\nabla p}{p_r} = \frac{64lh^3}{3d^4} \cdot \frac{1}{\ln(R/R_0)}. \tag{4.68}$$

Usually, we can set $\nabla p/p_r = 1/2$, which is equivalent to $\nabla p/p_s \approx 1/3$. Therefore, the geometric parameter of a capillary throttle can be determined according to Equation 4.68.

As for a constant pressure system, the film stiffness of a capillary throttle for a circular thrust plate is equal to

$$K = \frac{3W}{h}\left(\frac{h^3 Qk_c}{p_r + h^3 Qk_c}\right), \tag{4.69}$$

where $k_c = 128l/\pi d^4$ is the throttling coefficient.

Equation 4.69 shows that the stiffness of a capillary throttle system is less than that of a constant flow system.

4.3.3.3 Thin Walled Orifice Throttle

Such a throttle is used in a thin metal sheet to open a hole, and the diameter of the hole is greater than the thickness of the sheet to control flow. Suppose the diameter of the orifice is *d*, the flow through it will be

$$Q = \frac{\pi d^2}{4}\sqrt{\frac{2\nabla p}{\rho}}c_d, \tag{4.70}$$

where c_d is the flow coefficient; ρ is the density; ∇p is the pressure difference between the two ends of the throttle, $\nabla p = p_s - p_r$.

Then, according to the condition that the flow through the throttle is equal to that flow into the bearing, we have

$$\frac{\pi d^2}{4}\sqrt{\frac{2\nabla p}{\rho}}c_d = \frac{\pi h^3 p_r}{6\eta} \cdot \frac{1}{\ln(R/R_0)}. \tag{4.71}$$

Therefore, the parameters of the throttle can be calculated using Equation 4.71.

The above analysis shows that the thin walled orifice throttle has a slightly higher stiffness than that of the capillary throttle. However, the orifice throttle stiffness is related to the viscosity of lubricant, that is, stiffness will be influenced by temperature significantly.

4.4 Squeeze Bearing

If the carrying load changes in the thickness direction, the lubricant will be squeezed. Because the lubricant between the bearing surfaces is usually not squeezed out of the bearing at once, the film can produce a load to balance the load. For instance, in an aero-engine piston pin bearing, the squeezed pressure is larger than 35 MPa and in a shearing machine or punch crank bearing, the pressure may reach up to 55 MPa.

In order to analyze squeezing lubrication, we assume that there is no relative sliding between the two bearing surfaces and lubricant viscosity is constant. Therefore, Reynolds equation becomes

$$\frac{\partial}{\partial x}\left(h^3\frac{\partial p}{\partial x}\right)+\frac{\partial}{\partial y}\left(h^3\frac{\partial p}{\partial y}\right)=12\eta\frac{\partial h}{\partial t}. \tag{4.72}$$

By solving the equation we can determine the relationship between the load W and film thickness h.

4.4.1 Rectangular Plate Squeeze

As shown in Figure 4.11, there are two plates, acted on by load W. The clearance between the two plates is h, filled with viscous lubricant.

If the width of the plates is infinite, the Reynolds equation of Equation 4.72 becomes

$$\frac{d}{dx}\left(h^3\frac{dp}{dx}\right)=12\eta\frac{dh}{dt}. \tag{4.73}$$

The boundary conditions of the Reynolds equation are

$$p|_{x=\pm l/2}=0. \tag{4.74}$$

Because h is a constant, and not varying with x, we can obtain the following pressure distribution:

$$p=\frac{6\eta}{h^3}\frac{dh}{dt}\left(x^2-\frac{l^2}{4}\right). \tag{4.75}$$

Equation 4.75 shows that the pressure distribution is parabolic, and the maximum pressure is equal to $p_{\max}=(-3\eta l^2/2h^3)\,dh/dt$ in the middle of the plate. The carrying load of the plate per unit width is equal to

$$\frac{W}{b}=\int_{-l/2}^{l/2}pdx=-\frac{\eta l^3}{h^3}\frac{dh}{dt}. \tag{4.76}$$

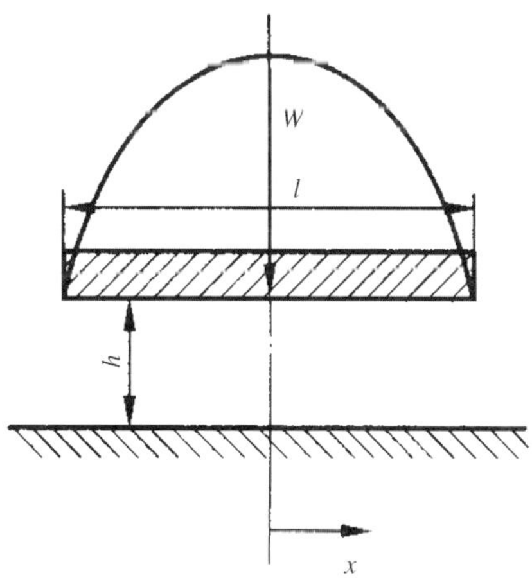

Figure 4.11 Rectangular plate squeezes.

From the above equations we know that when the film thickness gradually becomes thinner under the load, dh/dt is negative and a positive pressure is produced to carry the load. When the load is in an opposite direction, the film thickness will increase so that the compression effect disappears. If we continuously add lubricant into the clearance, the lubricant film can form again as the load presses down.

For a finite rectangular plate squeezing lubrication, its load-carrying capacity is equal to

$$\frac{W}{b} = -\beta \frac{\eta l^3}{h^3} \frac{dh}{dt}, \tag{4.77}$$

where β is the leakage coefficient, which depends on the ratio of l/b.

For a finite length rectangular plate, as the film thickness decreases from h_1 to h_2, the elapse time Δt is

$$\Delta t = -\int_{h_1}^{h_2} \beta \frac{\eta l^3 b}{W} \frac{1}{h^3} dh = \beta \frac{\eta l^3 b}{2W} \left(\frac{1}{h_2^2} - \frac{1}{h_1^2} \right). \tag{4.78}$$

4.4.2 Disc Squeeze

In order to analyze the disc squeeze lubrication with radius R, Equation 4.72 is first rewritten in the polar form:

$$\frac{\partial}{\partial r} \left(r h^3 \frac{\partial p}{\partial r} \right) + \frac{\partial}{r \partial \theta} \left(h^3 \frac{\partial p}{\partial \theta} \right) = 12 r \eta \frac{\partial h}{\partial t}. \tag{4.79}$$

Because the problem is polar symmetry, that is, $\partial p / \partial \theta = 0$, and h is independent to r, we have:

$$\frac{d}{dr} \left(r h^3 \frac{dp}{dr} \right) = \frac{12 \eta r}{h^3} \frac{dh}{dt}. \tag{4.80}$$

The boundary conditions of the above equation are: $dp/dr|_{r=0} = 0$; $p|_{r=R} = 0$. If we substitute the boundary conditions into the solution of Equation 4.80, the pressure becomes:

$$p = \frac{3\eta (r^2 - R^2)}{h^3} \frac{dh}{dt} \tag{4.81}$$

Integrating the pressure, the carrying load is equal to:

$$W = \int_0^R 2\pi p r dr = -\frac{3\pi \eta R^4}{2h^3} \frac{dh}{dt}. \tag{4.82}$$

Similarly, the elapse time the plate decants from h_1 to h_2 is equal to:

$$\Delta t = \frac{3\pi \eta R^4}{4W} \left(\frac{1}{h_2^2} - \frac{1}{h_1^2} \right). \tag{4.83}$$

For an ellipse plate squeeze with the long axis a and the short axis b, the carrying load is

$$W = -\frac{3\pi \eta}{h^3} \frac{a^3 b^3}{a^2 + b^2} \frac{dh}{dt}. \tag{4.84}$$

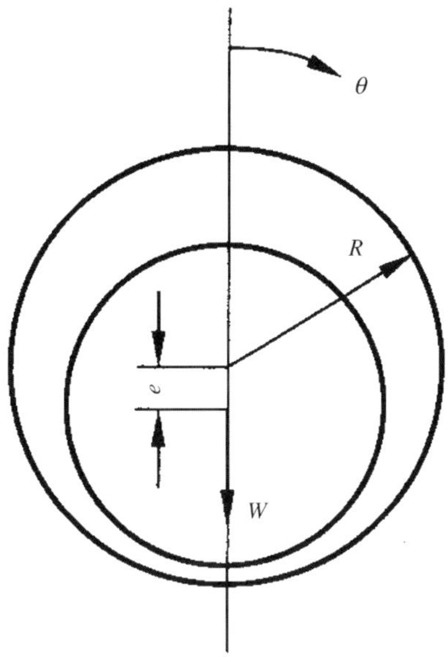

Figure 4.12 Journal bearings squeeze.

4.4.3 *Journal Bearing Squeeze*

As shown in Figure 4.12, a journal bearing is under load W in the squeeze lubrication. If the shaft center has a downwards speed $de/dt = c\, d\varepsilon/dt$, the film thickness variation is as follows:

$$\frac{dh}{dt} = c\cos\theta\frac{d\varepsilon}{dt}.$$

Between $\pi/2 < \theta < 3\pi/2$, dh/dt is negative, so that it can produce a positive pressure. Set $x = R\theta$, and for an infinite wide bearing, that is, $\partial p/\partial y = 0$, Equation 4.72 becomes

$$\frac{d}{d\theta}\left[(1+\varepsilon\cos\theta)^3\frac{dp}{d\theta}\right] = 12\eta\frac{R^2}{c^2}\cos\theta\frac{d\varepsilon}{dt}. \tag{4.85}$$

Integrate the above equation and substitute into the Sommerfeld boundary conditions: $p|_{\theta=0} = 0$ and $dp/d\theta|_{\theta=\pi} = 0$, the pressure distribution will be:

$$p = 6\eta\frac{R^2}{c^2}\frac{1}{\varepsilon}\left[\frac{1}{(1+\varepsilon\cos\theta)^2} - \frac{1}{(1+\varepsilon)^2}\right]\frac{d\varepsilon}{dt}. \tag{4.86}$$

Because the pressure distribution is symmetric to the axis $\theta = \pi$, the load-carrying capacity per width of the squeeze film is

$$\frac{W}{b} = 2\int_0^{\pi} pR\cos\theta d\theta = \frac{\eta R^3}{c^2}\frac{12\pi}{(1-\varepsilon^2)^{3/2}}\frac{d\varepsilon}{dt}. \tag{4.87}$$

From Equation 4.87, we can obtain Δt corresponding to the eccentricities varying from ε_1 to ε_2:

$$\Delta t = \frac{12\pi\eta bR^3}{Wc^2}\left(\frac{\varepsilon_2}{\left(1-\varepsilon_2^2\right)^{1/2}} - \frac{\varepsilon_1}{\left(1-\varepsilon_1^2\right)^{1/2}}\right). \tag{4.88}$$

It should be pointed out that Equations 4.87 and 4.88 give a film thickness in the full bearing, that is, 360°, because Sommerfeld boundary conditions are used in derivation. On the other hand, for semi-Sommerfeld conditions, pressure only exists between $\theta = \pm\,\pi/2$, and elsewhere $p = 0$, the solutions are as follows:

$$\frac{W}{b} = \frac{12\eta R^3}{c^2}\left[\frac{\varepsilon}{1-\varepsilon^2} + \frac{2}{(1-\varepsilon^2)^{3/2}}\arctan\left(\frac{1+\varepsilon}{1-\varepsilon}\right)^{1/2}\right]\frac{d\varepsilon}{dt}, \tag{4.89}$$

$$\Delta t = \frac{24\eta bR^3}{Wc^2}\left[\frac{\varepsilon_2}{\left(1-\varepsilon_2^2\right)^{1/2}}\arctan\left(\frac{1+\varepsilon_2}{1-\varepsilon_2}\right)^{1/2} - \frac{\varepsilon_1}{\left(1-\varepsilon_1^2\right)^{1/2}}\arctan\left(\frac{1+\varepsilon_1}{1-\varepsilon_1}\right)^{1/2}\right]. \tag{4.90}$$

Similarly, for a hemispherical bearing squeeze problem, the results will be:

$$W = \frac{6\pi\eta R^4}{c^2}\left[\frac{1}{\varepsilon^2}\ln(1-\varepsilon) + \frac{1}{\varepsilon^2(1-\varepsilon)} - \frac{1}{2\varepsilon}\right]\frac{d\varepsilon}{dt}, \tag{4.91}$$

$$\Delta t = \frac{3\pi\eta R^4}{Wc^2}\left[\frac{\varepsilon_2-\varepsilon_1}{\varepsilon_1\varepsilon_2} + \frac{1+\varepsilon_1^2}{\varepsilon_1^2}\ln(1-\varepsilon_1) - \frac{1+\varepsilon_2^2}{\varepsilon_2^2}\ln(1-\varepsilon_2)\right]. \tag{4.92}$$

4.5 Dynamic Bearing

As discussed earlier, a bearing on which the magnitude and direction of the load acting do not change is called a static load bearing. If the operating parameters are fixed, the axis of a journal bearing or thrust bearing will be at a certain position to keep unchanged. Therefore, the parameters of the bearing do not change with time.

However, the loads acting on most bearings change all the time in magnitude, direction or speed. Such bearings are called nonstable bearings or dynamic bearings. A dynamic bearing axis or thrust plate moves in a certain track. If the condition parameters are periodical, the axis track is closed.

Typical dynamic bearings can be found in the crankshaft, the connecting rod, the piston pin bearings and so on in an internal combustion engine, where the size and direction of loads vary periodically. For an unbalanced rotor, the direction of the load changes in rotary, but its magnitude does not change. Steady bearings also suffer dynamic loads during starting, stopping or under the impact of vibration.

Usually, dynamic bearings can be divided into two categories on its working principles. One is the journal which does not rotate around its own axis, that is, no relative sliding. For such a bearing, the journal center moves along a certain track so that squeeze effect is aroused. The other is the journal which not only rotates around its axis, but also moves. Therefore, the dynamic and squeeze pressures are included together.

The Reynolds equation for a dynamic bearing can be written as follows:

$$\frac{\partial}{\partial x}\left(\frac{h^3}{\eta}\frac{\partial p}{\partial x}\right) + \frac{\partial}{\partial y}\left(\frac{h^3}{\eta}\frac{\partial p}{\partial y}\right) = 6U\frac{\partial h}{\partial x} + 12W, \tag{4.93}$$

where $W = w_h\text{-}w_0$.

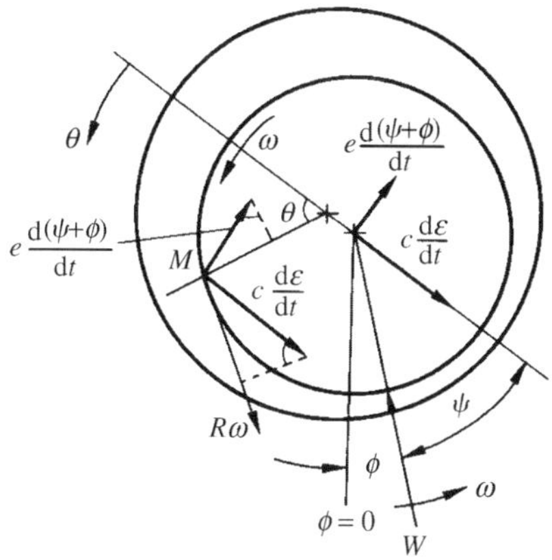

Figure 4.13 Dynamic bearing.

On the right-hand side of Equation 4.93, the first item is the dynamic pressure effect and the second is the squeeze one. If $w_h - w_0 = 0$, the equation will be the Reynolds equation for a bearing without squeezing.

Using Equation 4.93 to calculate the axis track of a dynamic bearing is an initial value problem. According to the given initial position of the axis, we can determine the axis track step-by-step.

4.5.1 Reynolds Equation of Dynamic Journal Bearings

In Figure 4.13 there is the relationship between the movements of a dynamic journal bearing. The journal rotates at the angular velocity ω around its center. Besides, the axis under the load W also moves in a certain track. The journal axis movement can be divided in the direction of the central line and perpendicular to the line. Then, the axis velocity components are equal to $c\, d\varepsilon/dt$ and $e\, d(\psi + \phi)/dt$. Here, we select the perpendicular axis as a reference axis; ϕ is equal to the load position angle, and ψ is the deviation angle, see Figure 4.13.

Each point M on the journal surface has a tangential velocity and a normal velocity. If we set the journal surface as the coordinate θ, the tangential velocity U and the normal velocity of V are equal to:

$$\begin{aligned} U &= R\omega + c\frac{d\varepsilon}{dt}\sin\theta - c\varepsilon\frac{d(\psi+\phi)}{dt}\cos\theta \\ W &= c\frac{d\varepsilon}{dt}\cos\theta + c\varepsilon\frac{d(\psi+\phi)}{dt}\sin\theta. \end{aligned} \tag{4.94}$$

Substitute U and V into Equation 4.93, and taking into account $c/R << 2$, $c/R << 2\cos\theta$ and $x = R\theta$ and $h = c(1 + \varepsilon\cos\theta)$, we can obtain the following Reynolds equation to analyze a dynamic journal bearing.

$$\frac{\partial}{\partial\theta}\left(h^3\frac{\partial p}{\partial\theta}\right) + R^2\frac{\partial}{\partial y}\left(h^3\frac{\partial p}{\partial y}\right) = 6\eta R^2\left[\left(\omega - 2\omega_L - 2\frac{d\psi}{dt}\right)\frac{dh}{d\theta} + 2c\frac{d\varepsilon}{dt}\cos\theta\right] \tag{4.95}$$

where $\omega_L = d\phi/dt$ is the angular velocity of the applied load W; $dh/d\theta$ in the right-hand side of Equation 4.95 is the dynamic pressure effect item; $d\varepsilon/dt$ is the squeezing effect item caused by the journal movement.

In order to obtain an analytical solution of Equation 4.95, the infinitely wide or infinitely narrow approximation must be used. Otherwise, numerical methods should be used.

For an infinitely wide bearing, $\partial p/\partial y = 0$. Substitute it into Equation 4.95 and use the Sommerfeld boundary conditions, $p|_{\theta=0} = 0$ and $p|_{\theta=2\pi} = 0$, we have:

$$p = 6\eta\left(\frac{R}{c}\right)^2\left\{\frac{2+\varepsilon\cos\theta}{(1+\varepsilon\cos\theta)^2}\frac{\varepsilon\sin\theta}{2+\varepsilon^2}\left(\omega - 2\omega_L - 2\frac{d\psi}{dt}\right) + \frac{1}{\varepsilon}\left[\frac{1}{(1+\varepsilon\cos\theta)^2} - \frac{1}{(1+\varepsilon)^2}\right]\frac{d\varepsilon}{dt}\right\}. \quad (4.96)$$

It can be seen that when $\omega_L = d\psi/dt = d\varepsilon/dt = 0$, Equation 4.96 has the same solution as a stable bearing as shown before.

Integrating Equation 4.96 along the central line and the perpendicular line respectively, we can obtain the load-carrying capacity as follows:

$$\frac{S\sin\psi}{12\pi^2} = \frac{\varepsilon}{(2+\varepsilon^2)(1-\varepsilon^2)^{1/2}}\frac{1}{\omega}\left(\omega - 2\omega_L - 2\frac{d\psi}{dt}\right), \quad (4.97)$$

$$\frac{S\cos\psi}{12\pi^2} = \frac{\varepsilon}{(1-\varepsilon^2)^{3/2}}\frac{1}{\omega}\frac{d\varepsilon}{dt}, \quad (4.98)$$

where S is the Sommerfeld number, $S = p(c/R)^2/N\eta$, which is often used as an indication of the load-carrying capacity of the dimensionless parameters; p is the average pressure, $p = W/bD$.

When $\omega_L = d\psi/dt = d\varepsilon/dt = 0$, Equation 4.98 becomes a steady bearing having a Sommerfeld analytical solution.

For an infinite narrow bearing, $\partial p/\partial\theta = 0$. Substitute it into Equation 4.95 and integrate twice, the pressure distribution can be obtained as follows:

$$p = 6\eta\left(\frac{R}{c}\right)^2\left[\left(\frac{b}{D}\right)^2 - \left(\frac{y}{R}\right)^2\right]\frac{1}{(1+\varepsilon\cos\theta)^3}\left[\frac{\varepsilon}{2}\left(\omega - 2\omega_L - 2\frac{d\psi}{dt}\right)\sin\theta - \frac{d\varepsilon}{dt}\cos\theta\right] \quad (4.99)$$

Again, integrate Equation 4.99, the load-carrying capacity of an infinite narrow bearing is:

$$\frac{S\sin\psi}{4\pi^2} = \left(\frac{b}{D}\right)^2\frac{\varepsilon}{2(1-\varepsilon^2)^{3/2}}\frac{1}{\omega}\left(\omega - 2\omega_L - 2\frac{d\psi}{dt}\right), \quad (4.100)$$

$$\frac{S\cos\psi}{4\pi^2} = \left(\frac{b}{D}\right)^2\frac{1+2\varepsilon^2}{(1-\varepsilon^2)^{5/2}}\frac{1}{\omega}\frac{d\varepsilon}{dt}. \quad (4.101)$$

The relationship between the load and the motion parameters is given by Equations 4.97 and 4.98 or the infinitely wide bearing and Equations 4.10 and 4.101 for an infinitely narrow bearing. They are the basic equations for dynamic bearing calculation. When the axis motion parameters $d\varepsilon/dt$ and $d\psi/dt$ are known, then by substituting them into Equations 4.97 and 4.98 or Equations 4.100 and 4.101 we will obtain algebraic equations that will allow us to carry out calculations easily.

However, if the axis motion parameters are not known but the load variation is known, it is much more difficult to obtain the axis motion track and the maximum eccentricity. The differential equations including ψ and ε should be solved. Especially for problems that S, ω and ω_L are the functions of time t, it is necessary to integrate Equations 4.97 and 4.98 or Equations 4.100 and 4.101 (which is difficult) in order to obtain analytical solutions.

4.5.2 Simple Dynamic Bearing Calculation

In this section, we will discuss some simple problems of dynamic journal bearings and give the calculations for lubrication properties [2].

4.5.2.1 A Sudden Load

For a stable load bearing, the journal axis is fixed at the equilibrium position relative to the bearing. When we suddenly add a load, the axis begins to move away from the equilibrium position. Its track curve depends on the direction and the magnitude of the load.

If the direction of the load is fixed, $\omega_L = d\phi/dt = 0$. For an infinitely wide bearing, Equations 4.97 and 4.98 becomes

$$\frac{S\sin\psi}{12\pi^2} = \frac{\varepsilon}{(2+\varepsilon^2)(1-\varepsilon^2)^{1/2}}\left(1-\frac{2}{\omega}\frac{d\psi}{dt}\right), \tag{4.102}$$

$$\frac{S\cos\psi}{12\pi^2} = \frac{\varepsilon}{(1-\varepsilon^2)^{3/2}}\frac{1}{\omega}\frac{d\varepsilon}{dt}. \tag{4.103}$$

The relationship between ε and ψ can be determined by the above two equations. Eliminating d_t and integrating the obtained equation, the axis track under a sudden load is

$$\sin\psi = \frac{12\pi^2}{5\varepsilon(1-\varepsilon^2)^{1/2}}\frac{1}{S} + K\frac{(1-\varepsilon^2)^{3/4}}{\varepsilon}. \tag{4.104}$$

Equation 4.104 describes a family of close track curves for different values of K.

Usually, a stable axis location is called the polar, denoted as ε_0 and ψ_0. From Equations 4.97 and 4.98, set $\omega_L = d\psi/dt = d\varepsilon/dt = 0$, we have the track in the pole coordinates as follows:

$$\frac{S}{12\pi^2} = \frac{\varepsilon_0}{(2+\varepsilon_0^2)(1-\varepsilon_0^2)^{1/2}}, \tag{4.105}$$

$$\psi_0 = \frac{\pi}{2}. \tag{4.106}$$

For an infinite narrow bearing, with the same method, the track and polar axis can be solved from Equations 4.100 and 4.101 as follows:

$$\sin\psi = \frac{\pi^2\varepsilon}{S(1-\varepsilon^2)^{3/2}}\left(\frac{b}{D}\right)^2 - K^2\frac{(1-\varepsilon^2)^{3/2}}{\varepsilon}, \tag{4.107}$$

$$\frac{S}{2\pi^2}\left(\frac{D}{b}\right)^2 = \frac{\varepsilon_0}{(1-\varepsilon_0^2)^{3/2}}, \tag{4.108}$$

$$\psi_0 = \frac{\pi}{2}. \tag{4.109}$$

In Figure 4.14, the results of Equations 4.104 and 4.107 are given to show the track curves while the eccentricity of the polar $\varepsilon_0 = 0.7$. There are two extreme states among these curves. One is only the point at $\varepsilon_0 = 0.7$; the other is a circle, which corresponds to $\varepsilon = 1$, which is practically impossible because the journal and the bearing will be in contact with each other.

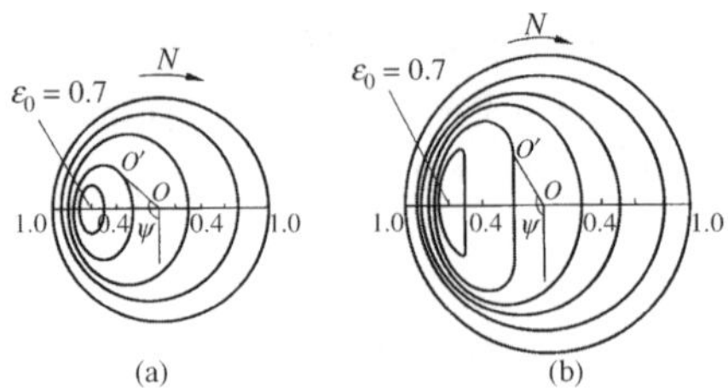

Figure 4.14 Tracks with a sudden load: (a) an infinitely wide bearings; (b) an infinitely narrow bearing.

It should be pointed out that the above analysis is based on the physical model of a nondamping free vibration system. In fact, due to lubricant film having damping, the axis will gradually tend to the pole's position and finally reach a steady state.

4.5.2.2 Rotating Load

Another typical dynamic bearing is that the magnitude of the load is fixed with a constant angular velocity. If the bearing works in a stable state, we can assume that the phase and amplitude of the axis track are constants, namely, $d\psi/dt = d\varepsilon/dt = 0$. Therefore, for an infinitely wide bearing, we have

$$\frac{S}{12\pi^2} = \frac{\varepsilon}{(2+\varepsilon^2)(1-\varepsilon^2)^{1/2}}\left(1 - 2\frac{\omega_L}{\omega}\right), \tag{4.110}$$

$$\psi = \frac{\pi}{2}. \tag{4.111}$$

It can be known that the carrying load of Equation 4.110 includes two parts. One is the unrotational load capacity, which makes the journal rotate around its own axis in the angular velocity ω; the other is the rotational load at the angular ω_L. The two loads are of difference phases of 180°. The sum is the total carrying capacity given by Equation 4.110.

Equation 4.110 also shows that the carrying load depends on ω and ω_L. When $\omega_L = 0$, that is, the $\omega_L/\omega = 0$, it is a stable bearing and this gives the maximum value of S. When $\omega_L = 2\omega$, $S = 0$ and the journal moves in the half-frequency whirl, which may bring about a severe vibration.

4.5.3 General Dynamic Bearings

For general dynamic bearings, the magnitude and direction changes with time. Furthermore, the journal speed may be a function of time. For such a bearing, because of complexity, a numerical method must be used to solve the load and the velocity according to the axis track.

As mentioned above, solving the axis movement is an initial value problem. From the start, as the initial position, take each instantaneous load as a steady load. Then find the next axis position. Finally, connect the points to obtain the axis moving track.

Several methods for solving general dynamic bearings are given below.

4.5.3.1 Infinitely Narrow Bearing

In 1962, Milne proposed an infinitely narrow bearing theory to calculate the axis track of a journal bearing [3]. This is based on directly integrating Equation 4.95 to obtain the pressure distribution as follows:

$$p = \frac{6\eta c}{h^3}\left(y^2 - \frac{b^2}{4}\right)\left[\frac{d\varepsilon}{dt}\cos\theta - \varepsilon\left(\frac{\omega}{2} - \omega_L - \frac{d\psi}{dt}\right)\sin\theta\right] \tag{4.112}$$

Then, the load components W_ε along the central line direction and W_a perpendicular to the central line can be integrated as shown in Figure 4.13. The circumferential border is known from the solution of infinitely narrow bearing, that is, $p = 0$. From Equation 4.112 we can obtain:

The film starting point $\theta_1 = \arctan\left[(2d\varepsilon/dt)/(\varepsilon(\omega - 2\omega_L - 2d\psi/dt))\right]$.

The film ending point $\theta_2 = \theta_1 + \pi$.

In order to determine θ_1 and θ_2, we must know $d\varepsilon/dt$, $d\psi/dt$, while $d\varepsilon/dt$ and $d\psi/dt$ depend on θ_1, θ_2, W_ε, W_a and so on. The relationships between them can be written as follows:

$$\frac{d\varepsilon}{dt} = \frac{k(I_2 W_\varepsilon + I_3 W_a)}{I_1 I_3 - I_2^2}, \tag{4.113}$$

$$\frac{d\psi}{dt} = \frac{\omega}{2} - \omega_L - \frac{k(I_1 W_\varepsilon + I_2 W_a)}{I_1 I_3 - I_2^2}, \tag{4.114}$$

where k is a constant related to the initial position and bearing parameters; I_1, I_2 and I_3 are the integral values containing θ and ε; θ_1 and θ_2 are the upper and lower limits.

Therefore, in order to obtain $d\varepsilon/dt$ and $d\psi/dt$, the above relationships need to be obtained by the simultaneous iterations.

An axis track can be obtained by the following steps. First, divide time into many intervals, and each interval of time should be very short so that the approximated $d\varepsilon/dt$ and $d\psi/dt$ can be considered as constants. This means that the axis moves with a constant velocity. Then, according to the given method before, determine the bearing load components W_a and W_a. Subsequently, solve the corresponding $d\varepsilon/dt$ and $d\psi/dt$ by a numerical method. And, from the initial position of the axis, determine the second axis position by the first $d\varepsilon/dt$ and $d\psi/dt$, the third position by the second $d\varepsilon/dt$ and $d\psi/dt$, and so on. Finally, with this step by step process, the whole axis track is obtained.

4.5.3.2 Superimposition Method of Pressures

In 1957, Hahm proposed the pressure superimposition method to consider the hydrodynamic effect and squeezing effect [4]. He first gave the dimensionless Reynolds equation with following parameters:

$$Y = \frac{2y}{b};\quad \alpha = \left(\frac{2R}{b}\right)^2;\quad P = \frac{p}{\eta\left(\omega - 2\omega_L - 2\frac{d\psi}{dt}\right)}\left(\frac{c}{R}\right)^2;\quad Q = \frac{2\frac{d\varepsilon}{dt}}{\omega - 2\omega_L - 2\frac{d\psi}{dt}}. \tag{4.115}$$

Then, Equation 4.95 becomes:

$$\frac{\partial}{\partial\theta}\left((1+\varepsilon\cos\theta)^3\frac{\partial P}{\partial\theta}\right) + \alpha(1+\varepsilon\cos\theta)^3\frac{\partial^2 P}{\partial Y^2} = -6\varepsilon\sin\theta + 6Q\cos\theta. \tag{4.116}$$

Because Equation 4.116 is a linear partial differential equation, the solutions of each right-hand side can be superimposed as follows:

$$P = P_1 + QP_2, \tag{4.117}$$

where P_1 and P_2 satisfy the following equations:

$$\frac{\partial}{\partial \theta}\left((1+\varepsilon\cos\theta)^3\frac{\partial P_1}{\partial \theta}\right)+\alpha(1+\varepsilon\cos\theta)^3\frac{\partial^2 P_1}{\partial Y^2}=-6\varepsilon\sin\theta, \tag{4.118}$$

$$\frac{\partial}{\partial \theta}\left((1+\varepsilon\cos\theta)^3\frac{\partial P_2}{\partial \theta}\right)+\alpha(1+\varepsilon\cos\theta)^3\frac{\partial^2 P_2}{\partial Y^2}=6\cos\theta. \tag{4.119}$$

From the above two equations, under the same boundary conditions P_1 and P_2 can be obtained so that from Equation 4.117 P can be obtained.

Integrate P along the positive pressure region to obtain the relationships between the carrying loads W_e, W_a and $d\varepsilon/dt$, $d\psi/dt$. Then, the loads can be used to calculate the corresponding velocities $d\varepsilon/dt$ and $d\psi/dt$. Then, step by step, determine the journal track curve.

4.5.3.3 Superimposition Method of Carrying Loads

To overcome difficulties to solve dynamic Reynolds equation, Holland in 1959 proposed a simplified calculation method [5]. The key points of the method are: (1) separately calculate the rotational and squeezing movements of the journal and solve them according to the respective boundary conditions. Then, calculate the sum of the hydrodynamic and squeezing carrying vectors to balance the external load to establish the relationship of the load and velocity.

As shown in Figure 4.15, because pressure distributions are calculated according to different boundary conditions, if we ignore interactions and negative pressure, the Holland method is actually a simplified calculation.

By Equation 4.95, the Reynolds equation only with a rotation movement is:

$$\frac{\partial}{\partial \theta}\left((1+\varepsilon\cos\theta)^3\frac{\partial p}{\partial \theta}\right)+R^2\frac{\partial}{\partial y}\left((1+\varepsilon\cos\theta)^3\frac{\partial p}{\partial y}\right)=-\frac{6\eta R^2\varepsilon}{c^2}\left(\omega-2\omega_L-2\frac{d\psi}{dt}\right)\sin\theta, \tag{4.120}$$

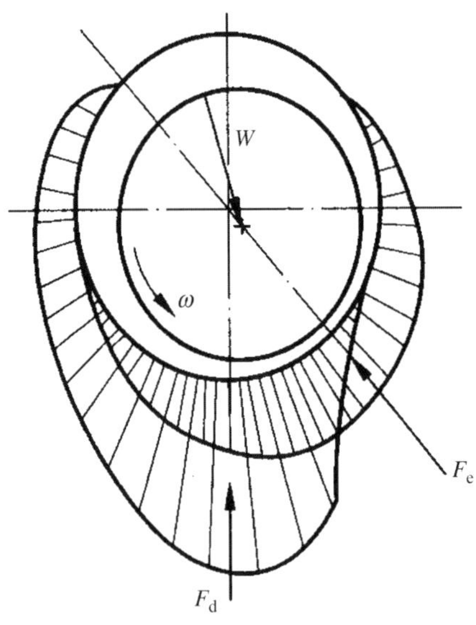

Figure 4.15 Rotation and squeeze film forces.

while only with a squeezing movement, the Reynolds equation is:

$$\frac{\partial}{\partial \theta}\left[(1+\cos\theta)^3\frac{\partial p}{\partial \theta}\right]+R^2\frac{\partial}{\partial y}\left[(1+\cos\theta)^3\frac{\partial p}{\partial y}\right]=\frac{12\eta R^2}{C}\frac{d\varepsilon}{dt}\cos\theta. \tag{4.121}$$

Based on the above two equations the load-carrying capacities F_d with the rotation movement and F_e with the squeezing movement can be obtained. To balance the external load W, the relationships between the load, $d\varepsilon/dt$ and $d\psi/dt$, can be established to obtain the journal track step by step.

The dynamic bearing lubrication is very complex, even for a smooth surface and isothermal lubrication problem. However, with the rapid development of computer technology, it is now possible to consider most factors affecting dynamic bearing lubrication. Wang in her doctoral thesis studied the dynamic bearing of the internal combustion engines to consider both the surface morphology and the thermal effect [6].

4.6 Gas Lubrication Bearings

Air is mainly used as a gas lubricant for high speed bearing operation. The sliding speeds may be higher than 100 m/s with an extremely low coefficient of friction and heat. For example, an air bearing with the diameter $D=34$ mm, length $L=40$ mm and rotational speed $n=21\ 000$ r/min, the temperature rise is only about 3 °C and the friction power loss is only 0.01 hp. In some working conditions of high or low temperature working environment, such as atomic energy, textile and food processing industries, the bearings are especially suitable for gas lubricated. However, the applications of gas lubrication are limited because the carrying capacity is low and manufacturing precision is high.

Similar to liquid lubrication, gas lubrications are also of two types: gas dynamic lubrication and gas static lubrication. Because the load-carrying capacity of the aerodynamic lubrication is low, more attention is drawn by gas static lubrication.

4.6.1 Basic Equations of Gas Lubrication

The main feature of gas in lubrication is compressible so that the density of gas is treated as a variable. Therefore, the Reynolds equation becomes

$$\frac{\partial}{\partial x}\left(\frac{\rho h^3}{\eta}\cdot\frac{\partial p}{\partial x}\right)+\frac{\partial}{\partial y}\left(\frac{\rho h^3}{\eta}\frac{\partial p}{\partial y}\right)=6\left[U\frac{\partial}{\partial x}(\rho h)+2\frac{\partial}{\partial t}(\rho h)\right] \tag{4.122}$$

Usually, viscosity of gas is low. For example, at 20 °C, the viscosity of air is only 4000th of the viscosity of the spindle oil. Therefore, under normal operating conditions, the friction power loss of gas lubrication can be negligible. Furthermore, not as a liquid, the viscosity of gas increases slightly with increase of temperature. Therefore, the thermal effect is significant only at high speed for a gas lubrication problem. Usually, gas lubrication is isothermal and the viscosity of gas is also considered as a constant.

The relationship of the density of gas with temperature and pressure is often given as follows:

$$\frac{p}{\rho}=RT, \tag{4.123}$$

where T is the absolute temperature; R is the gas constant.

As mentioned above that a typical gas lubrication problem is usually regarded as an isothermal process, Equation 4.122 becomes:

$$p=k\rho, \tag{4.124}$$

where k is the proportional constant.

When the speed in a gas lubrication process is very fast, the produced heat cannot be conducted easily. So, we can also take such a process as a diabatic process. Therefore, Equation 4.122 can be written as:

$$p = k\rho^n, \tag{4.125}$$

where n is the gas specific heat ratio. It is related to the structure of the gas molecule. For air, $n = 1.4$.

For an isothermal process, substituting Equation 4.124 into the Reynolds equation, it will be:

$$\frac{\partial}{\partial x}\left(h^3 p \frac{\partial p}{\partial x}\right) + \frac{\partial}{\partial y}\left(h^3 p \frac{\partial p}{\partial y}\right) = 6\eta\left[U \frac{\partial}{\partial x}(ph) + 2\frac{\partial}{\partial t}(ph)\right]. \tag{4.126}$$

Equation 4.126 is the basic equation used for calculation of gas lubrication.

4.6.2 *Types of Gas Lubrication Bearings*

Because the gas film thickness is very thin, gas lubricated bearings must be manufactured precisely. The main disadvantages of a gas lubrication bearing are both the load-carrying capacity and the low stability. In order to increase load-carrying capacity and to improve stability, many kinds of structures are designed, the principles of which can be also applied to fluid lubricated bearings.

Some aerodynamic journal bearings are shown in Figure 4.16. The bush of the porous bearing (Figure 4.16b) is made of porous material. Its stability can be improved on the bearing of Figure 4.16a. An elastic spring can also be used as in Figure 4.16c to improve the bearing stability further.

Some multi-wedge aerodynamic journal bearings are shown in Figure 4.17. The titling pad bearing is a typical multi-wedge bearing, which consists of a number of titling pads as shown in Figure 4.17a. The titling angle of the pad can be automatically adjusted with the load. Its stability is good, but it is complex.

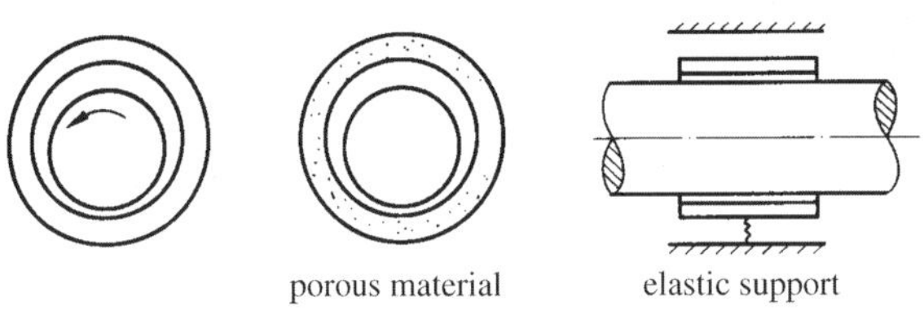

Figure 4.16 Some aerodynamic journal bearings.

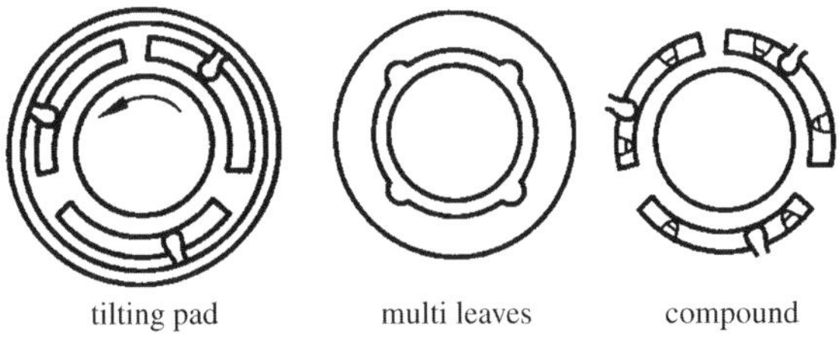

Figure 4.17 Multi-wedge journal bearings.

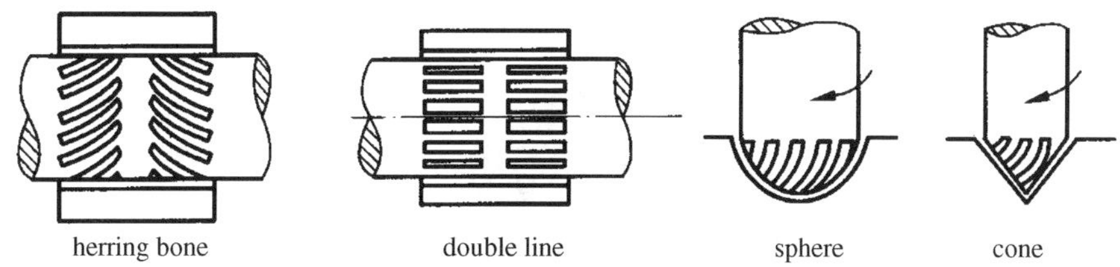

Figure 4.18 Grooved gas lubrication bearings.

Porous materials can also be used in pads. Another multi-wedge bearing is the multi-leave bearing as shown in Figure 4.17b, in which three or four leaves are usually used. In addition, compound bearings are also widely used, as shown in Figure 4.17c.

Sometimes, grooves can be manufactured on the bearing surface of an aerodynamic lubrication so that gas pressure can be effectively generated with the rotation of the journal, as shown in Figure 4.18. The load-carrying capacity and stability of these bearings are well improved and widely used in small-scale and high-speed rotating machineries. The shape of grooves may be herringbone or straight lines for journal bearings, or sphere or cone for thrust bearings.

In Figure 4.19, some flexible pad bearings are shown, which can be used in ultra high speed rotating machineries. These bearing surfaces are made of flexible foils. The surfaces are generally of different shapes meeting with particular requirements. The stability of the bearings is good, especially suitable for situations with small size and strict alignment.

In Figure 4.20 some thrust gas bearings are shown. The structures of these bearings are ladder-style, tilting-pad and flexible surface. A ladder bearing is simple with good stability, but the load-carrying capacity is low. The grooved bearing is of a high load-carrying capacity, but less stable. A tilting pad bearing is a high precision bearing, and the flexible surface is a little complex. These bearings are of different characteristics and can be selected according to actual needs.

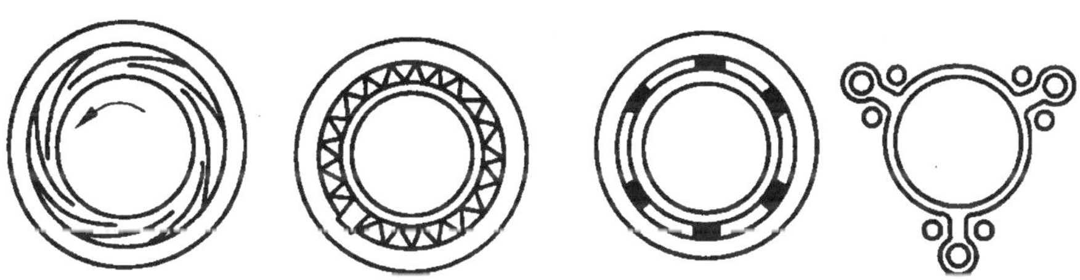

Figure 4.19 Flexible surface journal bearings.

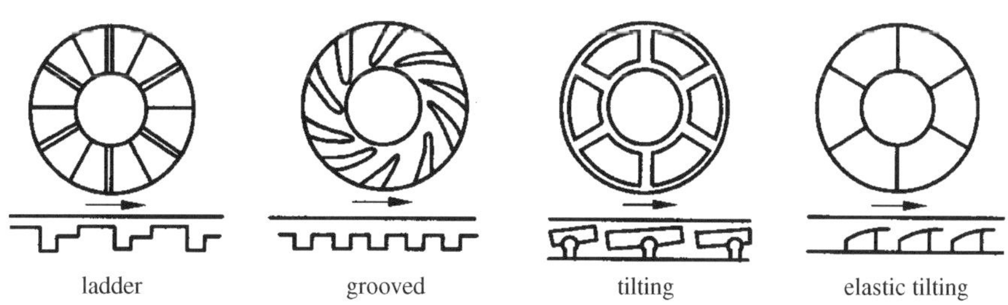

Figure 4.20 Gas thrust bearings.

Figure 4.21 From floating gas thrust bearings.

A floating gas thrust bearing is shown in Figure 4.21. Several holes and spiral slots are symmetrically manufactured on the thrust plate. The supplied gas with a static pressure can avoid surface scratches while the bearing starts or stops operation.

In Figure 4.22 two kinds of tilting-pad aerodynamic journal bearings are shown with floating supply structures. The shaft weight may cause an offset that results in side contact of the bearing. This makes the starting of the bearing difficult. While the journal is suspended by an external air supply, the gas can be easily supplied, called the floating supply. The gas is usually imported through the pivot shaft, as shown in Figure 4.22.

The load-carrying capacity of an aerostatic bearing is strongly influenced by the bearing structure. For example, the author studied two static air pressure bearing structures with a throttle, as shown in Figure 4.23 [7]. The cone-type bearing is with the cone angles of the rotor and stator 100° and 90° respectively. The throttle hole diameter is 0.4–1 mm; six small holes are uniformly designed to be parallel to the direction of the rotor axis (B) or with an angle of 44° (A), and the planar bearing end is made with an angle of 2–3° to control the flow in order to increase the load-carrying capacity.

Experimental results showed that in the same area, the load-carrying capacity of the cone-type bearing is higher than that of the planar. The greater the gas pressure, the larger the load-carrying capacity. For example, while $p_s = 0.6$ MPa, the load-carrying capacity of the cone-type bearings is 2.17 times that of the planar one.

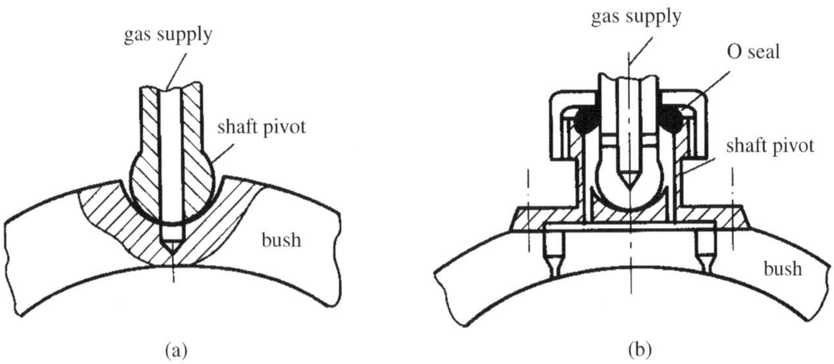

Figure 4.22 Structures of floating supply.

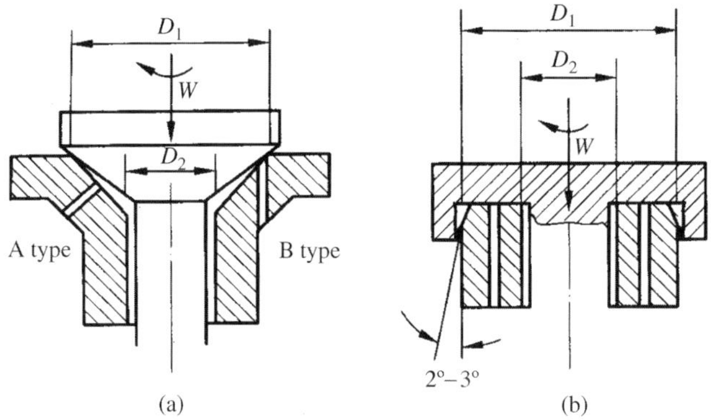

Figure 4.23 Two structures of aerostatic thrust bearings.

4.7 Rolling Contact Bearing

EHL theory has provided a theoretical basis for lubrication design of mechanical elements in line and point contacts. However, applications of EHL theory are still carried on because the formulas of EHL are used only under certain conditions, and contact situations of mechanical elements are very complex. Therefore, simplified analysis is often used.

It has been proved practically that a thin elastohydrodynamic oil film exists between the rolling bodies and the raceway while rolling contact bearings are rotating, such as the gyro motor bearings, aircraft engine bearings, spindle bearings, and precision machine tool spindle bearings. Because there is an EHL film, the contact fatigue life of a rolling contact bearing can be at least increased more than once as regulated by the American Bearings Manufacturers Association (ABMA). In order to carry out EHL calculation for a rolling contact bearing, the motion relationship between the rolling bodies, the rings and the force must be pre-determined. However, the dynamics analysis of a rolling contact bearing is very complex because the internal component movements of the bearing are closely related to the lubrication state. Dowson *et al.* on the roller bearing analysis showed that there exists a considerable relative sliding between the rollers and the rings while they consider lubricant as iso-viscous and solid surfaces as rigid. If EHL theory is adopted, the apparent sliding does not occur. Therefore, for elastohydrodynamic lubrication, the motion of a rolling contact bearing can be regarded as pure rolling.

Figure 4.24 shows a part of a roller bearing. The relationships between geometry and motion can be deduced to obtain the equivalent radius of curvature, the average speed and the carrying load.

4.7.1 *Equivalent Radius R*

If the radius of the inner ring is R_1, the outer ring is R_2, the roller's diameter is $d = 2r$, and let $\lambda = d/D_m$, where D_m is the average diameter $D_m = R_1 + R_2$, the equivalent radius for the roller and the inner ring at the contact point will be:

$$R = \frac{R_1 r}{R_1 + r} = \frac{\left(\frac{D_m}{2} - \frac{d}{2}\right)\frac{d}{2}}{\left(\frac{D_m}{2} - \frac{d}{2}\right) + \frac{d}{2}} = \frac{d}{2}(1 - \lambda). \tag{4.127}$$

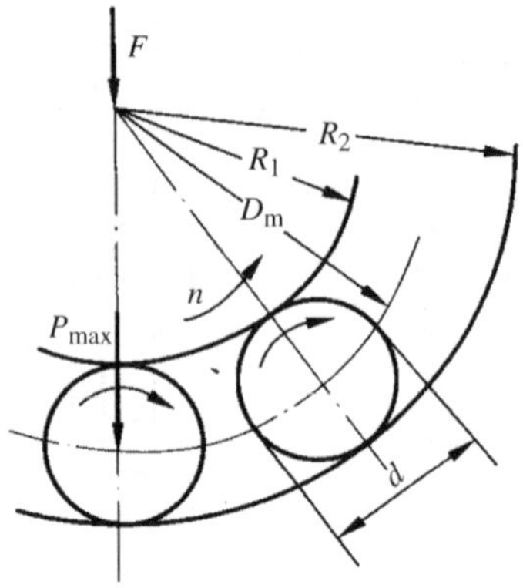

Figure 4.24 Roller bearing.

The equivalent radius for the roller and the outer ring at the contact point is:

$$R = \frac{R_2 r}{R_2 - r} = \frac{\left(\frac{D_m}{2} + \frac{d}{2}\right)\frac{d}{2}}{\left(\frac{D_m}{2} + \frac{d}{2}\right) - \frac{d}{2}} = \frac{d}{2}(1 + \lambda). \tag{4.128}$$

4.7.2 *Average Velocity U*

If n is the bearing inner ring speed, according to the pure rolling condition, the movement analysis is as follows:

$$\text{roller rotation speed } n_0 = \frac{1 + 2s}{2s(1 + s)} n,$$

$$\text{roller revolution speed } n_c = \frac{1}{2(1 + s)} n,$$

where $s = r/R_1 = \lambda/(1 - \lambda)$.

Therefore, the surface average velocity at the contact point is:

$$U = \frac{\pi}{30}(n - n_c)\left(\frac{D_m}{2} - \frac{d}{2}\right) = \frac{\pi n}{30}\frac{d}{4}\frac{1 - \lambda^2}{\lambda}. \tag{4.129}$$

4.7.3 *Carrying Load Per Width W/b*

Because the load distribution on the bearing components is related to deformation, the load acting on each rolling body is different. The maximum load should be found by calculating the minimum film thickness. Dowson *et al.* showed that if the total number of roller bearings is z and the sum load is F, the maximum load is equal to

$$P_{max} = \frac{4F}{z}. \tag{4.130}$$

If the effective roller contact length is l, the load per width is

$$\frac{W}{b}=\frac{4F}{zl}. \tag{4.131}$$

Substitute the above relationships into Dowson-Higginson formula, the minimum film thickness can be found.

1. Between the roller and the inner ring raceway:

$$h_{\min}=0.336\left[\frac{d}{2}(1-\lambda)\right]^{1.13}\alpha^{0.54}\left[\frac{\eta_0 n}{\lambda}(1+\lambda)\right]^{0.7}\times\frac{1}{E^{0.03}}\left(\frac{zl}{4F}\right)^{0.13}. \tag{4.132}$$

2. Between the roller and the outer ring raceway:

$$h_{\min}=0.336\left[\frac{d}{2}(1+\lambda)\right]^{1.13}\alpha^{0.54}\left[\frac{\eta_0 n}{\lambda}(1-\lambda)\right]^{0.7}\times\frac{1}{E^{0.03}}\left(\frac{zl}{4F}\right)^{0.13}. \tag{4.133}$$

The film thickness between the rollers and the outer ring raceway is usually thicker than that between the rollers and the inner ring raceway so that it is generally needed to calculate the latter film thickness.

For ball bearings, the steel ball and the rings are in point contacts. After the relationship between the bearing geometry and the motion is obtained, the EHL film thickness formula for point contacts can be used to find the minimum film thickness similar to the line contacts.

It should also be noted that for the rolling contact bearing in practical work, the movement and the force are always changing. Therefore, it does not always exist in the full film lubrication condition [8, 9].

4.8 Gear Lubrication

Gear lubrication is important and is closely related to elastohydrodynamic lubrication theory emergence and rapid development. In 1916, Martin first used the Reynolds equation to analyze gear lubrication. EHL has been continuously improved over the several decades and modern EHL lubrication theory is able to analyze practical gear lubrication problems accurately. AGMA (American Gear Manufacturers Association) has suggested that the EHL film thickness calculation of gear design is an important step.

It should be pointed out that the current EHL formula is established only for the steady state, that is, all physical variables do not vary with time. However, the working conditions of gear transmission are very complicated. Because the contact geometry, surface speed and load vary with time, the film thickness also varies all the time. For each tooth cycle, the engagement time is more than that for oil to flow through the Hertzian contact zone. Therefore, gear lubrication can be thought of as quasi steady. This means that the teeth in engagement can be treated as two equivalent cylinders in line contacts so that the film thickness of EHL contacts can be calculated on the instantaneous curvature radius, the relative speed and the load at the contact point without considering these parameters' variation. In this way, the EHL formula derived from the steady-state conditions can still be applied.

4.8.1 Involute Gear Transmission

First, let us calculate the oil film thickness for a pair of involute gears in an engaging cycle. As shown in Figure 4.25, if the center distance of the gears is $a=r_1+r_2$, the ratio $i=r_2/r_1>1$, the radii in a pitch

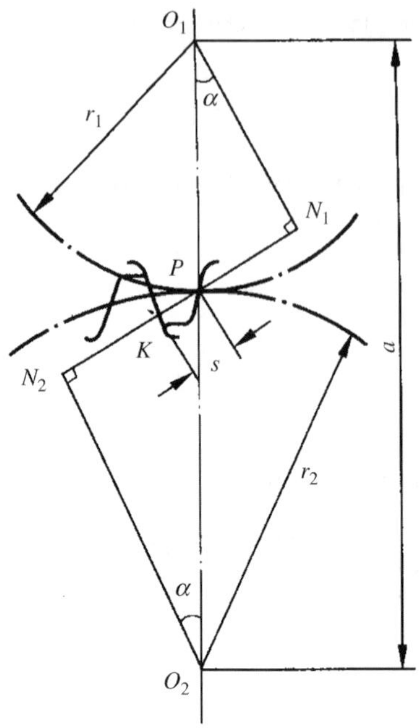

Figure 4.25 Involute gear engagement.

circle are:

$$r_1 = \frac{a}{i+1}$$
$$r_2 = \frac{ai}{i+1}. \tag{4.134}$$

If the contact point is at K and the centers of the two equivalent cylinders are N_1 and N_2 respectively, the center distance is equal to $N_1N_2 = (r_1 + r_2)\sin\alpha$, where α is the pressure angle of the gears, for the standard gears, $\alpha = 20°$.

The two cylinder radii are respectively equal to:

$$R_1 = r_1 \sin\alpha + s$$
$$R_2 = r_2 \sin\alpha - s, \tag{4.135}$$

where s is the distance from the contact point K to the node point P.

Then, the equivalent curvature radius R is:

$$R = \frac{R_1R_2}{R_1 + R_2} = \frac{(r_1 \sin\alpha + s)(r_2 \sin\alpha - s)}{(r_1 + r_2)\sin\alpha}. \tag{4.136}$$

The relative velocities of the two surfaces at contact point K are:

$$u_1 = \frac{\pi n_1}{30}(r_1 \sin\alpha + s)$$
$$u_2 = \frac{\pi n_2}{30}(r_2 \sin\alpha - s). \tag{4.137}$$

Therefore, the equivalent velocity is:

$$U = \frac{1}{2}(u_1 + u_2) = \frac{\pi n_2}{30}\left[r_2 \sin\alpha + \frac{s}{2}(1-i)\right]. \tag{4.138}$$

The engagement line coincides with the common tangent line of the two gear base circles. The contact length mainly depends on the gear module and pitch. The contact region of a pair of teeth can be expressed as s. If the roof heights of the two gear teeth are h_1 and h_2 respectively, s will vary from $-\left[\sqrt{(r_2+h_2)^2 - r_2^2\cos^2\alpha} - r_2\sin\alpha\right]$ to $+\left[\sqrt{(r_1+h_1)^2 - r_1^2\cos^2\alpha} - r_1\sin\alpha\right]$. Substitute Equation 4.76 and Equation 4.77 into the minimum film thickness of line contact EHL formula, the film thickness can be obtained as shown in Figure 4.26.

The parameters used in calculation are: the center distance $a = 0.3$ m; the speed of the larger gear $n_2 = 1000$ r/min; the transmission ratio $i = 5$; the load $W/b = 0.4 \times 106$ N/m; and the viscosity of lubricant $\eta_0 = 0.075$ Pa·s. The two diametric pitches $D_p = 5$ and 10 are given in the figure, which is equivalent to the modulus $m = 5.08$ and 2.54 mm. In the figure, the tooth roof heights are also shown. The standard height is equal to the multiplicative inverse of the diametric pitch.

Figure 4.26 shows that under the isothermal condition, when the small gear tooth top come into contact with the large gear root, the thickest film is the thickest, but when the big gear top come into contact with the small gear root, the oil film thickness is the thinnest. At the node, the film thickness is in the middle range. In addition, Martin suggested that the pressure-viscosity effect and the surface elastic deformation should not be considered. This will result in the viscous lubrication formula, but its results are quite different from practice.

The lubricant film thickness at the node is of a certain representation because the teeth surfaces are in the pure rolling at this time so that the EHL calculation is simple, and the isothermal EHL solutions is of high accuracy. Therefore, the lubrication of gear transmission design is usually based on the middle film thickness.

In 1974, Akin used the minimum film thickness formula of Dowson-Higginson in line contacts to analyze some involute gear lubrication problems [10] and he proposed the following film thickness formula.

4.8.1.1 Equivalent Curvature Radius *R*

According to the geometric simulation, the equivalent curvature radius has a relationship with two teeth contact radii as follows:

$$R = \frac{R_1 R_2}{R_1 \pm R_2}, \tag{4.139}$$

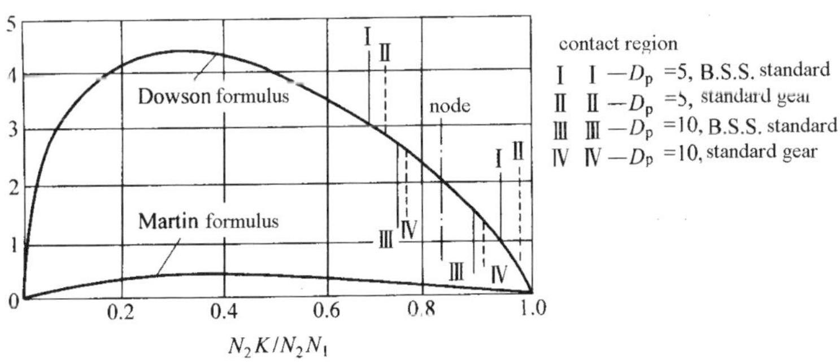

Figure 4.26 Film thickness changes in an engagement cycle.

where R_1 and R_2 are the node radii of the two teeth respectively; " + " is for the external engagement, and " − " is for the internal engagement.

From the principles of involute gear engagement, the equivalent radii of the typical gears at the node are:

$$\text{Spur gear} \qquad R = a \sin \alpha_n \frac{i}{(i \pm 1)^2},$$

$$\text{Helical gear} \qquad R = \frac{a \sin \alpha_n}{\cos^2 \beta} \frac{i}{(i \pm 1)^2},$$

$$\text{Straight bevel gear} \qquad R = L_m \sin \alpha_n \frac{i}{i^2 + 1},$$

$$\text{Arc bevel gear} \qquad R = \frac{L_m \sin \alpha_n}{\cos^2 \beta_m} \frac{i}{i^2 + 1}.$$

In the above formulas, α_n is the pressure angle; β is the pitch helix angle; L_m is the width of a bevel gear tooth at the midpoint of the pitch cone; β_m is the helix angle of the pitch circle at the midpoint arc of a bevel gear tooth, see Figure 4.27.

4.8.1.2 Average Velocity *U*

According to the relative speed on the normal surface of the node, the average velocities of the typical gears are:

$$\text{Spur gear} \qquad U = \frac{\pi n_1}{30} \left(\frac{a}{i \pm 1} \right) \sin \alpha_n,$$

$$\text{Helical gears} \qquad U = \frac{\pi n_1}{30} \left(\frac{a}{i \pm 1} \right) \frac{\sin \alpha_n}{\cos \beta},$$

$$\text{Straight bevel gear} \qquad U = \frac{\pi n_1}{30} \frac{L_m}{i} \sin \alpha_n,$$

$$\text{Arc bevel gear} \qquad U = \frac{\pi n_1}{30} \frac{L_m}{i} \frac{\sin \alpha_n}{\cos \beta_m},$$

where n_1 is the small gear rotational speed.

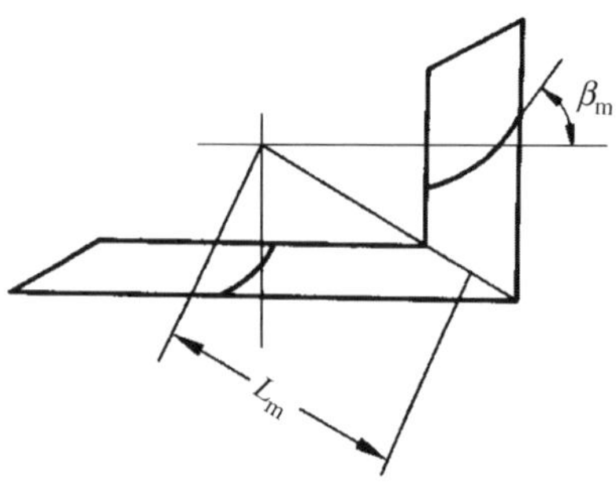

Figure 4.27 Bevel gear.

4.8.1.3 Load Per Width *W/b*

If F_t is the force in the circumference at the node; b is the tooth width; β_b is the helix angle, the loads per width on the contact point are:

Spur and straight bevel gear $$\frac{W}{b} = \frac{F_t}{b\cos\alpha_n},$$

Helical gear $$\frac{W}{b} = \frac{F_t\cos\beta_b}{b\cos\alpha_n\cos\beta},$$

Arc bevel gear $$\frac{W}{b} = \frac{F_t\cos\beta_b}{b\cos\alpha_n\cos\beta_m}.$$

Finally, substitute the above derived equivalent radii R, the average speed U and the unit load W/b into the minimum film thickness formula in line contacts, the minimum film thickness of the typical gears can be obtained.

For spur or helical gears:

$$h_{\min} = \frac{2.65\,\alpha^{0.54}}{E^{0.03}(W/b)^{0.13}}\left(\frac{\pi n_1\eta_0}{30}\right)^{0.7}\times\frac{(a\sin\alpha_n)^{1.13}}{\cos^{1.56}\beta}\times\frac{i^{0.43}}{(i\pm1)^{1.56}}. \tag{4.140}$$

For the straight-tooth bevels or arc bevels:

$$h_{\min} = \frac{2.65\alpha^{0.54}}{E^{0.03}(W/b)^{0.13}}\left(\frac{\pi n_1\eta_0}{30}\right)^{0.7}\times\frac{(L_m\sin\alpha_n)^{1.13}}{\cos^{1.56}\beta}\times\frac{i^{0.27}}{(i^2\pm1)^{0.43}}. \tag{4.141}$$

It should be noted that the above calculations are simplified except for involute spur gears. For most of the gears, such as space gears, they are of three-dimensional surfaces, not belonging to Hertzian contact. Therefore, not only is the contact area shape irregular, but also the sliding direction does not always coincide with the main plane of the contact area, even for existing rotation. Therefore, the classical EHL film thickness formula may result in significant errors. In the following, we will discuss arc bevel gear EHL [11–13].

4.8.2 Arc Gear Transmission EHL

Well lubricated gears are related to both tooth parameters and the profile. According to the characteristics of the relative movement of the two teeth, the formation of the oil film for the arc gears has very favorable conditions which the involute gears cannot realize. Especially in the high speed transmission, its advantages are obvious.

Figure 4.28 shows the engagement of a pair of arc gears.

In engaging, the arc gear teeth are equivalent to two cylinders, but the relationship between motion and geometry is very complete. For the present arc gear lubrication design, there is no systematic study. Liu Ying [14] has simplified EHL calculation of arc gears for engineering applications.

As shown in Figure 4.28, the two gear pitch radii are R_{01} and R_{02} respectively, the pressure angle is α_s, and the distance away from the contact point is l. As the arc radius of the convex tooth in the side profile (usually taken as l) is usually slightly smaller than that of the concave tooth, theoretically the arc gears engagement can be thought of as the point contact. However, after run-in the contact area extends rapidly to reach the total tooth length of 60–80% or more. Therefore, the engaging gears can actually be treated as two circular cylinders in the line contact. For this situation, the radii of the cylinders are R_{K1} and R_{K2}, the width is $2l$, and the arc surface radius is ρ.

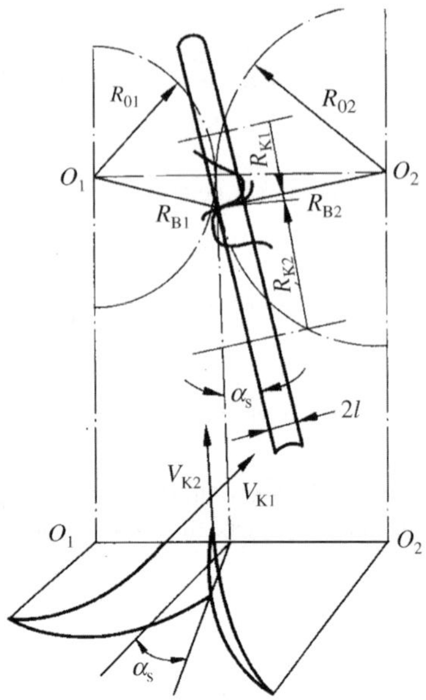

Figure 4.28 Arc gear transmission.

According to the engaging gear principles, the curvature radii R_{K1} and R_{K2} of the arc gears can be derived respectively to be:

$$R_{K1} = \frac{R_{B1}^2 + R_{01}^2 \cot^2 \beta}{R_{01} \sin \alpha_s + \rho} \sqrt{1 + \left(\frac{\cos \alpha_s}{\cot \beta}\right)^2}$$

$$R_{K2} = \frac{R_{B2}^2 + R_{02}^2 \cot^2 \beta}{R_{02} \sin \alpha_s - \rho} \sqrt{1 + \left(\frac{\cos \alpha_s}{\cot \beta}\right)^2}, \tag{4.142}$$

where β is the helix angle; R_{B1}, R_{B2} are the cylinder radii. And from the geometric relationship of Figure 4.28, they are equal to:

$$R_{B1} = \sqrt{\rho^2 \cos^2 \alpha_s + (R_{01} + \rho \sin \alpha_s)^2}$$

$$R_{B2} = \sqrt{\rho^2 \cos^2 \alpha_s + (R_{02} - \rho \sin \alpha_s)^2}. \tag{4.143}$$

The tooth contact length along the tooth height direction is $2l$, which is generally selected as:

$$l = \rho(\alpha_s - \delta) = Km_n(\alpha_s - \delta), \tag{4.144}$$

where K is coefficient, usually $K = 1.4$; m_n is the normal modulus of the gear; δ is the artificial angle.

The circumferential velocities V_{K1} and V_{K2} of the two circular cylinders are equal to:

$$V_{K1} = V_0\sqrt{1+\cot^2\beta+\left(\frac{\rho}{R_{01}}\right)^2+\frac{2\rho\sin\alpha_s}{R_{01}}} = V_0\sqrt{\cot^2\beta+\left(\frac{R_{B1}}{R_{01}}\right)^2}$$

$$V_{K2} = V_0\sqrt{1+\cot^2\beta+\left(\frac{\rho}{R_{02}}\right)^2-\frac{2\rho\sin\alpha_s}{R_{02}}} = V_0\sqrt{\cot^2\beta+\left(\frac{R_{B2}}{R_{02}}\right)^2}, \tag{4.145}$$

where V_0 is the velocity of the pitch circle.

The relationship of the normal F_n and the circumferential force F_t is

$$F_n = \frac{F_t}{\cos\alpha_s}\sqrt{1+\left(\frac{\cos\alpha_s}{\cot\beta}\right)^2}. \tag{4.146}$$

Therefore, the load per width is equal to $F_n/2l$.

Substitute the geometric sizes, velocities and loads of the two equivalent cylinders for arc gear engagement into the EHL minimum film thickness formula, combined with the arc gear simplified situation, we can get the minimum EHL film thickness for the steel arc gear lubrication by using mineral oil as follows:

$$h_{\min} = 0.8663\eta_0^{0.7}\left(\frac{2l}{F_n}\right)^{0.13}\left[(V_{k1}+V_{k2})-\frac{l^2}{2\rho}\left(\frac{V_{k1}}{R_{k1}}-\frac{V_{k2}}{R_{k2}}\right)\right]^{0.7}\left[\frac{R_{k1}R_{k2}}{R_{k1}+R_{k2}}-\frac{l(R_{k2}-R_{k1})}{2\rho(R_{k1}+R_{k2})}\right]^{0.43}. \tag{4.147}$$

In approximation, some items in the two square brackets of the above formula can be negligible, and if we take $R_{B1}=R_{01}$ and $R_{B2}=R_{02}$, the above equation is simplified as follows:

$$h_{\min} = 0.8663\eta_0^{0.7}\left(\frac{2l}{F_n}\right)^{0.13}\left[\frac{2V_0}{\sin\beta}\right]^{0.7}\left[\frac{1+\cot^2\beta}{\left(\frac{1}{R_{01}^2}-\frac{1}{R_{02}^2}\right)\rho+\left(\frac{1}{R_{01}}+\frac{1}{R_{02}}\right)\sin a_s}\sqrt{1+\left(\frac{\cos\alpha_s}{\cot\beta}\right)^2}\right]^{0.43}. \tag{4.148}$$

4.9 Cam Lubrication

The cam and follower are the main sliding friction pairs in line or point contacts. The contact stress of the cam surface is very high. For example, the maximum contact stress of the cam in an internal combustion engine is generally 0.7–1.4 GPa. Therefore, it used to be thought that a mixed lubrication condition exists between the cam and follower. However, with the development of EHL theory, it has become known that a cam and EHL film can be formed between the cam and the follower. The film thickness is an indicator in designing the profile of a cam.

In the working cycles of a cam mechanism, the curvature radius of the contact point, the speed and the load vary so that the film thickness will also vary accordingly. Similar to gear transmission, the cam EHL calculations can be taken as a quasi-steady problem in engineering design.

Here, only the EHL calculations of the cam and tappet are introduced. The EHL calculations of other types of cam mechanism are similar.

Deschler and Wittmann simplified the EHL analysis of the cam and tappet [15]. They took into account a steel cam mechanism lubricated by the mineral oil in line contacts. Because E and α vary little, they considered these parameters as constants, while W/b has little influence on $h_{\min}$ so it can be ignored.

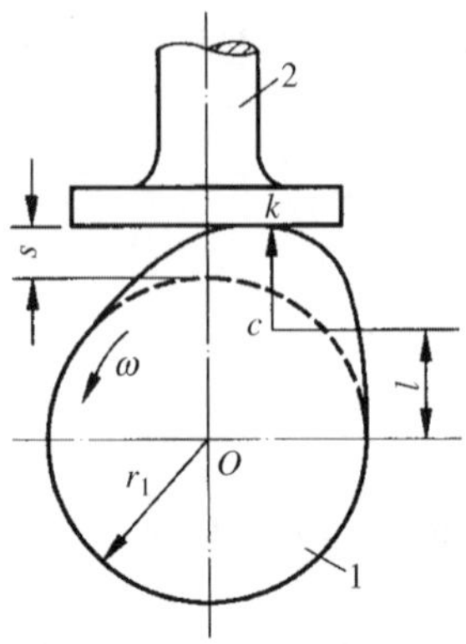

Figure 4.29 Cam and tappet.

Therefore, the minimum film thickness formula becomes

$$h_{\min} = 1.6 \times 10^{-5}\sqrt{\eta_0 UR}. \tag{4.149}$$

In Figure 4.29, a cam-tappet machinery is given. Suppose the tappet moves only along the vertical direction without rotation, the absolute velocity u_2 at the contact point K at the surface along the tangent (horizontal) direction is equal to 0. The absolute velocity of the cam surface at the contact point K along the tangent (horizontal) direction is equal to

$$u_1 = \omega(r_0 + s) = \omega(l + \rho), \tag{4.150}$$

where ω is the angular rotating velocity of the cam; r_0 is the base circle radius of the cam; s is the lift distance of the tappet; l is the vertical distance from the contact point to the cam curvature center c; and ρ is the curvature radius of the cam at the contact point.

Obviously, in the working process, the contact point K moves continuously. The entrainment velocity to form an effective hydrodynamic lubrication film is the relative velocity of the two surfaces at the contact point. Because the circumferential velocity of K is equal to the circumferential velocity of the curvature center c, that is, $u_c = \omega l$. Therefore, the entrainment velocity U is equal to

$$U = \frac{1}{2}[(u_1 - u_c) + (u_2 - u_c)] = \frac{\omega}{2}[2\rho - (r_0 + s)]. \tag{4.151}$$

Because the equivalent radius of curvature R is known to be equal to ρ, substitute the above equation into the simplified formula 4.149, we can obtain the minimum film thickness between the cam and tappet at the contact point K as follows:

$$h_{\min} = 1.6 \times 10^{-5}\sqrt{\frac{\omega\eta_0}{2}}(r_0 + s) \times \sqrt{|2N^2 - N|}, \tag{4.152}$$

where the dimensionless geometric parameter $N = \rho/(r_0 + s)$ is used, which indicates the geometry of the cam, known as the EHL characteristic number of a cam. From Equation 4.159 we can deduce that when

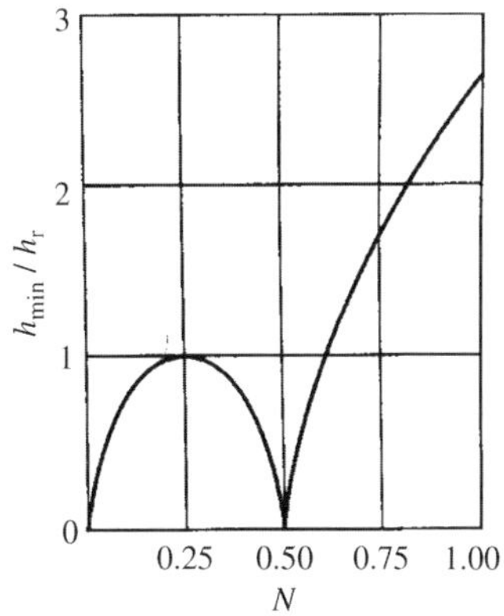

Figure 4.30 h_{min}/h_r via N.

$N = 0$ or $N = 0.5$, $h_{min} = 0$. h_{min} will be the maximum while $N = 0.25$. The relative thickness h_{min}/h_r via N can be plotted as shown in Figure 4.30.

From the figure we can see that the thin film region is $0 < N < 0.5$. When $N > 0.5$, the film thickness will increase continuously with increase of N. Therefore, in order to reduce wear of cam, the $0 < N < 0.5$ region should be avoided, but choose a larger N. Anyway, to increase N, other factors of a cam may limit the design.

After using Equation 4.152 to determine h_{min} and its variation in the working cycle at each position of the cam angle, the distribution of cam surface wear can be analyzed.

References

1. Wen, S.Z. (1990) *Principles of Tribology Theory*, Tsinghua University Publishing, Beijing.
2. Pinkus, O. (1980) *Sterntícht B Forward. Xi'an Jiaotong University Research Team Bearing Translation. Hydrodynamic Lubrication Theory*, Mechanical Industry Press, Beijing.
3. Milne, A.A. (December, 1962) Theoretical studies of the performance of dynamically loaded journal bearings, NEL Rept. No.70.
4. Hahn, H.W. (1957) Dynamically loaded journal bearings of finite length, *Proc. Conf. on Lubrication and Wear*, p. 100 (Instn Mech. Engrs, London).
5. Hooland, J. (1959) *Beitrag zur Erfassung der Schmierverhaltnisse in Verbrennungs Kraftmaschinen*, VDI-Forschungsheft.
6. Wang, X.L. (1999) Effect of surface topography into account internal combustion engine main bearing hydrodynamic lubrication analysis of thermal. PhD thesis, Tsinghua University.
7. Wen, S.Z., Wu, K., and Yu, D.Q. (1962) Hydrostatic air bearing test and analysis. *Journal of Mechanical Engineering*, **10** (3), 1–16.
8. Liu, J.H. and Wen, S.Z. (1992) Analysis of lubricating oil film bearing part of the state of two kinds of computational model. *Journal of Tribology*, **12** (2), 116–211.
9. Dowson, D. and Higginson, G.R. (1997) Elasto-Hydrodynamic Lubrication, Pergamon Press, London.
10. Akin, LS. (1974) EHD lubricant film thickness formulas for power transmission gears. *Journal of Tribology-Transactions of the ASME*, **96** (7), 426–431.
11. Huang, C.H. and Wen, S.Z. (1993) Elastohydrodynamic lubrication of wide elliptical contacts under heavy load. *Chinese Journal of Mechanical Engineering*, **6** (2), 145–152.
12. Huang, C.H., Wen, S.Z., Huang, P. *et al.* (1993) Multilevel solution of the elastohydrodynamic lubrication of elliptical contacts with rotational lubricant entrainment. *Proc. Lst International Symposium on Tribology. Beijing*, 1, pp. 124–131.

13. Huang, C.H., Wen, S.Z., and Huang, P. (1993) Multilevel solution of the elastohydrodynamic lubrication of concentrated contacts in spiroid gears. *Journal of Tribology-Transactions of the ASME*, **115** (3), 481–486.
14. Liu, Y. (2000) Arc gears Elastohydrodynamic lubrication calculation, in Tribology Design (eds Zhong.-Rong Zhouand Xie Bai), Southwest Jiaotong University Press, Chengdu, pp. 108–111.
15. Deschler, G. and Wittmann, D. (1978) Nockenauslegung fur Flachstopel unter Beachtung elastohydrodynamischer Schmierung. *MTZ Magazine*, **39** (3), S:123–S:127.
16. Wen, S.Z. (1982) Boundary element equation in lubrication problem: Rayleigh step bearings. *Lubrication and Sealing*, 3, 10–16.

5

Special Fluid Medium Lubrication

Oil and grease are the most commonly used lubricants. However, in special situations, special fluid media should be used as lubricants. This chapter, we will consider lubrication mechanism and the theory of several special lubricants, including magnetic hydrodynamic lubrication, micro-polar hydrodynamic lubrication, liquid crystal lubrication and the electric double layer effect of water lubrication.

5.1 Magnetic Hydrodynamic Lubrication

Magnetic fluid materials were discovered in the eighteenth century. It was found that ferromagnetic particles can be dispersed in liquid. In 1938, Elmore made a colloidal solution of ferromagnetic fluid containing 200 Å particle of Fe_3O_4 by using chemical methods. Forty years later, ferromagnetic fluid was used as a seal. The rapid development of magnetic fluid technology started in the mid-1960s. Magnetic fluids have been widely studied. These studies have prompted the usages of magnetic fluids in machinery, printing, processing, acoustic optical devices and medical areas. Applications of magnetic fluids are increasingly widespread.

5.1.1 Composition and Classification of Magnetic Fluids

Magnetic fluid is a kind of liquid with dispersed magnetic particles. It is sensitive to a magnetic field and flowable. It is composed of a base liquid (such as oil, water), dispersed materials (magnetic solid particles) and the dispersant coating the surfaces of the scattered material. According to the size of magnetic particles, magnetic fluids are divided into suspension, colloidal or real solutions. Suspension and colloidal solutions are composed of strong magnetic fine powder (such as Fe, Fe_3O_4), which is dispersed in the base liquid. A true solution is a saturated aqueous solution of a paramagnetic salt, such as $MnCl_2$, $Mn(NO_3)_2$.

A dispersed material is a ferromagnetic material. In this section, it refers to the ferromagnetic fluid unless otherwise specified. Water or oil (such as kerosene, esters, ethers or other organic solvents) is usually used as a base liquid. Sometimes a liquid metal or alloy (e.g. Hg, K, Na) can also be used as a base liquid to form a magnetic fluid. If water serves as a base fluid, it is known as the water-based magnetic fluid. For oil, we call the oil-based magnetic fluid.

The properties of a magnetic fluid are related to dispersed material and its concentration, while the fluid properties are based on the based liquid type and its concentration. The applications of a magnetic fluid depend on its magnetic properties, viscosity and stability. In order to get a stable magnetic fluid, the dispersed particles must be very fine, and furthermore, a surface-active agent must be used to have the

Principles of Tribology, First Edition. Wen Shizhu and Huang Ping.

particle surface treated to enable it not to be cohered and segregated under the action of gravitation, centrifugal force and magnetic field.

Some paramagnetic salts can form magnetic fluids while dispersed in water. The stability of such a fluid is good, but the magnetic properties are usually very poor. Therefore, its applications are limited. The usage of a magnetic fluid lies in its magnetic effect. Although the magnetic properties of a suspension magnetic fluid are better, its stability is poor. Precipitation and cohesion easily occur and they limit its application.

5.1.2 Properties of Magnetic Fluids

The essential components of a ferromagnetic fluid are the base liquid and the suspended ferromagnetic particles. Usually, it is assumed to be a uniform two-phase mixture. The main properties of a fluid are its density and viscosity. Because a magnetic fluid will be a controlled magnetic field, its magnetic property is also important.

5.1.2.1 Density of Magnetic Fluids

The weight and volume of a magnetic fluid are the sum of the base liquid and solid particles, that is

$$\begin{aligned} V_f &= V_c + V_p \\ G_f &= G_c + G_p, \end{aligned} \tag{5.1}$$

where V_f, V_c and V_p are the total volume of magnetic fluid, the base fluid volume and solid particle size respectively; G_f, G_c and G_p are the total weight of magnetic fluid, the base fluid weight and the weight of solid particles respectively.

The density of a ferromagnetic fluid is equal to

$$\rho_f = \frac{1}{g}\frac{G_c + G_p}{V_c + V_p} = \frac{V_c}{V_c + V_p}\rho_c + \frac{V_p}{V_c + V_p}\rho_p = (1 - \phi)\rho_c + \phi\rho_p, \tag{5.2}$$

where ϕ is the volume fraction, $\phi = V_p/V_c + V_p$, g is the gravity acceleration; $\rho_c = G_c/gV_c$; and $\rho_p = G_p/gV_p$.

5.1.2.2 Viscosity of Magnetic Fluids

The solid phase magnetic particle in a magnetic fluid can be thought of as a small magnetic ring of current. These small rings are subjected to a magnetic moment under an external magnetic field in the same direction. The influences of the external magnetic field on the magnetic fluid viscosity can be expressed as follows:

$$\eta_H = \eta_0 + \Delta\eta, \tag{5.3}$$

where η_H is the magnetic fluid viscosity; η_0 is the magnetic fluid viscosity with the external magnetic field strength equal to 0; $\Delta\eta$ is the additional viscosity generated by the external magnetic field.

If μ_0 is the vacuum magnetic conductivity, $\boldsymbol{M}$ is the magnetization, $\boldsymbol{H}$ is the magnetic field strength in the x and y directions, $\boldsymbol{\omega}$ is the fluid vortex rate, the additional viscosity $\Delta\eta$ will be:

$$\boldsymbol{\omega}\Delta\eta = \mu_0 \boldsymbol{M} \times \boldsymbol{H}. \tag{5.4}$$

As shown above, when the external magnetic field vector is perpendicular to the vortex vector, the viscosity of the ferromagnetic fluid generated by the magnetic field increases. When the external magnetic

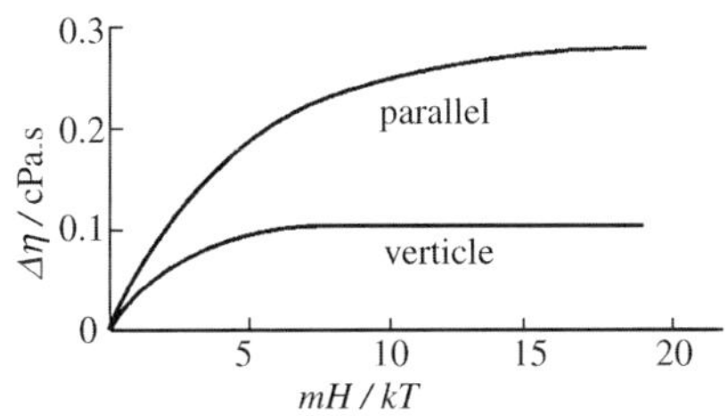

Figure 5.1 Variation of the ferromagnetic fluid viscosity with the magnetic field strength in the different flow directions.

field vector is parallel to the vortex vector, the magnetic field does not change the viscosity of the ferromagnetic fluid.

The influences of the magnetic field on a magnetic fluid with different flowing directions are different. Figure 5.1 shows the differences. When the magnetic field is parallel to the flow direction of the magnetic fluid, its viscosity increases considerably. In the lubrication calculation, for simplicity, the additional viscosity is generally not considered.

5.1.2.3 Magnetization Strength of Magnetic Fluids

Magnetic fluids belong to super-paramagnetic fluids. However, they follow the basic theory of paramagnetic substances, that is, the classical Langevin's theory still applies to the magnetization problems of magnetic fluids.

Suppose the volume of each magnetic particle is V_{pl} and the particle is of a dipole, then the magnetic moment $\boldsymbol{L}_m$ within the unit volume of magnetic particles, the magnetic field strength $\boldsymbol{H}_0$ and the magnetization $\boldsymbol{M}$ have the following relationship:

$$\frac{\boldsymbol{L}_m}{V_{pl}} = \mu \boldsymbol{M} \times \boldsymbol{H}_0. \tag{5.5}$$

5.1.2.4 Stability of Magnetic Fluids

The stabilities of magnetic fluids include thermal stability and colloidal stability. Thermal stability refers to the working performances of the ferromagnetic particles with variation of temperature under a magnetic fluid. For long-time work under high temperatures, the polymer synthetic lubricant is commonly used as a base fluid for magnetic fluids. Figure 5.2 gives the coagulation time of six different oil-based

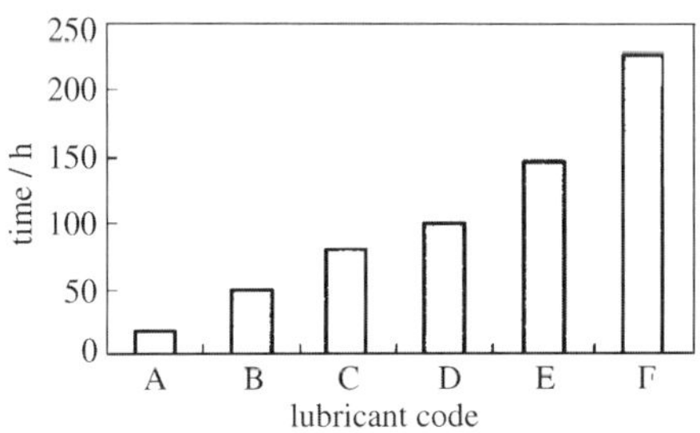

Figure 5.2 Thermal stabilities of magnetic fluids.

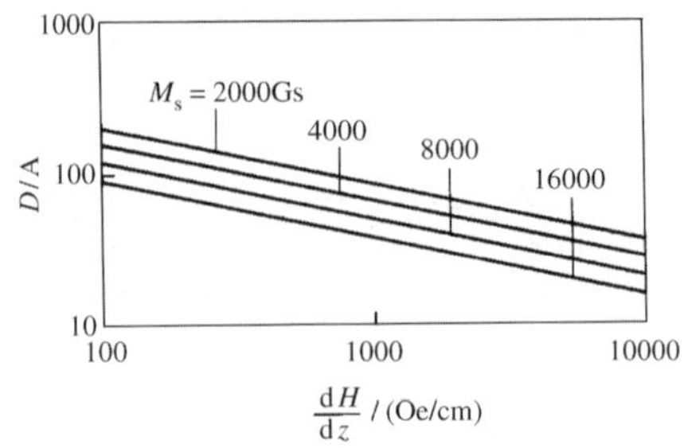

Figure 5.3 Magnetic field stability under the magnetic fluid.

magnetic fluids (with codes A, B, C, D, E and F) at 175 °C, showing the relative thermal stabilities of these magnetic fluids. Obviously, the longer the coagulation time, the better the thermal stability of a magnetic fluid.

The colloidal stability of a magnetic fluid is the anti-precipitation or anti-segregation capacity of a magnetic fluid with ferromagnetic particles under a magnetic field gradient to remain suspending in the base fluid. Generally, the percentage of the magnetic particles suspending in the colloid is the ability after a fixed magnetic field is added for a certain time. The higher the percentage, the more stable the fluid is.

Without considering the gravity, the gradient of the particle number along the height direction is equal to:

$$\frac{dn}{dz} = -\frac{D^3 M_s n}{24\,kT}\frac{dH}{dz}, \tag{5.6}$$

where n is the particle number; z is the height; D is the particle size including the coating layer; M_s is the saturation magnetization; k is the thermodynamic coefficient; T is the temperature; and H is the magnetic field strength.

The stability criterion of a magnetic fluid in a magnetic field is as follows:

$$\frac{1}{n}\frac{dn}{dz} \leq 1. \tag{5.7}$$

As the upper limit of the particle diameter is the function of the saturation magnetization and the gradient of a magnetic field, the particle diameter must be under the curves of the magnetization in Figure 5.3 so that the stability is guaranteed. The smaller the particle diameter, the higher stability is.

Table 5.1 shows the main physical characteristics of the magnetic fluids commonly used.

5.1.3 Basic Equations of Magnetic Hydrodynamic Lubrication

The Reynolds equation of magnetic hydrodynamic lubrication is derived from the fundamental equations of fluid mechanics, including the continuity equation, the momentum equations and the constitutive equations.

The simplified continuity equation is:

$$\frac{\partial \rho u}{\partial x} + \frac{\partial \rho v}{\partial y} + \frac{\partial \rho w}{\partial z} = 0. \tag{5.8}$$

Table 5.1 Main physical characteristics of commonly used magnetic fluids

Base liquid	Material code	Saturation magnetization strength (A/m)	Density (g/cm 3)	Viscosity (10^{-3} Pa·s)	Freezing (C°)	Boiling (133.3Pa)/°C	Initial susceptibility (M/H)	Surface tension (10^{-3} N/m)	Thermal conductivity (10^{-2} W/mC°)	Heat (kJ/kg· C°)	CTE (10^{-4}/C°)
2-ester	D01	2.51	1.185	75	−37	148.9	0.5				
Alkyl	H01	2.51	1.05	3	4.4	76.7	0.4	28	14.6	1.72	2.78
	H02	5.02	1.25	6	7	76.7	0.8	28	14.6	1.84	2.67
Carbon-fluorine	F01	1.26	2.05	2500	−34.4	182.2	0.2	18	8.4	1.97	3.28
Ester	E01	7.54	1.40	35	−63.2	40	1.0	21	13	3.73	2.5
	E02	5.02	1.30	30	−56.7	148.9	0.8	26	13	3.73	2.5
	E03	2.51	1.15	14	−56.7	148.9	0.4	26	13	3.73	2.5
Water	A01	2.51	1.18	7	0	25.6	0.6	26	58.6	4.2	1.61
	A02	5.02	1.38	100	0	25.6	1.2	26	58.6	4.2	1.56
Polyphenylene ether	V01	1.26	2.05	7500	10	260	0.2				

Note: In the table, viscosity is suitable for the shear rate $>10\,s^{-1}$; and the coefficient of expansion is suitable for the average temperature between 25–93 °C.

If the assumptions are the same as in Chapter 2, the Navier-Stokes equation can be simplified as follows:

$$\frac{\partial p}{\partial x}=\frac{\partial}{\partial z}\left(\eta_H\frac{\partial u}{\partial z}\right)+\frac{\partial}{\partial x}\int_0^{H_x}\mu_0 M_x dH_x$$

$$\frac{\partial p}{\partial y}=\frac{\partial}{\partial z}\left(\eta_H\frac{\partial v}{\partial z}\right)+\frac{\partial}{\partial y}\int_0^{H_y}\mu_0 M_y dH_y, \tag{5.9}$$

where μ_0 is the magnetic conductivity in vacuum; M_x and M_y are the magnetizations in the x and y directions; H_x and H_y are the magnetic field strengths in the x and y directions; η_H is the magnetic fluid viscosity.

If we consider that the magnetic fluid is isotropic and the magnetic field strengths of all directions are equal, the integral of Equation 5.9 can be expressed as:

$$p_M=\int_0^H \mu_0 M dH, \tag{5.10}$$

where p_M is the pressure induced by the magnetic field.

Then, Equation 5.9 becomes:

$$\frac{\partial p'}{\partial x}=\frac{\partial}{\partial z}\left(\eta_H\frac{\partial u}{\partial z}\right)$$

$$\frac{\partial p'}{\partial y}=\frac{\partial}{\partial z}\left(\eta_H\frac{\partial v}{\partial z}\right), \tag{5.11}$$

where $p'=p-\int_o^H \mu_0 M dH$ is called the equivalent pressure.

The method of derivation similar to Chapter 2 can be used to obtain the general Reynolds equation of magnetic hydrodynamic lubrication:

$$\frac{\partial}{\partial x}\left(\frac{\rho h^3}{\eta_H}\frac{\partial p'}{\partial x}\right)+\frac{\partial}{\partial y}\left(\frac{\rho h^3}{\eta_H}\frac{\partial p'}{\partial y}\right)=6\left[\frac{\partial(U\rho h)}{\partial x}+\frac{\partial(V\rho h)}{\partial y}+2\rho(w_h-w_0)\right]. \tag{5.12}$$

It should be pointed out that in Equation 5.12, the pressure p' is the hydrodynamic pressure but not the total pressure. The boundary conditions remain the usual Reynolds boundary conditions given in Chapter 2. However, unlike the general hydrodynamic lubrication, the difference is that the load-carrying capacity of a magnetic hydrodynamic film is the sum of hydrodynamic pressure and magnetic field pressure, and the magnetic field is in the whole region. Therefore, the area solved for the magnetic hydrodynamic lubrication equations is different from the area of hydrodynamic lubrication.

Therefore, the magnetic fluid load is equal to:

$$W=\iint_{\Omega\prime} p'dxdy+\iint_{\Omega}\left(\mu_0\int_0^H MdH\right)dxdy, \tag{5.13}$$

where W is the load, Ω is the hydrodynamic lubrication region; and Ω is the whole lubrication area.

From Equation 5.13 we can see that under the same load, due to a magnetic field existence, hydrodynamic pressure will be reduced accordingly. Or, for the same film thickness of a magnetic fluid lubricant, the load-carrying capacity will be larger than that of a common hydrodynamic lubricant.

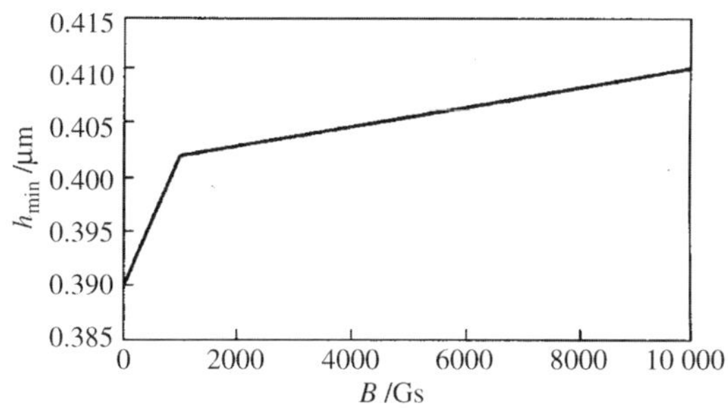

Figure 5.4 Magnetic induced intensity via the minimum film thickness.

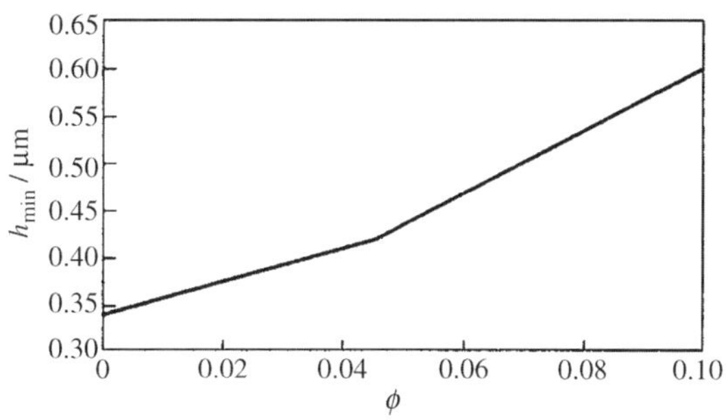

Figure 5.5 Volume fraction via the minimum film thickness.

5.1.4 Influence Factors on the Magnetic EHL

Wang [1] calculated the magnetic fluid characteristics of the magnetic field intensity and volume fraction in EHL.

1. Influence of the magnetic induced strength on the minimum film thickness:
   Figure 5.4 shows that when the external magnetic field increases, the minimum film thickness also increases because of the influence of the magnetic particles in the magnetic fluid. For a thinner film, the thickness increases quickly. When the film thickness is thick enough, the increase becomes slow.
2. Influence of the volume fraction ϕ on the minimum film thickness:
   Figure 5.5 shows the influence of the volume fraction on the minimum film thickness.

From the figure we can see that with an increase in the volume fraction, the minimum film thickness increases slowly, and then quickly, i.e. the larger the volume fraction, the thicker the film is. However, due to saturation, it is impossible for the volume fraction to be too large.

5.2 Micro-polar Hydrodynamic Lubrication

In hydrodynamic lubrication theory, a basic assumption is that a lubricant is a continuous media, that is, we do not need to take account of its internal micro-structure. However, polymers with long chain molecules or a lubricant within solid particles are not fully suitable for the assumption. Here, a fluid known as the micro-polar fluid is discussed. Each particle composing the group structure has a quality and speed.

In the micro-polar fluid model, the deformation of the micro-particle is ignored, but its movement remains. Therefore, a continuous medium theory is still suitable. However, because a particle has a length, the rotation must be added into the movement analysis so that a micro-polar fluid model has significant non-Newtonian fluid characteristics.

In 1982, Singh and Sinha [2] derived the Reynolds equation of micro-polar fluid lubrication. Based on their derivation, the basic theory of micro-polar hydrodynamic lubrication and some calculations are introduced in this section.

5.2.1 Basic Equations of Micro-polar Fluid Lubrication

5.2.1.1 Basic Equations of Micro-polar Fluid Mechanics

The basic equations of micro-polar hydrodynamic lubrication are derived under incompressible fluid conditions. In the Cartesian coordinates, the time-dependent three-dimensional expressions of the micro-polar fluid mechanics are as follows:

$$\frac{\partial u}{\partial x}+\frac{\partial v}{\partial y}+\frac{\partial w}{\partial z}=0, \tag{5.14}$$

$$\begin{aligned}
&\frac{1}{2}(2\mu+\chi)\left(\frac{\partial^2 u}{\partial x^2}+\frac{\partial^2 u}{\partial y^2}+\frac{\partial^2 u}{\partial z^2}\right)+\chi\left(\frac{\partial \omega_3}{\partial y}-\frac{\partial \omega_2}{\partial z}\right)-\frac{\partial p}{\partial x}=\rho\left(u\frac{\partial u}{\partial x}+u\frac{\partial u}{\partial y}+u\frac{\partial u}{\partial z}\right)\\
&\frac{1}{2}(2\mu+\chi)\left(\frac{\partial^2 v}{\partial x^2}+\frac{\partial^2 v}{\partial y^2}+\frac{\partial^2 v}{\partial z^2}\right)+\chi\left(\frac{\partial \omega_1}{\partial z}-\frac{\partial \omega_3}{\partial x}\right)-\frac{\partial p}{\partial y}=\rho\left(v\frac{\partial v}{\partial x}+v\frac{\partial v}{\partial y}+v\frac{\partial v}{\partial z}\right)\\
&\frac{1}{2}(2\mu+\chi)\left(\frac{\partial^2 w}{\partial x^2}+\frac{\partial^2 w}{\partial y^2}+\frac{\partial^2 w}{\partial z^2}\right)+\chi\left(\frac{\partial \omega_2}{\partial x}-\frac{\partial \omega_1}{\partial y}\right)-\frac{\partial p}{\partial z}=\rho\left(w\frac{\partial w}{\partial x}+w\frac{\partial w}{\partial y}+w\frac{\partial w}{\partial z}\right)\\
&\gamma\left(\frac{\partial^2 \omega_1}{\partial x^2}+\frac{\partial^2 \omega_1}{\partial y^2}+\frac{\partial^2 \omega_1}{\partial z^2}\right)+\chi\left(\frac{\partial w}{\partial y}-\frac{\partial v}{\partial z}\right)-2\chi\omega_1=\rho J\left(u\frac{\partial \omega_1}{\partial x}+v\frac{\partial \omega_1}{\partial y}+w\frac{\partial \omega_1}{\partial z}\right)\\
&\gamma\left(\frac{\partial^2 \omega_2}{\partial x^2}+\frac{\partial^2 \omega_2}{\partial y^2}+\frac{\partial^2 \omega_2}{\partial z^2}\right)+\chi\left(\frac{\partial u}{\partial z}-\frac{\partial w}{\partial x}\right)-2\chi\omega_2=\rho J\left(u\frac{\partial \omega_2}{\partial x}+v\frac{\partial \omega_2}{\partial y}+w\frac{\partial \omega_2}{\partial z}\right)\\
&\gamma\left(\frac{\partial^2 \omega_3}{\partial x^2}+\frac{\partial^2 \omega_3}{\partial y^2}+\frac{\partial^2 \omega_3}{\partial z^2}\right)+\chi\left(\frac{\partial v}{\partial x}-\frac{\partial u}{\partial y}\right)-2\chi\omega_3=\rho J\left(u\frac{\partial \omega_3}{\partial x}+v\frac{\partial \omega_3}{\partial y}+w\frac{\partial \omega_3}{\partial z}\right),
\end{aligned} \tag{5.15}$$

where u, v, w are the flow velocities in the x, y, z directions respectively; ω_1, ω_2, ω_3 are the rotational angular velocities in the x, y, z directions respectively; μ is the Newtonian fluid viscosity; χ is the rotational viscosity of the micro-polar fluid; ρ is the density; J is the inertia coefficient of micro-polar fluid; and γ is the material constant of the micro-polar fluid.

Equation 5.14 is the continuity equation. Because the fluid is assumed to be incompressible, the density ρ can be neglected. As the micro-polar fluid molecules have a characteristic length l, Equation 5.15 consists of three translational momentum equations, but also adds the three rotational momentum equations.

5.2.1.2 Reynolds Equation of Micro-polar Fluid

First, the dimensionless variables are as follows:

$$\begin{aligned} X &= x/a; \quad Y = y/b; \quad Z = z/h \\ \bar{u} &= u/U; \quad \bar{v} = v/U; \quad \bar{w} = w/U \\ \bar{\omega}_i &= \omega_i h/U; \quad P = ph_1^2/(\mu + \chi/2)Ua \\ \delta_1 &= h/a; \quad \delta_2 = h/b; \quad \delta_3 = h_1/a; \quad \delta_4 = b/a \\ \xi &= h_1/h; \quad L = h_1/l; \quad l = (\gamma/4\mu)^{1/2}; \quad N = \{\chi/(2\mu + \chi)\}^{1/2}, \end{aligned} \tag{5.16}$$

where a and b are the characteristic lengths of the lubrication region in the x and y directions; h is the thickness, h_1 is the minimum film thickness; U is the sliding velocity of the solid surface; l is the characteristic length of micro-polar fluid; L is the dimensionless characteristic length; and N is the coupling coefficient.

According to assumptions deriving from the Reynolds equation, that flow is laminar, the volume forces are neglected, the film thickness is much thinner compared with the length and width directions of the interface, sliding does not exist, the surface is smooth, and the surfaces are free from pores; therefore the following dimensional analysis will be given.

Compared with the unequal formulas below:

$$\begin{aligned} &\mathrm{Re} = 2\rho hU/(2\mu + x) << 1; \quad \mathrm{Re}' = \rho jUh/4\mu l^2 << 1 \\ &\frac{\partial}{\partial x} << \frac{\partial}{\partial z}; \quad \frac{\partial}{\partial y} << \frac{\partial}{\partial z} \\ &w << u; \quad w << v; \quad \delta_1 << 1; \delta_2 << 1; \delta_3 << 1 \\ &\xi \approx O(1); \quad \delta_4 \approx O(1), \end{aligned} \tag{5.17}$$

omitting the small amounts the dimensionless Equation 5.15 can be simplified as follows:

$$\begin{aligned} &\frac{\partial^2 \bar{u}}{\partial Z^2} - 2N^2 \frac{\partial \bar{\omega}_2}{\partial Z} - \frac{1}{\xi^2}\frac{\partial P}{\partial X} = 0 \\ &\frac{\partial^2 \bar{v}}{\partial Z^2} + 2N^2 \frac{\partial \bar{\omega}_1}{\partial Z} - \frac{1}{\xi^2}\frac{\partial P}{\partial Y} = 0 \\ &\frac{\partial P}{\partial Z} = 0 \\ &\frac{\partial^2 \bar{\omega}_1}{\partial Z^2} \quad \frac{N^2 L^2}{2(1 - N^2)\xi^2}\frac{\partial \bar{v}}{\partial Z} - \frac{N^2 L^2}{(1 - N^2)\xi^2}\bar{\omega}_1 = 0 \\ &\frac{\partial^2 \bar{\omega}_2}{\partial Z^2} + \frac{N^2 L^2}{2(1 - N^2)\xi^2}\frac{\partial \bar{u}}{\partial Z} - \frac{N^2 L^2}{(1 - N^2)\xi^2}\bar{\omega}_2 = 0 \\ &\frac{\partial^2 \bar{\omega}_3}{\partial Z^2} - \frac{N^2 L^2}{(1 - N^2)\xi^2}\bar{\omega}_3 = 0. \end{aligned} \tag{5.18}$$

From the third equation of Equation 5.18 we can see that $p = p(x, y)$ without z. From the sixth equation of Equation 5.18 and the boundary conditions available, we have $\omega_z = 0$. Therefore, six equations in Equation 5.18 are reduced to four equations as follows:

$$\left(\mu+\frac{1}{2}\chi\right)\frac{\partial^2 u}{\partial z^2}-\chi\frac{\partial\omega_2}{\partial z}-\frac{\partial p}{\partial x}=0$$
$$\left(\mu+\frac{1}{2}\chi\right)\frac{\partial^2 v}{\partial z^2}+\chi\frac{\partial\omega_1}{\partial z}-\frac{\partial p}{\partial y}=0$$
$$\frac{\partial^2\omega_1}{\partial z^2}-2\chi\omega_1-\chi\frac{\partial v}{\partial z}=0$$
$$\frac{\partial^2\omega_2}{\partial z^2}-2\chi\omega_2+\chi\frac{\partial u}{\partial z}=0. \tag{5.19}$$

With the given boundary conditions, we can derive the flow velocity from the corresponding expression. Usually, the boundary conditions of lubrication flow velocity are as follows:

$$As_4\ z=0,\quad u=U_{11}, v=U_{12}, \omega_1=0, \omega_2=0$$
$$As_4\ z=h,\quad u=U_{21}, v=U_{22}, \omega_1=0, \omega_2=0. \tag{5.20}$$

The following expressions can be verified to be the solutions of Equation 5.19 with the above-mentioned velocity boundary conditions:

$$u=\frac{1}{\eta}\left(\frac{\partial p}{\partial x}\frac{z^2}{2}+A_{11}z\right)-\frac{2N^2}{m}\{A_{21}\sinh(mz)+A_{31}\sinh(mz)\}+A_{41}$$
$$v=\frac{1}{\eta}\left(\frac{\partial p}{\partial y}\frac{z^2}{2}+A_{12}z\right)-\frac{2N^2}{m}\{A_{22}\sinh(mz)+A_{32}\sinh(mz)\}+A_{42}$$
$$\omega_1=\left\{-\frac{1}{2\eta}\left(\frac{\partial p}{\partial z}z+A_{12}\right)-A_{22}\cosh(mz)+A_{32}\sinh(mz)\right\}$$
$$\omega_2=\left\{-\frac{1}{2\eta}\left(\frac{\partial p}{\partial z}z+A_{11}\right)-A_{21}\cosh(mz)+A_{31}\sinh(mz)\right\}, \tag{5.21}$$

where

$$A_{1j}=\frac{1}{A_5}\left\{(U_{2j}-U_{1j})\sinh(mh)-\frac{h}{2\eta}\frac{\partial p}{\partial x_j}\left[h\sinh(mh)+\frac{2N^2}{m}(1-\cosh(mh))\right]\right\}$$
$$A_{2j}=\frac{1}{\eta A_5}\left\{(U_{2j}-U_{1j})\frac{1-\cosh(mh)}{2}-\frac{h}{2\eta}\frac{\partial p}{\partial x_j}\left[\frac{h}{2}\left\{\cosh(mh)-1\right\}+\left\{h-\frac{2N^2}{m}\sinh(mh)\right\}\right]\right\}$$
$$A_{3j}=\frac{A_{1j}}{2\eta}$$
$$A_{4j}=U_{1j}+\frac{2N^2}{m}A_{3j}$$
$$A_5=\frac{h}{\eta}\left[\sinh(mh)-\frac{2N^2}{mh}\left\{\cosh(mh)-1\right\}\right], \tag{5.22}$$

where subscript j takes 1 or 2; $x_1=x$; $x_2=y$; $m=N/l$; and $\eta=\mu+\chi/2$ is the viscosity of micro-polar fluid.

Integrating Equation 5.21, we can obtain the flow as follows:

$$q_x = \int_0^h \rho u dz = \frac{\rho h}{2}(U_{11} + U_{21}) - \frac{\rho f(N,l,h)}{12\eta}\frac{\partial p}{\partial x}$$
$$q_y = \int_0^h \rho v dz = \frac{\rho h}{2}(U_{12} + U_{22}) - \frac{\rho f(N,l,h)}{12\eta}\frac{\partial p}{\partial y}, \tag{5.23}$$

where $f(N,l,h) = h^3 + 12l^2h - 6Nlh^2\coth\left(\frac{Nh}{2l}\right)$.

Substitute Equation 5.23 into the integrated continuity equation of Equation 5.14, the general micro-polar fluid Reynolds equation is as follows:

$$\frac{\partial}{\partial x}\left(\frac{\rho f(N,l,h)}{12\eta}\frac{\partial p}{\partial x}\right) + \frac{\partial}{\partial y}\left(\frac{\rho f(N,l,h)}{12\eta}\frac{\partial p}{\partial y}\right) = \frac{1}{2}\frac{\partial}{\partial x}\{\rho(U_{11} + U_{21})h\} + \frac{1}{2}\frac{\partial}{\partial y}\{\rho(U_{12} + U_{22})h\}$$
$$- U_{21}\frac{\partial \rho h}{\partial x} - U_{22}\frac{\partial \rho h}{\partial y} + \rho V_h - \rho V_0. \tag{5.24}$$

It is noted that the micro-polar Reynolds equation is equivalent to modifying the viscosity of the classical Reynolds equation. When the characteristic length l of micro-polar fluid tends to 0 ($L \to \infty$), because $f(N, l, h) \to h^3$ and $\eta \to \mu$, Equation 5.24 will be reduced to the classical Reynolds equation.

It is also noted that in the derivation process, Singh and Sinha used the assumption of incompressible fluid so that Equation 5.24 can only be used for incompressible hydrodynamic lubrication. Here, the density ρ remains in Equation 5.24 in order to compare it with the classical Reynolds equation.

5.2.2 Influence Factors on Micro-polar Fluid Lubrication

5.2.2.1 Influence of Load

Figures 5.6 and 5.7 give the dimensionless pressure distribution and load of a journal bearing at the eccentricity $\varepsilon = 0.5$ with different B/D. Here, set the dimensionless characteristic length L to be the ratio of the bearing clearance c and micro-polar fluid characteristic length l. Let $L^2 = (c/l)^2 = 12$, and the squared coupling coefficient $N^2 = 1/3$.

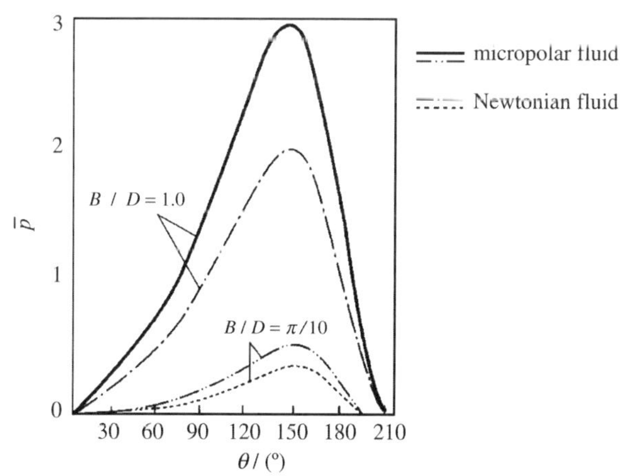

Figure 5.6 Dimensionless pressure distribution $\bar{p}$ via θ.

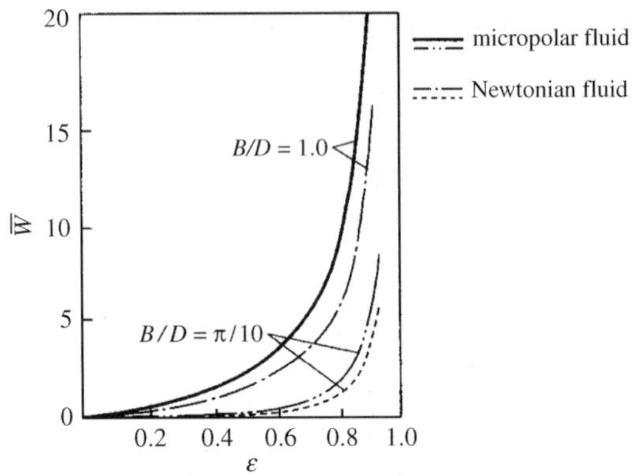

Figure 5.7 Dimensionless load $\bar{W}$ via ε.

As shown in the figures: under the same working conditions, the pressure and load of the micro-polar fluid significantly increase, compared with the Newtonian fluids.

5.2.2.2 Main Influence Parameters of Micro-polar Fluid

In the micro-polar fluid model there are two main parameters, the coupling coefficient N and the characteristic length l or the dimensionless characteristic length L. The following solutions of the journal bearing are discussed on these two parameters.

Figures 5.8–5.10 give the dimensionless load, the dimensionless frictional force and the friction coefficient respectively, varying with L or N as a reference parameter.

It can be seen from Figure 5.8 that (1) the smaller the dimensionless coupling coefficient, the greater the load-carrying capacity; (2) the smaller the dimensionless characteristic length L or the larger the micro-polar fluid characteristic length l, the further the load deviates from the load of Newtonian hydrodynamic lubrication. This means that the characteristic length significantly influences the non-Newtonian property of micro-polar fluid. When the clearance c is close to the micro-polar fluid characteristic length l (i.e. $L = 1$), the load-carrying capacity will quickly rise. If the coupling coefficient

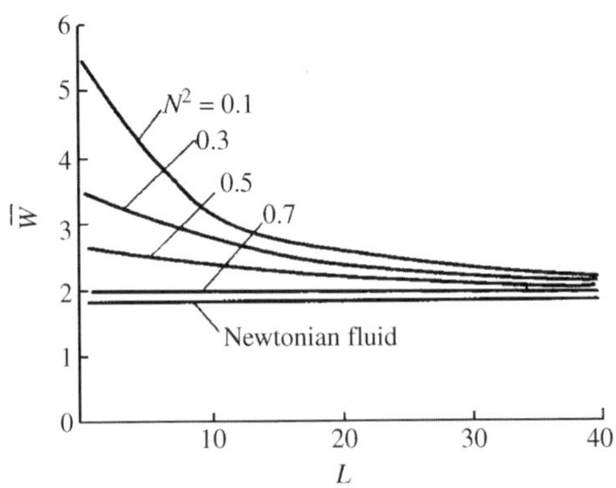

Figure 5.8 Dimensionless load via the dimensionless characteristic length.

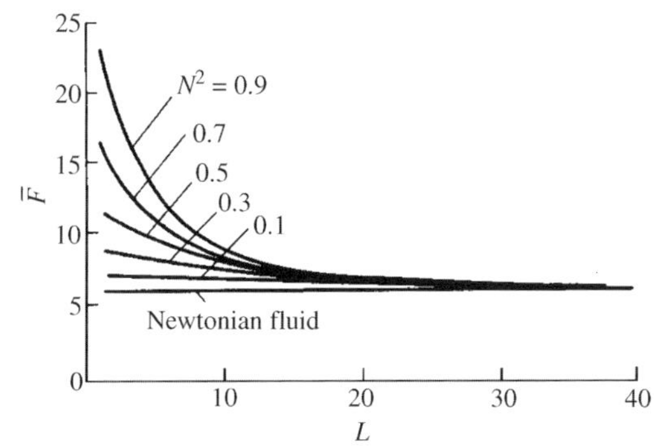

Figure 5.9 Dimensionless frictional force via the dimensionless characteristic length.

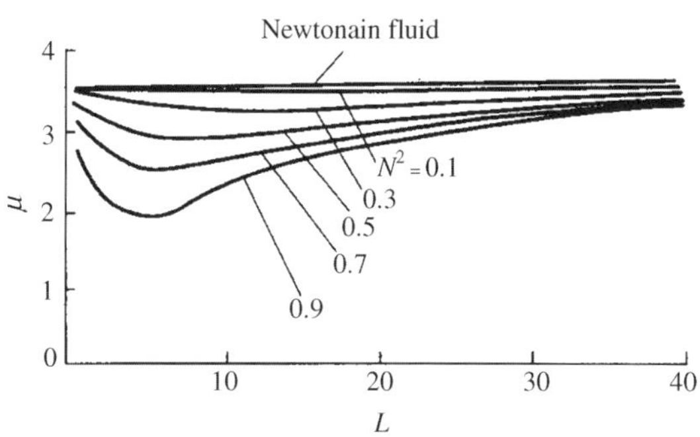

Figure 5.10 Friction coefficients via the characteristic length.

is small, the solution will deviate far away from Newtonian flow solutions. When L is very large, the micro-fluid tends to the ideal fluid so that the solution is also close to the solution of Newtonian fluid.

Figure 5.9 shows that when the dimensionless characteristic length l increases, the dimensionless frictional force tends to the Newtonian fluid solution. Figure 5.10 shows that the friction coefficient reaches the minimum as the dimensionless micro-polar fluid characteristic length is between 1 and 10. This is because the load increases faster than that of the frictional force so that the friction coefficient continues to drop. After the critical point, the friction will increase faster than the load. This will cause the friction coefficient to increase. However, the micro-polar fluid friction coefficient is always smaller than that of the Newtonian fluid. With continuous decrease of the coupling coefficient N, the friction coefficient tends to the Newtonian fluid friction coefficient.

5.3 Liquid Crystal Lubrication

Liquid crystal (LC), a material state between solid and liquid, is generally considered the fourth state of material beyond solid, liquid and gas. Molecules of liquid crystal are of long-range order arranged in one dimension or two dimensions. In the direction perpendicular to the surface of liquid crystal molecules it has the solid feature, to prevent direct contact between the surfaces; thus it has a strong load-carrying capacity. While in the tangent direction, LC possesses a low viscosity as a liquid, which can effectively reduce the friction coefficient [3–5].

The lubrication behaviors of liquid crystal are different from ordinary Newtonian fluids. The relationship of the friction coefficient, the velocity, the load and viscosity do not fit the Stribeck law. Compared with a Newtonian fluid under a high load and a low speed, the layered liquid crystal maintains a low friction coefficient. Furthermore, its friction coefficient almost does not vary with velocity and is lower than the minimum of a Newtonian fluid. Using liquid crystal as a lubricant, the tribological properties should be the environment stability (especially humidity). Generally speaking, the polar compounds are usually unstable, while the 2-alkyl derivatives are of good stability.

The structure of LC has a significant influence on the tribological properties. The structure of the columns or the smectic liquid crystal molecules can be divided into two parts: (1) the alkyl part is of a flexible part of the liquid crystal molecules; (2) the cyano and phenyl groups are the rigid part of the liquid crystal molecules. The other types of LC are: CB liquid crystal (4,4′-alkyl amino biphenyl), ECB liquid crystal (4,4′-alkoxy cyano biphenyl) and the CPC liquid crystal (alkyl cyclohexyl cyano benzene). The rigidity of CB is the lowest, but that of CPC is the highest. Studies have shown that the friction coefficient of CB is 0.035, CPC is 0.045, and ECB is 0.04. It can be seen that if the rigid part of the molecules of liquid crystal is strong, and the coefficient of friction is low.

The arrangement of liquid crystal molecules on the surface obeys the well-known FCK rule: if the molecules on the surface are in the high energy the arrangement is parallel, if in the low energy the arrangement is perpendicular. Normally, LC molecules are mainly arranged parallel to the surface. When the surface is processed by an active agent, the liquid crystal compounds can be of a lower friction coefficient. This is because the presence of the surface active agent induces liquid crystal molecules to arrange themselves perpendicular to the surface. In this situation, the share takes place between alkyl chains.

5.3.1 Types of Liquid Crystal

Although the crystal liquid performs as a liquid, it has the property of an orderly medium. LC can be divided into two categories: lyotropic liquid crystal and thermotropic liquid crystal.

5.3.1.1 Tribological Properties of Lyotropic Liquid Crystal

Lyotropic liquid crystal is a special solution. In a certain concentration range it has the properties of a crystal. Its structure is shown in Figure 5.11.

Lyotropic liquid crystal is composed of long chain molecules and solvent (usually water). Its long chain molecules are mainly compounds of soaps or chilly agents. A nonaqueous solvent lyotropic liquid crystal can be used as a lubricant. Its advantages are: (1) it possesses low shear strength; (2) perpendicular to the layered structure, it is similar to a solid in possessing a significant rigidity and a load-carrying capacity;

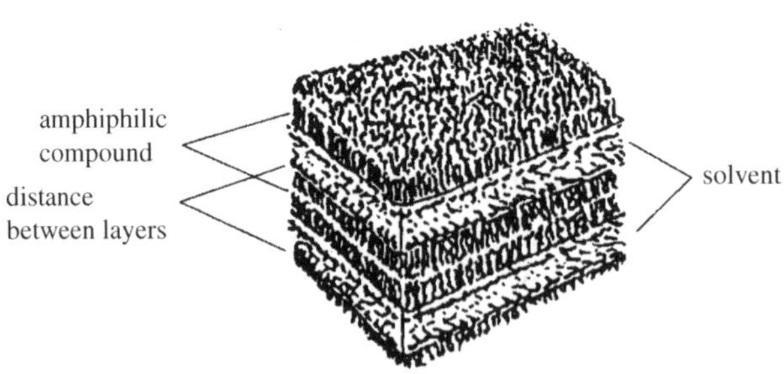

Figure 5.11 Structure of lyotropic liquid crystals.

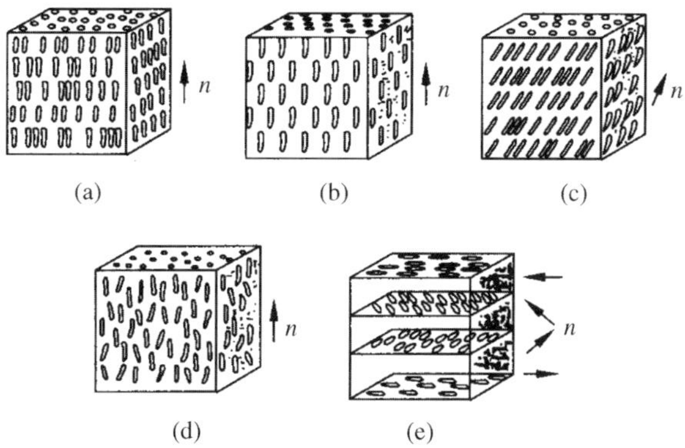

Figure 5.12 Types of thermotropic liquid crystals.

and (3) it is compatible with hydrophobic and hydrophilic additives. The disadvantage is that the temperature range of its crystalline structure is narrow and its flow is unstable.

Lyotropic liquid crystals are convenient and relatively cheap so they are widely used, in particular as a metal processing lubricant.

5.3.1.2 Tribological Properties of Thermotropic Liquid Crystal

Thermotropic liquid crystal is a non-isotropic solid. Compared with lyotropic liquid crystal, its properties and applications have been extensively studied. The structures of thermotropic liquid crystal are shown in Figure 5.12.

Thermotropic liquid crystal presents crystal properties at a certain temperature range. Some structures of thermotropic liquid crystal are as follows:

1. the micro-structure of smectic liquid crystal is oriented in the layer;
2. the micro-structure of the nematic liquid crystal is oriented in one direction;
3. the micro-structure of the cholesteric liquid crystal is oriented along the screw direction.

It is found that most thermotropic liquid crystal compounds have a low coefficient of friction at the room temperature.

Temperature is an important factor in the performance of liquid crystal lubrication. As shown in Figure 5.13, below the intercourse transition temperature T_{IN}, LC can be an effective lubricant and its friction coefficient is independent of temperature. If temperature is higher than T_{IN}, the orient structure of liquid crystal molecules has been disrupted so that their lubricating property decreases gradually. However, a common oil lubricant has no such characteristic, as the curve of MS20, a kind of aero lubricating oil, shows in Figure 5.13. In addition, the study also shows that when the thickness of liquid crystal MBBA is less than 100 nm, no structural change occurs. It may be that the surface energy of the liquid crystal film has a limited action on the orderly structure. By mixing different liquid crystal compounds, a low melting point mixture can be obtained, which will extend the temperature range of liquid crystal.

While the orient structure of a liquid crystal compound is subjected to an external field, such as an electric or magnetic field, its tribological properties will vary. Under the influence of a DC or AC electric field outside, the friction coefficient of liquid crystal may reduce down to 25%. However, this variation may be influenced by the nonisotropic transition temperature T_{IN}. If temperature is higher than T_{IN},

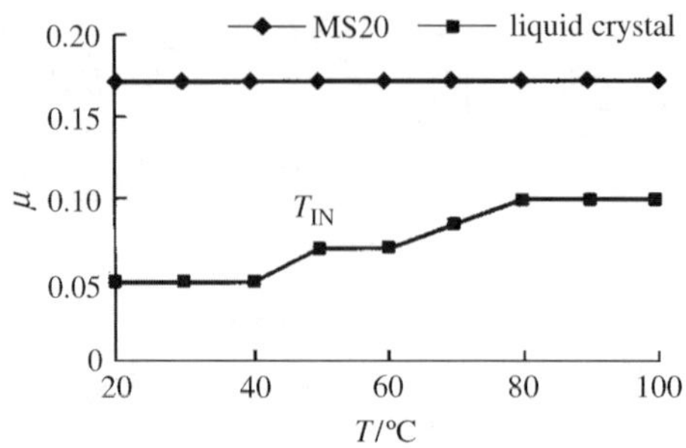

Figure 5.13 Relationship between friction coefficient and temperature of LC and MS20 (an aero lubricating oil).

although temperature has no significant influence on the friction coefficient of liquid crystal, the influence of an electric field on the friction coefficient thereby may be weakened.

Because the electrical field effect to liquid crystal is effective and reversible, it may be useful in the friction drive and friction control.

5.3.2 Deformation Analysis of Liquid Crystal Lubrication

Because the liquid crystal has a layered structure, its lubrication analysis method is different from the usual lubrication theory. So far, there has been no good LC lubrication theory. Rhim *et al.* set a simple model of liquid crystal lubrication for numerical deformation analysis [3]. The following discussion is limited to the situation that the orientation of liquid crystal is aligned to the normal direction of the layer as shown in Figure 5.12a.

As shown in Figure 5.14, the layers of the liquid crystal before entering the wedge-shaped gap are parallel to each other. When they enter the wedge zone, the liquid crystal layers inside the two surfaces are still maintained in parallel. Therefore, at the smaller end LC will inevitably be squeezed. If the wedge angle is 2θ, the wedge shaped length is l, the larger film thickness is $2h$, the smaller thickness is $2h_1$, Δa is the average thickness of liquid crystal layer, and b is the Burgers vector of values (i.e. LC the atomic spacing), the displacement on boundary is $L = b\Delta a/\theta$.

Figure 5.15 is a liquid crystal rheological model proposed by Rhim *et al.* [3]. They idealized a liquid crystal plane as a porous layer, connected by springs between the liquid crystal layers. LC can be thought of as the same as a Newtonian fluid flowing parallel to the plane. In the thickness direction, the flow of liquid crystal will be resisted by the porous layer in the same way as the penetration. The resistance is

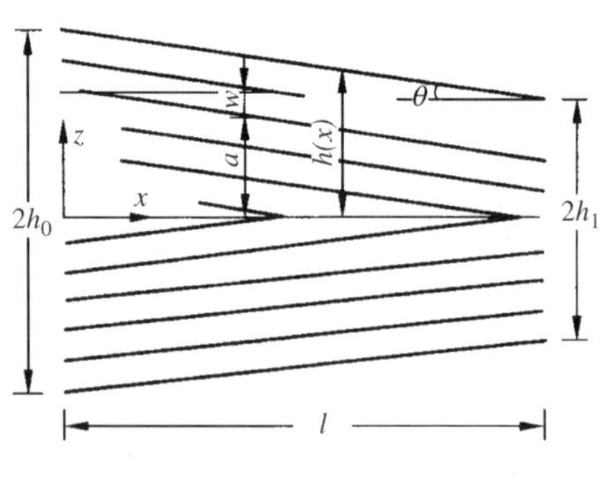

Figure 5.14 Schematic diagram of liquid crystal lubrication.

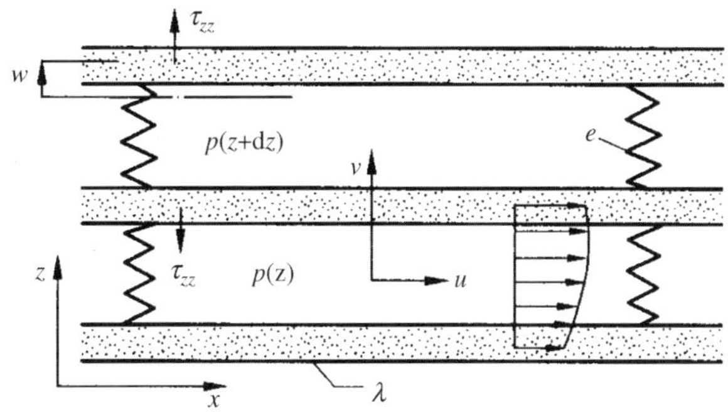

Figure 5.15 Rhim's rheological model of LC.

proportional to the pressure gradient. With the isothermal and incompressible assumptions, and without regard to the body force, the elastic deformation of the liquid crystal layers can be described only by one displacement component w (x, y, z), where w is the local displacement of a liquid crystal layer, and it obeys the following differential equation:

$$e\frac{\partial^2 w}{\partial z^2} - k\left[\frac{\partial^4 w}{\partial x^4} + 2\frac{\partial^4 w}{\partial x^2 \partial z^2} + \frac{\partial^4 w}{\partial z^4}\right] = 0, \tag{5.25}$$

where z is the coordinate perpendicular to the local liquid crystal layer; x and y are the coordinates along the local layer, and perpendicular to each other; e and k are the compression elastic modulus and the separation modulus of liquid crystal material, while e has a stress dimension, and k has a power dimension.

For simplicity, the following discussions are only limited to the two-dimensional infinite sliding problem. The dimensionless coordinates and displacements are as follows.

$$X = x/l; Z = z/h_0; W = w/h_0, \tag{5.26}$$

where l is slider width; h_0 is the half thickness at the larger end; W is the displacement of deformation.

Equation 5.25 can be written as:

$$\frac{\partial^2 W}{\partial Z^2} - \lambda^2\frac{\partial^4 W}{\partial X^4} = 0, \tag{5.27}$$

where λ is the liquid crystal permeability coefficient, for unit width, $\lambda = \sqrt{kh_0^2/el^2}$.

If liquid crystal only has one boundary dislocation, the local deformation W_i of Equation 5.27 has the following solution:

$$W_i = -\frac{\Delta a(Z - Z_0)}{|Z - Z_0|}\left[\mathrm{erf}\left(\frac{X - X_0}{\sqrt{4\lambda|Z - Z_0|}}\right) + \mathrm{erf}\left(\frac{X_0}{\sqrt{4\lambda|Z - Z_0|}}\right)\right], \tag{5.28}$$

where X_0 and Z_0 are the boundary coordinates of X and Z; Δa is the average thickness of liquid crystal layer; erf() is the error function.

When there are multiple internal dislocations, and as long as they are far enough apart from each other, the liquid crystal layer displacement solution can be obtained by superimposing all the single dislocation solutions. If in Figure 5.14 LC is limited to a thin, wedge-shaped area in the rigid boundaries, the deformation of liquid crystal layer can be obtained from Equation 5.28.

$$W = \sum_{i=1}^{n} [W_i(X, Z, X_{0i}, Z_{0i}) + \delta W(X, Z)]. \tag{5.29}$$

With the assumption that liquid crystal layers are parallel to the wedge surface, the corresponding boundary conditions are as follows:

$$\delta W[X, -H(X)] = 1 - H(X) - \sum_{i=1}^{n} W_i[X, -H(X)], \tag{5.30}$$

$$\delta W[X, H(X)] = 1 + H(X) - \sum_{i=1}^{n} W_i[X, H(X)], \tag{5.31}$$

$$\delta W(0, Z) = 0$$
$$\frac{\partial^2 W(0, Z)}{\partial X^2} = 0,$$

$$\delta W(1, Z) = -[1 - H_1]\mathrm{sign}(Z) - \sum_{i=1}^{n} W_i[1, Z]$$
$$\frac{\partial^2 W(1, Z)}{\partial X^2} = 0, \tag{5.32}$$

where sign(Z) is the function. While $Z \geq 0$, it takes a positive sign, and while $Z < 0$ a negative sign; H is the dimensionless thickness, $H = h/h_1$.

In fact, under these boundary conditions, an analytical solution of Equation 5.27 still cannot be directly obtained, such as Equation 5.28. Therefore, numerical methods must be used to solve $\delta W(X, Z)$. The usual approach is to use finite element method or finite difference method. Substitute the deformation of the nodes as the unknown quantities into Equation 5.27, and then solve them according to the above boundary conditions. For example,

$$\delta W(X, Z) = N_i(\xi, \eta)\delta W_i \tag{5.33}$$

where $N_i(\xi,\eta)$ is the element function, known as the shape function; δW_i is the unknown deformation on Node i.

Rhim *et al.* gave the analysis below on the finite element method. The parameters used in the calculation are: $l = 0.01$ m, $h = 10^{-4}$ m, $h_0 = 10^{-9/2}$ m, $e = 10^7$ Pa, $\lambda^2 = hh_0^2/el^4 = 10^{-18}$.

Figure 5.16 shows that if there is no dislocation, while the liquid crystal layer enters a wedge-shaped region, its shape is twisted to be a quadrilateral. In order to adapt to the wedge-shaped area, it gradually reduce its thickness so that the liquid crystal layer becomes thinner as a spring is gradually pressed. Clearly, the deformation would be impossible if the amount of compression was greater than the thickness of the layer.

Figure 5.17 shows three sub-images. They are the total displacement, node displacements, and side shape. The boundary dislocation is $L = b\Delta a = 0.1$ ($H_1 = 0.8$). In Equation 5.20, there exists a singular point, which can be clearly seen in Figure 5.17. Near the solid surface, LC is almost parallel to the surface, keeping a constant. However, the dislocations near the center occur abruptly rather than gradually (see Figure 5.17a). From the node displacements (Figure 5.17b) there exist many small wrinkles, which cause the liquid crystal layer to reverse distortion which, however, is quite different from Figure 5.16.

Figure 5.18 shows the displacement W of the small Burgers vectors with dislocations. Different from Figure 5.17, the figure has very small Burgers vectors, or the distance between the liquid crystal layers

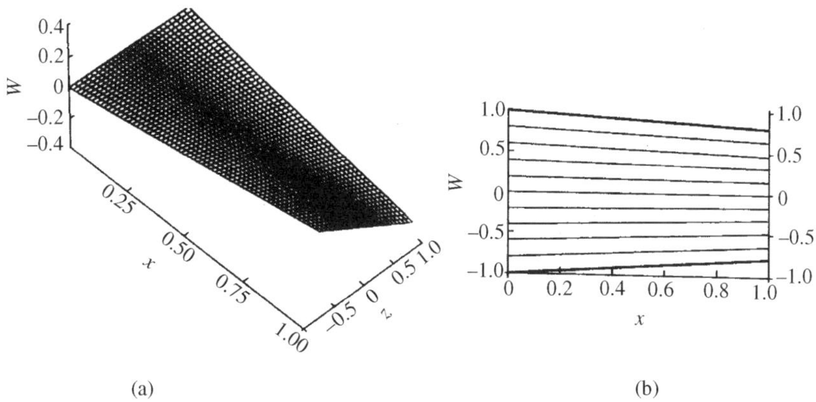

Figure 5.16 W distribution without dislocation ($H_1 = 0.8$); (a) displacement of liquid crystal layer; (b) side shape.

are larger, $L = b\Delta a = 0.02$. It can be seen from Figure 5.18 that the minimum thickness H_1 is equal to 0.8 and there are 20 dislocations per length at the boundary. In addition to dislocations, tiny wrinkles can also be seen near the dislocations, but wrinkles are hardly seen a little further away from the dislocations. This is mainly because the Burgers vectors are too small. In addition, the amplitudes also reduce in the central area.

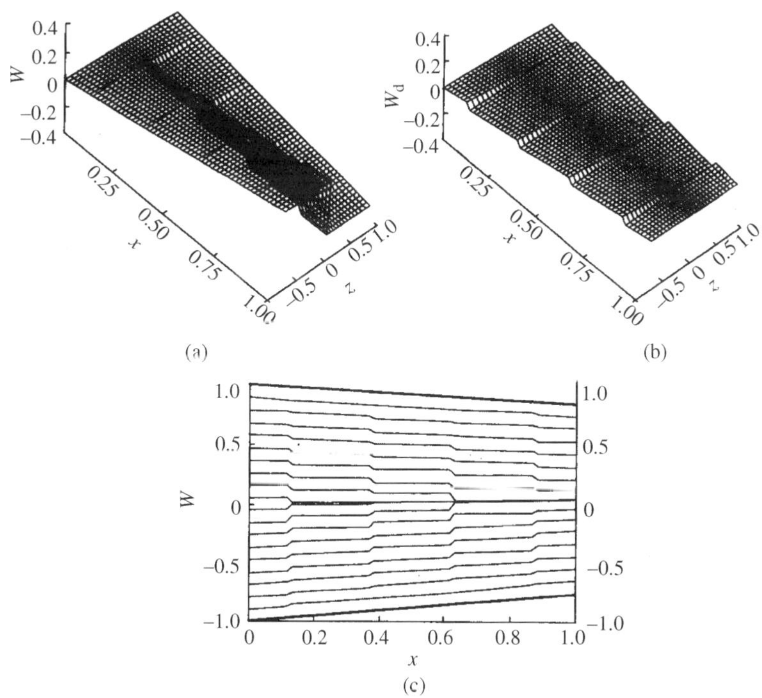

Figure 5.17 W distribution with dislocations ($H_1 = 0.8, L = 0.1$). (a) Total displacement; (b) node displacements; (c) side shape.

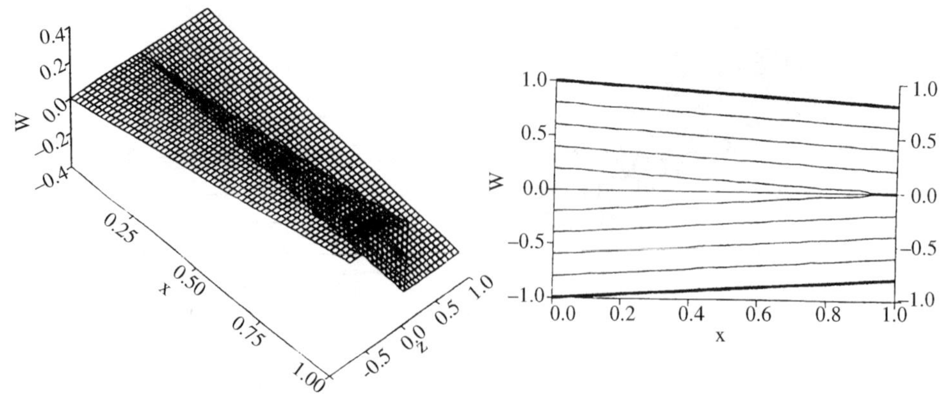

Figure 5.18 W distribution with dislocation of ($H_1 = 0.8$, $L = 0.02$).

5.3.3 Friction Mechanism of Liquid Crystal as a Lubricant Additive

The price of thermotropic liquid crystal compounds is expensive so simple usage of thermotropic liquid crystal as a lubricant is limited. However, they have wide applications as a lubricant additive because sometimes boundary lubrication is concerned. Adding liquid crystal compounds into the traditional oil-based lubricants has been proven to improve lubrication performances. The modified base oils include petroleum jelly, mineral oil, silicone oil, ester oil, artificial joints and some synovial fluids. It is also reported that cholesteryl ester has been used to improve the lubrication of MS20 aviation oil. A liquid crystal additive, 4,4′-pentyl-cyanobiphenyl (K15), is added into polyethylene glycol ester to make it more effective in reducing friction than using a pure liquid crystal compound. Moreover, liquid crystal as lubricant additives can also overcome deficiencies in stability and temperature adaptability for pure liquid crystal.

5.3.3.1 Tribological Mechanism of 4-pentyl-4′-cyanobiphenyl

Yao *et al.*, based on their research on the friction properties of 4-pentyl-4′-cyanobiphenyl (5CB) with different basic liquid, pointed out that 5CB possesses good friction performances in hexadecane [5], while in pentaerythritol ester, only 5CB does not reduce friction. If some inducer for parallel orientation is added, for example, sebacate ester, it will enable the friction coefficient to reduce significantly after the induction period of friction.

Usually, the concentration of 5CB in a lubricant is low so it can only form a thin boundary lubrication film. Under the action of surface forces, the film can maintain an orderly structure. Under friction, the structure is more easily formed. In the film, the long axis of liquid crystal molecules is parallel to the direction of friction. This structure is shown in Figure 5.19.

In the formation of liquid crystal film of 4-pentyl-4′-cyanobiphenyl, it presents a stiff performance in the load direction, which improves the load-carrying capacity and anti-wear ability. In the shear direction, it presents low viscosity. It may easily be sheared to reduce friction.

5.3.3.2 Tribological Mechanism of Cholesteryl Oleylcarbanate

Cholesteryl oleylcarbanate forms an ordered thin film with a helical structure in the base oil, as shown in Figure 5.20. In the film, the long axles of most liquid crystal molecules are along the direction of friction. Therefore, it shows a high stiffness in the load direction, but low viscosity in the shear direction. As an additive to the base oil, it can reduce the friction of the oil.

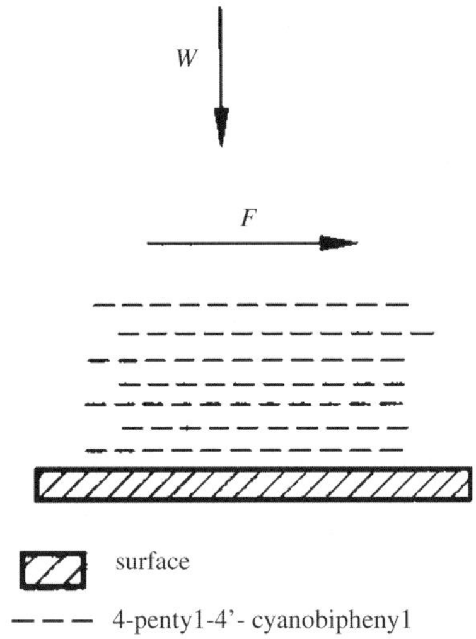

Figure 5.19 Film structure of 4-pentyl-4′-cyanobiphenyl.

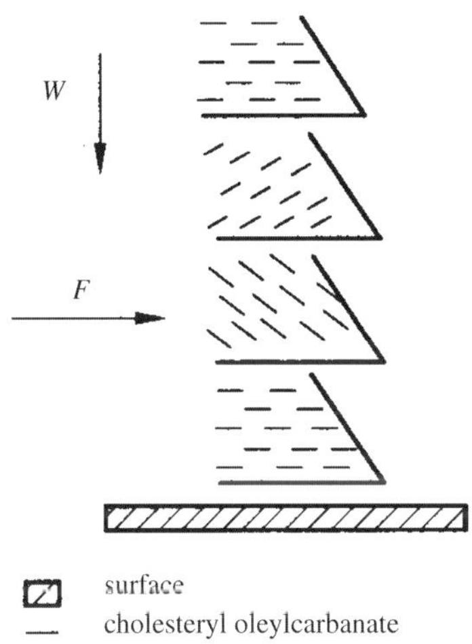

Figure 5.20 Film structure of cholesteryl oleylcarbanate.

Another application of liquid crystal additive is to improve the performance of synovial fluid in artificial joints. In laboratory conditions, the usage of liquid crystal additives can simulate the tribological properties of synovial fluid in human joints. Chromatography analysis shows that the components of the synovial fluid in human joints contain large amounts of cholesteryl esters.

5.4 Electric Double Layer Effect in Water Lubrication

Because the hardness of ceramic is even higher than steel, and it has chemical inertness and high temperature stability, it is a good friction material for water lubrication. Tomizawa and Fisher found that the friction coefficient of ceramic Si_3N_4 for water lubrication is less than 0.002, and the hydrodynamic lubrication film thickness is 70 nm. Their subsequent experiments also found that the friction coefficient of SiC is also very low, less than 0.01. Water lubrication of silicon ceramic is generally considered to have a low friction coefficient due to the chemical reaction to form a smooth layer, which makes hydrodynamic lubrication of a very thin water film possible.

Further studies have shown that the electric double layer effects in water film on ceramic surface plays an important role in lubrication. Here, the rheological characteristics of a thin water film on the ceramic surface are focused and differences with the traditional lubrication theory are discussed.

On solid surfaces, due to the existence of a very thin water film, the viscosity induced by electric double layer will significantly change the viscosity. Macauley measured the viscosity of water parallelly existing between glass planes with only 2.5 Å–0.1 μm distance, which is about 0.11P·s, while in the large tube, the viscosity of water is 0.01 P. It is estimated that the viscosity of adsorbed water film is 5–10 times larger than that of ordinary water. Elton thought that for a very thin water film the increase in viscosity is because of formation of ionic liquids on the solid surfaces caused by electric double layer of interface. Then he proposed a theoretical model which predicts the results to be consistent with the experimental results. Zhang and Umehara [6] derived the Reynolds equation on the electric double layer hydrodynamic lubrication, and carried out some calculations and experiments. This introduction is mainly based on their research in this area.

5.4.1 Electric Double Layer Hydrodynamic Lubrication Theory

5.4.1.1 Electric Double Layer Structure

The electric double layer occurs in the contact interface of an ion fluid and a solid. Figure 5.21 shows the diagram of electric double layer structure. The electric double layer is composed of the Stern layer and the diffusion layer. The Stern layer is an absorbing layer of water on a solid surface, and has a potential ψ. The macro-movement of fluid is believed to occur in the intersecting plane, with a few molecules of the Stern layer and the diffusion layer. The plane is called the slip plane. The electric potential on the slip plane is called ζ electric potential, which can be determined according to the characteristics of water and ceramics theoretically.

5.4.1.2 Hydrodynamic Lubrication Theory of Electric Double Layer

Figure 5.22 shows the hydrodynamic lubrication model of electric double layer, based on electric double layer existing on the two solid surfaces. The lower surface $z = 0$ has a velocity U along the x direction, the upper surface is static. Potential distribution of electric double layer along the z direction is determined by

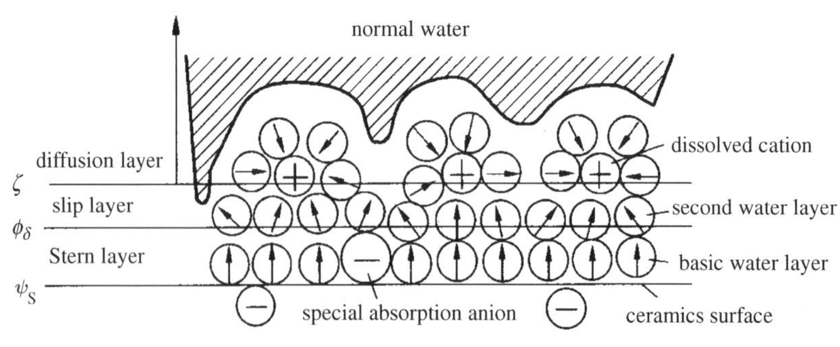

Figure 5.21 Schematic diagram of electric double layer.

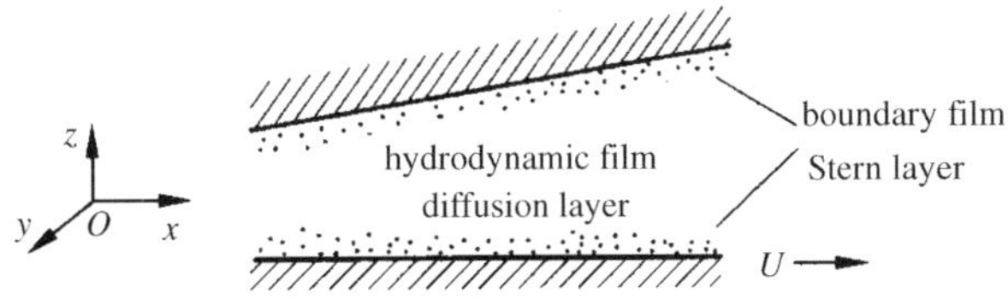

Figure 5.22 Electric double layer hydrodynamic lubrication model.

the following formula:

$$\psi = \begin{cases} \zeta \exp(-\chi z) & 0 < z < h/2 \\ \zeta \exp(-\chi(h-z)) & h/2 < z < h, \end{cases} \tag{5.34}$$

where h is the lubricant film thickness; χ^{-1} is the Debye double layer thickness; ζ is the electric potential of the slip plane.

In classical lubrication theory, only the viscous force and pressure are considered. However, due to electric double layer, we must consider the electric field. The viscous forces on a micro-element along the x and y directions are:

$$dF_x = \frac{\partial \tau_{zx}}{\partial z} dxdydz$$
$$dF_y = -\frac{\partial \tau_{zy}}{\partial z} dxdydz. \tag{5.35}$$

According to Newtonian viscosity law, the shearing stresses in the x and y directions are equal to:

$$\tau_{zx} = -\eta \frac{\partial u_x}{\partial z}$$
$$\tau_{zy} = -\eta \frac{\partial u_y}{\partial z}, \tag{5.36}$$

where u_x and u_y are the velocities of the fluid along the x and y directions; η is the fluid viscosity.

Substituting the above equations into Equation 5.35, we have:

$$dF_x = \eta \frac{\partial^2 u_x}{\partial z^2} dxdydz$$
$$dF_y = \eta \frac{\partial^2 u_y}{\partial z^2} dxdydz. \tag{5.37}$$

The pressure differences dP_x and dP_y caused by dx and dy along the x and y directions are:

$$dP_x = -\frac{\partial p}{\partial x} dxdydz$$
$$dP_y = -\frac{\partial p}{\partial y} dxdydz. \tag{5.38}$$

Along the x and y directions, the electric forces dR_x and dR_y are equal to:

$$\begin{aligned} dR_x &= E_x\,\rho dxdydz \\ dR_y &= E_y\,\rho dxdydz, \end{aligned} \tag{5.39}$$

where E_x and E_y are the fluid electric potentials inside the electric double layer along the x and y directions respectively, which are generated by flow; ρ is the density.

The relationships between the flow potential E_x and E_y, and pressure p are given by the Helmholtz-Smoluchowski formula as follows:

$$\begin{aligned} E_x &= -\frac{\zeta\varepsilon}{4\pi\eta_a\lambda}\frac{\partial p}{\partial x} \\ E_y &= -\frac{\zeta\varepsilon}{4\pi\eta_a\lambda}\frac{\partial p}{\partial y}, \end{aligned} \tag{5.40}$$

where η_a is the apparent viscosity of the fluid; λ is the conductivity; ε is the dielectric constant.

Equation 5.40 is derived under the assumption of capillary. Because the capillary diameter is much larger than the thickness of electric double layer, the macroscopic viscosity η in the original formula uses the micro-apparent η_a to consider the electric double layer effect. The force balance conditions for a micro-element are:

$$\begin{aligned} dF_x + dP_x + dR_x = 0 \\ dF_y + dP_y + dR_y = 0. \end{aligned} \tag{5.41}$$

Substituting Equations 5.37–5.39 into Equation 5.41, we have:

$$\begin{aligned} \eta\frac{\partial^2 u_x}{\partial z^2} - \frac{\partial p}{\partial x} + E_x\rho = 0 \\ \eta\frac{\partial^2 u_y}{\partial z^2} - \frac{\partial p}{\partial y} + E_y\rho = 0. \end{aligned} \tag{5.42}$$

In the electric double layer, the electric potential ψ is given by the following equation:

$$\nabla^2\psi = -\frac{4\pi\rho}{\varepsilon}. \tag{5.43}$$

Substitute Equation 5.43 into 5.42 and consider that the sizes of electric double layer in the x and y directions are much larger than that in the z direction, that is, $\partial^2/\partial x^2 \ll \partial^2/\partial z^2$ and $\partial^2/\partial y^2 \ll \partial^2/\partial z^2$; after simplification, the following equations are available:

$$\begin{aligned} \eta\frac{\partial^2 u_x}{\partial z^2} - \frac{\partial p}{\partial x} - \frac{E_x\varepsilon}{4\pi}\frac{\partial^2\psi}{\partial z^2} = 0 \\ \eta\frac{\partial^2 u_y}{\partial z^2} - \frac{\partial p}{\partial y} - \frac{E_y\varepsilon}{4\pi}\frac{\partial^2\psi}{\partial z^2} = 0. \end{aligned} \tag{5.44}$$

Suppose the pressure across the thickness is constant, integrating Equation 5.44 to z twice, we have:

$$\eta u_x - \frac{E_x \varepsilon}{4\pi}\psi = \frac{1}{2}\frac{\partial p}{\partial z}z^2 + Az + B, \tag{5.45}$$

where A and B are the integral constants to be determined by the following boundary conditions:

$$\begin{cases} u_x = 0 \\ \psi = \zeta \end{cases} \qquad \text{for } z = 0 \text{ and } z = h. \tag{5.46}$$

Substituting the boundary conditions into Equation 5.33, we obtain the following expressions:

$$A = -\frac{h}{2}\frac{\partial p}{\partial x}$$

$$B = -\frac{E_x \varepsilon}{4\pi}\zeta. \tag{5.47}$$

Substituting the above expressions of A and B into Equation 5.45, it became:

$$\eta u_x = \frac{z^2}{2}\frac{\partial p}{\partial z} - \frac{hz}{2}\frac{\partial p}{\partial x} + \frac{E_x \varepsilon}{4\pi}(\psi - \zeta) - \eta\left(1 - \frac{z}{h}\right)U. \tag{5.48}$$

With the same method to the y direction, we have:

$$\eta u_y = \frac{z^2}{2}\frac{\partial p}{\partial z} - \frac{hz}{2}\frac{\partial p}{\partial y} + \frac{E_y \varepsilon}{4\pi}(\psi - \zeta). \tag{5.49}$$

In derivation of Equation 5.49, the following boundary conditions are used:

$$u_x = -U; \quad \psi = \zeta \quad z = 0$$

$$u_x = 0; \quad \psi = \zeta \quad z = h. \tag{5.50}$$

The flow along the x direction is equal to:

$$Q_x = \int_0^h u_x dz. \tag{5.51}$$

Then, substituting Equation 5.48 into the above equation, we have:

$$Q_x = \frac{1}{\eta}\left\{-\frac{h^3}{12}\frac{\partial p}{\partial x} - \frac{E_x \varepsilon \zeta}{4\pi}\left[h - \frac{2}{\chi}\left(1 - e^{-\chi h/2}\right)\right]\right\} - \frac{hU}{2}. \tag{5.52}$$

We can also get the flow along the y direction as follows:

$$Q_y = \frac{1}{\eta}\left\{-\frac{h^3}{12}\frac{\partial p}{\partial y} - \frac{E_y \varepsilon \zeta}{4\pi}\left[h - \frac{2}{\chi}\left(1 - e^{-\chi h/2}\right)\right]\right\}. \tag{5.53}$$

Under the incompressible assumption, the integral continuity equation becomes:

$$\frac{\partial Q_x}{\partial x} + \frac{\partial Q_y}{\partial y} = 0. \tag{5.54}$$

Substituting Equations 5.52 and 5.53 into Equation 5.54, we have:

$$\frac{\partial}{\partial x}\left(\frac{h^3}{12\eta}\frac{\partial p}{\partial x}\right)+\frac{\partial}{\partial y}\left(\frac{h^3}{12\eta}\frac{\partial p}{\partial y}\right)=-\frac{U}{2}\frac{\partial h}{\partial x}-\frac{\partial}{\partial x}\left\{\frac{E_x\varepsilon\zeta}{4\pi\eta}\left[h-\frac{2}{\chi}\left(1-e^{-\chi h/2}\right)\right]\right\}$$
$$-\frac{\partial}{\partial y}\left\{\frac{E_y\varepsilon\zeta}{4\pi\eta}\left[h-\frac{2}{\chi}\left(1-e^{-\chi h/2}\right)\right]\right\}. \tag{5.55}$$

Substitute E_x and E_y in Equation 5.40 into Equation 5.55, and modify it in the form of Reynolds equation as follows:

$$\frac{\partial}{\partial x}\left(\frac{h^3}{12\eta_a}\frac{\partial p}{\partial x}\right)+\frac{\partial}{\partial y}\left(\frac{h^3}{12\eta_a}\frac{\partial p}{\partial y}\right)=-\frac{U}{2}\frac{\partial h}{\partial x}. \tag{5.56}$$

In Equation 5.56, η_a is the apparent viscosity as follows:

$$\eta_a=\eta+\frac{3\varepsilon^2\zeta^2\left\{h-\frac{2}{\chi}\left(1-e^{-\chi h/2}\right)\right\}}{4\pi^2\lambda h^3}. \tag{5.57}$$

Equation 5.56 is called the Reynolds equation considering electric double layer effect. The apparent viscosity expression Equation 5.57 shows that due to the existence of electric double layer, the increase of η_a is proportional to the square of ζ, and is inversely proportional to the cube of h. In other words, the thin film thickness will have a significant influence on the electric double layer hydrodynamic lubrication. In addition, ζ increases the load-carrying capacity of the lubrication film. This conclusion may help us choose surface materials and lubricants of thin film lubrication.

5.4.2 *Influence of Electric Double Layer on Lubrication Properties*

As shown in Figure 5.22, take a slider as an example to discuss the influence of electric double layer on the minimum film thickness and friction coefficient in a very thin water film with ceramic surfaces [7, 8].

5.4.2.1 Pressure Distribution

For an infinitely wide problem, variation of the pressure in the y direction can be omitted so that Equation 5.56 is simplified as:

$$\frac{\partial}{\partial x}\left(\frac{h^3}{12\eta_a}\frac{\partial p}{\partial x}\right)=-\frac{U}{2}\frac{\partial h}{\partial x}. \tag{5.58}$$

Integrating the above equation to x, we have:

$$\frac{h^3}{12\eta_a}\frac{\partial p}{\partial x}=-\frac{U}{2}(h-h_0), \tag{5.59}$$

where h_0 is the integral constant which can be determined by the boundary condition.

Substitute Equation 5.57 into the above equation, it becomes:

$$\frac{\partial p}{\partial x}=-\frac{6U(h-h_0)}{h^3}\left\{\eta+\frac{3\varepsilon^2\zeta^2\left[h-\frac{2}{\chi}\left(1-e^{-xh/2}\right)\right]}{4\pi^2\lambda h^3}\right\}. \tag{5.60}$$

For simplification, assume the film thickness h is much greater than electric double layer, that is, $\chi h >> 1$, then, Equation 5.60 becomes:

$$\frac{\partial p}{\partial x} = -\frac{6U(h-h_0)}{h^3}\left\{\eta + \frac{3\varepsilon^2\zeta^2}{4\pi^2\lambda h^3}\right\}. \tag{5.61}$$

Integrate the above equation to x, we have the pressure distribution of the infinitely wide slider:

$$p = \frac{6U\eta}{\alpha}\left\{\frac{1}{h} - \frac{1}{h_1} - \frac{h_0}{2}\left(\frac{1}{h^2} - \frac{1}{h_1^2}\right) + \frac{3\varepsilon^2\zeta^2}{4\pi^2\lambda\eta}\left[\frac{1}{3h^3} - \frac{1}{3h_1^3} - \frac{h_0}{4}\left(\frac{1}{h^4} - \frac{1}{h_1^4}\right)\right]\right\}, \tag{5.62}$$

where h_2 and h_1 are the inlet and outlet thicknesses of the slider respectively; α is the inclining angle of the slider, equal to $\alpha = (h_2 - h_1)/l$.

In Equation 5.62, the following boundary conditions have been used.

$$\begin{aligned} p|_{h=h_1} &= 0 \\ p|_{h=h_2} &= 0. \end{aligned} \tag{5.63}$$

Therefore, the integral constant h_0 is given as follows:

$$h_0 = \frac{\eta\left(\frac{1}{h_2} - \frac{1}{h_1}\right) + \frac{\varepsilon^2\zeta^2}{4\pi^2\lambda}\left(\frac{1}{h_2^3} - \frac{1}{h_1^3}\right)}{\frac{\eta}{2}\left(\frac{1}{h_2^2} - \frac{1}{h_1^2}\right) + \frac{3\varepsilon^2\zeta^2}{16\pi^2\lambda}\left(\frac{1}{h_2^4} - \frac{1}{h_1^4}\right)}. \tag{5.64}$$

Rewriting Equation 5.62 in a dimensionless form, we have:

$$p^* = h_1\left\{\frac{1}{h} - \frac{1}{h_1} - \frac{h_0}{2}\left(\frac{1}{h^2} - \frac{1}{h_1^2}\right) + \frac{3\varepsilon^2\zeta^2}{4\pi^2\lambda\eta}\left[\frac{1}{3h^3} - \frac{1}{3h_1^3} - \frac{h_0}{4}\left(\frac{1}{h^4} - \frac{1}{h_1^4}\right)\right]\right\}, \tag{5.65}$$

where p^* is the dimensionless pressure defined as follows:

$$p^* = p\left(\frac{\alpha h_1}{6\eta U}\right), \tag{5.66}$$

where α is the slider slope.

5.4.2.2 Load-carrying Capacity

The load-carrying capacity per unit width can be obtained by the integrating p along the length l of the slider:

$$\begin{aligned} W = \int_0^l p dx = \frac{1}{a}\int_{h_1}^{h_2} p dh = \frac{6U\eta}{\alpha}\int_{h_1}^{h_2}\left\{\frac{1}{h} - \frac{1}{h_1} - \frac{h_0}{2}\left(\frac{1}{h^2} - \frac{1}{h_1^2}\right)\right. \\ \left. + \frac{3\varepsilon^2\zeta^2}{4\pi^2\lambda\eta}\left[\frac{1}{3h^3} - \frac{1}{3h_1^3} - \frac{h_0}{4}\left(\frac{1}{h^4} - \frac{1}{h_1^4}\right)\right]\right\} dh \end{aligned} \tag{5.67}$$

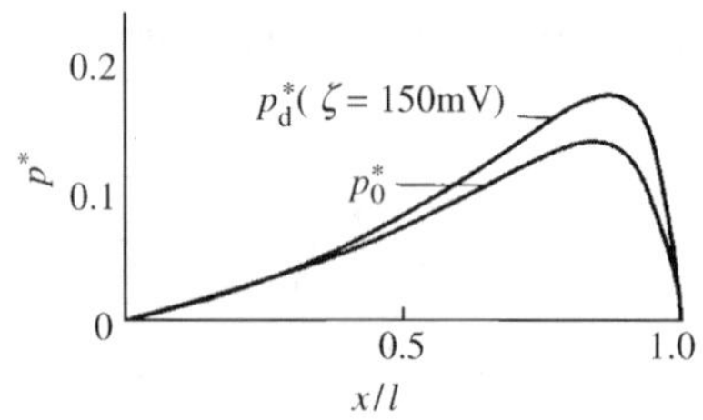

Figure 5.23 Pressure distributions with and without considering electric double layer effect.

The dimensionless load is:

$$W^* = \ln\frac{h_2}{h_1} - \frac{h_2 - h_1}{h_1} - \frac{h_0}{2}\left(\frac{1}{h_1} - \frac{1}{h_2} - \frac{h_2 - h_1}{h_1^2}\right)$$

$$+ \frac{3\varepsilon^2\zeta^2}{4\pi^2\lambda\eta}\left[\frac{1}{6}\left(\frac{1}{h_1^2} - \frac{1}{h_2^2}\right) - \frac{h_2 - h_1}{3h_1^3} - \frac{h_0}{4}\left(\frac{1}{3h^3} - \frac{1}{3h_1^3} - \frac{h_2 - h_1}{h_1^4}\right)\right], \tag{5.68}$$

where the dimensionless hydrodynamic load W^* is defined as $W^* = W\alpha^2/6\eta U$.

5.4.2.3 Friction Coefficient

Friction coefficient can be obtained by calculating the load and the frictional forces. Substitute Equation 5.40 into Equation 5.48, the velocity in the x direction is equal to

$$\eta u_x = \left\{\frac{z^2}{2} - \frac{hz}{2} - \frac{\varepsilon^2\zeta(\psi - \zeta)}{16\pi^2\lambda\eta_a}\right\}\frac{\partial p}{\partial x} - \eta\left(1 - \frac{z}{h}\right)U. \tag{5.69}$$

Because the viscous shear stress is equal to:

$$\tau_{zx} = -\eta\frac{\partial u}{\partial z} = \left(z - \frac{h}{2} - \frac{\varepsilon^2\zeta}{16\pi^2\lambda\eta_a}\frac{\partial\psi}{\partial z}\right)\frac{\partial p}{\partial x} + \eta\frac{U}{h}, \tag{5.70}$$

setting $z = 0$, the shear stress on the bottom surface is:

$$\tau_{zx} = -\eta\frac{\partial u}{\partial z} = -\left(\frac{h}{2} + \frac{\varepsilon^2\zeta}{16\pi^2\lambda\eta_a}\frac{\partial\psi}{\partial z}\right)\frac{\partial p}{\partial x} + \eta\frac{U}{h}. \tag{5.71}$$

Note: the right-hand side of the above equation contains two parts. They are induced by the fluid viscosity and the macro-apparent viscosity, considering with and without electric double layer effect.

Table 5.2 Electric double layer effects on the load

Minimum film thickness h_0 (nm)	W_0 ($\zeta = 0$ mV)	W ($\zeta = 150$ mV)	W_0/W
70	0.30	0.36	1.20
30	0.73	1.08	1.48

Integrating Equation 5.71 and using the pressure boundary conditions of $p(0)=0$ and $p(l)=0$, the frictional force per unit width is equal to:

$$F=\int_0^l \tau_{zx}dx=\int_0^l\left(-\frac{h}{2}\frac{\partial p}{\partial x}-\frac{\varepsilon^2\zeta}{16\pi^2\lambda\eta_a}\frac{\partial\psi}{\partial z}\frac{\partial p}{\partial x}+\eta\frac{U}{h}\right)dx=\frac{\alpha}{2}W+\frac{\eta U}{\alpha}\ln\frac{h_2}{h_1}. \tag{5.72}$$

Therefore, the friction coefficient can be written as:

$$\mu=\frac{F}{W}=\frac{1}{W}\left(\frac{\alpha}{2}W+\frac{\eta U}{\alpha}\ln\frac{h_2}{h_1}\right), \tag{5.73}$$

or

$$\mu=\frac{\alpha}{2}+\frac{\alpha}{6W*}\ln\frac{h_2}{h_1}. \tag{5.74}$$

5.4.2.4 An Example

By using water as a lubricant, its dielectric constant $\varepsilon=7.08\times10^{-10}$; the conductivity $\lambda=1.9\times10^{-4}\,\mu$F/m; the viscosity $\eta=0.001$Pas. If we take the slider slope $\alpha=0.003$; the length $l=1\times10^{-4}$ m, under the condition of the minimum thickness equal to 70 mm, the dimensionless pressure distribution p_d^* and p_0^* with and without considering electric double layer influence is as shown in Figure 5.23.

Clearly, when considering the influence of electric double layer, the pressure increases significantly. Therefore, the load-carrying capacity is significantly increased. The larger the potential ζ and especially the thinner the film thickness, the more remarkable is the pressure increase.

Table 5.2 gives the load ratio under different film thickness with and without considering the influence of the electric double layer. The calculation conditions are the same as mentioned above.

From the table, we can see that the smaller the lubricant film thicknesses, the greater the influence of the potential ζ of the electric double layer on the load-carrying capacity of the lubricant film.

References

1. Wang, R.Y. (1992) Magnetic fluid elastic hydrodynamic lubrication, Tsinghua University, a master's degree thesis.
2. Singh, C. and Sinha, P. (1982) The three-dimensional Reynolds equation for micropolar-fluid-lubricated bearing. *Wear*, **76** (2), 199–209.
3. Biresaw, Girma (ed.) (1990) *Tribology and Liquid-Crystalline State*, ACS Symposium Series 441, America Chemical Society, Washington DC.
4. Yao, J.B., Wen, S.Z., Wang, Q.L. *et al.* (2000) Antifriction additives 5CB liquid crystal properties. *Lubrication and Sealing*, 3, 24 27.
5. Yang, X.Y., Lu, C.F., and Wen, S. Z. (1992) Mineral oil Lubrication Characteristics of Liquid Crystal on Impact, National Tribology Conference Proceedings, Wuhan, Volume II, pp. 80–87.
6. Zhang, B. and Umehara, N. (1988) Hydrodynamic lubrication theory considering electric double layer for very thin water lubrication of ceramics. *JSME, Series C*, **41** (2), 285–290.
7. Huang, P., Huang, B.L., and Meng, Y.G. (2002) Study on influence of electric double-layer on the lubrication film thickness and pressure. *Journal of Mechanical Engineering*, **38** (8), 9–13.
8. Bai, S.X. and Huang, P. (2004) Study on influences of electric double layer viscosity on the lubrication properties. *Journal of Tribology*, **24** (2), 168–171.

6

Lubrication Transformation and Nanoscale Thin Film Lubrication

The Reynolds equation proposed in 1886 created a revolutionary hydrodynamic lubrication theory. The equation, based on viscous fluid mechanics, established a hydrodynamic lubrication theory that is now widely used in the design of mechanical parts where the lubricant film thickness is usually 100 μm or more. In the twentieth century, researchers combined hydrodynamic lubrication theory with elastic contact theory to develop EHL theory, which has been successfully used in the design of gear drives, rolling bearings and other mechanical elements with line and point contacts.

Because a concentrated load acts on the contact surfaces in the line or point contacts, the lubricant film thickness is significantly thin, usually within a range of 0.1 –1 μm. Although an EHL film thickness is thin, it still belongs to hydrodynamic lubrication so it follows the basic law of continuum mechanics.

In 1919, Hardy proposed boundary lubrication, that is, lubricating oil is adsorbed on metal surfaces through a physical or chemical process to form an absorption film. A boundary lubrication film is very thin, with a thickness of 0.005–0.01 μm. It is usually composed of a single or several molecular orientation layers. The boundary lubrication theory is based primarily on physical chemistry and the surface adsorption theory.

Obviously, whether using a film thickness or a physical characteristic such as friction coefficient to classify lubrication, it can be found that there is a transition zone between the boundary lubrication and the hydrodynamic lubrication (including EHL). We may ask: what kind of lubrication exists between boundary lubrication and hydrodynamic lubrication? And, how does the transition state happen between the absorption boundary film and the hydrodynamic film?

6.1 Transformations of Lubrication States

6.1.1 Thickness-Roughness Ratio λ

The classical lubrication theories believe that as the load increases, the lubricant film thickness gradually drops. As the direct contact of asperities occurs, lubrication fails. Therefore, a thickness-roughness ratio λ is used as the criterion for the state division:

$$\lambda = h/\sigma = h/\sqrt{\sigma_1^2 + \sigma_2^2}, \tag{6.1}$$

Principles of Tribology, First Edition. Wen Shizhu and Huang Ping.

where h is the film thickness; σ is the variance of the equivalent surface roughness; σ_1 and σ_2 are the variances of the two surface roughnesses respectively.

Generally, if $\lambda \geq 3$, it is full-film lubrication (hydrodynamic lubrication or EHL), and if $1 \leq \lambda < 3$, the asperities will contact so it belongs to mixed lubrication.

However, further study shows that the EHL lubricant film is of a heavy load-carrying capacity. Therefore, the above criterion is not suitable any more.

6.1.2 Transformation from Hydrodynamic Lubrication to EHL

In 1885, Thurston observed that with increase of the load, a journal bearing was found to reach the minimum friction coefficient. He believed that this is the conversion point from hydrodynamic lubrication to mixed lubrication.

Later, Gümbel combined the phenomenon with the Stribeck experimental curve and proposed the lubrication state diagram, as shown in Figure 6.1. According to variation of the friction coefficient μ with the dimensionless parameter $\eta\omega/p$, Gümbel proposed three lubrication states. They are hydrodynamic lubrication, mixed lubrication and boundary lubrication. Here, η is the lubricant viscosity; ω is the angular velocity of the journal; p is the average pressure or the load acting on the unit projected area.

It is generally belived that the minimum friction coefficient of hydrodynamic lubrication is in the order of 10^{-3}, and that of boundary lubrication is about 0.1. Because mixed lubrication is between hydrodynamic lubrication and boundary lubrication, with an increase in the proportion of the hydrodynamic film, the friction coefficient will gradually decrease.

EHL is a special form of lubrication state, found in 1960s. If the film thickness is h and the friction coefficient is μ, their relationships with the working condition parameter $\eta\omega/p$ can be expressed as follows:

$$h \propto \frac{\eta\omega^a}{p}$$
$$\mu \propto \frac{\eta\omega^b}{p}, \tag{6.2}$$

where a and b are the constants related to the contact surface shapes.

According to the results of the line contact EHL, Dowson proposed the curve in Figure 6.2. It is similar to the Stribeck curve to show the influence of $1/p$ on h and μ [1]. From the figure, it can be seen that in the

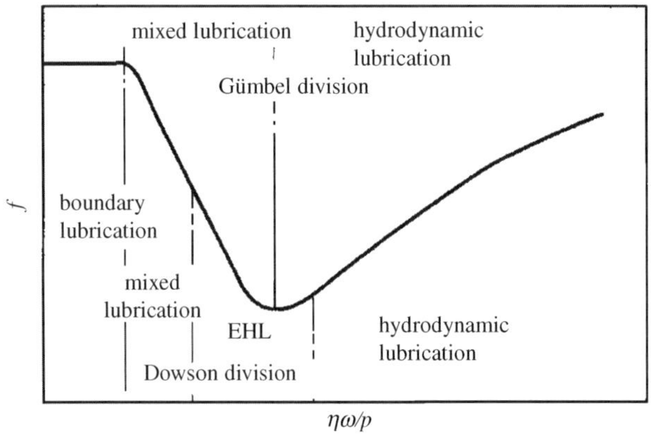

Figure 6.1 Lubrication state diagram.

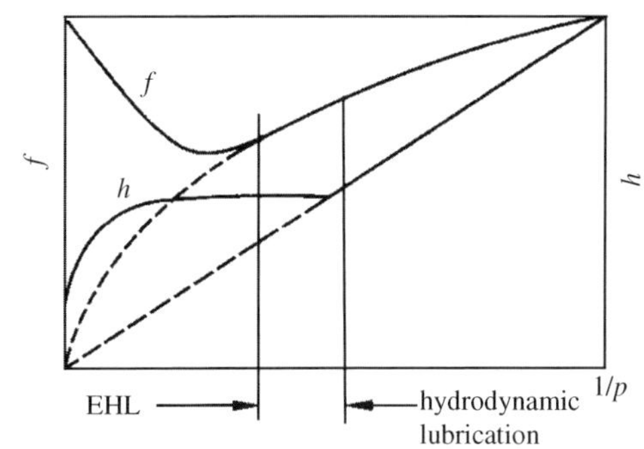

Figure 6.2 Dowson's division [1].

EHL region, the elastic deformation of the surface prevents the friction coefficient and the lubricant film thickness from decrease with increase of the load.

Based on Figure 6.2, Dowson proposed his lubrication state division as shown at the bottom of Figure 6.1.

Going from thin to thick film, the four states are: boundary lubrication, mixed lubrication, EHL and hydrodynamic lubrication. Dowson believed that when the fluid film decreases to the oil molecule thickness, the boundary film occurs on the roughness peaks and enters mixed lubrication state. He proposed 25 nm as the conversion film thickness for EHL to transform to mixed lubrication.

In concluding the development of lubrication technology, Dowson pointed out that due to the improvement of lubrication design and manufacturing technology, the lubricant film thickness will continuously drop and he gave Table 6.1 to illustrate this trend.

It should be noted that the above analysis was given in the earlier study of EHL theory. The authors believe that for the precision machining surfaces, the thin-film lubrication should be a new lubrication state among EHL, boundary lubrication and mixed lubrication.

6.1.3 Transformation from EHL to Thin Film Lubrication

For some well-lubricated gears or rolling bearings, the film thicknesses calculated by EHL formula are about a dozen layers of molecules of a lubricant. Obviously, such a thin film will not meet the conditions of the continuous medium mechanics, which means that assumptions of the EHL theory are not available for its correct application.

Table 6.1 Varying trend of the minimum lubrication film thickness

Age	Typical examples	Minimum thickness/m
1900	General bearing	10^{-4}–10^{-5}
1950	Steady journal bearing	10^{-5}
1980	Crankshaft bearings	10^{-5}–10^{-6}
	Gears, rolling bearing	10^{-6}–10^{-7}
1990–2000	Lubrications of asperity, low elastic modulus of the surface, magnetic recording devices and plastic hydrodynamic lubrication	10^{-7}–10^{-8}, even 10^{-9}

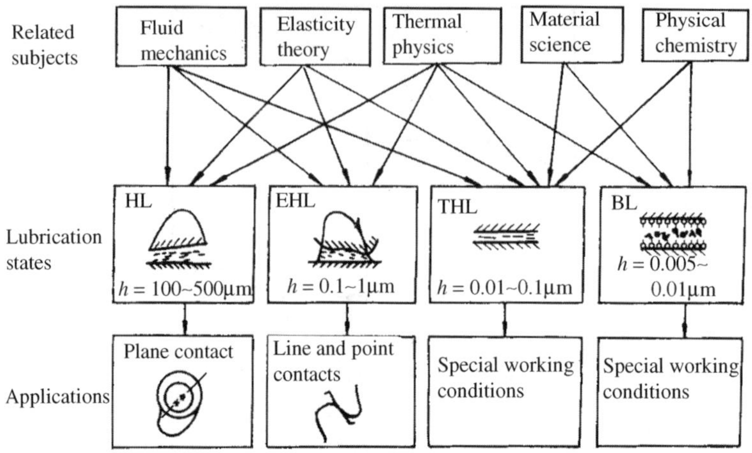

Figure 6.3 Lubrication states and their applications.

This raises two questions: first, how do we determine the minimum thickness limit in order to correctly use EHL theory? Second, how do we establish a physical mathematical model to evaluate thin film lubrication if EHL is a failure? For the first question, the experiments of Johnston *et al.* showed that when the film thickness is less than 15 nm, the variation of the thickness with the velocity deviates from the EHL theory prediction such that the film thickness is lower than that of the EHL theory predicted [2]. The experiments of Luo Jianbin showed that the minimum film thickness of EHL is about 26 nm [3], while the Streator and Gerhardstein results showed that the limit of the minimum thickness is about 23 nm [4]. Because lubricant molecules are usually about 0.5–3 nm long, the above experiments mean that the EHL theory can be applied to a limit that the minimum thickness is approximately 10–50 molecular layers thick. More research needs to be carried out on thin film lubrication in order to answer the latter question.

With the development of EHL, in the early 1990s the author proposed a new lubrication state, thin film lubrication. This is used to describe the transition between the boundary lubrication and EHL lubrication [5]. Later, both the theoretical and experimental studies showed that the lubrication state with a submicron or nanometer thickness exists. Research on thin lubrication has made significant progress [6]. Figure 6.3 briefly describes the features and applications of the lubrication states [7].

The transformation from EHL to thin film lubrication mainly depends on the lubricant film thickness. As the EHL film thickness drops to a certain value, the variation rule of the film thickness will deviate from that of EHL. This thickness should be the conversion film thickness.

Johnston *et al.* experimentally investigated the transformation from EHL to thin film lubrication [2]. Some experimental results can be seen in Figure 6.4.

Based on EHL theory, the film thickness formula in point contacts can be simplified to be:

$$h_c = ku^{0.7}\eta_0^{0.7}\alpha^{0.7}, \tag{6.3}$$

where h_c is the thickness in the central contact area; u is the entrainment velocity; η_0 is the viscosity under the atmospheric pressure; α is the viscosity-pressure coefficient; k is a constant.

It can be seen that if η_0 and α are known, the EHL film thickness h_c is proportional to the entrainment velocity u on the logarithmic coordinates.

Figure 6.4 shows that lubricants at very low velocities can keep a stable and reliable thin film thickness of 5 nm. Furthermore, as the film thickness is less than 15 nm, the variations of the central film thickness

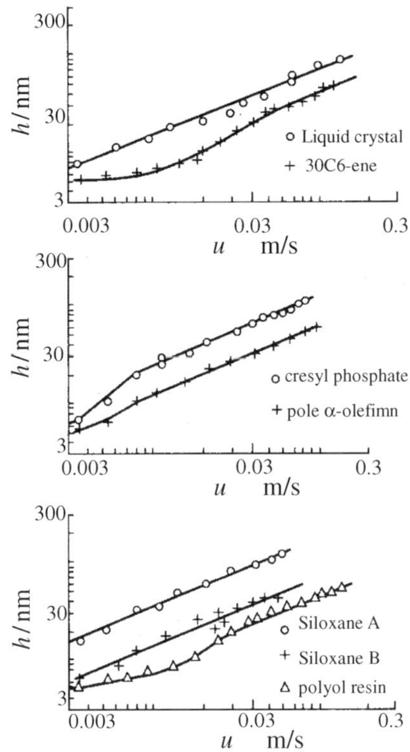

Figure 6.4 Relationship between the central thickness h and the velocity u.

with the velocity for the other lubricants, except for LC and a low viscosity silicone, have deviated from the rule of EHL, that is, the film thickness is proportional to the velocity. This shows that the lubrication state is not EHL any more, but thin film lubrication.

When the thickness is less than 15 nm, the continuum mechanics hypothesis does not exactly apply. Because the lubricant molecules near the solid surface present an ordered arrangement, the effective viscosity of the lubricant is higher than its body viscosity so as to reduce the lubricant film thickness. With the air method to measure the viscosity near the solid surface, it was found that the viscosity of 5-sebacic acid having a chemical polarity is quite different while the distance away from the solid surface is less than 10 nm. This indicates that the ordered structure of the lubricant molecules near the solid surface has a resulting difference from the bulk viscosity.

Because the state transformation is related to the structure of the lubricant molecules, the conversion conditions should be connected with the properties of a lubricant, such as the film thickness. Furthermore, the transformation conditions should also be related to the other properties of the lubricant. Their conversion film thicknesses are also different. We used the principle of optical interference in relative intensity to develop a nano-film thickness measurement device for point contact EHL and to systematically study a variation of thin lubrication film thicknesses [8]. Figure 6.5 shows some experimental results.

The figure shows that in the high entrainment velocity region, the central film thickness is proportional to the entrainment velocity. That is the typical nature of an EHL film.

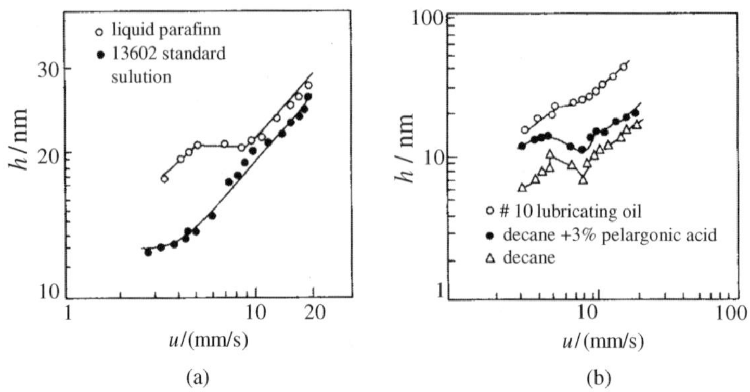

Figure 6.5 Relationship between the central film thickness and the velocity of different lubricants.

For no. 13602 standard solution, when the film thickness is reduced to 15 nm, the velocity index gradually decreases less than that in Equation 6.3. Therefore, the influence of the entrainment velocity is weakened such that the film thickness decreases and transits to thin film lubrication.

However, the conversion film thicknesses are different for different lubricants. In Figure 6.5, the conversion film thickness of the liquid paraffin is 20 nm, while the conversion film thicknesses of 10# mechanical lubricating oil is 24 nm. For decane with a low viscosity, its conversion thickness not only reduces to about 12 nm but the relationship between the thickness and velocity is also abnormal. The film thickness increases with decrease of the entrainment velocity in a certain range. If 3% pelargonic acid is added into decane, the film thickness increases more significantly, but the variation of thickness with the velocity is similar to that of the decane without pelargonic acid.

Therefore, we believe that the thin film lubrication is a feature of the ordered molecular film, and its thickness is closely related to the interface adhesion energy and liquid molecular structure. The interface adhesion energy is related to the liquid molecular structure, the molecule weight and the ambient temperature similar to the equivalent viscosity. Therefore, an ordered film thickness of liquid should vary with change of the equivalent viscosity. Figure 6.6 shows the relationship between the film thickness and the lubricant equivalent viscosity while EHL begins to transform to thin film lubrication [9].

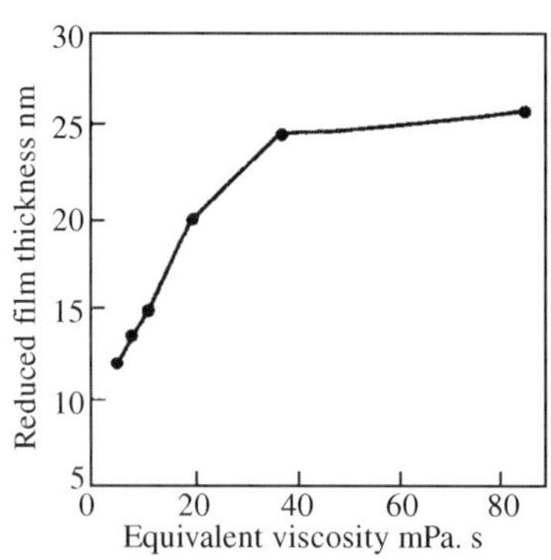

Figure 6.6 Relationship between the reduced film thickness and the equivalent viscosity.

It can be seen from the figure that the conversion lubricant film thickness increases with increase of the viscosity. As the thickness reaches about 26 nm, the variation of the thickness slows down significantly. This is connected with the effective region of the adhesion energy.

6.2 Thin Film Lubrication

According to the survey, the lubricant film thickness ranges from a few molecular layers to a few hundred micron meters. This includes the lubrication of different states. However, a variety of lubrication states have their typical characteristics and thickness ranges. The classical Stribeck curve gives the variation of the entire lubrication regions. Fluid film lubrication (including hydrodynamic lubrication and EHL) and boundary film lubrication have been well understood. However, the middle lubrication state between these well-known states is usually referred to mixed lubrication, and its phenomena are still not fully studied. The modern lubrication theory study should be focused on this field.

As early as 1971, Roberts and Tabor experimentally discovered that within the friction pairs composed of rubber and glass, a 10-nm thick water film exists. In the late 1980s, Spikes found a way to improve the optical interferometer to measure the EHL film thickness, and the nanometer thin film thickness had been measured. At a very low entry velocity and in point contacts, the EHL film thicknesses they measured are between 5 and 10 nm, which are equivalent to 3–6 layers of the lubricant molecules. Subsequently, they used this device to study different materials and lubrication additives. The results showed that zinc dialkyldithiophosphates (ZDDP) forms a viscous lubricant film with thickness of only 10–50 nm. They also studied the properties of nano-thin film lubrication of nonhydrocarbon lubricants, emulsion and so on. The author successfully measured the nano film thicknesses based on the principle of optical interference in the relative intensity [10]. The developed testing device improved the measurement accuracy and resolution and the measured results also confirmed the formation of a complete nano thin lubrication film on a point contact surface.

Practice has shown that water-based lubrication materials are widely used in industrial settings. Because their viscosities are low, the film is very thin. Some mechanical components work under high temperature with a nanometer lubrication film due to the oil viscosity dropping drastically with rising temperature. Some anti-wear additives can form a very thin lubricant film on the element surface. Furthermore, with friction surfaces under ultra low velocity or a very heavy load, a thin film lubrication also exists.

Thin-film lubrication study is of great significance in developing lubrication and wear theory, and meets the needs of engineering applications. Thin film lubrication may be a key technology to ensure that high-tech equipment and ultra-precision machine work properly. In addition, as manufacturing techniques improve, so should the general machinery. Because of the flattening effects of micro-EHL on the asperities of the surface topography, thin film lubrication will see increasingly wider applications.

6.2.1 Phenomena of Thin Film Lubrication

Luo Jianbin studied the performances and mechanisms of nanoscale thin film with an optical interference device based on the relative intensity method of point contact [8,9]. Figure 6.7 gives three light interference fringe images of the contact area under a load of 6.05 N, temperature 20 °C, the steel ball with a diameter of 20 mm and different entrainment velocities by using 10# lubricating oil. It can be seen from the diagram in the static contact that while the entrainment velocity is zero, the interference fringes are a set of concentric circles. For the dynamic contacts, the necking phenomenon is shown in the outlet of the contact area. The higher the velocity, the more intense the necking shrick is, demonstrating typical characteristic of EHL.

Figure 6.8 shows the thickness measured curves with 13 602 standard solution under the load 4 N, temperature of 25 °C, and ball diameter of 20 mm. For the static contact, the thickness is thin and stable, the film shape is flat and there are no hydrodynamic effect and side leakage phenomena. The lubricant film can be an adsorption molecule film in the ordered arrangement. While entrainment velocity increases

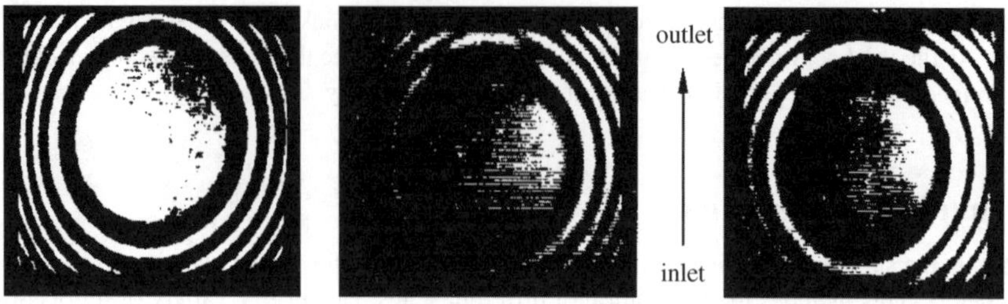

(a) $u = 0$ mm/s (b) $u = 3.12$ mm/s (c) $u = 5.0$ mm/s

Figure 6.7 Interference fringes of the contact area of thin film lubrication.

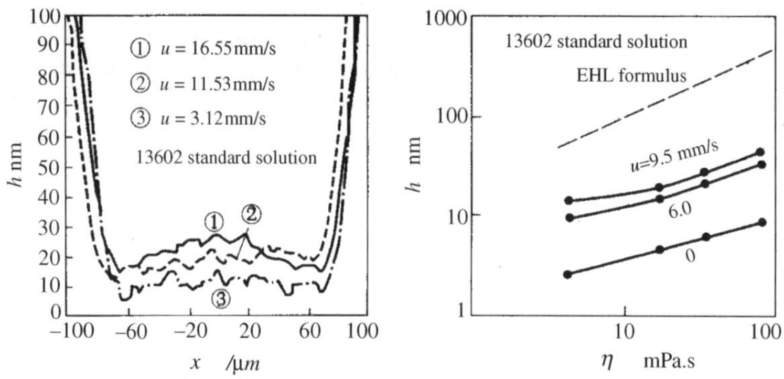

Figure 6.8 Experimental results of thin film lubrication. (a) Central film shape; (b) central thickness via viscosity.

gradually, the film thickness rises. When the velocity is low, the thickness is smoother and the hydrodynamic pressure effect is weaker. This shows that most films also belong to an adsorption film.

With analysis of Figure 6.9, Luo Jianbin proposed a physical model of thin film lubrication [9]. He believed that the sub-microns or nanoscale structures and properties of the lubricant film consist of three different films, the adsorption film, the orderly liquid film and the viscous fluid film. The main points of the model include: (1) Close to the solid surface, the film consists of two parts. One part is the adsorption film formed during the static contact. The other is the orderly liquid film formed during lubrication. This part is more regular and compact due to shearing. The total thickness of the adsorption film is of several

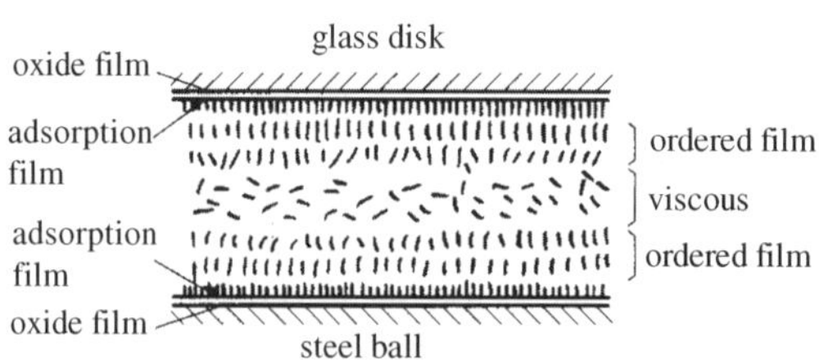

Figure 6.9 Thin film lubrication model.

molecular layers of the lubricant, connected firmly with the surface. Because it is not of the property of a liquid, the adsorption film does not flow. The absorption film displays boundary lubrication characteristics so it can also be called the boundary lubrication film. (2) In the central part, the formation of the lubricant film relies on the hydrodynamic effect. Because it is of EHL characteristics, it is known as EHL film. (3) The orderly liquid film is between the viscous fluid film and the adsorption film. It is promoted to form because the friction shears the liquid molecules and the surface energy exists. Under normal circumstances, the thin film lubrication has a thickness of a few dozen molecular layers.

6.2.2 Time Effect of Thin Film Lubrication

The authors observed that the lubricant film thickness gradually rises to a stable value after a continuous shearing process in thin film lubrication conditions [3,11]. The thixotropic property of a lubricant cannot be used to explain the time effect of thin film lubrication, because the thixotropic effect gradually reduces the thickness to achieve stability.

The experimental results show that the time effect has relationships with the load, the entrainment velocity and the viscosity of the lubricant.

Figure 6.10 shows that under the conditions of the steel ball diameter $d = 20$ mm, temperature $T = 27\,°C$, the entrainment velocity $u = 4.49$ mm/s, the load $W = 4$ N and 7 N respectively, the central lubrication film thickness of liquid paraffin varies with the run-time. By comparison, it can be seen that when the load is 4 N and other parameters remain unchanged, the film thickness increases about 6 nm after 80 min of continuous running. The thickness is 7 nm after stopping so it increases 2 nm more than that of the beginning. When the load is increased to 7 N, the film thickness increment is more than 13 nm after 70 min running and 7 nm after stopping. Thus, the load increases the film thickness increment.

Further experimental studies showed that the time effect of the entrainment velocity on the lubricant film thickness is more complex. The lower the entrainment velocity, the stronger the time effect is. However, the static contact film thickness does not change with time so there is no time effect, nor is there any time effect with high entrainment velocity. Therefore, we can conclude that the time effect only exists at a certain range of velocity.

Figure 6.11 gives the time effect of different lubricants, with the load $W = 4$ N, the entrainment velocity $u = 3.12$ mm/s, the steel ball diameter $d = 20$ mm, the central film thicknesses of 10#, 30# and 40# mechanical oils, the continuous operation time $t = 70$ min. The film thickness of 10# rises significantly, but that of 40# hardly rises at all.

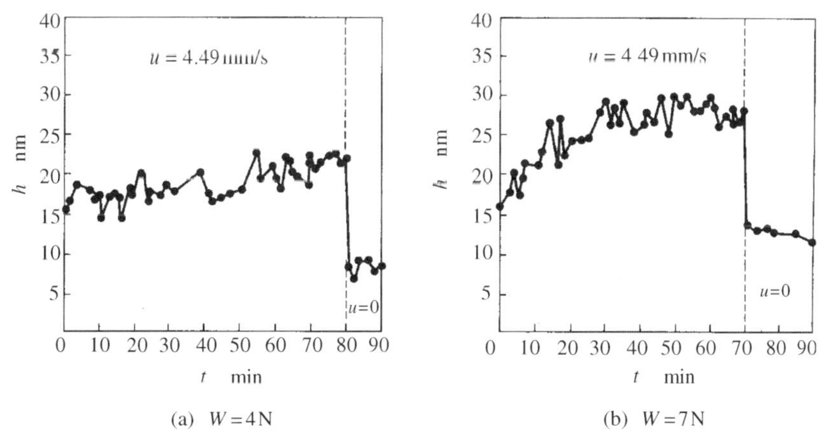

Figure 6.10 Film thickness via running time.

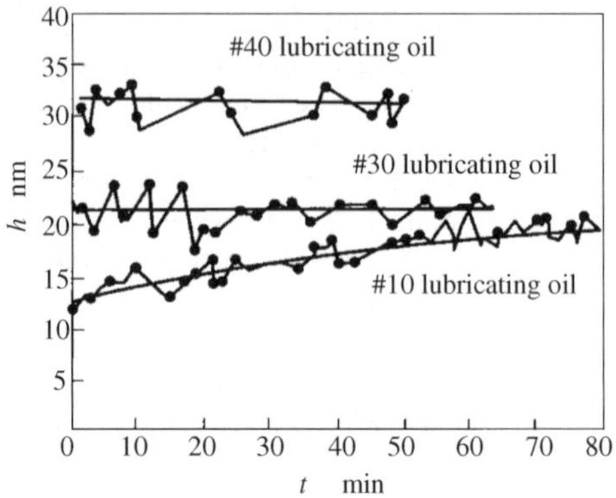

Figure 6.11 Time effect on three lubricants.

Luo Jianbin summarized a large number of experimental results and concluded that the relationship between the load and the time effect is as shown in Figure 6.12 and the relationship between the velocity times viscosity and the time-effect is as shown in Figure 6.13 [9].

The above study shows that in a certain velocity region, the time-effect is stronger for small viscosity and large load. That means the lubricant film thickness will increase as the running time increases. However, according to the theory of hydrodynamic lubrication, these changes of the working condition parameters for viscous fluid are unfavorable factors in reducing the film thickness. So the film-forming mechanism of the thin film lubrication is different from that of hydrodynamic lubrication as the main factors are surface energy and structure of lubricant film molecular ordering.

Experiments show that during operation the increase in thin film thickness slowly drops after the thickness rise to about 30 nm. This thickness stabilized, and is related to the surface force range. This shows that the time effect is related to the action of surface energy in the friction interface.

The thin lubricant film is confined in a narrow gap between the friction surfaces. Because of shearing of molecules, the lubricant film will be restructured. First, the liquid molecules near the surface are arranged vertically to form an adsorption film. Then, the adsorption film passes the force and the adsorption potential to the neighboring molecules, gradually in an orderly arrangement. As the running time increases, the orderly arrangement of the molecules increases so that the film thickness

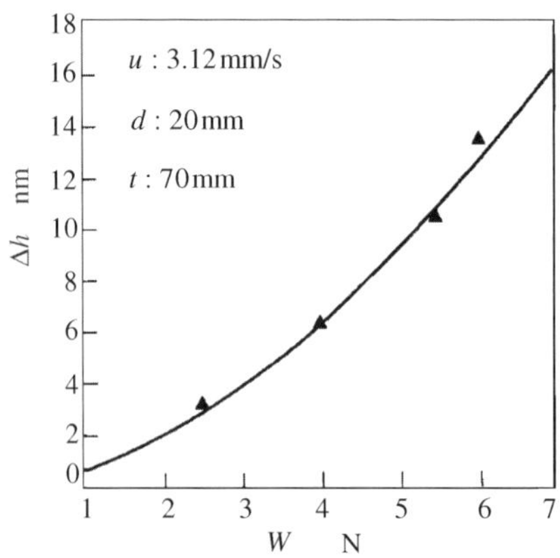

Figure 6.12 Time effect under different loads.

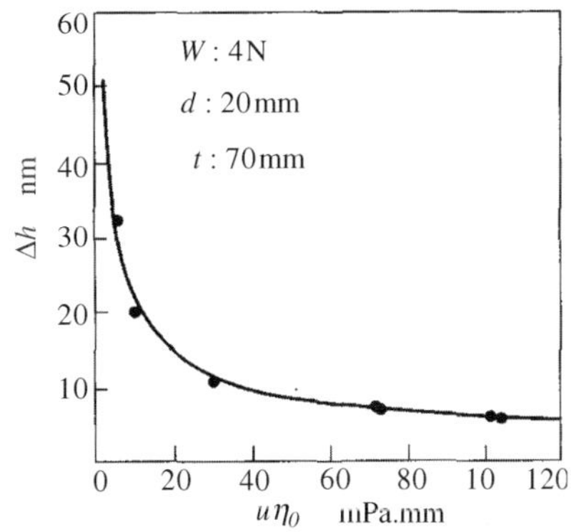

Figure 6.13 Time effect with different $u\ \eta_0$.

increases until it reaches the range where the surface force is balanced. Then, the ordered film thickness becomes stable.

The above conclusions can be verified by the research of Alstern and Granick (1990), and Thompson *et al.* [12]. They used molecular dynamics to simulate the shearing process of molecular behaviors of thin film lubrication, and found that the structure of the original phase will be broken so the molecule phase will be transited or re-crystallized. The spherical molecules being squeezed between the two surfaces are much faster than the unconstrained crystallization of molecules. The chain-like molecules in the liquid state are much longer than the spherical molecules, and it is much easier to form an ordered arrangement. The simulation also showed that high pressure contributes to the occurrence of the phase transition.

The above analysis shows that in the thin film lubrication state, the time effect changes with the molecular structures. With decrease of the lubricant viscosity, increase of the load and decrease of the velocity, the time effect will be strengthened such that the film thickness increases with shearing time and then stabilizes at a certain value. Furthermore, in the static contact the time effect does not appear, and the time effect is connected with the history of shearing.

6.2.3 Shear Strain Rate Effect on Thin Film Lubrication

Streator and Gerhardstein experimentally studied nano-lubricant film properties in different shear rates [4]. Their experiments were completed by a modified computer hard disk. The disk surface was coated with an amorphous carbon film. The arithmetic mean deviation of the surface roughness is $R_a = 3–5$ nm, the head material is Al_2O_3-TiC ceramics, and the contact area is 3.1 mm^2. With four kinds of PFPE as a lubricant, using dip-coating technique to coat one on the surface serves as the hard disk lubricant. When the hard disk rotates, the magnetic head and disk consist of a wedge slider lubrication problem. The performances of four lubricants are listed in Table 6.2.

Table 6.2 PFPE lubrication properties (26 °C)

Lubricant	Kinematic viscosity/mm^2/s	Density/g/cm^3	Surface tension/10^{-5} N/cm
143AZ	29.4	1.9	16
143AY	104	1.9	18
143AX	300	1.9	18
133AD	1030	1.9	19

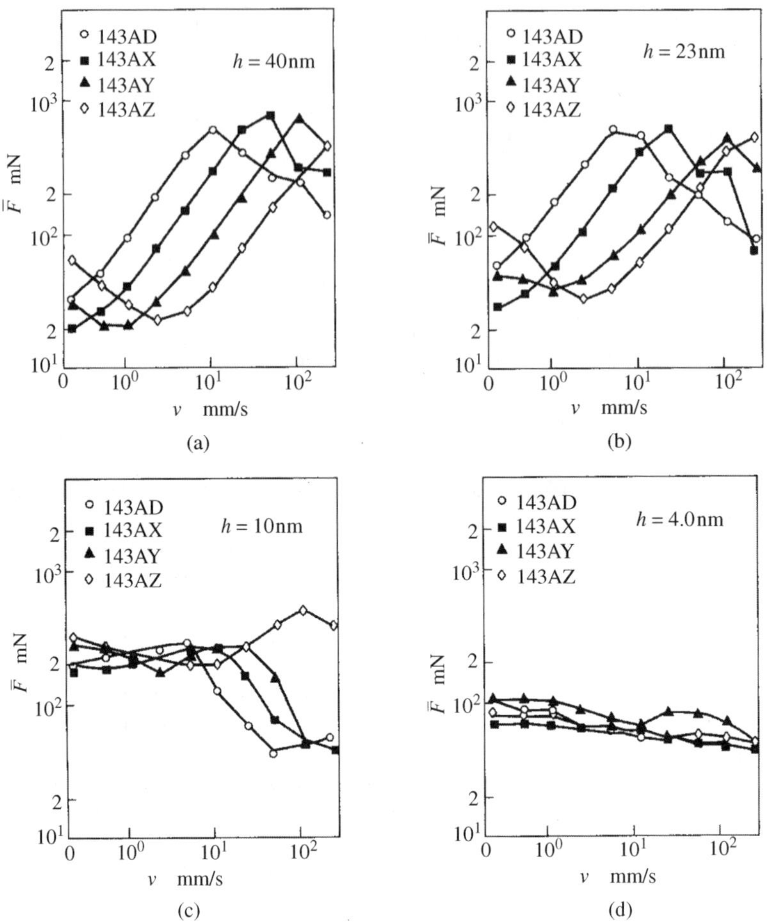

Figure 6.14 Average frictional force via velocity.

Streastor *et al.* measured variation of the frictional force with the sliding velocity under the different thicknesses of 2.3–80 nm with a load of 150 mN, a sliding velocity 0.25–250 mm/s. They found that there are three nano-lubricant film lubrication states, adhesive lubrication, hydrodynamic lubrication and shear-thinning lubrication.

Figure 6.14 shows that when the film thickness h is equal to 40, 23, 10 and 4 nm, the average friction force $\bar{F}$ varies with the sliding velocity under the above conditions. The relationship of the frictional force and the shear rate is the same.

Figure 6.14a is for the film thickness $h = 40$ nm. In the middle velocity, the frictional force varies linearly with the sliding velocity for the four lubricants. The relative frictional force is corresponding to the viscosity values in Table 6.2. It can be seen that in the middle velocity (or shear rate), the lubrication state belongs to hydrodynamic lubrication. Figure 6.14b shows that when the film thickness $h = 23$ nm, the variation of the frictional force is similar to that of the film thickness $h = 40$ nm.

Experiments show that the thickness $h = 23$ nm is probably the minimum thickness for the continuum media mechanics. Below the thickness, no hydrodynamic lubrication state appears, as shown in Figure 6.14c and d. Only for 143AZ with the lowest viscosity, the film thickness $h = 10$ nm belongs to the hydrodynamic lubrication. For the other lubricants, the frictional force decreases with increase of the sliding velocity. Such a feature is not compatible with the boundary lubrication mechanism so it is called the adhesive lubrication. This is because at a low sliding velocity there is a high friction force. For example, when the film thickness $h = 10$ nm, the low-velocity frictional force is as much as 318 mN, which is five times more than the typical frictional force without lubrication, that is 50 mN. The boundary lubrication frictional force is always lower than the solid frictional force of dry lubrication. Furthermore, according to the boundary lubrication mechanism, the frictional force decreases with the boundary film thickness. In Figure 6.14c and d, however, the frictional force of the thickness of 10 nm at low velocities is larger than that of the thickness of 4 nm.

The adhesion lubrication mechanism is much complex. Simply, the acting force is related to Van der Waals force of the surface molecules. The state conversion between the adhesive lubrication and the hydrodynamic lubrication depends on the shear rate in the lubricant film. The shear rate is determined by the film thickness and the sliding velocity.

From Figure 6.14a and b, it can be seen that in the high velocity sliding area, the frictional force decreases with increase of the velocity. This is due to a strong shear-thinning phenomenon because the high shear rate significantly decreases the viscosity of the lubricant and result in friction dropping. For example, for 143AD, when the sliding velocity is 250 mm/s, the frictional force is 130 mN, corresponding to the viscosity being only 6.7×10^{-3} Pa·s, which is much smaller than its normal viscosity 1.95 Pa·s, at almost 300 times lower. This is called the shear-thinning lubrication.

It should be noted that the shear rate the lubricant film withstands is actually the flow velocity gradient across the thickness. Because the film thickness is at the nanometer scale, the small sliding velocity can achieve a high shear rate, as high as 10^7–$10^8\ s^{-1}$. Thus, the shear-thinning effect cannot be ignored in the thin-film lubrication.

6.3 Analysis of Thin Film Lubrication

6.3.1 Difficulties in Numerical Analysis of Thin Film Lubrication

It should be noted that the numerical calculation of sub-micron and nano-thin film lubrication will face many complex phenomena. These include the molecular structure of lubricant film, non-Newtonian properties, thin-film shearing, phase change, the limit shear stress, yield failure, surface roughness, solid particles contained in a lubricant film and so on. Some are not studied yet and are not well understood.

As mentioned above, when a lubricant film is thin to the molecular scale, traditional assumptions no longer apply. It is now widely recognized that the lubricant film thickness should be at least larger than the molecular length of the liquid so that the continuum mechanics can effectively be applied to analyze the lubrication behaviors.

For a molecular smooth surface, the Reynolds equation based on continuum mechanics can be applied to film thicker than 30 nm. Some people believe that for the thin lubricant film, only a few correction factors simply need to be added and thus the Reynolds equation is still able to be used for analysis.

Under high pressure, phase transition of a liquid will take place, that is, solidification. The solidified pressure of a lubricant is related to the lubrication film thickness and the shear rate. Hu Yuanzhong calculated using molecular dynamics simulation that the equivalent viscosity will increase with decrease of a nano liquid film thickness [13]. As the film thickness drops to a certain extent, the liquid will lose the mobility, that is, the liquid-solid phase transition appears, and the phase transition pressure drops with a decrease in film thickness. Granick (1991) pointed out that the continuous shearing movement prevents liquid from solidification. The higher the shear rate, the higher the solidified pressure. This shows that the lubricant film solidification brings great difficulty for numerical analysis on thin film lubrication.

In addition to solidification, the rheology of a lubricant is another important feature. The existence of the limit of shear stress is a kind of rheological property. As shown in Figure 6.15, various rheological

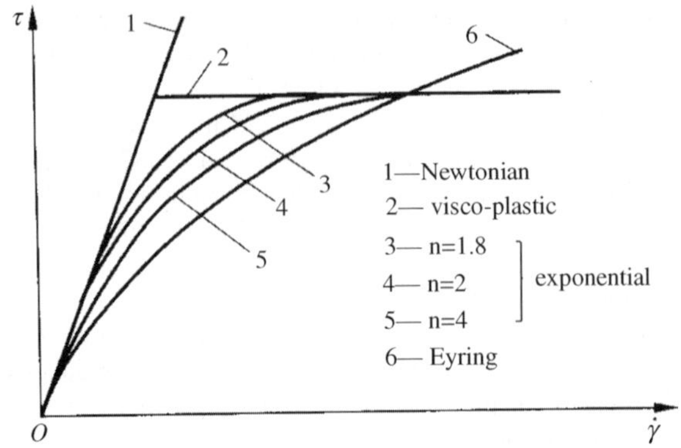

Figure 6.15 Several rheological models.

models proposed for the lubricant film are of the shear stress limit. The visco-plastic limit and the exponential limit are the common two. In the exponential limit models, the different index n is also proposed, usually $n = 1.8$ or 2.

As a rheological property, the limit shear stress should be related to pressure, temperature and so on. Hu Yuanzhong *et al.* showed that the limit shear stress will increase with decrease in film thickness [13]. The flow of thin liquid lubrication film is divided into two regions. In the region near the surface, liquid withstands much high shear stress. As the stress reaches the limit, a slip occurs in the liquid–solid interface.

Huang and Wen studied lubrication failure caused by the limit shear stress [14]. They pointed out that as the shear stress in the lubricant film reaches its limit, the fluid begins to yield and a slip exists inside or in the liquid–solid interface. With increase of the shear rate, the slip zone is gradually expanded, which leads to the loss of the load carrying capacity of lubricant film that cause lubrication failure.

In addition, with decrease of the film thickness, the surface topography becomes very important. For thin film lubrication, whether to use the Reynolds equation or the simplified Navier-Stokes equations to analyze rough surface lubrication is still unclear. Elrod (1973) distinguished two kinds of roughness, they are: Reynolds roughness and Stokes roughness. If l is the rough peak wavelength, h is the film thickness, when $h/l \ll 1$ it is the Renault roughness, suitable for thin film lubrication conditions. At this time Reynolds equation can be applied to calculate and analyze the lubrication problem. When $h/l \gg 1$ the Stokes roughness applies. If so, Navier-Stokes equations must be used. However, some people suggest using Navier-Stokes equations to analyze thin film lubrication of rough surfaces.

A lubricant film contains more or less solid particles, which may also cause difficulties in analysis of thin film lubrication. Experimental results show that even after a fine filtering, the solid particles contained in a lubricant are usually several times larger than the thickness. These solid particles affect the thin film lubrication performances and can even cause surface damage. A lubrication model with solid particles has not yet been established.

In short, although the physical model and numerical analysis of thin film lubrication have gained wide attention because of its complexity, there is limited understanding of this phenomenon and its laws so far.

6.3.2 *Tichy's Thin Film Lubrication Models*

In the 1990s, Dr Tichy of Rensselaer Polytechnic Institute studied thin film lubrication and proposed three physical mathematical models and gave some numerical results for thin film lubrication.

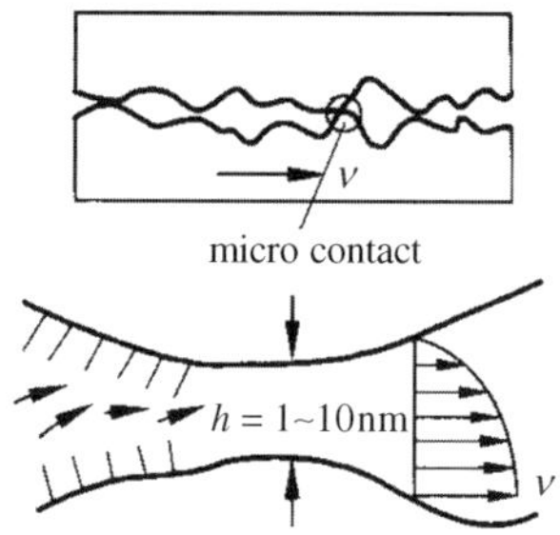

Figure 6.16 Direction factor model.

6.3.2.1 Direction Factor Model

In 1993 in a symposium on tribology at Tsinghua University, Tichy presented the direction factor model and numerical solutions [15]. He believed that thin film lubrication is composed of micro-contact areas, as shown in Figure 6.16. The film thickness of the micro-contact areas is about 1–10 nm, where the lubricant molecular size is about 1nm. The micro-contact structure is made of the polar molecules of the lubricant adsorbed on the surface. Near the solid surface the lubricant layer is of quasi solid characteristics. Far away from the surface, the lubricant still has liquid characteristics. Tichy proposed that the direction factor is a unit vector whose direction follows the direction of molecular orientation. It can represent the structure of the lubricant film. He then proposed three parameters of material to describe the rheological performance, as follows:

1. **Conventional viscosity:** this characterizes the viscous resistance of the liquid part of the lubricant film. Its role in lubrication is the same as common viscosity in conventional lubrication.
2. **Direction viscosity:** this causes a flow resistance due to the direction variation of a direction factor.
3. **Elastic modulus:** in this case a lubricant layer near the surface appears to have the characteristics of quasi solid due to a direction factor attaching to the surface.

With the rheological model of Figure 6.16, Tichy calculated the lubrication problems of a wedge-shape slider. The calculation results show that when the lubricant film was perpendicular to the moving direction of the direction factors, viscous resistance is quite large. When the lubricant film is fixed along the moving direction of the direction factors, there is a significant drop in the viscous resistance. Then, the resistance can be expressed by the common viscosity.

6.3.2.2 Surface Layer Model

Tichy (1995) proposes a surface layer model of thin film lubrication according to the fact that the viscosity near the surface is several orders large than that of the buck lubricant. The model consists of conventional viscosity, the surface layer thickness and the viscosity of the surface layer of the three material parameters. In Tichy's view, the micro-structure of the lubricant molecules near the surface can be regarded as a high rigid viscosity surface layer. There are two kinds of viscosity along the direction of the lubricant film thickness.

If z is the coordinate across the film thickness, h is the film thickness, δ is the surface layer thickness, η_0 is the conventional viscosity, and η_s is the surface layer of viscosity, then

$$\begin{aligned} \eta &= \eta_s \quad z \le \delta \quad \text{or} \quad z \ge h - \delta \\ \eta &= \eta_0 \quad \delta < z < h - \delta, \end{aligned} \tag{6.4}$$

where η_0 and η_s can be measured with a viscometer, while δ value can be calculated based on the lubricant molecular structure and the surface energy.

According to the surface layer model, Tichy derived a modified Reynolds equation and applied it to solve the wedge-shaped slider problem of thin film lubrication. His calculations showed that the surface layer thickness and the surface layer viscosity are the main parameters to affect the performances of lubrication. Their rise will increase the load carrying capacity, but decrease the friction coefficient.

6.3.2.3 Porous Surface Layer Model

Tichy (1995) treated the molecular structure of the lubricant film close to the surface as a layer of porous coating, which is attached to the solid surface. Three material parameters used in the model are the conventional viscosity, the porous layer thickness and the pore-like parameter. Among them, the porous layer thickness can be calculated by the lubricant molecular structure and the surface energy, while the hole-like parameter is measured by the experiments.

According to Darcy law, Tichy determined the properties of the porous layer. He then derived a modified Reynolds equation and applied it to obtain the load carrying capacity, the friction coefficient and the relationship between the porous layer thickness and the pore-like parameter of a wedge-shaped slider lubrication.

It should be noted that the above several models are not perfected due to a lack of large experimental data.

Zhang Chaohui *et al.* proposed a corrected formula of the viscosity for the nano-thin film based on the experiments [16]. Furthermore, his computational results are generally consistent with his experimental results.

6.4 Nano-Gas Film Lubrication

The general equations for gas lubrication are based on compressible gas dynamics equations of the continuous medium. The compressible Reynolds equation is often used to solve the gas lubrication problems. However, in the environment of the near-vacuum conditions, or in the conditions of a very small gap at the room temperature and the atmospheric pressure, the common model based on gas dynamics theory no longer applies.

The conventional hydrodynamic theory cannot explain the phenomena of near-vacuum conditions or a very small clearance flow due to the rarefied gas effect. The thinning effect will appear, such as in micro-electro-mechanical systems, or the magnetic head/disk of a hard disk and so on.

Rarefied gas flows first occur in a high-altitude environment. When the flight of a high-velocity object occurs at high altitude, it is in the rarified conditions. An aerospace is related to the studies of the rarefied gas dynamics and heat transfer. With the development of nano-technology, rarified gas problems exist in many research fields, such as plasma material processing, micro-etching, micro-electro-mechanical systems (MEMS), fine chemicals, vacuum systems and so on. The following discussion focuses on the thin gas lubrication effect in the magnetic hard disk of the computer [17].

6.4.1 Rarefied Gas Effect

In order to describe the deviation extent of gas from the continuous medium, Knudsen defined a dimensionless parameter K_n, that is, Knudsen number. Its expression is

$$K_n = \frac{\lambda}{h}, \tag{6.5}$$

where λ is the molecular mean free path; h is the characteristic size of the gas flow; h is usually a channel height or a tube diameter.

When the free path of gas molecules changes only a little, the smaller the characteristic size h, the larger the Knudsen number K_n is, so that the problem is more deviated from the classical gas lubrication problems. Therefore, the rarified effect exists in a small sized device.

K_n expresses the extent of ratification of gas. The large number denotes significant rarified effect. According to the number K_n, the gas flow region can be divided into the following four regimes.

1. $K_n \leq 0.01$, the flow regime is considered to be a continuous medium region. In this region, the mobile is of no-slip boundary condition of the Navier-Stokes equation.
2. When $0.01 < K_n \leq 0.1$, it is the slip flow regime (slip flow). It is usually assumed that the no-slip boundary condition no longer applies. A mean free path from the bottom of the Knudsen layer began to play a control role between the wall and fluid. In the slip flow regime, flow is under the control of Navier-Stokes equations and the velocity slip and temperature jump conditions are used to establish the rarified effect model through the part of the wall slip.
3. When $0.1 < K_n \leq 10$, it is the transition flow regime (transition flow). In this regime the Navier-Stakes equations are no longer valid. The effect of molecules colliding with each other must be considered.
4. When $K_n > 10$, the fluid is considered to be in free molecular flow (free molecular flow). In the free molecular flow regime, each collision between molecules can be ignored compared with the role of collision between molecules and the boundary.

The inverse Knudsen number is also used afterwards as a rarified effect criterion. The expression of the inverse Knudsen number D is as follows:

$$D = \frac{\sqrt{\pi}}{2K_n}. \tag{6.6}$$

For a nano gas flow, it still has a relatively large Knudsen number even though it is not really thin because the characteristic size is small. When the Knudsen number K_n is between 0.01 and 10, the rarefied gas is neither an absolutely continuous medium nor a free molecular flow. This regime needs further classifying.

6.4.2 Bounardy Slip

6.4.2.1 Slip Flow

Since the Reynolds equation applies only to a large enough gas film thickness, we must consider the gas rarified effect for the sub-micron gap of gas flow. There is a layer of thin stream near the solid surface, called the Knudsen layer, as shown in Figure 6.17.

Although the continuous fluid theory no longer applies to the Knudsen layer, it still applies to other areas beyond the layer. A boundary slip velocity u_s can be introduced to describe the Knudsen layer. The velocity is a function of the shear stress. In the following, we will introduce several commonly used boundary slip models. In these models, the height h used as the characteristic size is

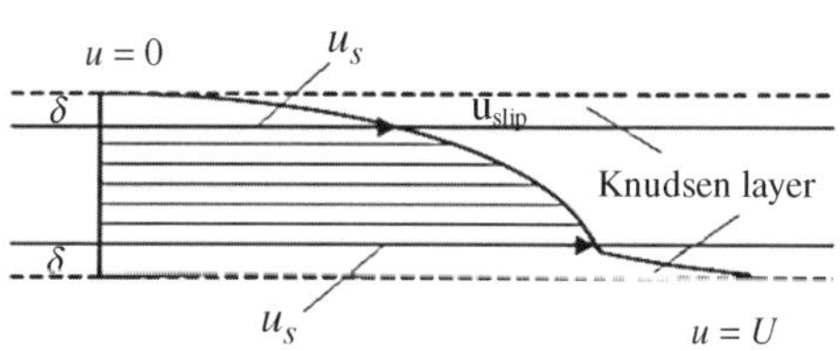

Figure 6.17 Knudsen layer and slip velocity.

the distance of the two parallel plates. The up plane is fixed and the down plate moves with a velocity U.

If K_n is much smaller than 0.01, it belongs to the continues media flow. Therefore, it can be solved by the Navier-Stokes equations. Because the slope of the static plate and the thickness are much smaller than the horizontal scale, the flow velocities are given as follows (see Chapter 2).

$$u = \frac{1}{2\eta}\frac{\partial p}{\partial x}\left(z^2 - zh\right) + (U_h - U_0)\frac{z}{h} + U_0, \tag{6.7}$$

$$v = \frac{1}{2\eta}\cdot\frac{\partial p}{\partial y}\left(z^2 - zh\right) + (V_h - V_0)\frac{z}{h} + V_0. \tag{6.8}$$

Integrate the above equations across the thickness, the no slip flow rates are equal to

$$q_x = -\frac{h^3}{12\eta}\cdot\frac{\partial p}{\partial x} + (U_0 + U_h)\frac{h}{2}$$
$$q_y = -\frac{h^3}{12\eta}\cdot\frac{\partial p}{\partial y} + (V_0 + V_h)\frac{h}{2}, \tag{6.9}$$

The original compressible Reynolds equation should be modified if the rarefied effect is considered. The velocity and flow should be re-derived, because of the presence of the slip.

6.4.2.2 Slip Models

1. The first-order boundary slip model

   In the first-order boundary slip model, assume the velocity boundary conditions of the up and down surfaces are:

$$u|_{z=h} = U_h - a\lambda\frac{\partial u}{\partial z}\bigg|_{z=h}, \quad v|_{z=h} = V_h - a\lambda\frac{\partial v}{\partial z}\bigg|_{z=h}$$
$$u|_{z=0} = U_0 + a\lambda\frac{\partial u}{\partial z}\bigg|_{z=0}, \quad v|_{z=h} = V_0 + a\lambda\frac{\partial v}{\partial z}\bigg|_{z=0}, \tag{6.10}$$

   where a is the surface adapted coefficient, which is used to consider the gas molecules reflecting on the boundary wall after collision.

   Then, integrating the velocities across the thickness direction again, the flow rates of the first-order slip model are:

$$q_x = -Q_1\frac{h^3}{12\eta}\cdot\frac{\partial p}{\partial x} + (U_0 + U_h)\frac{h}{2}$$
$$q_y = -Q_1\frac{h^3}{12\eta}\cdot\frac{\partial p}{\partial y} + (V_0 + V_h)\frac{h}{2}, \tag{6.11}$$

   where the dimensionless flow rate Q_1 is equal to

$$Q_1 = \frac{D}{6} + \frac{a\sqrt{\pi}}{2}. \tag{6.12}$$

2. The second-order boundary slip model
   The velocity boundary conditions of the up and down surfaces for the model are:

$$u|_{z=h} = U_h - a\lambda\frac{\partial u}{\partial z}\bigg|_{z=h} - \frac{1}{2}\lambda^2\frac{\partial^2 u}{\partial z^2}\bigg|_{z=h}, \; v|_{z=h} = V_h - a\lambda\frac{\partial v}{\partial z}\bigg|_{z=h} - \frac{a}{2}\lambda^2\frac{\partial^2 v}{\partial z^2}\bigg|_{z=h}$$
$$u|_{z=0} = U_0 + a\lambda\frac{\partial u}{\partial z}\bigg|_{z=0} - \frac{1}{2}\lambda^2\frac{\partial^2 u}{\partial z^2}\bigg|_{z=0}, \; v|_{z=0} = V_0 + a\lambda\frac{\partial v}{\partial z}\bigg|_{z=0} - \frac{a}{2}\lambda^2\frac{\partial^2 v}{\partial z^2}\bigg|_{z=0}, \quad (6.13)$$

   where the dimensionless flow rate Q_2 is as follows:

$$Q_2 = \frac{D}{6} + \frac{a\sqrt{\pi}}{2} + \frac{a\pi}{4D}. \quad (6.14)$$

3. The one and half (1.5) order boundary slip model
   The velocity boundary conditions of the model are:

$$u|_{z=h} = U_h - a\lambda\frac{\partial u}{\partial z}\bigg|_{z=h} - \frac{a}{2}\left(\frac{2}{3}\lambda\right)^2\frac{\partial^2 u}{\partial z^2}\bigg|_{z=h}, \; v|_{z=h} = V_h - a\lambda\frac{\partial v}{\partial z}\bigg|_{z=h} - \frac{1}{2}\left(\frac{2}{3}\lambda\right)^2\frac{\partial^2 v}{\partial z^2}\bigg|_{z=h}$$
$$u|_{z=0} = U_0 + a\lambda\frac{\partial u}{\partial z}\bigg|_{z=0} - \frac{a}{2}\left(\frac{2}{3}\lambda\right)^2\frac{\partial^2 u}{\partial z^2}\bigg|_{z=0}, \; v|_{z=0} = V_0 + a\lambda\frac{\partial v}{\partial z}\bigg|_{z=0} - \frac{1}{2}\left(\frac{2}{3}\lambda\right)^2\frac{\partial^2 v}{\partial z^2}\bigg|_{z=0}, \quad (6.15)$$

   where the dimensionless flow rate $Q_{1.5}$ is as follows:

$$Q_{1.5} = \frac{D}{6} + \frac{a\sqrt{\pi}}{2} + \frac{a\pi}{9D}. \quad (6.16)$$

4. Liu's first-order and second-order boundary slip models
   Liu derived his first-order and second-order boundary slip models with two-dimensional molecular dynamics simulation [18]. The dimensionless flow rates of Liu are as follows.

$$Q_{1,\,\mathrm{Liu}} = \frac{D}{6} + \left(\frac{2}{3}\right)\frac{a\sqrt{\pi}}{2}, \quad (6.17)$$

$$Q_{2,\,\mathrm{Liu}} = \frac{D}{6} + \left(\frac{2}{3}\right)\frac{a\sqrt{\pi}}{2} + \left(\frac{1}{2}\right)\frac{a\pi}{4D}. \quad (6.18)$$

   Figure 6.18 shows the relationships of the Poiseuille flow rate Q_p with the inverse Knudsen number D for the different models. As can be seen from the figure, in the larger D regime, the curves are close to the curve of the continuum theory. As D decreases, the flows rates Q of all the models are higher than that of the continuous one, but the curves of these models show significant differences. Q of the first-order model is small, Q of the second-order model large; Q of the 1.5 order model is in the middle. Q of Liu's first-order and second-order model are slightly smaller than the original ones.

Boltzmann Equation for Rarified Gas Lubrication

When a gas flow is in the nano transition region, Boltzmann equation should be used to solve the problem [19]. The Boltzmann equation with collision considered is as follows.

$$\frac{\partial f}{\partial t} + v\frac{\partial f}{\partial r} + F\frac{\partial f}{\partial v} = \int_{-\infty}^{+\infty}\int_0^{4\pi}(f'f'_1 - ff_1)g\sigma(g, r)d\Omega dv_1, \quad (6.19)$$

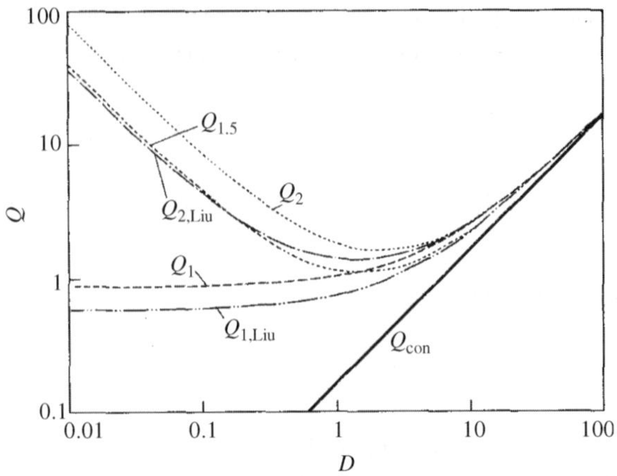

Figure 6.18 Q_p via D for the different models.

where $f = f(r, v, t)$, $f_1 = f(r, v_1, t)$, $f' = f(r, v', t)$, $f_1{}' = f(r, v_1{}', t)$; f gives the number of molecules within the velocity between v to $v + dv$ at time t and in the micro-volume element of r to $r + dr$; v and v_1, are the velocities of the two molecules before collision; v' and v_1' are the velocity after collision; the relative velocities of the two molecules before and after collision are unchanged, but the direction Ω changes; σ is the micro-collision cross-section, which is related to the force model between molecules; and F is the force per unit mass of molecules.

The Boltzmann equation is in the six-dimensional phase space, that is, three positions and three velocities. It gives the conservation relationship between the massive molecules. It is quite difficult to get a solution of such a complex integral equation. Currently only a few analytical solutions can be obtained; one is the Maxwell distribution solution as follows:

$$f = \frac{n}{\left(\frac{2\pi kT}{m}\right)^{2/3}} \exp\left(-\frac{v_x^2 + v_y^2 + v_z^2}{2kT/m}\right). \tag{6.20}$$

6.4.3 Reynolds Equation Considering the Rarefied Gas Effect

Flows between the magnetic head and disk in a computer hard disk significantly show the rarefied gas effect. The space flight height between the head and disk is less than 10 nm nowadays. Here we will discuss the influence of rarefied gas effect on the gas flow.

When consider the rarified effect, a velocity distribution is usually different so the Reynolds equation should be modified. Fukui used the Boltzmann equation to derive a modified compressible Reynolds equation [20].

The inverse Knudsen number D is usually a function of the pressure p and the thickness h. In addition to Equation 6.6, it can also be written as follows:

$$D = D_0 \frac{ph}{p_0 h_0}, \tag{6.21}$$

where D_0 is the characteristic inverse Knudsen number, determined by the minimum thickness h_0 and the ambient pressures p_0, $D_0 = p_0 h_0 / \eta\sqrt{2RT_0}$; R is the gas constant (=287 J/kg.K); T_0 is the ambient temperature, generally taken as 293 K; η is the gas viscosity.

Then, the modified dimensionless Reynolds equation by Fukui can be written as follows:

$$\frac{\partial}{\partial X}\left(PH^3Q\frac{\partial P}{\partial X}\right)+A^2\frac{\partial}{\partial Y}\left(PH^3Q\frac{\partial P}{\partial Y}\right)=\Lambda_x\frac{\partial(PH)}{\partial X}+\Lambda_y\frac{\partial(PH)}{\partial Y}+\sigma\frac{\partial(PH)}{\partial T}, \tag{6.22}$$

where P is the dimensionless pressure, $=p/p_0$, p_0 is the ambient pressure; H is the dimensionless film thickness, $=h/h_0$, h_0 is the minimum film thickness; X, Y are the dimensionless coordinates, $X=x/L$ and $Y=y/B$; A is the ratio of length and head width, $=L/B$; Λ is the bearing number, $\Lambda_x=6\eta UL/p_0/h_0^2$ and $\Lambda_y=6\eta VB/p_0/h_0^2$; η is the gas viscosity; U is the sliding velocity; Q is a dimensionless coefficient of flow; T is the dimensionless time, $T=t/\omega_0$; ω_0 is the characteristic resonance frequency; σ is the squeeze constant, $\sigma=12\eta\omega_0L^2/p_0/h_0^2$.

Fukui and Kaneko also gave the polynomial formulas fitting with the numerical calculation data for the flow coefficient by the inverse Knudsen number D [20]. The formulas are as follows:

$$\begin{aligned}
Q &= \frac{Q_p}{Q_c} \\
Q_c &= \frac{D}{6} \\
Q_p &= \frac{D}{6}+1.0162+\frac{1.0653}{D}-\frac{2.1354}{D^2} && 5\le D \\
Q_p &= 0.13852D+1.25087+\frac{0.15653}{D}-\frac{0.00969}{D^2} && 0.15\le D<5 \\
Q_p &= -2.22919D+2.10673+\frac{0.01653}{D}-\frac{0.0000694}{D^2} && 0.01\le D<0.15
\end{aligned} \tag{6.23}$$

6.4.4 *Calculation of Magnetic Head/Disk of Ultra Thin Gas Lubrication [21]*

The dimensionless Reynolds equation for the steady ultra-thin gas lubrication of the head/disk can be written as:

$$\Lambda_x\frac{\partial(PH)}{\partial X}+A\Lambda_y\frac{\partial(PH)}{\partial Y}=\frac{\partial}{\partial X}\left(PH^3Q\frac{\partial P}{\partial X}\right)+A^2\frac{\partial}{\partial Y}\left(PH^3Q\frac{\partial P}{\partial Y}\right). \tag{6.24}$$

In order to solve Equation 6.24, the following two problems must be solved first.

6.4.4.1 Large Bearing Number Problem

In the ultra thin gas lubrication, the bearing number Λ is very large, usually more than 1 million. This will cause the solution to be severely unstable. In order to solve the large bearing number problem in ultra-thin gas bearing lubrication, we analyzed Equation 6.24 as follows:

1. The bearing consists of two shear flow items, Λ_x and Λ_y. When h_0 is small, they are far greater than the other items. Therefore, if the traditional lubrication calculation is still used, the large number will make the iterative process instable.
2. In the incompressible Reynolds equation, the shear flow does not contain the pressure so that the pressure must be solved from the left hand items of Equation 6.24. However, due to gas compressibility, the right hand items contain pressure p now. This provides a new possibility in solving the equation.

Concerning the ultra-thin gas lubrication for the above two points, by using the upwind scheme, Equation 6.24 can be shown as follows:

$$
\begin{aligned}
&2\Lambda_x\left(P_{i,j}H_{i,j}-P_{i-1,j}H_{i-1,j}\right)/\Delta X_i+2A\Lambda_y\left(P_{i,j}H_{i,j}-P_{i,j-1}H_{i,j-1}\right)/\Delta Y_j\\
&=0.5\left[(QH^3)_{i+1/2,j}(P^2_{i+1,j}-P^2_{i,j})-(QH^3)_{i-1/2,j}(P^2_{i,j}-P^2_{i-1,j})\right]/\Delta X_i^2\\
&+0.5A^2\left[(QH^3)_{i+1/2,j}(P^2_{i,j+1}-P^2_{i,j})-(QH^3)_{i-1/2,j}(P^2_{i,j}-P^2_{i,j-1})\right]/\Delta Y_j^2.
\end{aligned}
\tag{6.25}
$$

After finishing, we have:

$$
\hat{P}_{i,j}=\frac{\left[\begin{array}{l}2\hat{P}_{i-1,j}\Lambda_x H_{i-1,j}/\Delta X_i+0.5\left((\hat{Q}H^3)_{i+1/2,j}\tilde{P}^2_{i+1,j}+(\hat{Q}H^3)_{i-1/2,j}\hat{P}^2_{i-1,j}\right)/\Delta X_i^2\\ +2\hat{P}_{i,j-1}A\Lambda_y H_{i,j-1}/\Delta Y_j+0.5A^2\left((\hat{Q}H^3)_{i,j+1/2}\tilde{P}^2_{i,j+1}+(\hat{Q}H^3)_{i,j-1/2}\hat{P}^2_{i,j-1}\right)/\Delta Y_j^2\end{array}\right]}{\left[\begin{array}{l}2\Lambda_x H_{i,j}/\Delta X_i+2A\Lambda_y H_{i,j}/\Delta Y_j+0.5\tilde{P}_{i,j}\left((\hat{Q}H^3)_{i+1/2,j}+(\hat{Q}H^3)_{i-1/2,j}\right)/\Delta X_i^2\\ +0.5A^2\tilde{P}_{i,j}\left((\hat{Q}H^3)_{i+1/2,j}+(\hat{Q}H^3)_{i-1/2,j}\right)/\Delta Y_j^2\end{array}\right]}.
\tag{6.26}
$$

where the variables with – above are the values before iteration, with ∧ above are the values after iteration.

6.4.4.2 Sudden Step Change Problem

In Chapter 3, we have introduced a method to deal with such cases, that is, the following formula is used:

$$
q_{i+1/2,j}=\frac{2q_{i+1,j}q_{i,j}}{q_{i+1,j}+q_{i,j}},
\tag{6.27}
$$

where $q=PHQ^3$.

The flow coefficient in the y direction has a similar formula as Equation 6.27. It should be noted: if we did not use the above conversion formula, the results obtained may have contained errors.

6.4.4.3 Solution of Ultra Thin Gas Lubrication of Multi-Track Magnetic Heads

In the actual multi-track heads, the main sizes, conditions and structures are given in Table 6.3 and Figure 6.19.

Table 6.3 Size and condition parameters

Parameters	Value	Unit
Head length l	1.235×10^{-3}	m
Head width b	0.7×10^{-3}	m
Minimum film thickness h_{min}	10×10^{-9}	m
Disk linear velocity U	25	m/s
Gas viscosity η	1.8060×10^{-5}	Pa.s
External pressure p_0	1.0135×10^{5}	Pa
Skew angle α_s	12.361	°
Rolling angle α_r	6.14653×10^{-6}	rad
Pitch angle α_p	1.87×10^{-4}	rad

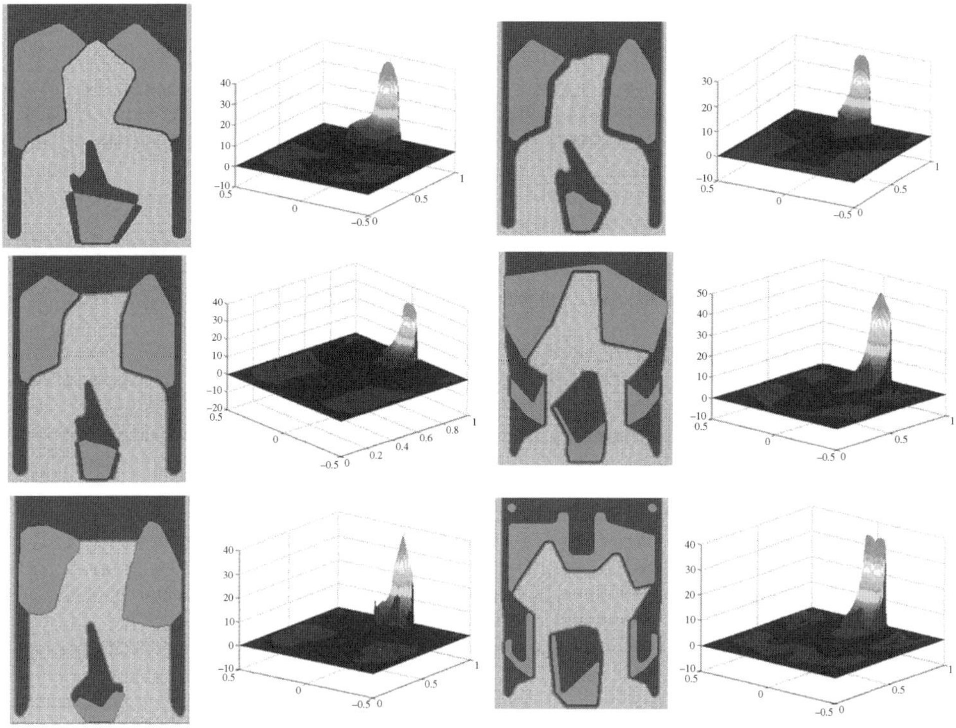

Figure 6.19 Head structures and pressure distributions.

Here, the studied magnetic heads have two steps, 114 nm and 1.77 μm with respect to the top surface of the head. Therefore, the film thickness is expressed as follows:

$$h(x, y) = h_0(x, y) + l(x_N - x)\sin(\alpha_p + \alpha_0) + b(y_M - y)\sin\alpha_r + (x + x_0 - l)(x - x_0)/(7.8125 \times 10^6) + h_{\min}, \tag{6.28}$$

where $h_0(x, y)$ is the original head height (i.e. original film thickness); item 2 and 3 of the right-hand side of Equation 6.28 is for the tilt; item 4 is a coupled arc with the maximum height of 33 nm; and $h_{\min}$ is the minimum film thickness, $h_{\min} = 10$ nm; x_0 and α_0 are the coordinate and initial inclination angle of the point of the minimum thickness; x_N and y_M are the coordinates of the head end.

In addition, the velocity components of the disk are calculated by the following formulas:

$$U_x = U\cos\alpha_s \\ U_y = U\sin\alpha_s. \tag{6.29}$$

By using Equations 6.26 and 6.27 to solve the problems of the large bearing number and the step sudden change, and combined with Equations 6.28 and 6.29, we can finally obtain the pressure of Equation 6.24 and then the other lubrication performances. Figure 6.19 gives the pressure distribution of six kinds of different structures of the actual heads. Compared with the approximate analytical solutions, the two solutions are very close to each other. This indicates that the results are reasonable.

References

1. Dowson, D. (1969) Chapter II, in *Boundary Lubrication – An Appraisal of World Literature*, ASME Press, USA.
2. Johnston, G.J., Wayte, R., and Spikes, H.A. (1991) The measurement and study of very thin lubricant films in concentrated contacts. *Tribology Transaction*, **34** (2), 187–194.
3. Luo, J.B., Wen, S.Z., Huang, P. *et al.* (1995) The characteristics of nano-thin film lubrication study. *Journal of Scientific Instrument*, **16** (1), 240–244.
4. Streator, J. L. and Gerhardstein, J.P. (1993) Lubrication regimes for nanometer-scale lubricant films with capillary effects, in *Thin Films in Tribology* (ed. D. Dowson), Elsevier, pp. 461–470.
5. Wen, S.Z. (1993) On thin film lubrication, in *Proc. 1st International Symposium on Tribology*, **1**, International Academic Publishers, Beijing, pp. 30–37.
6. Wen, S.Z. and Luo, J.B. (2001) Nano-thin film lubrication. *Journal of Tsinghua University*, **41** (4/5), 63–68, 76.
7. Wen, S.Z. (1993) From the elasto-hydrodynamic lubricant film to thin film lubrication - lubrication theory, new area of research. *Lubrication Engineering*, **6**, 48–56.
8. Huang, P., Luo, J.B., and Wen, S.Z. (1994) NGY-2-based nano-film thickness measuring instrument. *Journal of Tribology*, **14** (2), 175–179.
9. Luo, J. B. (1994) Experimental technology and characteristics of thin film lubrication study of, PhD thesis, Tsinghua University.
10. Wen, S.Z. (1992) *Elasto-Hydrodynamic Lubrication*, Tsinghua University Press, Beijing.
11. Wen, S.Z. (1998) *Nano-Tribology*, Tsinghua University Press, Beijing.
12. Thompson, P.A., Robbins, M.O., and Grest, G.S. (1993) Simulations of lubricant behavior at the interface with bearing solids, in *Thin Film in Tribology* (ed. D. Dowson), Elsevier, pp. 347–366.
13. Hu, Y.Z. (1997) Nano-fluid lubricant film of Molecular Dynamics Simulation: I ball molecular liquid simulation results. *Journal of Materials Research*, **11** (2), 131–136.
14. Huang, P. and Wen, S.Z. (1995) Study of visco-plastic fluid lubrication failure – sliding problem. *Progress in Natural Science*, **5** (4), 435–439.
15. Tichy, J.A. (1993) Ultra thin film structured tribology, in *Proc of 1st International Symposium on Tribology*, **1**, International Academic Publishers, Beijing, pp. 48–57.
16. Zhang, C.H. i., Luo, J.B., and Wen, S.Z. (2001) Nano-scale amendments to the viscosity of lubricant film and thin film lubrication calculation. *Journal of Mechanical Engineering*, **37** (1), 42–45.
17. Niu, R.J. (2007) Rarefied gas lubrication numerical simulation and experimental analysis, South China University of Technology doctoral dissertation.
18. Lin, W. (2001) *Physical Modeling and Numerical Simulations of the Slider Air Bearing Problem of Hard Disk Drives*, University of California, Berkley.
19. Bird, G.A. (1976) *Molecular Gas Dynamics*, Clarendon Press, Oxford.
20. Fukui, S. and Kaneko, R. (1988) Analysis of ultrathin gas film lubrication based on linearized Boltzmann equation: first report-derivation of a generalized lubrication equation including thermal creep flow. *Journal of Tribology-Transactions of the ASME*, **11**, 253–262.
21. Huang, P., Xu, L.G., Meng, Y.G., and Wen, S.Z. (2007) Solving head/disk ultra-thin gas film lubrication properties of the effective finite-difference algorithm. *Journal of Mechanical Engineering*, **43** (3), 43–49.

7

Boundary Lubrication and Additives

If hydrodynamic or elastohydrodynamic lubrication cannot be realized, adding lubricant containing some additives to the friction surfaces can also reduce friction and wear. This is boundary lubrication.

Boundary lubrication is important because it is widely applied to the actual machineries and equipment. Even if a machine works under hydrodynamic lubrication conditions, boundary lubrication may still exist, such as during starting, stopping, overloading or because of assembly errors.

Boundary lubrication is not well understood, because:

1. It is very difficult to study and analyze the properties of a very thin boundary lubrication film, even if under laboratory conditions.
2. There are a lot of uncontrollable factors, such as surface properties (geometrical, physical and chemical), the ingredients in lubricant and the environmental conditions (oxygen, temperature, humidity, etc.) that may affect the results.
3. Usually, several different kinds of lubricating mechanisms exist and interact with each other that made boundary lubrication research more complicated.

Because there is no well-accepted theory of boundary lubrication, its applications still depends on experiments and experience.

7.1 Types of Boundary Lubrication

Boundary lubrication generates a film below 0.1 μm on the friction surfaces, referred to as the boundary film. The properties of the film are different from those of the body lubricant. According to structures, boundary films can be divided into adsorption film, chemical reaction film and thicker viscosity film. According to lubrication mechanisms, boundary films are divided into polished film or softening film.

7.1.1 Stribeck Curve

For a journal bearing, the friction coefficient varies with load, speed and viscosity, as shown in Figure 7.1, which is the well-known Stribeck curve that shows the transformation process of lubrication states. In the figure, the abscissa is $S_0 = \eta v/w$, Summerfield number, where η is the viscosity of the lubricant; v is the velocity; w is the load. The ordinate is the friction coefficient f.

Principles of Tribology, First Edition. Wen Shizhu and Huang Ping.

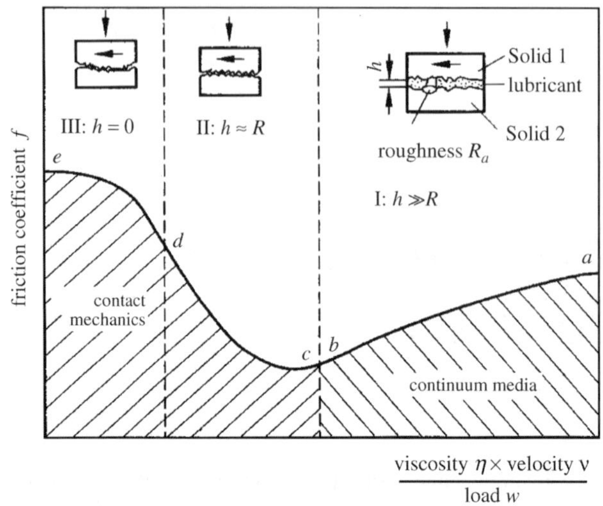

Figure 7.1 Stribeck curve of a journal bearing.

In the above figure, in District I the bearing is in the hydrodynamic lubrication state. The two friction surfaces are separated by a lubricant film. In this region, the film thickness h of the fluid is larger than the surface roughness R_a. The internal friction of the fluid is very small and the friction coefficient increases slightly with increase of S_0. In District II the bearing is in the mixed lubrication state. The film thickness between the friction surfaces is close to the surface roughness R_a. The load is carried by both the roughness and the noncontinuous fluid film. The friction is produced by the lubricant fluid shearing, and the asperity deformating and shearing. With decrease of S_0, the friction coefficient reaches the lowest and then rises again. In District III, the friction caused by the asperity interactions enhances to a stable value. The film thickness reduces to only one or two molecular layers. The friction is mostly determined by the physicochemical characteristics of the film and the surface materials.

It is generally believed that the friction coefficient of boundary lubrication or mixed lubrication consists of two parts: the fluid friction coefficient f_L and the solid friction coefficient f_s. f_L is caused by the shearing flow of the lubricant and is equal to $\eta\ du/dz$, where η is the viscosity of the lubricant; and du/dz is the velocity gradient across the film. In order to separate the two surfaces by the lubricant, the relative sliding velocity must be large enough. f_L increases with increase of the relative sliding velocity. We can see that along the abscissa, the friction coefficient of Curve *ab* in the figure decreases as the sliding velocity v decrease. As v decreases, the film thickness between the friction surfaces is also reduced. The surface contact produces the solid friction f_S. Because the heat generated by the friction reduces the local viscosity of the lubricant, f_L will decline. This makes the total friction coefficient f continuously decrease to the lowest point c. With further decrease of the sliding velocity v, the solid contact area will increase so that the solid friction f_S becomes the main part of the total friction coefficient f. Therefore, f increases along *cd* to arrive at the boundary lubrication state, that is, District III.

7.1.2 Adsorption Films and their Lubrication Mechanisms

7.1.2.1 Adsorption Phenomena and Adsorption Films

Lubricating oils usually contain some polar substances, for example, with 1–2% fatty acids ($C_nH_{2n+1}+COOH$) as shown in Figure 7.2a. The end with -COOH is called a polar group. A fatty acid

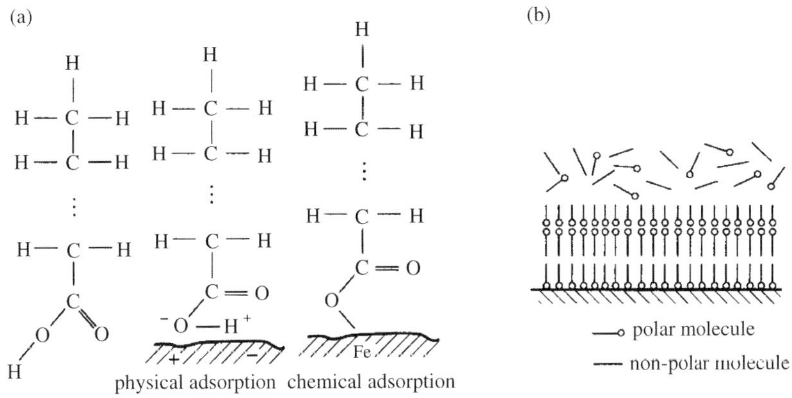

Figure 7.2 Polar molecular structure of fatty acid and adsorption film structure.

is of a long-chain structure. The straight lines are used to represent the nonpolar molecules, and the circles represent the polar groups, as shown in Figure 7.2b. Relying on the Van der Waals force between molecules or atoms, polar groups can be firmly adsorbed on the metal surface to form a monolayer or multi-molecular directional and orderly film. Such an adsorption is referred to as the physical adsorption. Except for some individual peaks of roughness, the adsorption film separates the two friction surfaces, providing a low shear resistance interface so that its friction coefficient is low and the metal surface adhesion can be avoided. Although the long-chain hydrocarbon has a physical adsorption capacity, its physical absorption film is relatively weak. Furthermore, the physical adsorption film can be reversibly desorbed.

As shown in Figure 7.2b, an adsorption film usually consists of three or four molecular layers. Each molecular layer is closely arranged by the cohesion force between molecules so that the molecular film has a load-carrying capacity to be able to separate the two friction surfaces. Friction occurs between neighboring adsorption layers during sliding.

When the temperature is high, the polar molecules adsorbed on the metal surface may form metal soap films, such as C_nH_{2n+1} + COOM. They are also polar molecules to be adsorbed on the metal surface through chemical combinations, called chemical adsorption. The metal ions in a chemical adsorption film do not stray from the original metal lattice, and the lubricant molecules are also retained in their original physical properties. Metal soap film is of a high melting point and thermal stability. A chemical adsorption film is not reversible and its friction coefficient is low. Compared with the physical adsorption film, the chemical adsorption film can work under heavy load, fast speed and high temperature conditions.

The number of adsorbed molecules per area is called the adsorption ratio, which is a measurement of the adsorption capacity. The adsorption ratio increases with increase of the concentration or the adsorbed polar molecule number inside the base fluid. The maximum adsorption capacity is called the saturation quantity. Figure 7.3 gives adsorption curves of several polar molecules. It is shown that the highest adsorption capacity is the stearic acid, and the lowest is the stearic acid acetate. It is also shown that a turning point can be found in the curve of the octadecanol. It is generally believe that the point is caused by the phase transformation.

Experimental results show that at the beginning adsorption is quick. For example, in the first five minutes, about 90% polar molecules may have been adsorbed. However, it will take a very long time to reach the saturation. This is because the adsorption capacity decreases with increase of the distance away from the metal surface. The adsorption of the second layer mainly depends on the adsorption force of the first layer. In saturation, the chemical adsorption film has good lubrication properties. For example, its friction coefficient maintains low and stable. Therefore, in order to get a good lubrication, the adsorption film should have enough adsorbed layers, as shown in Table 7.1.

Table 7.1 Adsorbed layers for good lubrication

	Platinum	Stainless	Silver	Nickel	Mo
Stearate	>10	3	7	3	3
Stearic acid soap	7–9	1	3	3	3

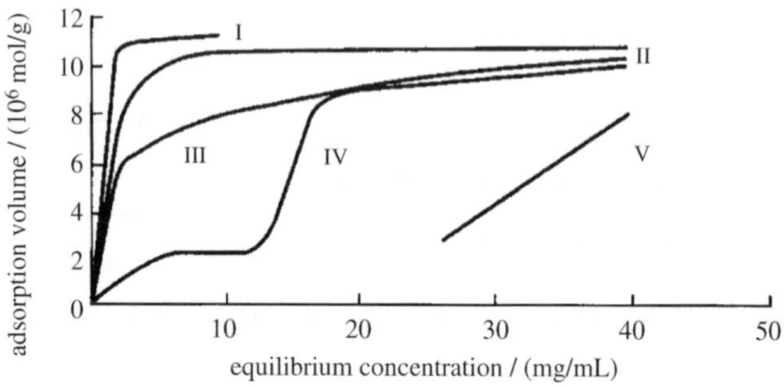

Figure 7.3 Adsorption curves of some polar molecules. I: stearic acid; II: palmitic acid; III: laurate; IV: octadecanol; V: stearic acid acetate.

The adsorption capacity and adsorption intensity of a lubricant on metal is called oiliness. Oiliness is a composite indicator related to the nature and state of lubricants and metal surfaces. Animal oils are of good oiliness, vegetable oils are fairly good, but mineral oils are less oily because they contain few fatty acids, but usually contain some unsaturated hydrocarbons, which also have some absorption capacities. Active metals, such as copper, iron, vanadium, and so on, has a good adsorption capacity, while the adsorption capacity of nickel, chromium, or platinum is poor.

7.1.2.2 Structure and Property of Adsorption Films

The viscosity of a boundary film can be determined by the air flow method as shown in Figure 7.4a. Under the action of an air flow, a liquid in the narrow gap presents a certain shape. According to the shape and Newton's viscosity law, viscosity can be obtained. Figure 7.4b gives some measured results.

For vaseline oil with nonpolar molecules, its curve is a straight line in 40 nm away from the solid wall. This indicates that the viscosity of this nonpolar lubricant is a constant, while for the amyl sebacic acid

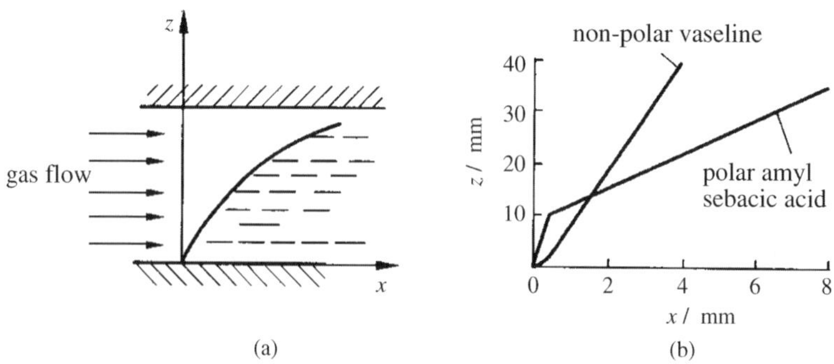

Figure 7.4 Boundary layer viscosity.

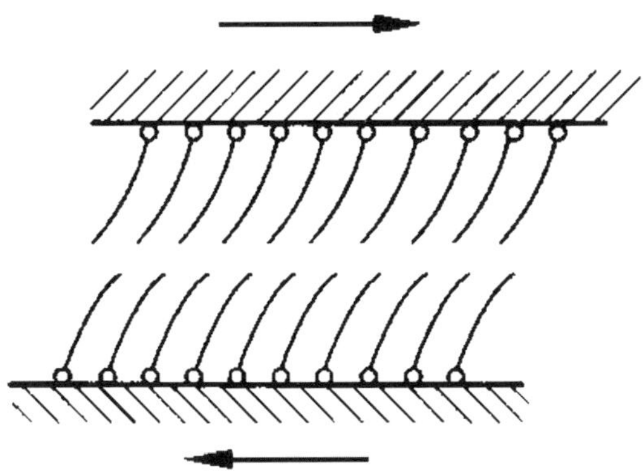

Figure 7.5 Surfaces sliding with an adsorption film.

with polar molecules, there is a turning point at 10 nm away from the wall. Its viscosity is much larger when close to the wall than that away from the wall. This shows that the properties of an adsorption film being of an alignment structure are quite different from the original liquid.

The polar molecules of the adsorption film are parallel to each other and perpendicular to the solid surface. Such an arrangement results in the maximum number of the adsorbed molecules. During sliding, the adsorbed molecules tilt and bend to form molecular brushes in order to reduce the resistance of friction so that the friction coefficient is low and this arrangement also effectively prevents direct contact of the two friction surfaces, as shown in Figure 7.5.

Molecules of a fatty acid are easily adsorbed. However, because the lengths of molecular chains are different, the lubrication abilities are different. The molecular chain of acetic acid is short, but that of the stearic acid is long. Usually, a long molecular chain forms a thicker adsorption film so that the two friction surfaces are separated easily. Usually, the friction coefficient of a boundary film decreases with increase of the chain length of the polar molecules and tends to be more stable. The number of carbon atoms determines the chain length. Therefore, the greater the number of carbon atoms of the polar molecules, the lower is the friction coefficient.

Molecules of an adsorption film are oriented and hierarchically structured to be capable of carrying loads. Because the molecules in the same layer are independent of the other layers so as to form many easily sliding interfaces, therefore, the boundary friction is the outer friction between the layers, and the film thickness presents in a ladder-style, as shown in Figure 7.6.

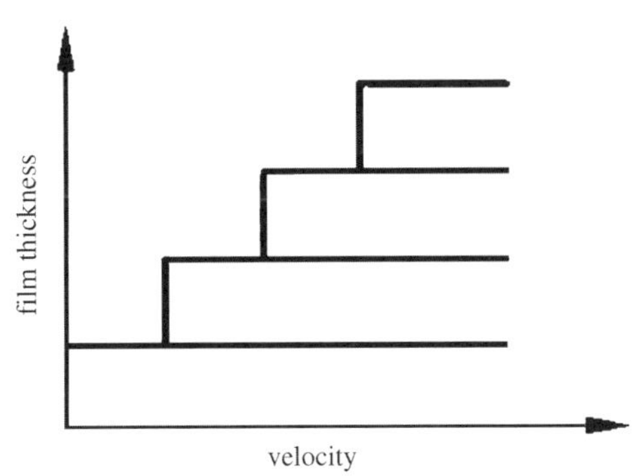

Figure 7.6 Velocity distribution.

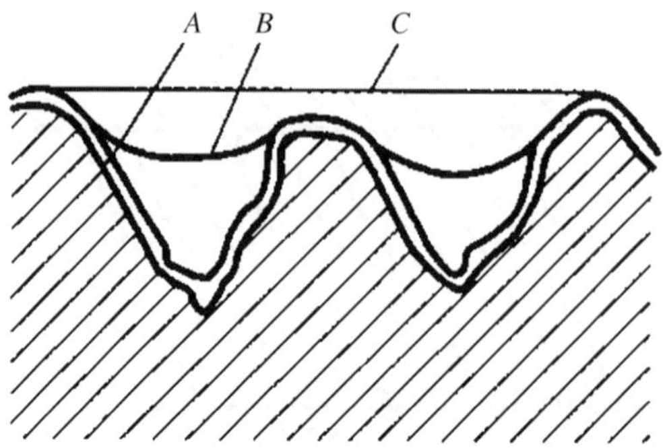

Figure 7.7 Adsorption film forms.

The effect of boundary lubrication is closely related to the quantity of lubricants, as shown in Figure 7.7. The solid surface is covered by an adsorption film which reduces the surface free energy. If lubricant is lacking, the adsorption film on the surface is only with a single molecular layer and the corresponding surface free energy is the minimum, as Shape *A* in Figure 7.7. If lubricant is abundant, the film is as Shape *C* in Figure 7.7. If the quantity of lubricant is between *A* to *C*, the film is Shape *B* in Figure 7.7. At this time, because friction mostly occurs on the peaks, the lubricant has no significant influence on the friction coefficient. During friction, once the adsorption film on the peaks is destroyed, the lubricant in the valley will be rapidly applied to recover the film quickly. When the lubricant is less than *A*, it cannot be applied again. Therefore, if the film on the peaks is destroyed, dry friction occurs. When lubricant exceeds Shape *C*, the friction coefficient will be unstable.

A polar molecular lubricant has two ways to influence friction and wear. One is to form an adsorption film to prevent metals from directly contacting so as to reduce friction and wear. The other way is that when a metal surface has some cracks, the polar molecule will expand the cracks. As shown in Figure 7.8a, in order to form an adsorption film, the polar molecules are forced to enter the crack tip from the outside so as to expand the crack. Such a phenomenon is called the wedge effect.

The wedge effect is also the reason that an adsorption film of a peak can carry a load. In Figure 7.8b, the polar molecules are forced to enter the contact area of two surfaces during adsorption. Because of the wedge effect, the increasing pressure separates the two surfaces away to prevent them from directly contacting.

It is very difficulty to obtain a sufficiently thick lubricant film if the load is too heavy, the impact is too large, starting or stopping is too frequent, the speed is too low, surface is too rough, or in an intermittent swinging operating condition. Under these circumstances, the polar substances can be added into a

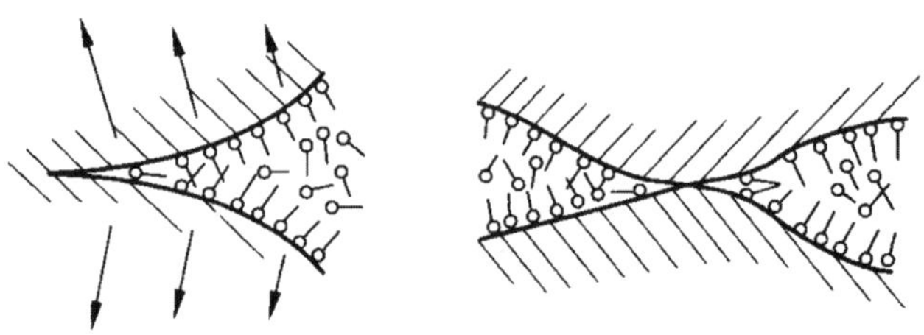

(a) (b)

Figure 7.8 Wedge effect of adsorption film.

lubricant to form an adsorption film on the metal surface to prevent from tear and to reduce wear. For the effectiveness of an adsorption film, the following notes are important.

1. Carefully select the surface materials and the lubricants as well as strictly control the surface roughness;
2. Add some necessary additives into the basic lubricant, because a pure mineral oil is nonpolar and a mineral oil not highly refined contains few fatty acids, only 1–2%. Therefore, their adsorption capacities are poor. Common oily additives include high fatty acids, esters, alcohols and metal soaps, such as oleic acid, dimer acid, aluminum stearate, oleic acid ester, dimer acid glycol monoester, zinc dialkyl dithio phosphate (ZDDP), formic acid phosphate ester, and chlorinated paraffins, and so on. Oily additive is usually added less than 10%. In addition, although the animal and vegetable oils have good adsorption capacities, they are easily oxidized.
3. Strictly control the working conditions. An adsorption film usually exists below 200 °C and under the slow relative velocity. In the light and middle load conditions, an adsorption film can effectively reduce the adhesive, fretting and oxidized wears of the surfaces, but its effectiveness for the abrasive wear depends a lot on the sizes of the abrasive particles.

7.1.3 Chemical Reaction Film and its Lubrication Mechanism

Under high speed, heavy load and middle temperature conditions, the ingredients in the extreme pressure additives, such as sulfur, phosphorus, chlorine, may quickly react with the metal surfaces to generate a thick inorganic film, as shown in Figure 7.9. The film, with a high melting temperature, a low shear strength and a strong linkage with the metal surface, can protect the surface from adhesive wear. During sliding, if the reaction film has been worn out, a new one will be quickly formed to effectively prevent the two friction surfaces from direct contacting.

The chemical reaction films are mainly used to prevent adhesive wear under the high temperature, high speed and heavy load conditions. They are widely used in gear drive lubrication, such as the heavy-loaded gears or worms, so that the dry friction on the pure metal surfaces hardly exists. The most common chemical reaction film is the oxide film. An oxide film can generally reduce friction, but it often causes metal wear.

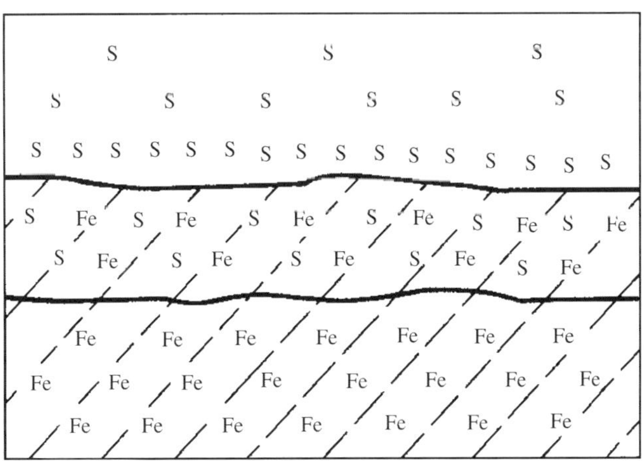

Figure 7.9 Chemical reaction film model.

A chemical reaction film is much more stable than an adsorption film. Its friction coefficient is between 0.10 and 0.25. Good lubrication performances of a reaction film require it to be of certain sufficient thickness. Usually, its thickness is between 1 and 10 nm. It is impossible for a chemical reaction film to be desorbed. However, during a friction process, the chemical reaction film continues to generate after it is worn away so the lubricating performance depend on the balance of generation and wearing. If a new chemical reaction film cannot be generated in time, the lubrication effects will be lost.

The properties of a chemical reaction film also depend on the connecting strength. If a reaction film firmly connects with the metal surfaces, it can protect the surfaces from being worn, otherwise it increases metal corrosion. Furthermore, if the chemical properties of the additives are quite active, it easily corrodes the metal. Therefore, the additives and the usage amount should be selected based on the friction materials and the working conditions.

The main role of a reaction film is to prevent surface from scoring under heavy load, high speed, and high temperature conditions. So, sometimes these additives are called extreme pressure additives. Such additives are corrosive so corrosion wear always exists.

Usually, the critical [*pv*] of a chemical reaction film is high. Therefore, extreme pressure additives can greatly improve the friction surfaces from being seized. If extreme pressure additives are added, the anti-scoring ability of a lubricant usually increases 2–4 times, or even more than 10 times. Therefore, research on the extreme pressure additives has been developing rapidly. In China a variety of extreme pressure additives have also been developed. For example, extreme pressure additives for the middle load industrial gear oils have been standardized and high extreme pressure gear oils have also been serialized.

7.1.3.1 Additives of Chemical Reaction Film

Commonly used additives of the chemical reaction films are:

1. Sulfur compounds including: vulcanized butyl, vulcanized trisobutylene, as well as dibenzyl disulfide, sulfide terpenes, ethylene dixanthate and so on. A sulfide film is of higher load-carrying capacity, higher anti-wear ability, higher hydrolysis stability and higher melting temperature than a chloride film. The organic phosphorus compounds include: tricresyl phosphate, triethyl phosphate (TEP), orthophosphite diethylester and so on.
2. The azotic derivatives of phosphate, ortho-phosphate and phosphate sulfide have developed rapidly as extreme pressure additives, such as phosphate amine salt, phosphate ester amide, dialkyl dithiophosphate amine salt and so on. These compounds not only possess the extreme pressure properties of the sulfur and phosphorus compounds, but also have introduced long-chain polar compounds to enhance properties of oiliness, anti-rust ability and anti-oxidation ability as well. The organic phosphides react with metals to form the low melting alloys to smooth the surfaces so as to reduce friction and wear.
3. Zinc dialkyl dithio phosphate (ZDDP) as a sulfur-phosphorus based extreme pressure additive is widely used in the extreme pressure gear oils and the hydraulic oils, but its water resistance is poor.
4. Borate as an extreme pressure additive has good lubricating performances under no water conditions.

7.1.3.2 Notes for Applications of Extreme Pressure Additives

1. Match the ratio of the extreme pressure additive and oily additives. Generally, the friction coefficient of a base lubricant is high and increases with temperature. The friction coefficient of a lubricant containing some fatty acid (oily additive) below the critical temperature is low. However, while temperature exceeds the critical temperature, friction and wear will sharply increase. Inversely, the friction coefficient of a lubricant containing some extreme pressure additives is relatively high below the chemical reaction temperature, but becomes low above reaction temperature. With a good ratio of

the oily additive and the extreme pressure additive, its friction coefficient can be kept stably low under both low temperature and high temperature.
2. Match the materials of the metal surfaces with the additives. The extreme pressure additives are corrosive to metals. The surface material match can avoid occurrence of a severe and progressing corrosion. For example, the copper alloy is easy to break if sulfur exists.
3. Be careful of corrosion of the extreme pressure additives on metals. Without any corrosion control measurement, the amount of an extreme pressure additive should be strictly limited, usually 0.5% or less. However, the usage of an extreme pressure additive currently tends to be higher, sometimes up to 10–20%.

7.1.4 Other Boundary Films and their Lubrication Mechanisms

7.1.4.1 High Viscosity Thick Film

In recent years, a number of organic polymer additives have been developed, which can form high viscosity thick films in a friction process due to chemical interactions between each other. The thickness may be up to 1–50 nm. Such a film is a continuous film with many spherical granules, as shown in Figure 7.10. Due to its high viscosity, the film can produce a hydrodynamic effect to carry a heavy load at a very low sliding speed. The combination of the film and the solid surfaces depends on the Van der Waals force.

This kind of film can reduce the adhesion and furrow simultaneously while under light or middle load and middle temperature.

7.1.4.2 Polishing Thin Film

Some compositions in lubricant reacting with metal surfaces may form a thin chemical reaction film. Its shear strength is low and easy to slide so it possesses an anti-adhesion ability. In addition, during the friction process, as the film is slowly worn out, a new film is formed to make the surfaces gradually smooth. Therefore, it not only polishes the surfaces, but alleviates surface plowing as well.

This type of boundary film is usually applied in medium temperature and medium load working conditions.

7.1.4.3 Surface Softening Effect

Some special chemical substances in lubricant may soften the metal surfaces, called the Rehbinder effect. Because furrows are reduced, the friction coefficient decreases. The lubricating mechanism of such an effect has not been well understood yet, but it may be important for precisely machining metal surfaces. Its applications in engineering need studying in depth.

Furthermore, the oxide film naturally forming on the pure metal surface can significantly reduce the dry friction coefficient, and its mechanism is the same as the chemical reaction film.

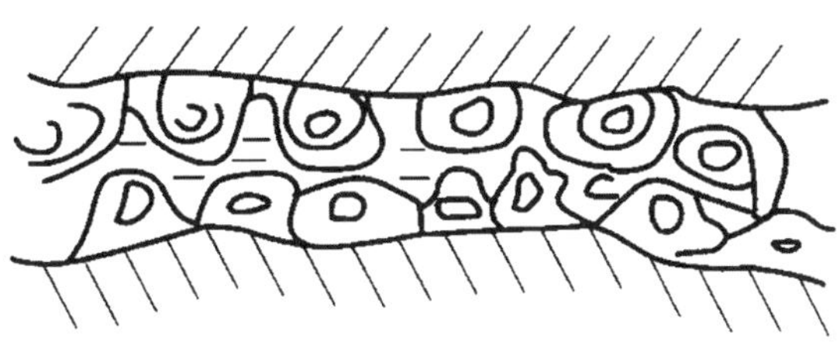

Figure 7.10 High viscosity thick film.

7.2 Theory of Boundary Lubrication

7.2.1 Boundary Lubrication Model

Generally, engineering surfaces are rough. Boundary lubrication is a very important form of lubrication for the rough surfaces. Figure 7.11 gives a model of boundary lubrication.

When two friction surfaces are loaded, because the pressure on some high asperities is large such as at *A* of Figure 7.11, the boundary film will break and the two solid surfaces will directly contact with each other, while at *B* of Figure 7.11, a boundary lubrication film exists. At *C* of Figure 7.11, however, an oil cavity is formed between two peaks without surface contact. And a thick lubricant film for relative motion is generated by the hydrodynamic pressure to carry part of the load at *S* of the figure. If the contact area is *A*, the frictional force F is expressed as follows:

$$F = A[\alpha_S\tau_S + (1-\alpha_S)\tau_L] + F_F, \tag{7.1}$$

where α_S is the percentage of a solid contact area A_S to the total area A, $\alpha_S = A_S/A$; τ_S and τ_L are the shear strengths of the solid and the liquid. F_F is the resistance force of furrows.

The total load is equal to:

$$W = A[\alpha_S p_0 + (1-\alpha_S)p_L], \tag{7.2}$$

where p_0 is the plastic pressure of the lower hardness metal; p_L is the lubricating pressure. If $\bar{p}$ is the averaged pressure, Equation 7.2 becomes

$$W = A\bar{p}. \tag{7.3}$$

Divide the above equation by Equation 7.1, we obtain the friction coefficient of boundary lubrication f_B as follows:

$$f_B = F/W = \alpha_S\left(\frac{\tau_S}{\bar{p}}\right) + (1-\alpha_S)\frac{\tau_L}{\bar{p}} + f_F, \tag{7.4}$$

where $f_F = F_F/\bar{p}A$.

Usually, f_F is relatively small and can be ignored.

Under the dry friction, the load *W* and the frictional force *F* are given as follows:

$$W = Ap_0, \tag{7.5}$$

$$F = A\tau_S + F_F, \tag{7.6}$$

where $p_0 = \bar{p}$.

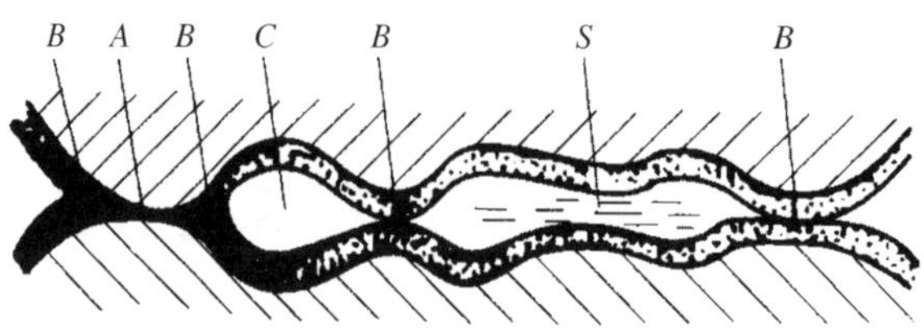

Figure 7.11 Boundary lubrication model.

If we ignore F_F, the friction coefficient f_D of dry friction is:

$$f_D = \tau_S/\bar{p}. \tag{7.7}$$

Ignoring f_F in Equation 7.4 and dividing it by Equation 7.7, we have:

$$f_{BL}/f_D = \alpha_S + (1-\alpha_S)\frac{\tau_L}{\tau_S}. \tag{7.8}$$

Since $\tau_L \ll \tau_S$, the second term in the right hand of Equation 7.8 can also be ignored. Therefore, we have:

$$f_B/f_D = \alpha_S. \tag{7.9}$$

Usually, $\alpha_S \ll 1$, that is, the friction coefficient of boundary lubrication is much smaller than that of dry friction.

Substituting Equations 7.7 to 7.4, we finally obtain:

$$f_B = \alpha_S f_D + (1-\alpha_S)\frac{\tau_L}{\bar{p}} + f_F. \tag{7.10}$$

For a good boundary lubrication, α_S is usually smaller than 0.01 or 0.001, or even less. Then, Equation 7.10 can be approximately expressed as:

$$f_{BL} = \frac{\tau_l}{\bar{p}} + f_F. \tag{7.11}$$

7.2.2 *Factors Influencing Performances of Boundary Film*

In order to understand the mechanism of boundary lubrication further, wetting and contact angle of lubricants are used to determine lubricating performances, and the critical temperature is calculated to determine the boundary lubrication failure. Although the theoretical analysis is not perfect, it plays an important role in the development of boundary lubrication study.

7.2.2.1 Internal Pressure Caused by Surface Tension

In Chapter 1 the concepts, measurement and calculation of the lubricant wetting and contact angle are described. It is generally believed that the wettability of a lubricant has an important influence on boundary lubrication. The adhesive energy is closely related to the surface wettability which forms an adsorption film on the surface. Under the usual circumstances, wetting can be determined through measuring the contact angle of lubricant on a solid surface. The wettability of a lubricant is of important significance for studying the phenomena of adsorption.

When there is a lubricant droplet between two parallel solid plates, its edges will form a meniscus. If the internal pressure is less than the ambient pressure, the concave meniscus forms (see Figure 7.12a), otherwise, the convex meniscus forms. The pressure difference inside and outside the meniscus is called the capillary pressure or Laplace pressure, which may be positive or negative, depending on whether the internal pressure is larger than the ambient pressure or not.

According to Laplace equation, the force between two parallel plates induced by the capillary pressure is as follows:

$$F_L = \frac{\pi R^2 \gamma_{\mathrm{lg}}(\cos\theta_1 + \cos\theta_2)}{h}, \tag{7.12}$$

where θ_1 and θ_2, are the contact angles of the lubricant on the upper and lower surfaces.

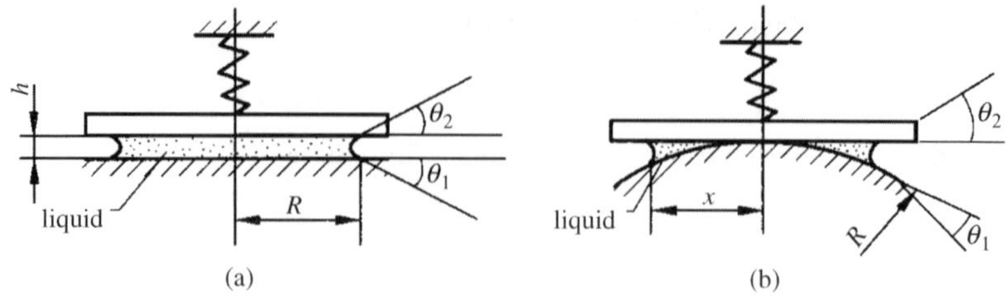

Figure 7.12 Meniscus force. (a) Flat-flat; (b) flat-sphere.

It can be seen from the cross sections that with increase of the surface radius R and decrease of the gap h, the meniscus force increases.

If the spherical surface contacts with a plane and the lubricant exists around the contact point (see Figure 7.12b), the meniscus force is as follows:

$$F_L = 2\pi R\gamma_{\rm lg}(\cos\theta_1 + \cos\theta_2). \tag{7.13}$$

When $\theta_1 = \theta_2 = \theta$, we have:

$$F_L = 4\pi R\gamma_{\rm lg}\cos\theta. \tag{7.14}$$

It can be seen that the meniscus force is related to the spherical radius R, the liquid surface tension $\gamma_{\rm lg}$ and the contact angle θ, but is not related to the lubricant volume. In the micro-scale boundary lubrication, the meniscus force is often important and cannot be ignored.

7.2.2.2 Adsorption Heat of Boundary Film

From the boundary lubrication model we can see that the actual contact area of the friction surfaces is composed of the direct contact area A_d and the boundary film area A_b. Set $\alpha \equiv A_d/(A_d + A_b)$, which is called the relative oil film deficiency, indicating the probability of occurrence of wear and tear of mixed lubrication, because wear and tear occur in direct contact.

According to Bowden and Tabor analysis, the relationship between friction coefficient of mixed lubrication f, the dry friction coefficient f_d and the friction coefficient of boundary lubrication f_b can be written as follows [1]:

$$f = \alpha f_d + (1-\alpha)f_b, \tag{7.15}$$

where

$$\alpha = \frac{f - f_b}{f_d - f_b} \quad \text{or} \quad 1 - \alpha = \frac{f_d - f}{f_d - f_b}. \tag{7.16}$$

Kingsbury proposes a formula to calculate α according to times ratio as follows [2]:

$$1 - \alpha = \exp\left(-\frac{t_x}{t_r}\right), \tag{7.17}$$

where t_x is the time of friction surface passing through the contact length x with the sliding velocity U_s, $t_x = x/U_s$; t_r is the average time for the adsorbed molecules occupying contact surface. Frenkel proposed the following equation to calculate t_r [3]:

$$t_r = t_0 \exp\left(\frac{E}{RT_s}\right), \tag{7.18}$$

where t_0 is the thermal vibration periodic time of the polar molecule perpendicular to the surface; E is the adsorption heat, which is a fundamental nature of a lubricant, cal/mol; R is the gas constant; and T_s is the surface contact temperature, K.

Substituting the above expression into Equation 7.17, we have

$$1 - \alpha = \exp\left\{-\left[\left(\frac{x}{U_s t_0}\right)\right]\exp\left(-\frac{E}{RT_s}\right)\right\} \tag{7.19}$$

or

$$\ln\left[\ln\left(\frac{1}{1-\alpha}\right)\right]^{-1} = \ln\left(\frac{U_s t_0}{x}\right) + \frac{E}{RT_s}. \tag{7.20}$$

From Equations 7.19 or 7.20 we can see that the greater the heat of adsorption of lubricant, the greater the proportion of the boundary lubricant film is, therefore, the better the effect of boundary lubrication.

According to the result given by Lindemann: [4]

$$t_0 = 4.75 \times 10^{-13}\left(\frac{MV^{2/3}}{T_m}\right)^{1/2}, \tag{7.21}$$

where M is the grams per mole of lubricant; V is the molar volume, cm^3; T_m is the melting temperature of the lubricant, K.

In 1971, Hansen and Beerbower proposed: [5]

$$x = 1.46 \times 10^{-8}\, V^{1/3} \tag{7.22}$$

And he suggested selecting $T_m = 0.4T_r$. Here, T_r is the critical temperature of the lubricant, which is the absolute temperature, above which scratches will occur. Substituting t_0, x, and T_m into Equation 7.20, it becomes:

$$\ln\left[\ln\left(\frac{1}{1\quad\alpha}\right)\right]^{-1} = \ln\left[5.15 \times 10^{-5}\, U_s\left(\frac{M}{T_r}\right)^{1/2}\right] + \frac{E}{RT_s}, \tag{7.23}$$

or

$$E = RT_s\left\{\ln\left[\ln\left(\frac{1}{1-\alpha}\right)\right]^{-1} \quad \ln\left[5.15 \times 10^{-5}\, U_s\left(\frac{M}{T_r}\right)^{1/2}\right]\right\} \tag{7.24}$$

Equation 7.24 is called the adsorption (desorption) heat equation, which establishes a relationship between the critical temperature, the adsorption or the desorption heat, the friction coefficient and the surface contact temperature.

7.2.2.3 Critical Temperature

Boundary lubrication failure can be analyzed by α-T_r curve of Kingsbury's [6]. The adsorption heat of Equation 7.23 can be used to obtain the critical temperature T_r. Because $0 < \alpha < 1$, $\alpha - T_r$ curve can be plotted.

Table 7.2 Friction coefficients and critical temperatures of some saturated fatty acids

Name	Melting point (°C)	Friction materials	Friction coefficient below the critical temperature	Critical temperature (°C)
Octanoic acid	16	steel–cast iron	0.11	70
		steel–steel	0.13	70
Capric acid	31.5	steel–steel	0.10	46
Dodecanoic acid	44	cast iron-cast iron	0.15–0.20	40–50
		zinc–zinc	0.04	94
		copper–copper	0.10	97
		steel–steel	0.09	84
Myristic acid	58	steel–steel	0.08	87
Hexadecanoic acid	64	steel–steel	0.08	91
		copper–copper		105

Usually, the critical point on the curve can be chosen as the criteria of the boundary lubrication failure. The critical point should be found by determining the maximum curvature on the curve, that is, the point that satisfies that $d^2\alpha/dT_r^2$ has reached the maximum. According to this condition, the critical temperature T_r' of the failure of boundary film should be:

$$T_r' = \frac{3\left(\frac{E}{RT_r'} - 2\right) - \left[5\left(\frac{E}{RT_r'}\right)^2 - 12\left(\frac{E}{RT_r'}\right) + 12\right]^{1/2}}{\frac{6.16}{U_s}\left(\frac{T_m}{M}\right)^{1/2}\left(\frac{E}{RT_r'}\right)\exp\left(-\frac{E}{RT_r'}\right)}. \tag{7.25}$$

An iterative method can be used to obtain T_r' of Equation 7.25.

Akin proposed a simple method to calculate the critical temperature [7]. He set $\alpha = 0.001$ as the failure criterion, and derived the critical temperature from Equation 7.20 as follows:

$$T_r' = \frac{E}{R}\left[\ln\left(\frac{x}{t_0}\right) - \ln U_s + 6.907\right], \tag{7.26}$$

or

$$T_r' = \frac{E}{R}\left\{\ln\left[3.08 \times 10^4\left(\frac{T_m}{M}\right)^{1/2}\right] - \ln U_s + 6.907\right\}. \tag{7.27}$$

Equation 7.27 is the called Akin equation. Theoretically, when the surface contact temperature T_s exceeds the critical temperature T_r', the boundary lubricant film will be broken so that wear and tear will occur severely.

Temperature is an important factor in boundary lubrication. An adsorption film only works in a certain temperature range. If temperature is higher than the critical, the adsorption film will be softened or desorbed to cause lubrication failure. Table 7.2 gives the friction coefficient and the critical temperatures of some saturated fatty acids.

7.2.3 *Strength of Boundary Film*

The ability of a boundary film to prevent failure is called strength. The cause of failure is very complicated. It depends on the strength of boundary film, connection strength with the metal surface, temperature, load and so on. Currently, the critical [pv], the critical temperature and the critical friction coefficient are used to estimate the boundary film strength.

In the boundary lubrication conditions, the sliding velocity v is kept to a constant and the load per unit area p is gradually increased; or p is kept as a constant and v is gradually increased. When the product of p and v reaches the critical value, the temperature, the friction coefficient and wear are dramatically increased. Therefore, the critical value [pv] can be obtained under this working condition.

The critical temperature is a main parameter to measure the strength of boundary lubrication film. When the friction surface temperature is high enough, the adsorbed molecules starts to disorientate and soften, the film begins to be desorbed, and the friction coefficient increases rapidly, but the film still has some lubricating ability. This temperature is called the first critical temperature, as given in Table 7.2. When the temperature rises higher to make the lubricant or grease decompose or aggregate, the boundary film will have completely failed and severe wear and tear occur. This temperature is called the second critical temperature. The second critical temperature of fatty acids is between 150 and 160 °C, while that of the soap lubricants can reach up to 300 °C.

The number of repeating friction times to cause boundary film failure is called the critical friction frequency. Under normal circumstances, critical friction frequency increases with increase of the sliding speed, but decreases with increase of the load and temperature. The longer the polar molecular chain of an adsorption film, the larger is the critical friction frequency.

A reasonable choice of friction materials and lubricant, and deliberately reducing surface roughness, can effectively enhance the strength of a boundary film. The most convenient way is to add some appropriate amount of oiliness or extreme pressure additives into the lubricant.

7.3 Lubricant Additives

In order to improve the performances of a lubricant effectively, we can add a small amount (for example, 1–2%) of additives into the base oil. In the friction process, lubricants, additives and metal friction surfaces chemically react. The chemical reactions are usually classified according to tribo-chemical reaction, frictional oxidation and mechanochemistry. From the chemistry viewpoint, a friction chemical reaction is a special form of chemical reactions. There are two ways for friction to promote a chemical reaction, that is, friction heat and friction surface activation. When the old metal is worn the new metal surface is exposed. The additives react with the active elements of the surface in what is called a mechanochemical reaction. According to the purposes of usage, the types of additives are mainly classified as follows [8].

7.3.1 Oily Additives

Oily additives are of very strong polar long-chain molecules so at room temperature the adsorption film can be formed on the metal surface. Under medium temperature and light load, an oily additive can form a thick high-viscosity film. A good oily additive should possess the polar groups which have a strong absorption energy on the metal surface. Besides, in order to completely separate the friction surfaces and obtain a low friction coefficient, the composition of the polar groups should contain more than 12–14 carbon atoms, as shown in Figure 7.13. For the polar molecular structures, the linear chain of Figure 7.14a is better than that of Figure 7.14b.

The types of oily additives include animal and vegetable oils, fatty acids, dimeric dibasic acids, polymers and organic sulfur compounds.

1. Animal and vegetable oils have good lubricity, adhesion, and emulsification; their disadvantages are of poor oxidation resistance, bad solubility and limited resources.
2. Fatty acids, such as oleic acid and stearic acid, have good lubricity. The disadvantages are poor oily solubility, precipitation or turbidity after long-time storage, and acidic corrosion. In order to

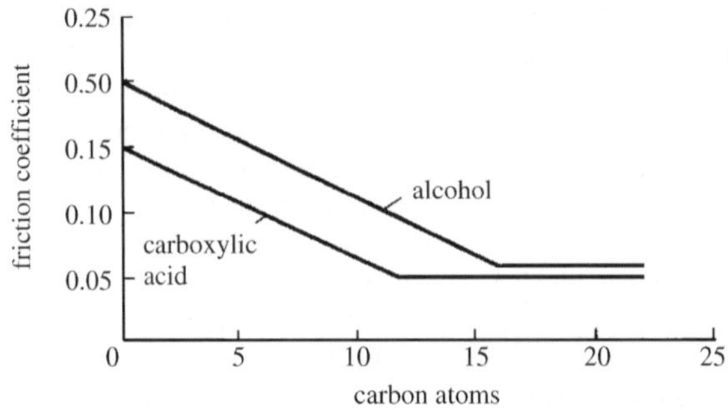

Figure 7.13 Effect of carbon atoms on friction coefficient.

compensate for the disadvantages of fatty acids, fatty acid salts, alcohols, esters and amines are developed as oily additives.

3. Dimeric acids are made from linolic acids with catalysts by thermal polymerization. It has advantages such as good performances under low temperature, noncrystalline, hydrocarbon soluble and so on.
4. Oily additives using high molecular polymers have been developed recently. The mixture of random polypropylene (molecular weight is larger than 20 000) and mineral oil can be used as a rolling oil. By adding 5% random polypropylene, the naphthenic lubricant can be used in rolled aluminum, magnesium and some other metals. Its best advantage is that there is no oil patch. Adding the low molecular weight polyisobutylene into a cutting emulsion can improve its lubricity. Polyethylene is also used as the cold-rolling lubricant additives.

In addition, because liquid crystal has a special structure, it is commonly used for anti-friction additives. Particularly under the low-speed conditions, liquid crystal can obtain a stable and low friction coefficient [9]. It can also be used as additives to prevent the precision guide rails from stick-slip.

7.3.2 Tackifier

Tackifier additives are polymers with a chain structure. At low temperature, their molecules curl like balls with little effect on viscosity. At high temperature, their structure stretches to resist the movement of lubricant molecules such that the viscosity increases. Tackifiers are used in engines to improve the viscosity-temperature characteristic of lubricant and make light lubricating oils tacky.

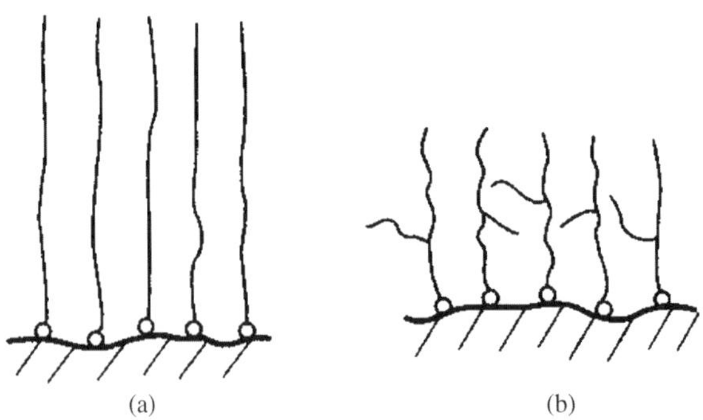

Figure 7.14 Polar molecular structures.

Tackifiers commonly include polymethacrylate gel, polyisobutylene, ethylene/propylene copolymer, styrene/diene copolymer and polyethylene-normalbutyl-ether and so on. Among them, polymethacrylate gel and polyisobutylene are of oil-soluble polymer chains, widely used in hydraulic systems of engines, aircrafts, CNC machine tools and so on.

The characteristics of tackifiers include:

1. Thermal oxidation stability
   Temperature in an engine cylinder is sometimes as high as 270 °C. Under this condition, a polymer tackifier will be decomposed to accumulate carbon. The thermal oxidative stability of a polymer depends largely on its structure. The polymethacrylate will decompose at about 200 °C, while the polyisobutylene will decompose at 250 °C.
2. Hypothermia viscosity and hypothermia starting performance
   The low temperature or hypothermia starting performance of an engine depends largely on the hypothermia viscosity of the tackifier oil, which is based on hypothermia viscosity of the base oil.
3. Hypothermia pumping capacity
   When an engine starts at a low temperature, the oil pressure of the lubrication system must reach the normal state at a very short period to ensure all key parts obtain adequate lubrication in time. The capacity that the oil pump sends lubricant to all the components of the engine is called the pumping capacity. The pumping capacity of the tackifier oil is closely related to its viscosity. In the commonly used tackifiers, the polymethacrylate has a good pumping capability under the low temperature.
4. Shear stability
   When a lubricant circulates in a lubricating system, under the mechanical shear, its viscosity thickening effect will decline because the polymer molecular chains are broken. This causes the viscosity to decrease. The ability to resist the mechanical shear of a tackifier is called the shear stability. Olefin copolymers have higher shear stability than those of polyethylene and polyethylene-normalbutyl-ether.

7.3.3 Extreme Pressure Additives (EP Additives)

Friction in metal under high speed and heavy load usually generates a lot of heat. The adsorption film on common anti-wear additive forms cannot withstand high temperature under heavy load conditions. Usually, such a harsh boundary lubrication is called extreme pressure lubrication. Extreme pressure additives can generate a stronger chemical reaction film to prevent metal surfaces from scratch and sinter.

The extreme pressure additive is an organic compound containing chlorine, sulfur, phosphorus, such as chlorinated paraffins, zinc dialkyl dithiophosphate, and so on. Under heavy load and high temperature conditions, the extreme pressure additive separates out some active elements to produce the low shear strength metal compound film to prevent metal surfaces from being severely worn.

By using an extreme pressure additive, the appropriate conditions should be carefully chosen to be able to form a chemical reaction film. If the extreme pressure additive is too active or its concentration is excessively high, corrosive wear may occur on metal surfaces.

7.3.4 Anti-Wear Additives (AW Additives)

If metal surfaces withstand a medium load in boundary lubrication, the anti-wear additive can be adsorbed on the metal surface or have chemical reactions with the surface metal to form an adsorption film or reaction film to prevent the metal surfaces from being severely worn.

Anti-wear additives are organic compounds containing sulfur and phosphorus. Their function is to form a thin sulfide or phosphating film on the metal surfaces. Some believe that the lubrication mechanism of the anti-wear additives is to form a thick film with a high viscosity. Compared with oily additives, anti-wear additives are capable of withstanding higher temperatures. In addition, the anti-wear additives have a polishing function to smooth a metal surface.

The lubrication performances of anti-wear additives are closely related to the material friction surfaces, known as compatibility. Therefore, it is very important to choose a suitable anti-wear additive [10].

7.3.5 Other Additives

Other types of additives include:

1. Anti-oxidants
   If a lubricant constantly contacts with air, oxidation reaction may occur. An anti-oxidant is used to delay the oxidation process to prolong the life of the lubricant.
2. Pour point depressant additive
   Paraffin in a lubricating oil may easily be solidified at low temperatures to form a mesh structure. The pour point depressant additives can prevent paraffin from forming a mesh structure, thus lowering the solidification temperature of a lubricant and improving its flow property under low temperatures.
3. Anti-foam additive
   If oil is used repeatedly, it may be mixed with air to form many bubbles so as to reduce its lubricity. Anti-foam additives are used to reduce the surface tension of the lubricant to prevent it from forming bubbles.
4. Anti-corrosion additives
   A lubricating oil due to heat is easily oxidized to form peroxides, which may generate acidic products harmful to the metal surface and causing corrosion. Anti-corrosion additives may decompose peroxides to reduce the acidic material, and form a chemical reaction film to protect the metal surface from corrosion.

Sometimes, different additives may have the effect of restraining each other. Therefore, when several additives are used together, one should pay attention to their integral effect. For example, polar substances, such as oily additives and anti-corrosion additives, may cause the extreme pressure additives to fail.

Figure 7.15 shows how to properly use different additives. The effects of different additives are listed in the figure. Friction coefficient is the coordinate and temperature is the abscissa. The friction coefficient of paraffin as a base oil is high and increases with temperature because it is nonpolar. If an oily additive is

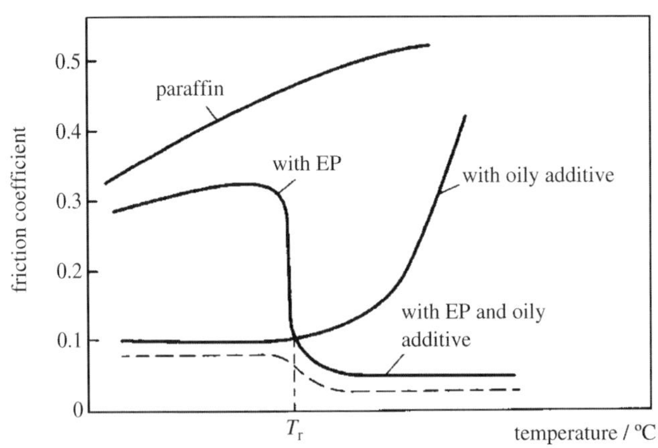

Figure 7.15 A combined effects of additives.

added, it has a good anti-friction effect at low temperature, but the friction coefficient increases quickly after temperature reaches the critical temperature T_r. If an extreme pressure additive is added, the friction coefficient is high at low temperature, but dramatically drops after temperature reaches the critical temperature T_r. By using both additives, we are able to make the friction coefficient stable and low for all temperatures, as shown by the dotted curve in Figure 7.15.

However, to obtain a good combined effect of several additives is often not so easy. There is a so-called temperature gap problem in some gear oils. That is, under medium temperatures, the adsorption film may be broken, but a chemical reaction film has not been formed in time by the extreme pressure additives. Therefore, surface damage will occur because all the additives are void.

References

1. Bowden, F.P. and Tabor, D. (1954) *The Friction and Lubrication of Solid*, Clarendon Press, Oxford.
2. Kingsbury, E.P. (1958) Some aspects of the thermal desorption of a boundary lubricant. *Journal of Applied Physiology*, **29**, 6–11.
3. Frenkel, J. (1924) Theorie der Adsorption und verwandter Erscheinungen. (Theory of adsorption and related phenomena). *Zeitschrift Fur Physik*, **26**, 117–138.
4. Lindemann, F.A. (1910) The calculation of molecular vibration frequencies. *Zeitschrift Fur Physik*, **11**, 609–612.
5. Hansen, C.M. and Beerbower, A. (1971) Solubility parameters, in *Kirk-Othmer Encyclopedia of Chemical*, Technology, Suppl. Vol., 2nd edn(ed. A. Standen), Interscience, New York, 889–910.
6. Kingsbury, E.P. (1960) The heat of adsorption of a boundary lubricant. *ASLE Transactions*, **3**, 30–33.
7. Akin, L.S. (1972) An interdisciplinary lubrication theory for gear. *ASME Transactions, Journal of Engineering for Industry, Series B*, **75** (4), 1178–1195.
8. Yan, Z.G. (ed.) (1999) *New Lubricating Materials and Lubrication Technology Practical Handbook*, National Defense Industry Press, Beijing.
9. Tang, X.Y., Lu, C.F., and Wen, S.Z. (1992) Liquid crystal on mineral oil lubrication characteristics. *Wuhan: Proceedings of the Fifth National Conference on Tribology, the Next Book*, pp. 80–87.
10. Zhou, C.H., Dong, J.X.u., and Wen, S.Z. (1992) Study on friction materials and compatibility of commonly used anti-friction additives. *Journal of Tribology*, **12** (1), 73–80.

8

Lubrication Failure and Mixed Lubrication

Lubrication failure is said to have occurred when surfaces are worn or the load-carrying capacity of the lubricant film is dramatically reduced or even totally lost. The main factors leading to lubrication failure are surface roughness, non-Newtonian behavior of lubricant and interface temperature rise. Although it has been recognized that lubrication failure is caused by these factors, it is not clearly understood how these factors contribute to the mechanism of lubrication failure.

This chapter will also discuss mixed lubrication, which is the most widespread lubrication in actual machineries. As the surface roughness of the friction surfaces affects the thickness of the lubricant film, a variety of lubrication conditions exist at the same time. Furthermore, the roughness changes randomly so mixed lubrication is very complicated and a satisfactory physical model and its related theory are yet to be established.

8.1 Roughness and Viscoelastic Material Effects on Lubrication

8.1.1 Modifications of micro-EHL

The numerical solutions of the conventional micro-EHL show that an EHL film is still able to prevent the solid surface from contacting with each other even if surfaces are very rough. This is because the fluid viscosity and the elastic deformation of solid material significantly increase with rise of pressure. The pressure on roughness peaks generates sufficient deformation to avoid asperity to contact and collide. Therefore, it is impossible to bring about wear. However, conventional EHL theory has used the following two assumptions that might possibly result in solution errors.

1. A lubrication problem of two moving surfaces is generally equivalent to a lubrication problem of a moving and rigid surface with the other stationary and elastic surface.
2. Solid materials are thought as perfectly elastic, that is, their elastic deformation needs no time.

To modify the former assumption, the elastic deformation should be separately added to the two surfaces. The configuration is significantly different from the original one.

The latter hypothesis should be modified to take into account the viscosity of solid materials. Introduction of the solid viscosity will lead to lubricated wear because the deformation cannot occur

Principles of Tribology, First Edition. Wen Shizhu and Huang Ping.

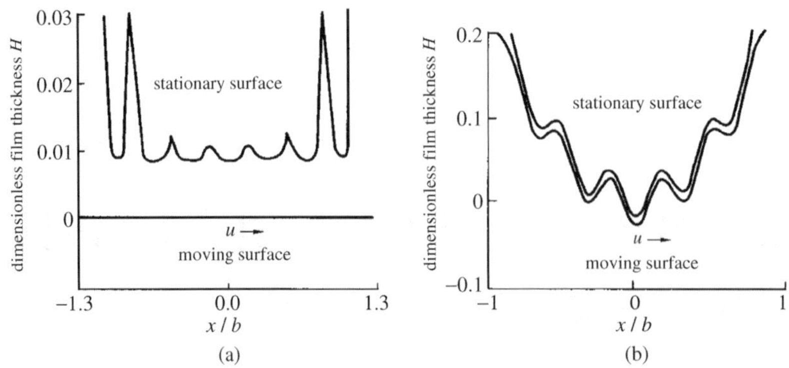

Figure 8.1 Differences between equivalent and actual deformations. (a) Equivalent EHL deformation (b) actual EHL deformation.

as soon as possible anymore. The author suggests the above modifications on the micro-EHL, having obtained the following results, which explain the reasons for lubricated wear [1].

For example, let us discuss a pure sliding micro-EHL problem for a smooth surface sliding beneath a stationary sinusoidal rough surface. We will compare the conventional solution with the solution using the former modification.

Suppose the materials of the two surfaces are the same, that is, $E_1 = E_2$ and $\nu_1 = \nu_2$. The equivalent solution of the film thickness is shown in Figure 8.1a. Although the rough peak amplitude ($\delta = 0.562\ \mu$m) is significantly higher than the average film thickness ($\bar{h} = 0.119\ \mu$m) in the Hertzian area, but the equivalent solution shows that the lubricant film is still thick enough to separate the two surfaces. From Figure 8.1a, it can be seen that because the velocity is parallel to the surface in the equivalent solution, it means that the relative motion cannot cause wear. However, Figure 8.1b gives the cylinder and plane contact situation, where $H = h/h_H$ is the dimensionless film thickness and $h_H = b^2/R$ is the Hertzian width.

If the elastic deformation is allocated to the two surfaces, the configuration of the stationary and moving surfaces determined by the following formula:

$$\text{Stationary surface} \qquad z_1 = h_0 + \frac{x^2}{2} + h\delta(x) + \upsilon_1, \tag{8.1}$$

$$\text{Moving surface} \qquad z_2 = -\upsilon_2, \tag{8.2}$$

where υ_1 and υ_2 are the elastic deformations of the stationary and moving surfaces respectively.

Although the film thickness is still $h = z_1 - z_2$, the lubrication gap region appears to have fluctuations in the shape as shown in Figure 8.1b. Because the two surfaces are no longer parallel to the velocity, the two surfaces embed in each other after deformation. Therefore, lubricated wear may occur even if thick lubricant film exists between the surfaces.

8.1.2 Viscoelastic Model

As mentioned above, if only actual deformation is considered, but the surface material is perfectly elastic, wear still does not occur. Because a load is added, deformation is fully realized as soon as possible.

However, any deformation will take time, which means that solid material has a viscosity. When the deformation time is shorter than the loading time, the material viscosity can be neglected. The loading

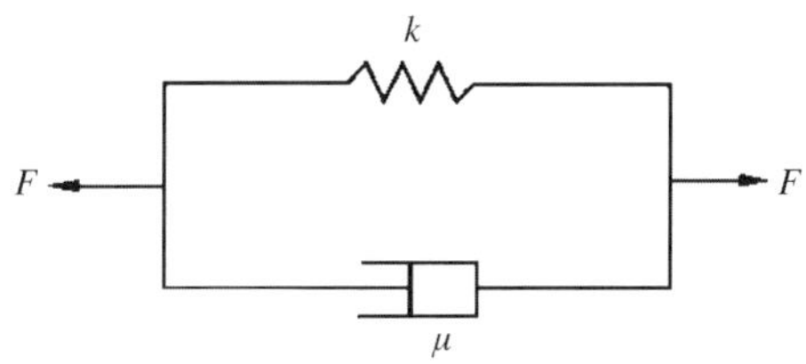

Figure 8.2 Kalvin-Voigt viscoelastic model.

time of the point or line contacts is usually very short. However, when the loading time is the same or less than the deformation time, material viscosity must be considered.

There are a number of viscoelastic models for solid materials. The most common and simplest two are the Maxwell model and the Kalvin-Voigt model. In the Maxwell model, the deformation is considered as the sum of the elastic and the viscous. A load exists with elastic deformation, and deformation tends to infinity with increase of time. The Kelvin-Voigt model is shown in Figure 8.2. Compared with the Maxwell model, it is more suitable for the relationship between deformation and time considered in micro-EHL.

The deformation in Kelvin-Voigt model is controlled together by both elasticity and viscosity. In the initial stages, due to larger velocity, viscosity plays a major role. In later stages, the main factor is the elastic. The total deformation is limited by the elastic. The relationship between the tension strain and time in the Kelvin-Voigt model is as follows:

$$\varepsilon = \varepsilon_0\left(1 - e^{-t/\tau}\right), \tag{8.3}$$

where ε_0 is the total tension strain; t is time; τ is the lag time, $\tau = \eta/G$, η is the viscosity of surface material, and G is the shear elastic modulus.

Although most materials are visco-elastic, it is still very difficult to accurately measure the viscosity of a high elastic material experimentally, because it only appears at the very beginning for a short time. Experiments show that the viscosity of steel is about 1 MPa·s. Because the steel shear modulus is 80 GPa, the lag time of steel is about 1.25×10^{-5} s. It should be pointed out that the experimental data are only obtained from the tensile tests, but the viscosity data of the contact deformation have not been reported.

Equation 8.3 gives the relationship between the strain and time. As the deformation is an integral of the strain, for convenience but without bringing about substantial deviations, a similar formula is used to describe the relationship between the deformation and time as follows:

$$\delta = \delta_0\left(1 - e^{-t/\tau}\right), \tag{8.4}$$

where δ_0 is the total elastic deformation.

If the wavelength of roughness is Δ; the relative sliding velocity of two friction surfaces is u, it takes time of $\tau_1 = \Delta/u$ for a moving peak to pass through a stationary peak, as shown in Figure 8.1b. When $\tau_1 \le \tau$, the deformation will be incomplete in τ_1. In the example of Figure 8.1b, $\tau = 8.62 \times 10^{-6}$ *s*. If the material is steel, the deformation cannot be accomplished in the load time τ. Therefore, the moving peak will collide with the stationary peak, and lubricated wear occurs.

8.1.3 Lubricated Wear

From the above analysis, we know that if an actual peak is much higher than the lubricant film thickness, the two friction surfaces will be embedded in each other. And, if the deformation is not complete, lubricated wear may occur. Based on the above analysis, we proposed a wear model for lubricated wear [1]. If a rough peak slides over a stationary surface with velocity u, and suppose:

1. wear caused by the asperity movement occurs only on the stationary surface;
2. both the moving peak and the stationary surface are viscoelastic;
3. the surface elastic deformation is calculated from Equation 8.4.

In the following, the conditions of lubricated wear are given first. Then, a corresponding model of lubricated wear is developed.

8.1.3.1 Lubricated Wear Criteria

If Δ is the wavelength of roughness and δ_0 is the total deformation, the passing time of a moving peak from the stationary rough starting edge to its center is $t_1 = \Delta/2u$. During this period, the material deformation δ' is equal to:

$$\delta' = \delta_0(1 - e^{-t_1/\tau}). \tag{8.5}$$

If the peak height of roughness after deformation is still higher than the lubricant film thickness h_0, that is, $\delta_0 - \delta' \geq h_0$, lubricated wear will occur. Substitute Equation 8.5 into the inequality, the relationship of five parameters: film thickness, roughness peak height, roughness wavelength, relative sliding velocity and lag time, is as follows:

$$\frac{\Delta}{2u\tau} \leq \ln(\delta_0/h_0). \tag{8.6}$$

8.1.3.2 Lubricated Wear Model

In dry friction, the wear equation of the cone-shaped asperities passing through a plane can be expressed as:

$$\frac{dV}{ds} = h^2 \tan\theta, \tag{8.7}$$

where h is the peak height of roughness; θ is the cone angle; W is the load; H is the material hardness; and k is the factor of wear.

If udt is used instead of the displacement increment ds, the formula of lubricated wear can be written in the following similar one:

$$\frac{dV}{dt} = u[\delta_0 e^{-\Delta/2u\tau} - h_0]^2 \Delta/\delta_0. \tag{8.8}$$

The application of Equation 8.8 is based on the satisfaction of Equation 8.6. Equation 8.8 shows that lubricated wear is related to the lubricant film thickness h_0, the wavelength of roughness Δ, the relative sliding velocity u, the material viscosity η and the lag time τ.

8.1.3.3 Lubricated Wear Example

Here is an example to calculate the lubricated wear with Equation 8.8 [2]. A line contact friction problem between a stationary smooth block and a rotating rough ring is considered. Although wear is a dynamic process, for convenience sake, a quasi steady solution is obtained. A step-by-step method is used to solve the pressure and the film thickness according to the steady EHL, and then used to

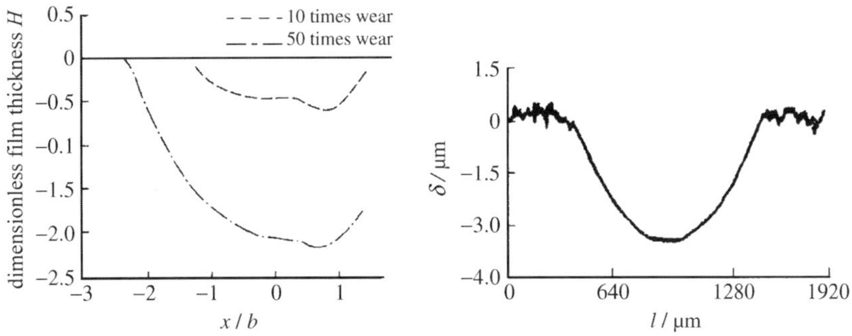

Figure 8.3 Comparison of calculated and experimental worn shapes. (a) Calculated worn surfaces; (b) experimental worn surface.

calculate the next step wear. In addition, assume that when the rough ring slides over the contact area, wear only occurs on the smooth block but not on the surface of the rough ring. The calculation data are: $E = 221$ MPa; $R = 0.02$ m; $u = 1$ m/s; and $\eta_0 = 0.05$ Pa·s.

Calculations show that if the peak height after deformation is 1.1 times of the initial center thickness of the film, the roughness larger than the film will be worn out during the sliding process. Here, the wear volume is obtained after taking into account the elastic deformation. The next step of solving EHL is on the basis of the former worn configuration.

Figure 8.3a shows the worn surface curves after the 10th step and the 25th step. It can be seen from the figure that in the wear process, the lubrication area continues to grow. The initial worn shape is similar to the EHL film thickness with obvious necking phenomenon. But with increase of wear, the contact area becomes a parabola shape. This is similar to the shape of the ring.

Figure 8.3b is the experimental results obtained by using a Timken tester [2]. It can be seen that the worn shape is similar to the calculated results.

In order to study further characteristics of lubricated wear, more experiments are carried out to compare with the calculated results as shown in Figure 8.4 [2]. The results show that the visco-elastic model is feasible to analyze lubricated wear.

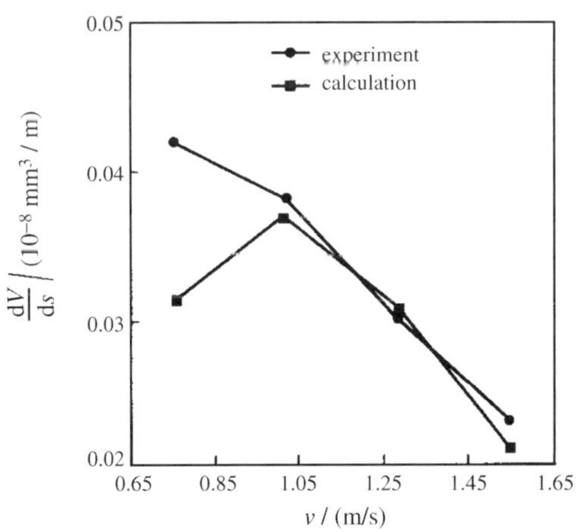

Figure 8.4 Relationship of lubricated wear and sliding velocity.

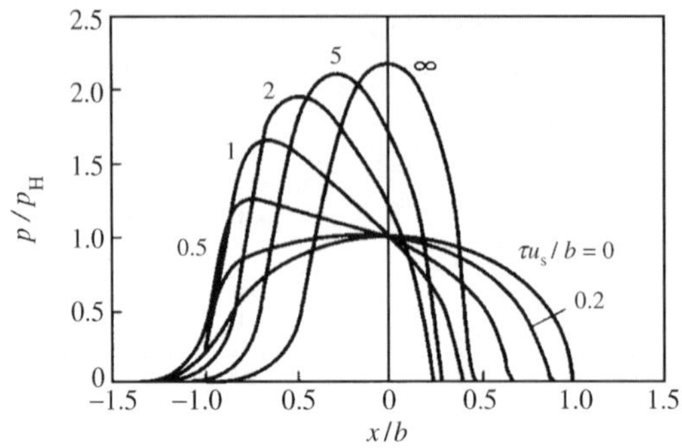

Figure 8.5 Pressure of rubber EHL.

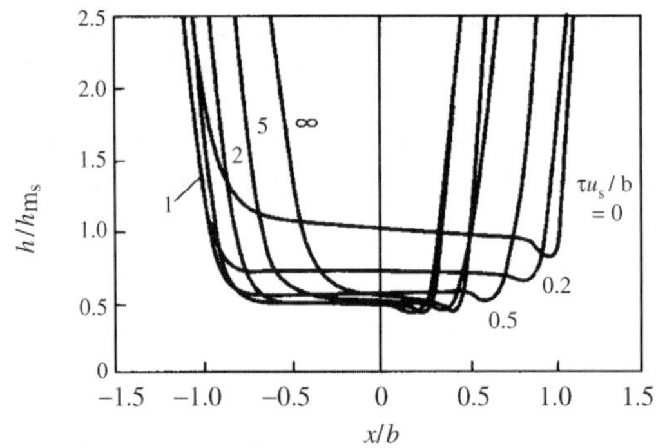

Figure 8.6 Film thickness of rubber EHL.

Further study has been carried out on EHL problems of rubber, a strong visco-elastic material [3,4]. Visco-elasticity not only brings about lubricated wear but also has a significant influence on the pressure and film thickness of EHL. Rubber has greater differences from the usual metallic materials. Figures 8.5 and 8.6 give the pressure distributions and film thickness curves of EHL of rubber. In the figures, $\tau u_s/b$ is an index viscosity parameter of material. When it is equal to zero, it is purely an elastic problem, that is, without considering deformation time.

From the above figures, it can be found that in the inlet there is a clear pressure peak which is different from that of the metal EHL. Although the film thickness in the outlet area still has some necking, there is no secondary pressure peak, and the pressure distribution throughout the area decreases continuously with increase in $\tau u_s/b$. The final lubricated wear area is only about half of the whole contact area. In addition, the greater the deformation of the material, the greater the difference in pressure at the inlet.

8.2 Influence of Limit Shear Stress on Lubrication Failure

Non-Newtonian characteristics of a lubricant have been used to explain the friction characteristics of a lubricant film. However, according to present lubrication theories, the hydrodynamic pressure is generated by the shear stress, so the non-Newtonian behaviors will inevitably influence the load-carrying capacity of the lubricant film. Based on this fact, we have systematically studied lubrication failure of the visco-plastic fluid [5–8].

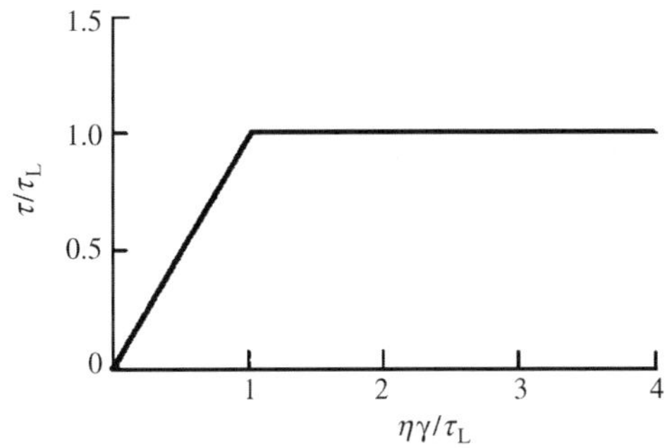

Figure 8.7 Visco-plastic fluid constitutive curve.

8.2.1 Visco-Plastic Constitutive Equation

Figure 8.7 is a simplified visco-plastic constitutive curve, where τ_L is the limit shear stress. Based on the constitutive equation, when the shear stress of the fluid reaches the limit its load-carrying capacity will drop significantly.

The visco-plastic fluid constitutive equation is:

$$\begin{aligned} \tau &= \eta \frac{du}{dz} \qquad \eta \left| \frac{du}{dz} \right| < \tau_L \\ \tau &= \tau_L \qquad \eta \left| \frac{du}{dz} \right| > \tau_L. \end{aligned} \tag{8.9}$$

The parameters of the visco-plastic fluid lubrication of a straight slider in calculations are: the slider width $b = 0.1$ m; the film ratio $K = h_1 - h_2/h_2 = 2.2$, where h_1 and h_2 are the maximum and minimum film thicknesses of the slider; the sliding velocity $u_0 = 5$ m/s; the viscosity of the lubricant $\eta = 0.1$Pa·s; and the limit shear stress $\tau_L = 6.9 \times 10^5$ Pa.

8.2.2 Slip of Fluid-Solid Interface

Figure 8.8 shows the pressure distribution and the interface shear stress curves, taking into account the limit shear stress [7,8]. In the figure, $P = p(h_1 - h_2)h_2/2u_0\eta b$ is the dimensionless pressure; $T = \tau/\tau_L$ is the dimensionless shear stress; and $X = x/b$ is the dimensionless coordinate. Slip between the fluid and solid

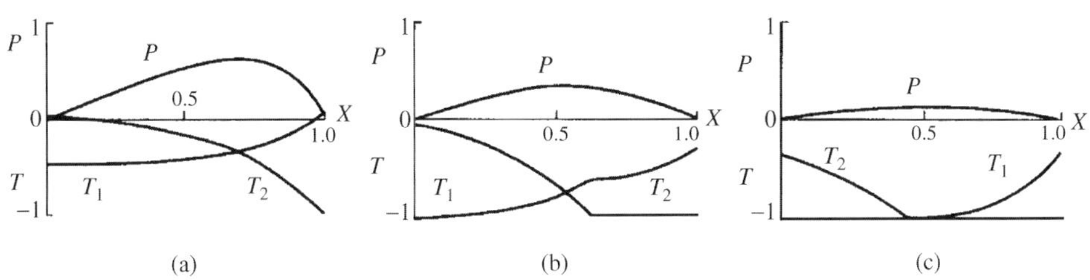

Figure 8.8 Shear stress and pressure distribution curves on the interface. (a) $H_2 = 2.1$; (b) $H_2 = 0.86$; (c) $H_2 = 0.6$.

interface occurs when the shear stress reaches the limit. It can be seen that the slip starts at the minimum film thickness. In Figure 8.8a, when the dimensionless minimum film thickness $H_2 = h_2/h_m$ is equal to 2.1 ($h_m = \eta u_0/\tau_L \approx 0.72\ \mu$m), the upper boundary shear stress T_2 of the minimum film thickness reaches the limit first, where the slip occurs. The slip results in a continuous drop of the film thickness.

It can be found from Figure 8.8 that the slip zone expansion can be divided into two steps:

1. After occurrence of slip, as the film thickness continues to fall, the slip extends from the minimum thickness to the contact center. At the same time, the shear stress at the maximum thickness increases continually.
2. When H_2 reduces to 0.86 (for $H_1 = 1.89$), the shear stress at the maximum thickness reaches the limit, too. Then the inlet area starts slip. The slip expands from both the outlet and inlet toward the contact center until $H_2 = 0.60$, (for $H_1 = 1.32$) when the slippage occurs in the whole contact area.

8.2.3 *Influence of Slip on Lubrication Properties*

Influence of slip caused by the limit shear stress on lubrication properties have been studied [9] and the parameters are the same as those of Figure 8.8. Figure 8.9 shows the calculation results of a straight slider considering the limit shear stress. The curves of the ratios of the carrying capacity w, the friction coefficient μ and the flow q of the visco-plastic fluid to w_0, μ_0 and q_0 of the Newtonian fluid are given, varying with the minimum film thickness H_2. In Figure 8.9, the thickness H_2 is chosen from the beginning to the global contact area of the slip. It can be seen from the figure that at the beginning ($H_2 = 2.2$), the load-carrying capacity, friction coefficient and flow of Newtonian fluid and visco-plastic fluid are very close. With the increase in the slip area, the differences become obvious. As the global area slips ($H_2 = 0.6$), the load-carrying capacity drops to 22%, the friction coefficient is 3.3 times as the original, and the flow increases 21%. The results show that if there is slip in the interface, using the classical lubrication theory to calculate the load-carrying capacity and other features will bring about significant errors.

The reasons for the differences of the load-carrying capacity, friction coefficient and flow of visco-plastic fluid lubrication are as follows.

1. While considering slip, the drop of pressure causes the load-carrying capacity to drop.
2. The friction coefficient depends on both the shear stress and load-carrying capacity. Although the shear stress has the limit, it drops less than the load-carrying capacity and result in increase in friction coefficient.
3. The increase of the flow is caused by increase in the fluid velocity because of slip.

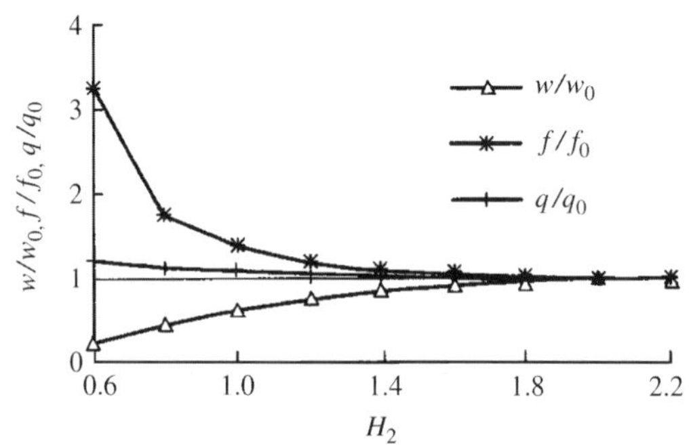

Figure 8.9 Comparison of visco-plastic fluid lubrication and Newtonian fluid lubrication.

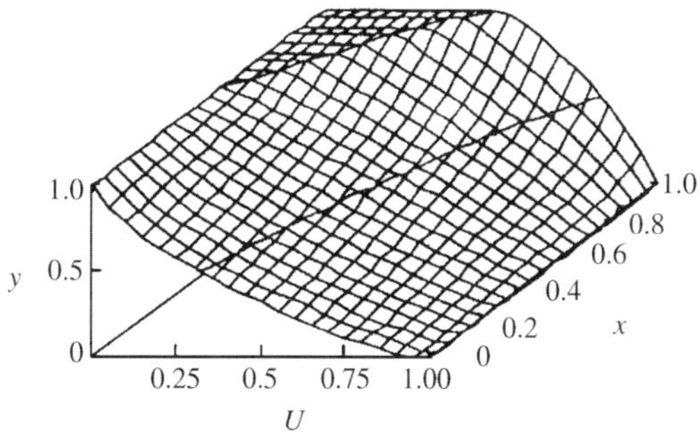

Figure 8.10 Velocity distribution along the directions of the length X and the thickness Z ($K = 2.2$, $H_2 - 0.7$).

In Figure 8.10, when $H_2 = 0.7$, the velocity distribution along the direction of the thickness and width is shown, where $U = u/u_0$ is the dimensionless velocity, $X = x/l$ is the dimensionless length, and l is the length of the slider, $Z = z/h$ is the dimensionless height, and h is the film thickness.

From the figure it can be known that in the inlet, the dimensionless velocity of the lower surface is no longer equal to 1, but less than 1. In the outlet, the dimensionless velocity of the upper surface is no longer 0. For example, when the global slip occurs, the largest dimensionless velocity of the upper surface $U = 0.6$.

Table 8.1 shows H_2 of the beginning slip, slip at both ends, and the global area slip, with the different film ratio K. From the table we can see that at the beginning, the larger K is, the larger slip the film of the inlet. While the slip occurs at both ends or global area, K is small.

Finally, the ratio of w/w_0, μ/μ_0 and q/q_0 varying with the film ratios K at the global slip are shown in Figure 8.11.

Influences of the limit shear stress on the lubricating properties can be summarized as follows:

1. The load-carrying capacity decreases obviously due to slip. At the same time, it also decreases with the film ratio K. When the film ratio K is the minimum, the relative load-carrying capacity is also the minimum.
2. The friction coefficient increases sharply to more than 10 times.
3. The flow ratio increases with most film ratios, but it decreases slightly when the film ratio is small.
4. The thinner the film, the larger the influence of slip on the lubrication properties.

Table 8.1 The minimum film thickness H_2 varying with the film ratio K

K	H_2		
	beginning slip	slip at the both ends	global slip
1.2	1.27	1.03	0.87
1.4	1.50	1.01	0.78
1.6	1.69	0.98	0.72
1.8	1.85	0.94	0.67
2.0	2.00	0.89	0.64
2.2	2.10	0.86	0.60
2.4	2.23	0.82	0.58
2.6	2.33	0.78	0.56
2.8	2.42	0.75	0.54
3.0	2.50	0.72	0.53

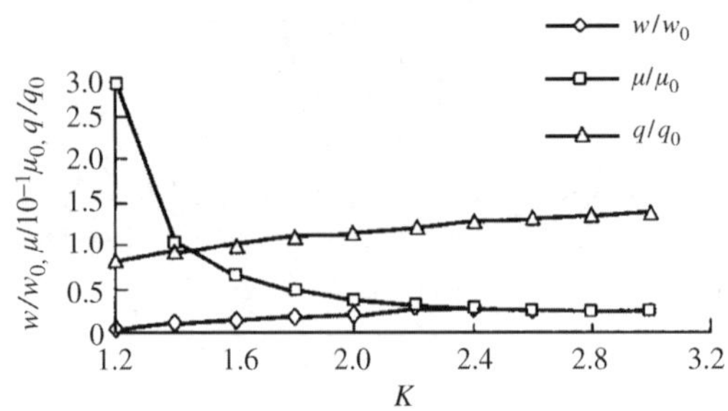

Figure 8.11 w/w_0, μ/μ_0 and q/q_0 varying with the film ratios K at the global slip.

According to the pressure distribution, film thickness, load-carrying capacity of the Newtonian fluid and the visco-plastic fluid, we will discuss the slip influences on hydrodynamic lubrication and elastohydrodynamic lubrication [5,6].

Figure 8.12 shows the pressure distribution of the Newtonian fluid and the pressure distribution of the visco-plastic fluid while the shear stress reaches the limit for the straight slider lubrication.

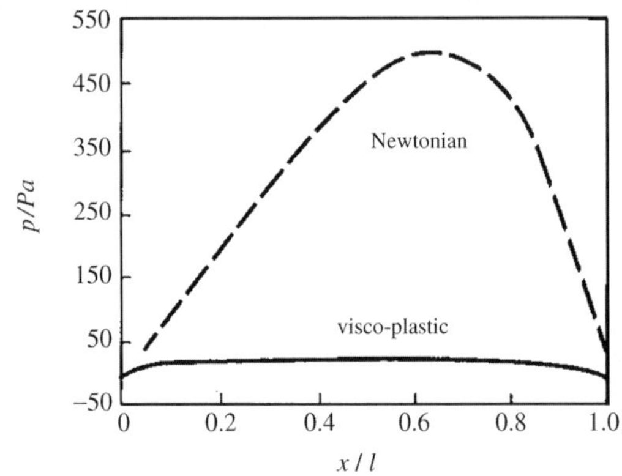

Figure 8.12 Pressure distributions of Newtonian fluid and visco-plastic fluid.

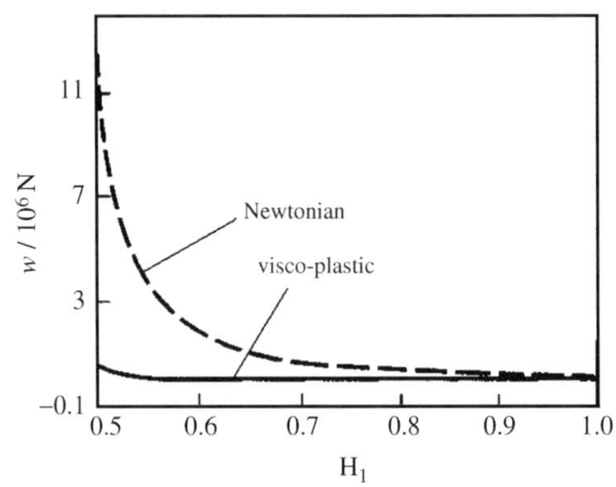

Figure 8.13 Load-carrying capacities of Newtonian fluid and visco-plastic fluid.

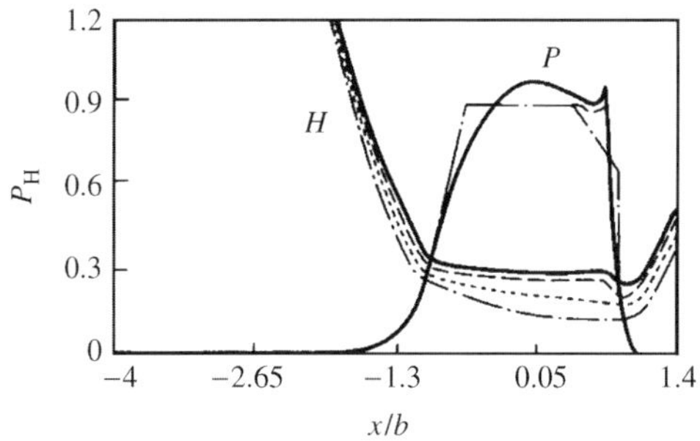

Figure 8.14 Pressure distribution and thickness of visco-plastic fluid EHL.

Figure 8.13 shows the load-carrying capacity. It can be seen that in the figure, the pressure distributions and the load curves of the Newtonian fluid and the visco-plastic fluid are significantly different. With the shear stress limit, the visco-plastic fluid pressure and the load-carrying capacity are much smaller than those of the Newtonian fluid.

Figure 8.14 gives the pressure and the film thickness curves of the visco-plastic EHL with a sliding-rolling ratio $s = 1$. The solid line shows the pressure distribution of the Hertzian pressure of 0.49 GPa. At this time the boundary slip has not happened yet. Therefore, there is no difference between the common EHL solutions, e.g. with an obvious secondary pressure peak and the necking. The dotted line, point line and dash-dotted line indicate that the maximum Hertz contact pressures are 0.5, 0.53 and 0.55 GPa respectively. Under these conditions, slip starts increase significantly with increase of the Hertz contact pressure. It can also be seen that after the interface slip, not only does the film thickness decrease significantly, but the shape of the pressure distribution is also clearly changed.

In summary, based on the fact that the modern lubricants containing a lot of polymer additives have a significant non-Newtonian characteristic and the shear stress limit, the lubrication theory of the visco-plastic fluid is established. As the interface slips, the load-carrying capacity reduces and the film thickness drops significantly.

8.3 Influences of Temperature on Lubrication Failure

With increase in the high speed and large-scale power machinery, more and more attention has been paid to lubrication failures. For example, lubrication failure of steam turbosets bearings sometimes results in severe accidents. According to statistics, it is shown that accidents resulting from lubrication failure of steam turbosets are more than 60% of the total number of accidents. As lubrication failure is a sudden accident, it is very important to study the problems.

Most of the accident bearings are designed according to lubrication theory to show they should work in good lubrication condition. Therefore, this raises the question whether the prediction of classical lubrication theory has any deficiency. Although external factors such as shocks, vibration, processing and installation errors may bring about bias from the results of theoretical analysis, the factors of the lubrication film should be the primary factor in causing lubrication failure, or the lubricant film itself will lose its load-carrying capacity under certain conditions.

In large heavy-load thrust bearings, temperature plays an important role in lubrication failure. As lubrication failure occurs, temperature is always significantly increased. Therefore, taking into account the influence of temperature is the key to analyzing lubrication failure. Although the thermal calculation of hydrodynamic lubrication is similar, it cannot fully reflect the influence of temperature on the load-carrying capacity. The authors have analyzed the influences of the thermal non-Newtonian constitutive on lubrication failure as follows [10–12].

It should be pointed out that a large thrust bearing carries a heavy load so it generally runs in a thin film thickness. As is well known, the thin lubrication film thickness can significantly increase load-carrying capacity, but it also significantly increases the shear strain rate in the lubricant film, leading to the temperature rising quickly. As temperature increases, the viscosity of lubricant will rapidly decrease. This greatly reduces the shear stress of the lubricant film to change its constitutive characteristic. Therefore, the thermal fluid constitutive equation has been deduced. The research shows that after considering the thermal effect, shear stress is at its maximum. When shear stress reaches the limit, the load-carrying capacity of the lubricant film is significantly reduced and even lost.

The following is an analysis of lubrication failure based on the thermal constitutive equation.

8.3.1 *Mechanism of Lubrication Failure Caused by Temperature*

According to the Reynolds equation, pressure is inversely proportional to the quadratics of the film thickness. In other words, with a decline in the film thickness, the pressure increases rapidly to increase the load-carrying capacity. When considering the thermal influence on the load-carrying capacity, temperature decreases the film thickness so as to increase the shear strain rate of the lubrication film. This leads to temperature rise. Temperature also significantly reduces the lubricant viscosity because pressure is proportional to viscosity. The viscosity decrease will result in the load-carrying capacity of the lubricant film decreasing. Therefore, the lubrication stability problem stems from the thermal effect. Their relationship is as follows:

$$\text{Film thickness}\downarrow$$

$$\Rightarrow\begin{cases}\text{pessure}(\Delta p_1)\uparrow\uparrow\Rightarrow\text{load carrying capacity}\uparrow\uparrow\\ \text{shear strain rate}\uparrow\Rightarrow\text{temperature}\uparrow\Rightarrow\text{viscosity}\downarrow\downarrow\Rightarrow\text{pressure}(\Delta p_2)\downarrow\downarrow\Rightarrow\text{load carrying capacity}\downarrow\downarrow\end{cases}$$

Based on the above analysis, we have proposed the concepts of local and global thermal instability [10]:

Local thermal stability: at any point of the lubrication region, if $\Delta p_1 > \Delta p_2$, with decrease of film, it is said to be thermally stable. If $\Delta p_1 = \Delta p_2$, it is critical. If $\Delta p_1 < \Delta p_2$, it is thermally unstable.

Global thermal stability: For the entire lubrication region, if there is a decrease in film thickness, $\sum\Delta p_1 > \sum\Delta p_2$, it is said to be globally thermally stable. If $\sum\Delta p_1 = \sum\Delta p_2$, it is critical. If $\sum\Delta p_1 < \sum\Delta p_2$, it is globally thermally unstable.

8.3.2 *Thermal Fluid Constitutive Equation*

From the above analysis, we can know that to determine the thermal stable state of a lubricant film is an important step in the analysis of lubrication failure. Therefore, we will derive the thermal fluid constitutive equation and propose the critical thickness to determine the thermal stable state of a lubricant film.

For classical lubrication theory to consider an incompressible Newtonian viscous fluid, its constitutive equation is:

$$\tau = \eta\frac{du}{dz}. \tag{8.10}$$

When considering the influence of temperature on viscosity, a simplified energy equation and Barus viscosity-temperature equation are used, as well as taking into account the operating conditions of the

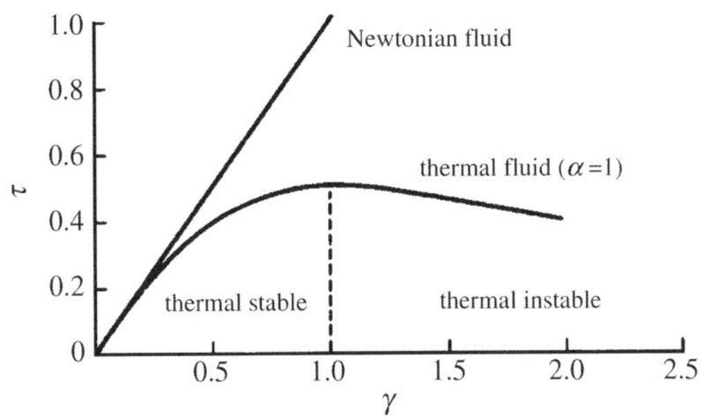

Figure 8.15 Constitutive curves of Newtonian and thermal fluids.

high shear strain rate, that is $\dot{\gamma} \approx U/h$. We introduce the thermal fluid constitutive equation as follows [10]:

$$\tau = \frac{\eta_0 \dot{\gamma}}{1 + \alpha \dot{\gamma}^2}, \tag{8.11}$$

where α is a coefficient, $\alpha = \sqrt{2\beta\eta_0 x/\rho c U}$.

Equation 8.11 shows that in the inlet, the lubricant is Newtonian, while in the outlet, it presents significant non-Newtonian behavior. Figure 8.15 gives the relationship between the dimensionless stress and shear strain rate of Newtonian and thermal fluids.

The difference in the thermal fluid constitutive equation from the other constitutive equations is that it has extreme shear stress. In Figure 8.15, the division line of the thermally stable and unstable zones is shown by the extreme shear stress. By using the thermal fluid constitutive equation, it is easy to obtain the corresponding film thickness where the shear stress reaches the extreme so that the occurrence of lubrication failure can be predicted easily.

To find the extreme value of Equation 8.11, the critical thickness can be obtained when setting the shear stress as the extreme. The expression is as follows:

$$h_c = \sqrt{\frac{2\beta\eta_0 U x}{\rho c}} \tag{8.12}$$

The critical lubricant film thickness can be used to judge whether the film at any point is stable or not. It is known that the shear strain rate is inversely proportional to the film thickness. Therefore, when the film thickness h is thicker than h_c, it is thermally stable. Otherwise, it is thermally unstable. And the larger the difference between critical film thickness and film thickness, the more instability there is.

8.3.3 Analysis of Lubrication Failure

Considering the thermal lubrication failure, the theoretical calculations and experimental results are given for a tilting-pad thrust bearing [11–13]. Table 8.2 lists the parameters of the bearing.

Figure 8.16 shows the numerical calculations and experimental results of the load-carrying capacity via the minimum film thickness, based on the thermal fluid constitutive equation. In addition, in the figure the result of the conventional theory for a Newtonian fluid is also given under the same conditions. According to the conventional theory, the lubrication pressure is inversely proportional to the quadratics of the film thickness, that is, $p \propto 1/h^2$. The result obtained from the method introduced by mechanical design handbook is also given for comparison [14].

Table 8.2 Parameters of thrust bearing

Parameters	Value
B/m	0.014
D_2/m	0.565
L/m	0.158
U/(m/s)	76.53
Z	8
α/rad	0.69
η/Pa.s	0.01436
β/(1/K)	0.03

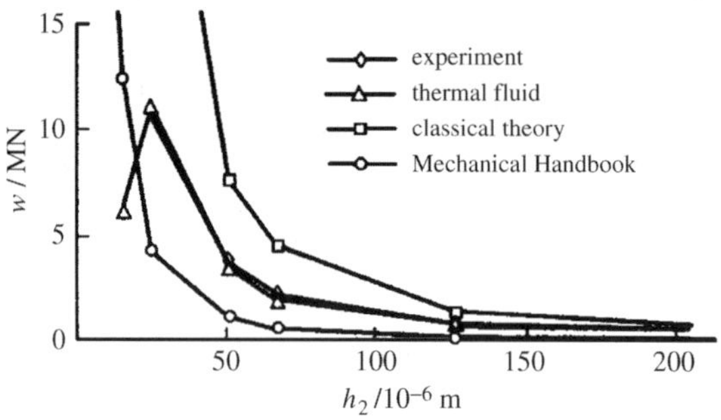

Figure 8.16 Load-carrying capacities via the minimum film thickness.

The following conclusions can be obtained from the results.

1. In conventional Newtonian fluid theory, with decrease of the film thickness, the load-carrying capacity increases infinitely and therefore lubrication failure cannot happen.
2. The calculation results of the load-carrying capacity according to the thermal fluid constitutive equation is consistent with experimental data.
3. With decrease of the film thickness, the load-carrying capacity of the thermal fluid and the experimental data are significantly lower than that of the conventional lubrication theory.
4. After the load-carrying capacity reaches the extreme, it begins to drop, that is, lubrication failure occurs.

The results also show that the conventional design methods indicate that the load-carrying capacity can increase infinitely with decrease in film thickness. This means that lubrication failure is impossible, and especially that load-carrying capacity rapidly increases when film thickness is thin. However, the thermal fluid constitutive theory is able to calculate the minimum film thickness h_2. When the film is thinner, lubrication starts to fail. By Equation 8.12, if $x = l$, the critical film thickness h_c causing thermal instability can be obtained to be equal to 91 μm. Therefore, when the outlet thickness $h_2 = 25.4\,\mu$m, the maximum carrying capacity has been reached. By using the film ratio $K = 4.2$ and Equation 8.12 again, we have $x = 0.57l$. This means that almost half of the bearing is thermally unstable so the thrust bearing has reached critical globally thermal instability.

8.4 Mixed Lubrication

Mixed lubrication is a combination of dry friction, boundary lubrication and hydrodynamic lubrication. Research has discovered that the micro-EHL film has an ability to flatten the surface roughness. It is

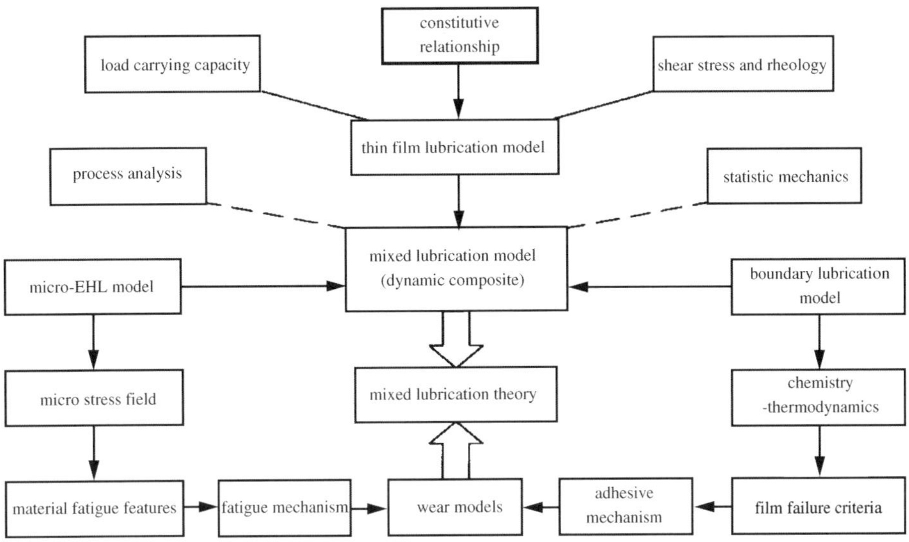

Figure 8.17 Mixed lubrication research and development diagram.

believed that mixed lubrication does not include dry friction, because the boundary film and the thin film cover the friction surfaces. The authors described the research and development of mixed lubrication as shown in Figure 8.17.

Clearly, except a very smooth surface, mixed lubrication widely exists. Study on mixed lubrication helps us understand the influence of surface morphology on lubrication. Kong *et al.* used totally optical reflection technology to realize the measurement of nano-film thickness, which has been applied to mixed lubrication [15,16]. They used a rough steel disk and a smooth sapphire to compose the sliding frictional pairs in line contacts to study the film thickness distribution along the contact line and the relationships of the average film thickness with the load, sliding velocity, oil viscosity and surface roughness.

The three kinds of roughnesses of the steel disc surface are obtained with a vertical texture after grinding or polishing. The root mean square deviation σ of roughness is 0.065 μm, 0.118 μm or 0.166 μm. The two mechanical lubricating oil used are No. 40 and No. 20 mechanical oils, with respect to the dynamic viscosities of 0.034 Pa·s and 0.017 Pa·s at 50 °C.

Figure 8.18 shows the relationships of the average thickness h_m along the contact line center with the sliding velocity v in the logarithmic coordinates, with No. 40 mechanical oil.

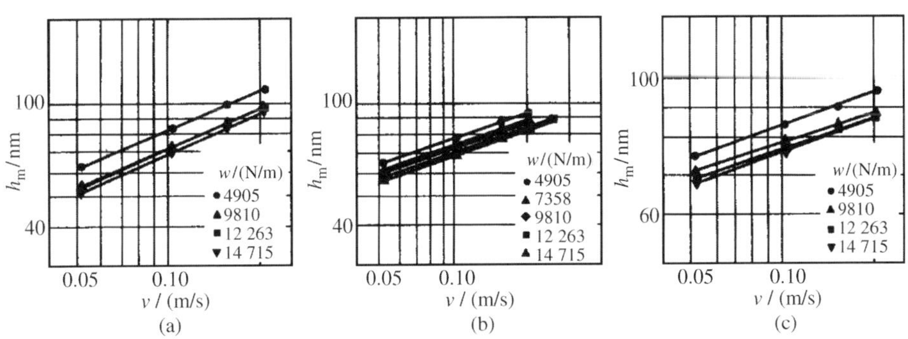

Figure 8.18 Influence of velocity on the average film thickness.

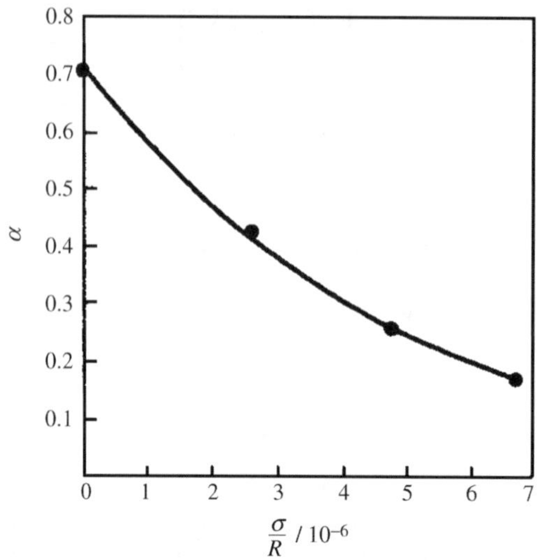

Figure 8.19 Relationship between α and σ.

The figure shows that similar to a smooth surface EHL, the average film thickness of the rough surface is an exponential function of velocity:

$$h_m \propto v^{\alpha}, \tag{8.13}$$

where α is the speed index, related to roughness as shown in Figure 8.19.

When $\sigma = 0.065\ \mu\text{m}$, $\alpha = 0.42$; $\sigma = 0.118\ \mu\text{m}$, $\alpha = 0.26$; and $\sigma = 0.166\ \mu\text{m}$, $\alpha = 0.16$. All these are lower than $\alpha = 0.71$ for a smooth surface. This shows that with increase of roughness, the influence of the velocity on film thickness decreases.

Figure 8.19 gives the relationship between the velocity index α and roughness σ, where R is the radius of the disk. According to the experimental data, the relationship between α and σ can be shown as follows:

$$\alpha = 0.71 \times 0.8^{\sigma/R}. \tag{8.14}$$

When $\sigma = 0$, the surface is smooth, then $\alpha = 0.71$ which is very close to the line contact EHL film thickness, with velocity index of 0.70.

By using No. 40 mechanical lubricating oil, the relationship between the average film thickness h_m and the load w per width is shown in Figure 8.20. The experimental data also obey an exponential relation:

$$h_m \propto w^{-\beta}, \tag{8.15}$$

where β is the load index, varying with roughness. When roughness σ is 0.065, 0.118 or 0.166 μm, the corresponding β is 0.14, 0.12 or 0.09. That is, with increase of roughness, the influence of load on the film thickness reduces.

Figure 8.21 gives the relationship between the load index β and roughness σ. Its fitting equation is:

$$\beta = 0.18 \times 0.91^{\sigma/R}. \tag{8.16}$$

For the smooth surface $\sigma = 0$, $\beta = 0.18$, which is same as the load index in the Dowson and Higginson formula of the line contact EHL.

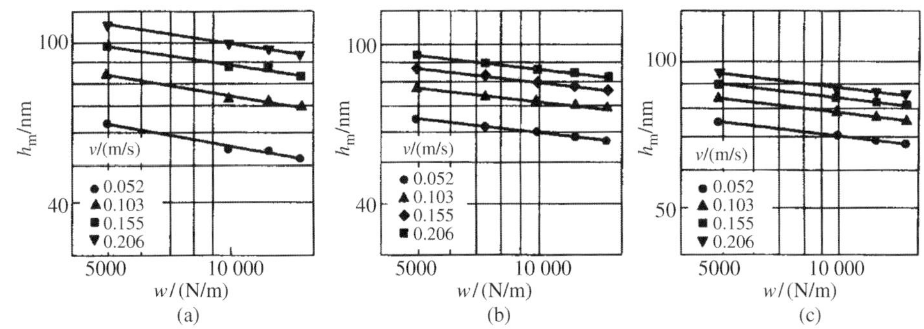

Figure 8.20 Influence of load on the average film thickness.

Figure 8.22 shows the influence of viscosity of the two mechanical lubricating oils on the average film thickness. The film thickness increases with increase in viscosity. In the figure, the relationships between the film thickness and speed or load are shown by the parallel lines. This indicates that for the same roughness, variation of the viscosity does not influence the speed and load. Equations 8.14 and 8.16 are applied to most mechanical lubricating oils.

The influence of roughness on the film thickness ratio h_m/σ is shown in Figure 8.23. From the figure we can see that the film thickness decreases with increase of roughness. At the same time, the influences of velocity and load on average film thickness are weakened with increase in roughness.

It should be noted that, because surface roughness is random, its influence on lubricant film thickness is very complicated. The existing research findings vary and further studies are needed. Based on the above results, it should be pointed out: Firstly, before experiments, the surface roughness measured is in an initial state. After the friction process, the actual roughness has been greatly changed. Secondly, the film thickness *hm* is the average film thickness along the center of the line contact. In fact, the actual film thicknesses of different points vary. According to our measurement, the influence of the viscosity of

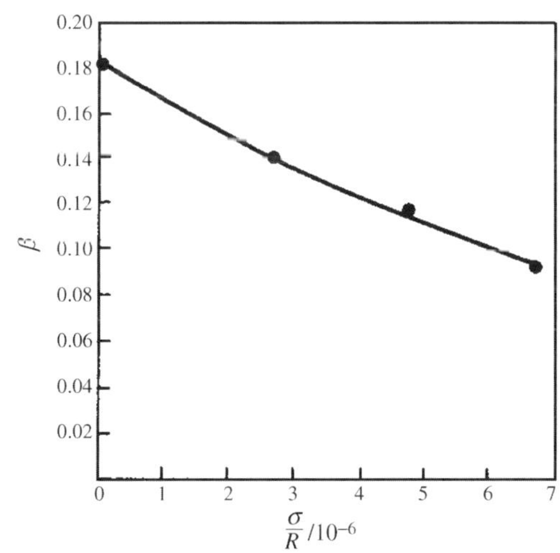

Figure 8.21 Relationship between β and σ.

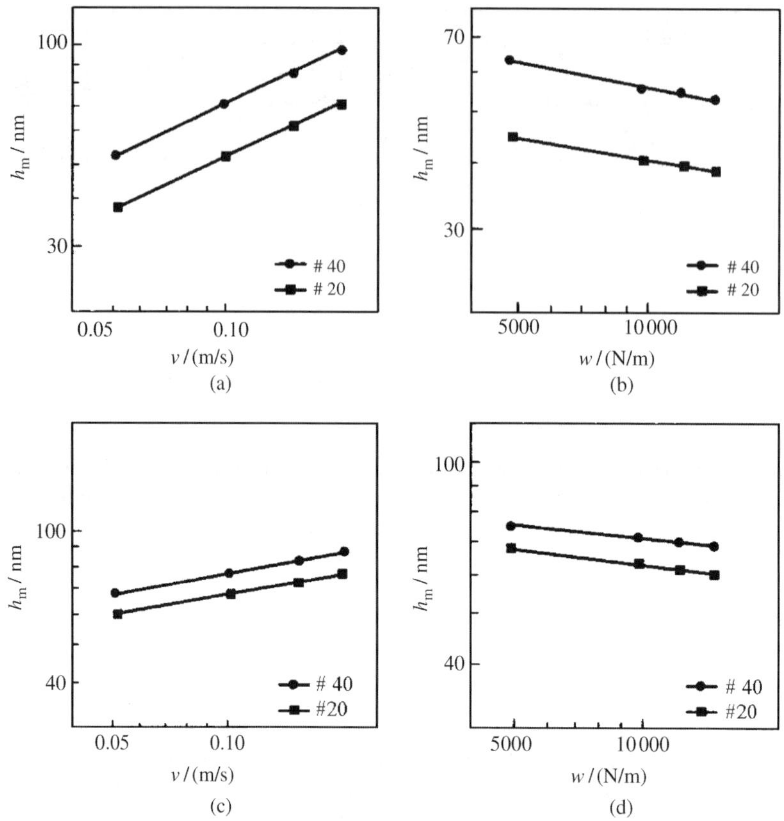

(a) $\sigma = 0.065\ \mu$m, $\omega = 14\ 715$N/m; (b) $\sigma = 0.065\ \mu$m, $v = 0.052$m/s;
(c) $\sigma = 0.166\ \mu$m, $\omega = 14\ 715$N/m; (d) $\sigma = 0.166\ \mu$m, $v = 0.052$m/s;

Figure 8.22 Influence of viscosity on the average film thickness.

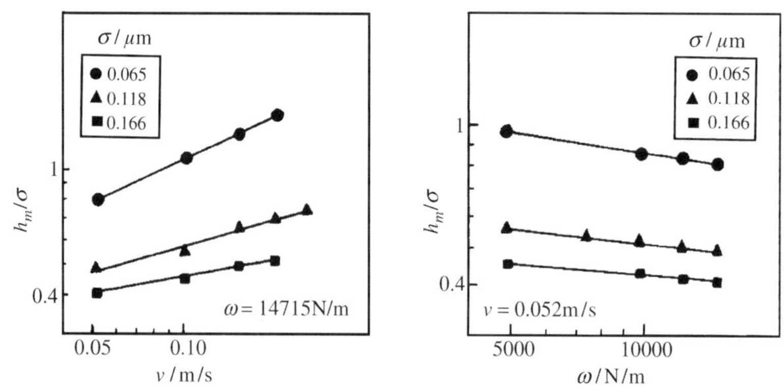

Figure 8.23 Influence of roughness on the film thickness ratio.

different points on the film thickness is around 10–110 nm. It clearly shows that it belongs to the mixed state of boundary lubrication, thin film lubrication and micro-elastohydrodynamic lubrication.

References

1. Huang, P. and Wen, S.Z. (1996) Visco-elastic hydrodynamic lubrication and lubrication wear. *Journal of Mechanical Engineering*, **32** (3), 27–34.
2. Zou, Q., Huang, P., and Wen, S.Z. (1996) Abrasive wear model for lubricated sliding contacts. *Wear*, **196**, 72–76.
3. Huang, P., Zheng, J., Hooke, C.J., and Wen, S.Z. (1996) Surface roughness on-line exposure to low elastic modulus properties of EHL. *Journal of Tsinghua University*, **36** (10), 50–56.
4. Hooke, C.J. and Huang, P. (1997) Elastohydrodynamic lubrication of soft viscoelastic materials in line contact. *Proceedings of the Institution of Mechanical Engineers, Part J: Journal of Engineering Tribology*, **211** (3), 185–194.
5. Huang, P. and Wen, S.Z. (1994) A new model of visco-plastic fluid lubrication for sliding problems. *ACTA Tribologyica of Romania*, **2** (1), 23–30.
6. Huang, P. and Wen, S.Z. (1995) Study of visco-plastic fluid lubrication failure sliding problem. *Progress in Natural Science*, **5** (4), 435–439.
7. Huang, P. and Wen, S.Z. (1996) Sliding under the conditions of elastohydrodynamic lubrication film thickness and yield the boundary. *Lubrication and Sealing*, **6**, 18–23.
8. Huang, P., Luo, J.B., and Wen, S.Z. (1999) Theoretically study on the lubrication failure for the lubricants with a limiting shear stress. *Tribology International*, **32** (10), 421–426.
9. Huang, P., Luo, J.B., and Wen, S.Z. (1999) Visco-plastic fluid interface slip on the lubrication properties of impact studies. *Acta Mechanica Sinica*, **31** (6), 745–752.
10. Huang, P., Chen, Y.Z., Luo, J.B., and Wen, S.Z. (2000) Thrust bearing failure analysis of hydrodynamic lubrication. *Journal of Mechanical Engineering*, **36** (1), 96–100.
11. Huang, P. and Wen, S.Z. (1996) Non-Newtonian effects of temperature and lubrication failure mechanism analysis. *Lubrication and Sealing*, **2**, 14–16.
12. Zhang, G.J., Huang, P., and Wen, S.Z. (1996) The temperature lubricant non-Newtonian effect and its influence on the lubrication properties. *Lubrication and Sealing*, **3**, 5–9.
13. Zhang, G.J., Meng, H.R., Huang, P., and Wen, S.Z. (1998) Lubricant boundary slip on the elastohydrodynamic lubrication characteristics. *Journal of Tribology*, **18** (3), 243–247.
14. Mechanical Engineering Handbook Committee (1980) *Mechanical Engineering Handbook*, Article 29 Articles, bearing. Mechanical Industry Press, Beijing.
15. Kong, F.R. (1993) Mixed lubrication characteristics of experimental techniques and, Tsinghua University doctoral thesis.
16. Zou, Q., Kong, F.R., and Wen, S.Z. (1995) A new method for measuring the film thickness of mixed lubrication in line contacts. *Tribology Transactions*, **38** (4), 869–874

Part II
Friction and Wear

9

Surface Topography and Contact

Tribology is the subject of studying the action and variation taking place between relatively moving surfaces. Therefore, to understand and study surface morphology and contact is the basis of analyzing friction, wear and lubrication. The geometrical features of surfaces play an important part in the interplay between lubrication, dry friction and wear.

9.1 Parameters of Surface Topography

Any friction surface is made up of many different shapes of peaks and valleys. From a macro viewpoint, a surface may be smooth and flat, but observed under a microscope, it is composed of a number of irregular peaks and valleys, that is, topography. This is because any manufacturing process makes a surface form different shapes of peaks and valleys, as shown in Figure 9.1.

Topography parameters are usually used to describe surface geometry. The most commonly used parameter of surface topography for tribological study is roughness, which is represented by the profile of a cross section of the surface. According to representation methods, topography parameters can be divided into one-dimensional, two-dimensional and three-dimensional.

One-dimensional topography parameters are usually represented only by height and are widely used in practice. As shown in Figure 9.2, we usually plot the outline of heights along the cross section and select the x axis to let the upper area and the bottom area be equal. The x axis is known as the centerline.

9.1.1 Arithmetic Mean Deviation R_a

After measuring the heights of n points along x direction, calculate the arithmetic mean as follows:

$$R_a = \frac{1}{l}\int_0^l |z(x)|dx = \frac{1}{n}\sum_{i=1}^{n} |z_i|, \tag{9.1}$$

where $z(x)$ is the height function; l is the measuring length; n is the measured point number; z_i is the measured height at the ith point.

9.1.2 Root-Mean-Square Deviation (RMS) σ or R_q

The formula to calculate σ or R_q is as follows:

$$\sigma = \sqrt{\frac{1}{l}\int_0^l [z(x)]^2 dx} = \sqrt{\frac{1}{n}\sum_{i=1}^{n} z_i^2}. \tag{9.2}$$

Principles of Tribology, First Edition. Wen Shizhu and Huang Ping.

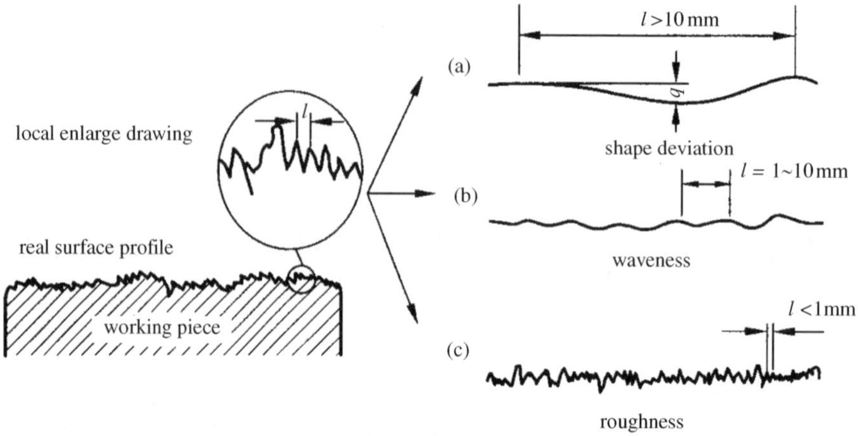

Figure 9.1 Surface topography.

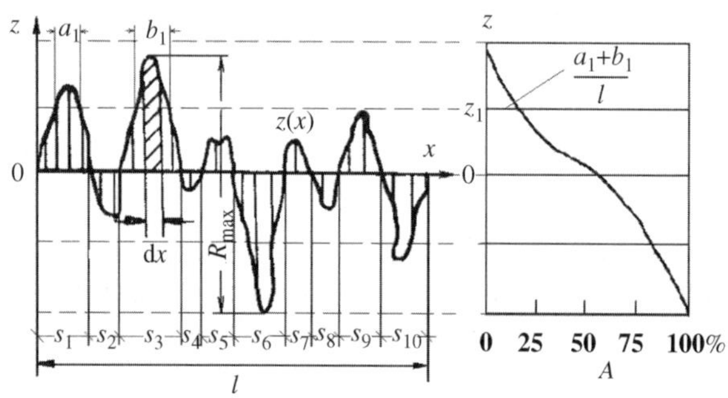

Figure 9.2 One-dimensional topography parameters. (a) Surface roughness and measuring parameters; (b) load-carrying area curve.

9.1.3 *Maximum Height R_{max}*

The maximum height R_{max} is the distance between the lowest valley and the highest peak within the measurement length l, as shown in Figure 9.2.

9.1.4 *Load-carrying Area Curve*

The load-carrying area curve is drawn according to the surface roughness profile. The theoretical carrying area curve is shown in Figure 9.2b. Suppose that rough surface is worn to the depth of z_1, form two planes with width as a_1 and c_1 as shown in Figure 9.2a. The sum of a_1 and c_1 divided by l is plotted corresponding to z_1 in Figure 9.2b. Thus, the carrying area curve can be obtained.

9.1.5 *Arithmetic Mean Interception Length of Centerline S_{ma}*

S_{ma} is the average intercept of all the n interceptions on the centerline within the measured length l. It reflects the irregularly undulating wavelength or of peak-valley spacing of the roughness of a surface.

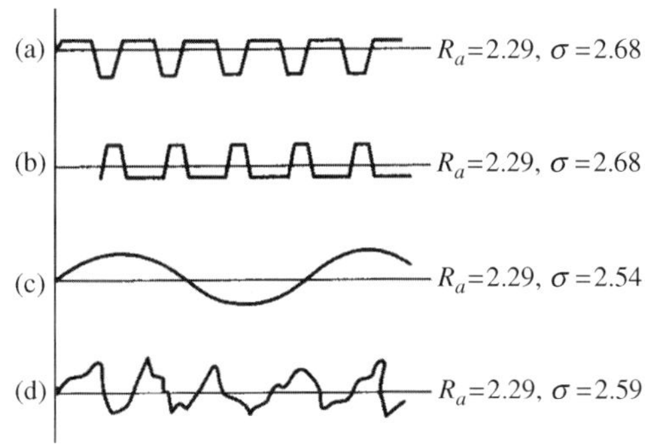

Figure 9.3 R_a and σ for different profile.

In Figure 9.2, ten intercepts, $s_1, s_2, \ldots, s_{10}$, are given in the roughness profile. To calculate their mean arithmetic value or the average, we can easily obtain its S_{ma}.

It should be pointed out that the one-dimensional topography parameters cannot illustrate a surface geometry very well. As shown in Figure 9.3, R_a of the four surface profiles is the same, but their profiles are quite different, or even completely opposite. σ of a profile is a little better than R_a. However, as shown in Figure 9.3, RMS of the two opposite profiles (a) and (b) are the same, but the results are unaccepted. Usually, the one-dimensional topography parameters are applied only to describe similar surface profiles manufactured by the same method. If we combine one-dimensional height parameters with one-dimensional wavelength parameters, we can roughly estimate the two-dimensional topography image of the surface.

The one-dimensional surface topography parameters cannot well represent the tribological properties of a surface because the slopes and curvatures are very important to friction, wear and lubrication. Therefore, to reflect the tribological and contact conditions of a rough surface, the following parameters of the two-dimensional topography are used.

9.1.5.1 Slope $\dot{z}_a$ or $\dot{z}_q$

The slope of a curve is a gradient of a profile, that is, $\dot{z} = \mathrm{d}z/\mathrm{d}x$. It is important in micro-EHL calculation. The arithmetic mean slope $\dot{z}_a$ or RMS slope $\dot{z}_q$ is often used in calculation.

9.1.5.2 Peak Curvature C_a or C_q

The peak curvature is the second derivative on a peak curve. It has an influence on lubrication and surface contacts. Because there are many peaks in a rough profile, the peak curvature is usually presented by the arithmetic mean curvature C_a or RMS curvature C_q.

However, the two-dimensional topography parameters are sometimes not enough to describe a complex rough surface. A better way is to use the three-dimensional topography methods to describe them. The three-dimensional topography can be represented by the following methods.

1. Two-dimensional family profiles
   As shown in Figure 9.4, a group of closed profiles are used to represent the three-dimensional topography. It gives a good description of the surface.
2. Contour map
   As shown in Figure 9.5, a contour map of the surface topography is often used to show the up and down roughness of a surface.

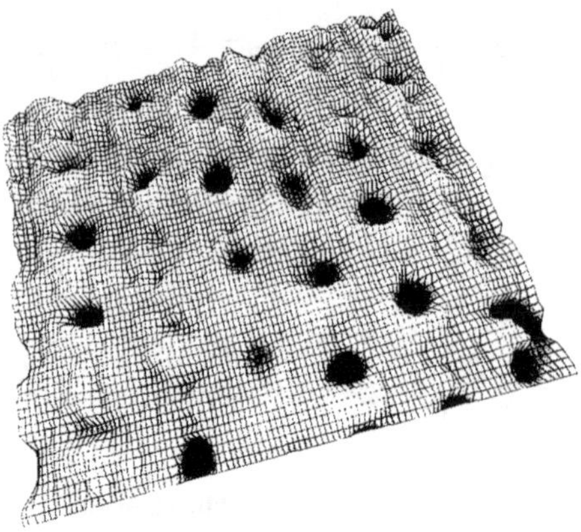

Figure 9.4 Two-dimensional family profiles.

Figure 9.5 Contour map.

9.2 Statistical Parameters of Surface Topography

A surface topography is usually composed of two components, the periodical variation and the random variation. Using statistical topography parameters instead of a single topography parameter to describe the surface geometry features is more accurate. The probability density function is used to describe variations of height, wavelength, slope or curvature at each point on a profile.

9.2.1 *Height Distribution Function*

As shown in Figure 9.6, if x is the centerline and z is the height of the profiles curve, the probability density distribution curve can be drawn as follows.

From the profile draw a line, as l in Figure 9.6. Calculate the sum length Σl_i of the segments of the peaks if above x or valley if below x. Obtain the ratio of $\Sigma l_i / l$ to draw a bar chart, as shown on the left side of the figure. A histogram can thus be obtained. Use a smooth curve to plot the bar chart, which is the probability density curve of the height distribution.

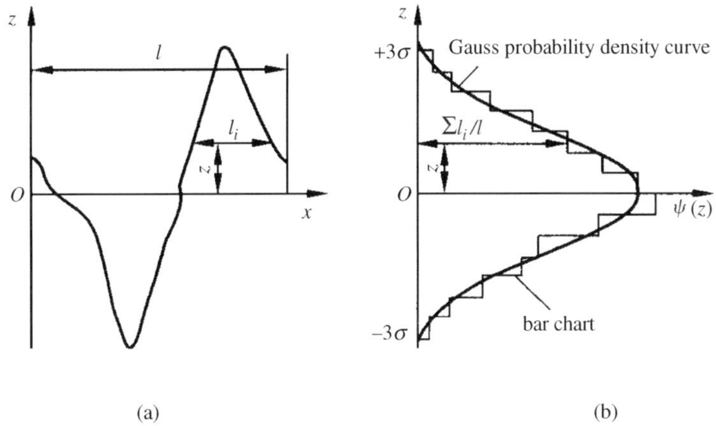

Figure 9.6 Roughness height distribution density curve.

Therefore, the rough heights of a cutting processed surface are close to the Gauss distribution. Its probability density function is as follows:

$$\psi(z) = \frac{1}{\sigma\sqrt{2\pi}}\exp\left(-\frac{z^2}{2\sigma^2}\right), \tag{9.3}$$

where σ is the RMS deviation of the heights, which is also called the standard deviation in the Gauss distribution; and σ^2 is called the variance of the Gauss distribution.

Equation 9.3 is the standard Gauss distribution curve. The probability density function $\psi(z)$ gives the probability of the height z. The range of the standard Gauss distribution is theoretically from $-\infty$ to $+\infty$, but it actually contains about 99.9% heights between -3σ and $+3\sigma$. Therefore, the heights beyond $\pm 3\sigma$ can be negligible in practice.

It should be noted that for the two-dimensional topography parameters such as slopes and peak curvatures can also be described by the statistical method.

For example, according to the surface profile, a slope of any point on a profile can be expressed as $\dot{z} = \mathrm{d}z/\mathrm{d}x$. Then, to draw the slope histogram similar to the height one above, we can obtain the probability distribution density function $\psi(\dot{z})$ of the slope as shown in Figure 9.7.

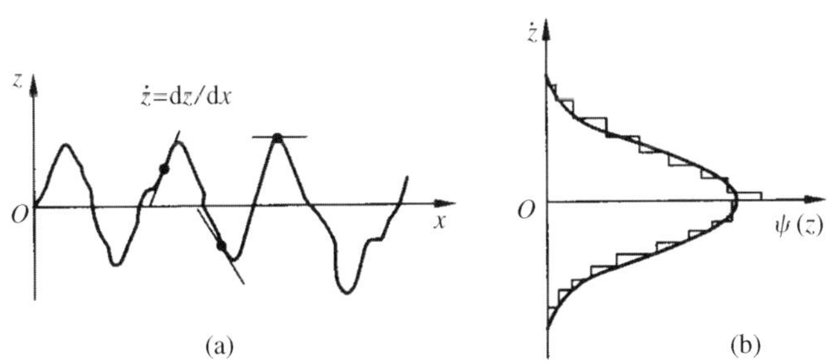

Figure 9.7 Distribution density curve of slope.

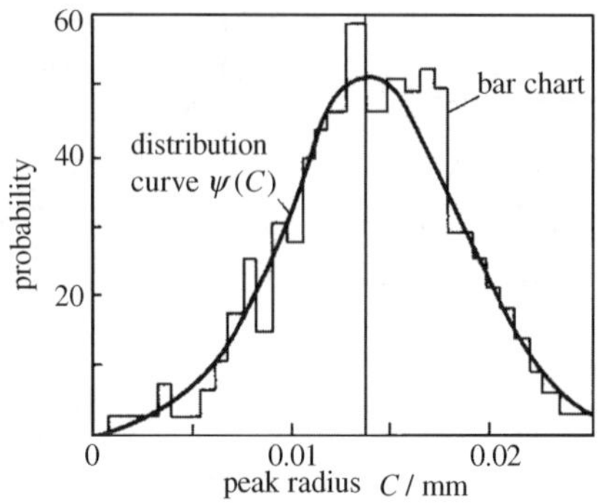

Figure 9.8 Peak curvature distribution density curve.

For the peak curvature C or the peak radius r ($r = 1/C$), by using the similar approach above, we can also obtain the probability density function $\psi(C)$ or $\psi(r)$. Figure 9.8 is given a peak curvature distribution density chart and curve based on the real cutting surface.

9.2.2 *Deviation of Distribution*

A distribution of a cutting surface topography is different from the standard Gauss distribution. For such a situation, some statistical parameters are usually used to show the difference. They are

1. Skewness S

   The skewness expresses the deviation from the symmetrical axle of a distribution curve. Its equation is as follows:

$$S = \frac{1}{\sigma^3}\int_{-\infty}^{+\infty} z^3\psi(z)dz. \tag{9.4}$$

   Substitute the standard Gauss distribution function into Equation 9.4, we have $S = 0$. The skewness of a distribution is equal to zero and is said to be normal or symmetrical. The skewness of a nonsymmetrical distribution is not equal to zero. It may be positive or negative. They are given in Figure 9.9.

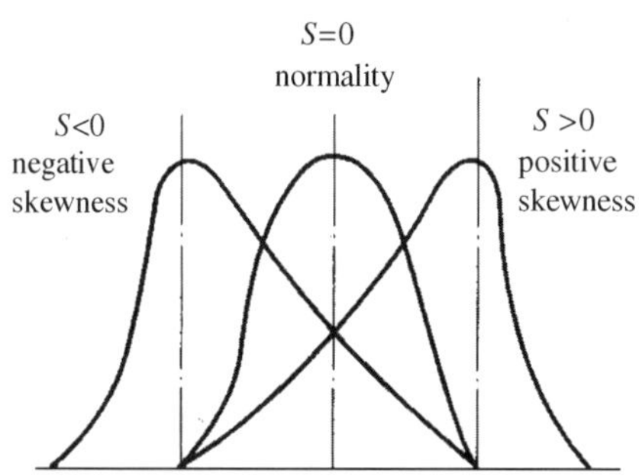

Figure 9.9 Skewness.

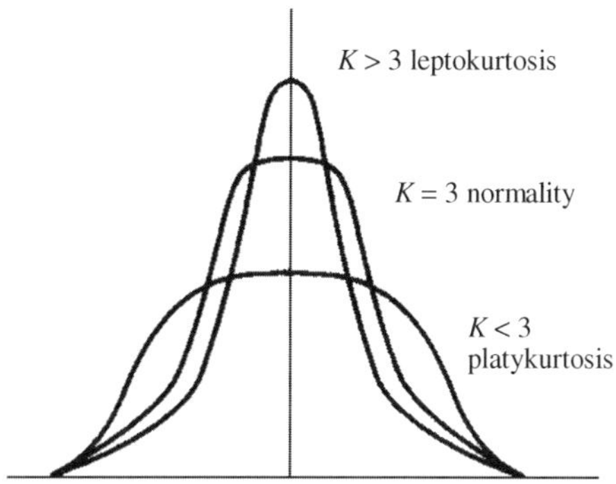

Figure 9.10 Kurtosis.

2. Kurtosis K
   The kurtosis shows the sharpness of a distribution curve. It is defined as:

$$K = \frac{1}{\sigma^4} \int_{-\infty}^{+\infty} z^4 \psi(z) dz. \tag{9.5}$$

   Substitute Equation 9.3 into Equation 9.5, it can be seen that the kurtosis K of the standard Gauss distribution is equal to 3 and is said to be normal. When $K < 3$, it is said that the distribution is platykurtosis, while $K > 3$, it is leptokurtosis. Figure 9.10 has shown these cases.

9.2.3 Autocorrelation Function of Surface Profile

The autocorrelation is the cross-correlation of a profile with itself. It is a mathematical tool for finding repeating patterns. An autocorrelation function is a mathematical expectation of the product between the profile heights of each point away from a fixed interval l:

$$R(l) = E[z(x) \times z(x + l)], \tag{9.6}$$

where E is the mathematical expectation.

If the measured point number within the total measured length L is n, and the coordinate is x_i, $R(l)$ is expressed as:

$$R(l) = \frac{1}{n-1} \sum_{i=1}^{n-1} z(x_i) \times z(x_i + l). \tag{9.7}$$

For a continuous profile curve, an autocorrelation function can be written in the integral form.

$$R(l) = \lim_{L \to \infty} \frac{1}{L} \int_{-L/2}^{+L/2} z(x) \times z(x + l) dx. \tag{9.8}$$

$R(l)$ is the function of the sampling interval l. Particularly, when $l = 0$, the autocorrelation function denoted by R_0, that is, R_0 is equal to the variance σ^2. The dimensionless autocorrelation function can be expressed as follows:

$$R^*(l) = \frac{R(l)}{R_0} = \frac{R(l)}{\sigma^2}. \tag{9.9}$$

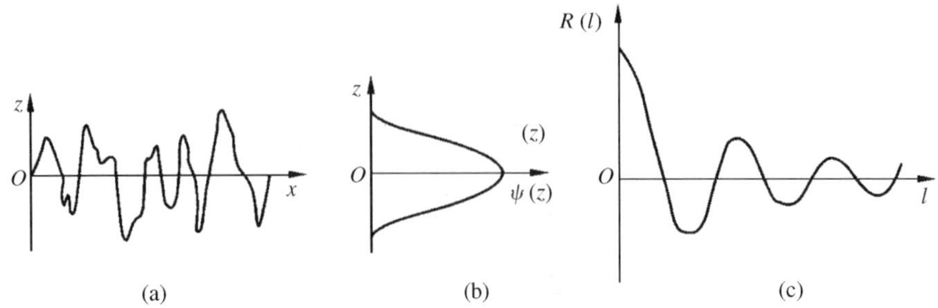

Figure 9.11 Probability distribution function and autocorrelation function of a profile.

Figure 9.11 gives a profile, its probability distribution curve, and its autocorrelation curve. An autocorrelation function can be decomposed into two components, the random and periodical. The random component shows that correlation decreases with increase of the interval l. The periodical component reflects the oscillating variation.

Calculation of an autocorrelation function needs to collect and deal with large amounts of data. For simplification, the random component is usually expressed as an exponential function and the periodical component as the trigonometric functions. Analysis shows that the component of a planed rough surface is mainly the periodical component, but that of a super-finished surface is the random component.

An autocorrelation function is very important for studying the surface topography. The characteristics of any surface topography can be described by a high degree probability density function $\psi(z)$ and an autocorrelation function $R^*(l)$.

9.3 Structures and Properties of Surface

Tribological properties of solid surfaces are directly affected by surface topography. In addition to geometric shapes, the surface physical and chemical states are also important for controlling tribological behaviors and process [1].

Machining processes greatly change the structure of a mechanical element surface. After being processed, a surface usually consists of a number of layers. The typical structure of the metal surface is as shown in Figure 9.12.

The deformed layers are on top of the metal basement. They are strengthened layers during processing with a total thickness of tens of microns, gradually transmitting from the small to large deformation layers. The Bainite layer is on top of deformed layers, which is produced due to surface melting, suddenly

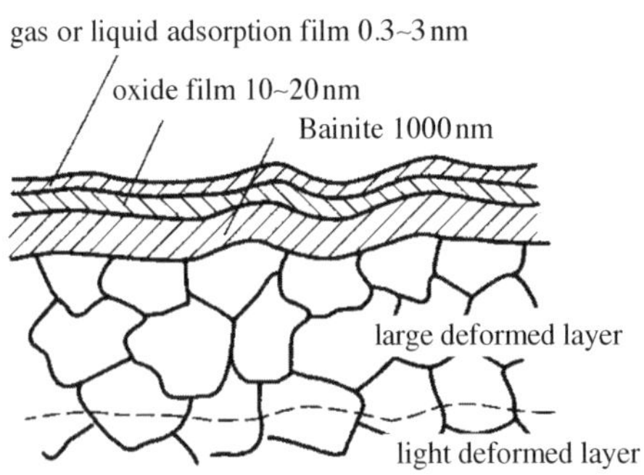

Figure 9.12 Metal surface structure.

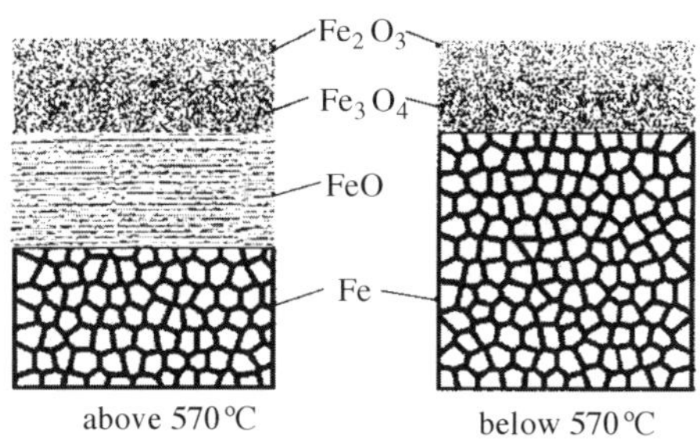

Figure 9.13 Oxide layer of Fe.

cooling, amorphously forming or micro-crystallining during the cutting process. Then, an oxide layer is formed due to surface contact with the atmosphere. Its organizational structure is related to oxygen density. The outside layer on the surface is an adsorption film or a contamination film, formed by gas or liquid polar molecules around the surface.

It is known that the organization of the structure of a metal surface changes with the processing method. Meanwhile, the surface mechanical properties and the substrate material are quite different. Strengthening, micro-hardness and residual stress have an important impact on the tribological properties of the metal surface.

In the process of machining or wear, once fresh metal surface is exposed to the atmosphere, it will quickly form an oxide layer. Under different temperatures, a metal usually forms different oxide layers. The oxide layer structures of Fe above and below 570 °C are shown in Figure 9.13. The oxidation rate depends on the rate of the inner diffusion of oxygen or the outer diffusion of the metal ions. Because the lattice constants of metal and oxide are different, this resists the oxygen to proliferate into the deeper inside of the metal. The oxide layer influencing tribological properties is related to the strength of the oxide layer. Usually, a thinner oxide film is of a higher strength and can prevent the metal from adhesion, such as Al_2O_3. If the oxide layer is thick, such as Fe_2O_3, the strength is weak and it is easily worn off so the wear process is accelerated. The surface energy, the adsorption effect and the surface oxidation are important to tribology.

In a friction process, due to the force and heat, a friction surface will be in a series of changes. The surface topography and micro-structure of the contact surface is constantly changing, while the adsorption film and the oxide film will break down, regenerate and metastasize. All these have a great impact on the friction and wear properties of a surface.

The metal friction pairs also produce a complex organizational structure change during a friction process. When temperature exceeds the transition temperature, due to high surface pressure and the environmental media, metal may have a special phase transition or structural change. For example, a steel surface will form a special kind of austenite layer during friction, which is not only very hard but also quite plastic. In addition, under the repeat actions of pressure and temperature of the friction process, the alloy atoms in the metal are prompted to diffuse toward the contact surface, making the sub-surface porous. Furthermore, the friction temperature will recrystallize the surface to alter the plasticity of it during the friction process. These changes in the friction process have a considerable impact on the mechanical properties of the friction surface.

9.4 Rough Surface Contact

When two solid surfaces contact each other, due to roughness, the real contact area is only a fraction of the apparent contact area. The real contact area and the distribution play a decisive influence on friction and wear.

The contact area shape of the roughness is usually like an ellipse. As the contact only occurs in quite a small part compared with the whole ellipse, it can be approximately regarded as an asperity sphere. Therefore, the two-surface contact can be regarded as the contact between a series of balls of different height. In analysis of rough surface contact, this model is often used.

As mentioned before, the contact of the two elastic surfaces can be converted into the contact of a rigid smooth plane and an equivalent elastic sphere with the equivalent radius R and the equivalent elastic modulus E.

9.4.1 Single Peak Contact

When the load W produces the deformation δ, the shape of the elastic sphere changes from the dotted curve to the solid curve as shown in Figure 9.14. Clearly, the real contact area is the circle with the real radius a rather than the nominal radius e.

According to the theory of elasticity, the deformation, radius and load have the following relationships with the load and its properties:

$$\delta = \left(\frac{9W^2}{16E^2R}\right)^{1/3}$$

$$a = \left(\frac{3WR}{4E}\right)^{1/3}$$

$$W = \frac{4}{3}ER^{1/2}\,\delta^{3/2}. \tag{9.10}$$

From the above, we have $a^2 = R\delta$. So, the real contact area A is equal to:

$$A = \pi a^2 = \pi R\delta. \tag{9.11}$$

According to the geometry relation, the nominal radius is equal to:

$$e^2 = R^2 - (R-\delta)^2 = 2R\delta - \delta^2 \approx 2R\delta. \tag{9.12}$$

Therefore, the relationship of the nominal contact area A_0 and the real area A is:

$$A_0 = \pi e^2 = 2\pi R\delta = 2A. \tag{9.13}$$

We can see that for the single elastic asperity contact, the real contact area is only half of the nominal contact area.

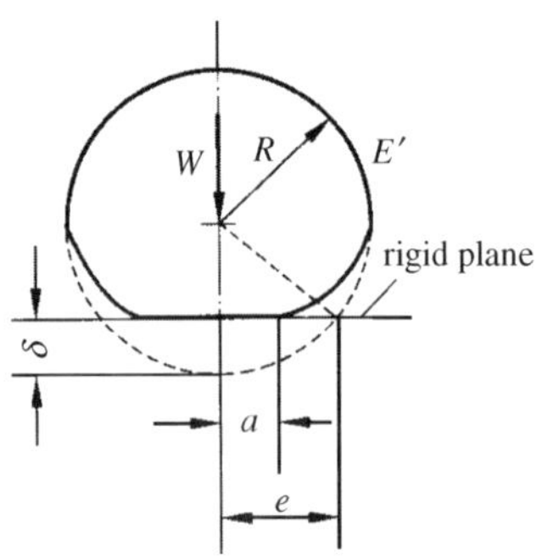

Figure 9.14 Single peak contact model.

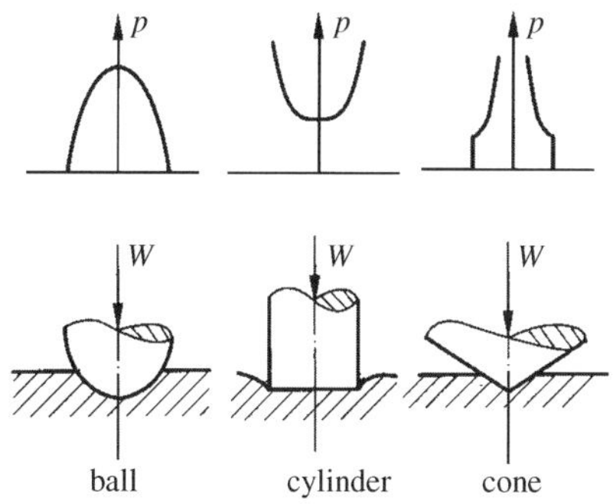

Figure 9.15 Different shape asperities contacts.

In addition to the ball shape asperity contact model, the commonly used models also include the cylinder shape or cone shape contact model as shown in Figure 9.15. In the figure, the pressure distribution of the ball model is an ellipse. Those of the cylinder or cone shaped model appear indefinite in part of the contact area, that is, the pressure at the edge or the central region tends to infinity. This will make calculation of elastic deformation difficult. The real contact area of the cylinder model that remains unchanged is equal to the real contact, while the contact area in the cone shaped model is close to the area.

9.4.2 Ideal Roughness Contact

Here, an ideal rough surface is composed of a row of rough peaks with the same radius and the same height. At the same time, the load and deformation of each peak are exactly the same and do not affect each other, as shown in Figure 9.16. If the maximum height of rough peak is h with a radius R, and when a smooth plane under a load W moves downwards, it generates the normal deformation equal to $(h-d)$. The distance between the rigid smooth surface plane and the rough surface centerline is d.

If there are n peaks and each carries the same load W_i, the total load W according to Equation 9.10 is equal to:

$$W = nW_i = \frac{4}{3} nER^{\frac{1}{2}}(h-d)^{\frac{3}{2}}. \tag{9.14}$$

The real contact area is equal to the sum of the real contact area A_{ei} of each asperity, that is:

$$A_e = nA_{ei} = n\pi R(h-d). \tag{9.15}$$

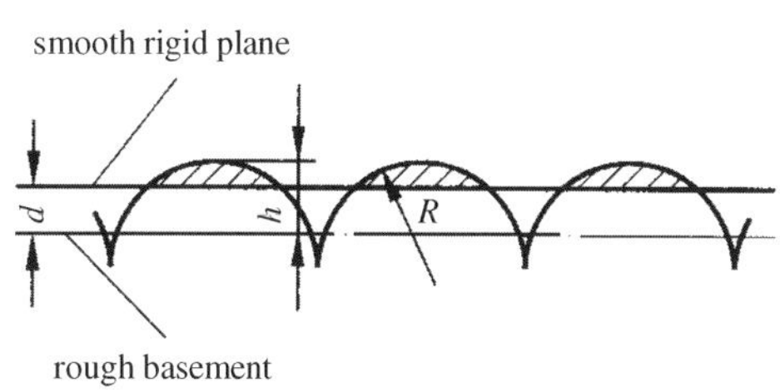

Figure 9.16 Ideal spherical roughness contact.

By using Equation 9.15 to eliminate $(h - d)$ in Equation 9.14, the total load becomes

$$W = \frac{4E}{3\pi^{3/2}\, n^{1/2}\, R} A_e^{3/2}. \tag{9.16}$$

It can be seen that for the elastic contact, the load is proportional to the 3/2 powers of the real contact area.

When a surface is in the plastic contact, if the individual asperity peak suffers the same yield stress σ_s and the deformation does not produce any horizontal expansion, the nominal contact area of a rough peak is $A_{pi} = 2\pi R(h - d)$. Therefore, the load each peak carries becomes:

$$W_{pi} = \sigma_s A_{pi} = 2\sigma_s A_{ei}. \tag{9.17}$$

So, the total load is:

$$W = nW_{pi} = 2\sigma_s A. \tag{9.18}$$

Equation 9.18 shows that for the plastic contact, the load is proportional to the real contact area.

Theoretically, it is commonly believed that the real contact area has a linear relationship with the load. The analysis of the model of the ideal roughness surface shows that only in the plastic contact is this relationship correct, but in the elastic contact, the relationship is not linear. This conclusion is because the model of the ideal roughness surface is too simple. By using the random roughness contact model as follows, we will find that in the elastic contact the relationship of the real contact area and the load will be linear.

9.4.3 Random Roughness Contact

The peak heights of roughness of a real surface are random so it should be expressed by the probability distribution density function. Therefore, the peak contact points should be calculated according to probability [2].

Figure 9.17a shows the contact of the two random roughness surfaces. The root mean square roughness of the two surfaces is σ_1 and σ_2 respectively, and the distance between the centerlines is h. The contact can be converted to the contact of a rigid smooth surface and an elastic rough surface with the equivalent root mean square roughness $\sigma = \sqrt{\sigma_1^2 + \sigma_2^2}$, as shown in Figure 9.17b.

The distance between two centerlines is h. Only the roughness belong $z > h$, the contact will take part. On the probability density distribution curve, the probability of $z > h$ is equal to:

$$P(z > x) = \int_h^\infty \psi(z)dz. \tag{9.19}$$

If the number of peaks in the rough surface is n, the number of the peaks participating in contact is m. Then,

$$m = n\int_h^\infty \psi(z)dz. \tag{9.20}$$

Because the deformation for each peak is equal to $(z - h)$, by using Equation 9.11 the real contact area A can be obtained as follows:

$$A = m\pi R(z - h) = n\pi R\int_h^\infty (z - h)\psi(z)dz. \tag{9.21}$$

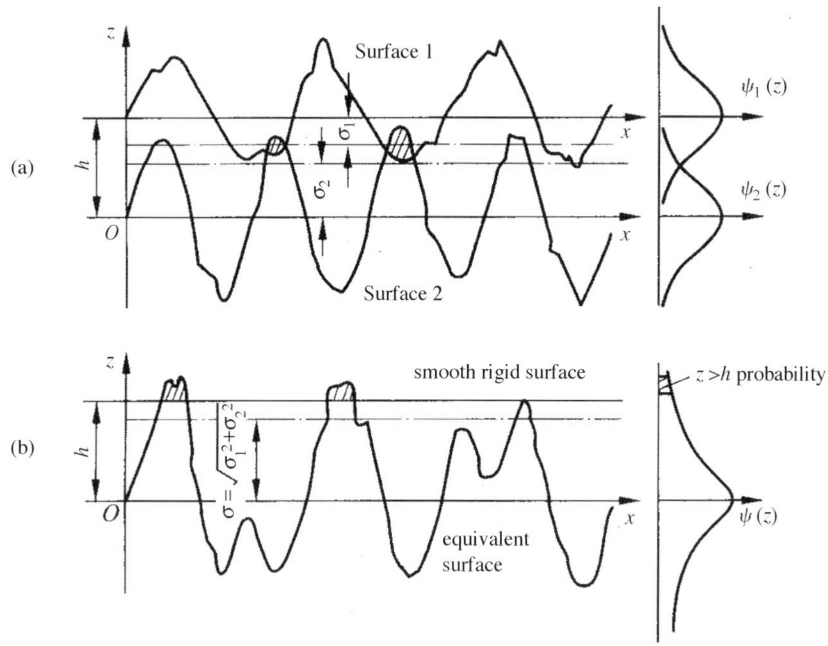

Figure 9.17 Two rough surfaces contact.

Therefore, the total load W is:

$$W = \frac{4}{3} m E R^{\frac{1}{2}} (z-h)^{\frac{3}{2}} = \frac{4}{3} n E R^{\frac{1}{2}} \int_h^{\infty} (z-h)^{\frac{3}{2}} \psi(z) dz \tag{9.22}$$

Usually, the profile heights of an actual surface obey the Gauss distribution. The distribution of the peaks is close to an exponential distribution. If we set $\psi(z) = \exp(-z/\sigma)$, we have

$$\begin{aligned} m &= n \exp(-h/\sigma) \\ A &= \pi n R \sigma \exp(-h/\sigma) \\ W &= \frac{4}{3} n E R^{\frac{1}{2}} \sigma^{\frac{3}{2}} \exp(-h/\sigma). \end{aligned} \tag{9.23}$$

From the above equations, we can conclude that $W \propto A$ and $W \propto m$. It can be seen that if the two rough surfaces are in the elastic contact, the load is proportional to the real contact area and the contact number of peaks.

When the two surfaces in the plastic contact, from the above analysis we have:

$$\begin{aligned} A &= 2\pi n R \int_h^{\infty} (z-h) \psi(z) dz \\ W &= \sigma_s A = 2\pi n R \sigma_s \int_h^{\infty} (z-h) \psi(z) dz. \end{aligned} \tag{9.24}$$

That means, the load has a linear relationship with the real contact area, but is not related to the distribution function $\psi(z)$.

To sum up, the relationship between the real contact area and the load depends on the surface profile and the contact state. When the contact is plastic, the relationship of the load and the real contact area is linear no matter how the distribution is. In the elastic contact, if the profile heights of the surface are close to the Gauss distribution, the load and the real contact area also have a linear relationship.

9.4.4 Plasticity Index

In fact, when two rough surfaces contact, a mixture of elastic and plastic deformations usually exists, that is, the higher peaks are in the plastic contact, but the lower peaks are in the elastic contact. With increase in the load, the normal deformations of the two surfaces increase and the peak number in the plastic contact also increases. So, the normal deformation can serve as a measuring scale to check the surface deformation state.

Greenwood and Williamson gave the following analysis on the contact problems [3]. From Equations (9.10) and (9.11), the average pressure on the contact area can be obtained as follows:

$$p_c = \frac{W}{A} = \frac{4E}{3\pi}\sqrt{\frac{\delta}{R}}. \tag{9.25}$$

Calculation of the plastic deformation shows that when the average pressure p_c reaches $H/3$, the plastic deformation has occurred, where H is the Brinell hardness (HB) of the material. When p_c is equal to H, the plastic deformation can be visible. Usually, $p_c = H/3$ is selected as the criterion for occurrence of the plastic deformation. Substitute $p_c = H/3$ into Equation 9.24 the plastic deformation δ can be obtained as follows:

$$\delta = \left(\frac{\pi H}{4E}\right)^2 R = \left(0.78\frac{H}{E}\right)^2 R. \tag{9.26}$$

Because transformation from the elastic deformation to the plastic deformation is a gradual process, an appropriate margin is used to change the above equation as follows:

$$\delta = \left(\frac{H}{E}\right)^2 R. \tag{9.27}$$

For convenience, a dimensionless parameter for the plastic deformation is used to determine whether the contact is elastic or plastic:

$$\Omega = \sqrt{\frac{\sigma}{\delta}} = \frac{E'}{H}\sqrt{\frac{\sigma}{R}}, \tag{9.28}$$

where Ω is the plasticity index.

When $\Omega < 0.6$, the contact is elastic. When $\Omega = 1$, a very few of peaks are in the plastic contact. When $1 < \Omega < 10$, it belongs a mixed contact with elastic and plastic deformations. The larger the Ω, the more the proportion of plastic deformation will be.

References

1. Wen, S.Z. (1998) *Nano-Tribology*, Tsinghua University Press, Beijing.
2. Halling, J. (1975) *Principles of Tribology*, McMillan Press Ltd.
3. Greenwood, J.A. and Williamson, J.B. (1966) Contact of nominally flat surface. *Proceedings of the Royal Society of London Series A-Mathematical Physical and Engineering Sciences*, **295**, 300–319.

10

Solid Friction and Control

The friction between two relative moving solid surfaces only related to the interaction of the contact surfaces is called the external friction. In boundary lubrication, the friction occurring on the adsorption film or other films is considered external friction. However, the friction induced by the relative movement of fluid molecules is known as the internal friction.

The common feature of external friction and internal friction is that an object or one part of it transmits the movement to another object or the other part so as to try to have both objects or parts to move at the same velocity and during this friction process, the energy converts.

The difference between external friction and internal frictions is that the internal movement is of different characteristics. For internal friction, the velocity of the neighbor particles in the fluid changes continuously. It possesses a certain velocity gradient. However, for external friction, the sliding velocity on the surface has an abrupt change. In addition, the internal frictional force is proportional to the relative sliding velocity and it disappears when the velocity is equal to zero. However, the relationship between the friction and the sliding velocity of external friction varies with working conditions and static frictional force still exists when velocity is equal to zero.

In this chapter we discuss the dry friction between solid surfaces. It is the sliding or rolling friction without lubricant.

10.1 Basic Characteristics of Friction

It is generally believed that Leonardo da Vinci (1452–1519) first proposed the concept of friction. Later, Amontons also carried out a series of experiments and established the friction laws. Subsequently, Coulomb, based on his further experiments, developed Amontons' work and derived four classical friction laws as follows.

Law 1: The frictional force is proportional to the load. Its mathematical expression is

$$F = fW, \tag{10.1}$$

where F is the frictional force; f is the friction coefficient; W is the normal load.

Equation 10.1 is often called the Coulomb's law, and it can be considered as the definition of the friction coefficient. Except for the case that the actual contact area is close to the apparent area under heavy load, the first law is correct.

Law 2: The friction coefficient is not related to the apparent contact area.

Principles of Tribology, First Edition. Wen Shizhu and Huang Ping.

The second law is generally suitable only for materials having a yield limit, such as metal. It does not apply to elastic or visco-elastic material.

Law 3: The static friction coefficient is larger than the dynamic friction coefficient.

This law does not apply to visco-elastic material, whether it has a static friction coefficient or not is not conclusive yet.

Law 4: The friction coefficient is not related to the sliding velocity.

Strictly speaking, the fourth law does not apply to any material, although metal is mostly consistent with this rule. The friction coefficient of a viscoelastic elastomer is obviously related to the sliding velocity.

Although according to recent study it has been found that most of the classical friction laws are not entirely correct, they basically reflect the mechanisms of the sliding friction, and therefore they are still widely used to solve many engineering problems.

In-depth studies on the sliding friction have shown that it also has the following features.

10.1.1 Influence of Stationary Contact Time

A tangential force is required to have one surface to slide relatively to another. The force is called the static frictional force. Furthermore, the force to keep the surface continuously moving is called the dynamic frictional force. Usually, for most engineering materials, the dynamic frictional force is smaller than the static frictional force, although the dynamic frictional force of a viscoelastic material is sometimes larger than the static frictional force.

It has also been found that the static friction coefficient is influenced by the contact time. As shown in Figure 10.1, the static friction coefficient increases with increase in contact time. For the plastic material, this effect is more significant.

Under the action of the normal load, the actual contact area of a friction surface increases because the roughness embedded in each other produces a high contact stress to induce the plastic deformation. With increase of the stationary contact time, the embedding depth and the plastic deformation increase such that the static friction coefficient also increases.

10.1.2 Jerking Motion

The sophisticated experiments have shown that a dry sliding friction movement is not continuous, but intermittent. Such a phenomenon is known as the jerking motion. When the surface is elastic, the jerking motion is more significant. The jerking motion is a typical phenomenon of the dry friction, different from the well-lubricated case.

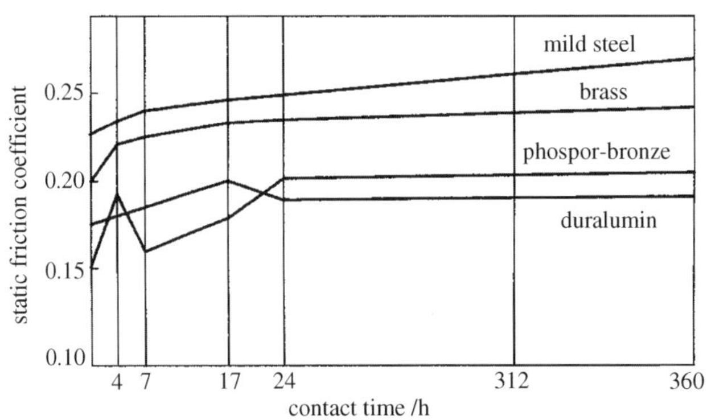

Figure 10.1 Relationship between the static friction coefficient and contact time.

The adhesion friction theory proposed by Bowden *et al.* can explain the jerking motion of metals, but it cannot explain the phenomena of non-metallic materials. Some believed that the static electricity causes the jerking motion, but it does not satisfy the condition.

A more satisfactory explanation of the jerking motion includes two aspects. One is that the motion is the result that, with increase of frictional force, sliding velocity decreases. The other is that the extension of the contact time is the reason of the jerking motion. Actually, both effects cause the motion. At high sliding velocity, the main effect is the former; while at the low velocity, the decisive factor is the latter.

The jerking motion in sliding friction is harmful to the smooth operation of machines. For examples, the vibration of a frictional clutch, the screaming noise during braking of a vehicle, the vibration and the creep on the sliding rail of a cutting tool are all connected with the jerking motion. Therefore, in order to improve the smoothness, an important way is to reduce the jerking motion during the friction process.

10.1.3 Pre-Displacement

When an external force makes an object trending to slide, if the tangential force is less than the static frictional force, the object does not move apparently but only produces a very small pre-displacement to a new stationary position. The pre-displacement increases with increase of the tangential force, and the maximum pre-displacement just before sliding is called the limit pre-displacement. The tangential force corresponding to the limit pre-displacement is the maximum static frictional force.

Figure 10.2 lists the pre-displacement curves of several metallic materials. From the figure we can see that only at the very beginning the pre-displacement is proportional to the tangential force. As it tends to the limit pre-displacement, the pre-displacements grow sharply to the limit, but the friction coefficient hardly increases.

The pre-displacement is partially reversible. That is, after the tangential force is removed, the displacement tends to disappear but not all. The greater the tangential force, the greater the residual pre-displacement. As shown in Figure 10.3, when we apply a tangential force, the point moves along *OlP* to *P*, having the pre-displacement *OQ*. When the tangential force is removed, the point moves along *PmS* to *S*, and the corresponding residual pre-displacement is *OS*. If the same tangential force is re-applied, the object will move to *P* again, but along *SnP*.

The pre-displacement is very important to the mechanical element design. The friction transmission and traction between the wheels and the tracks are based on the ability to generate a frictional force under the condition of the pre-displacement. The frictional force under the pre-displacement is also important to the reliability of the brake device.

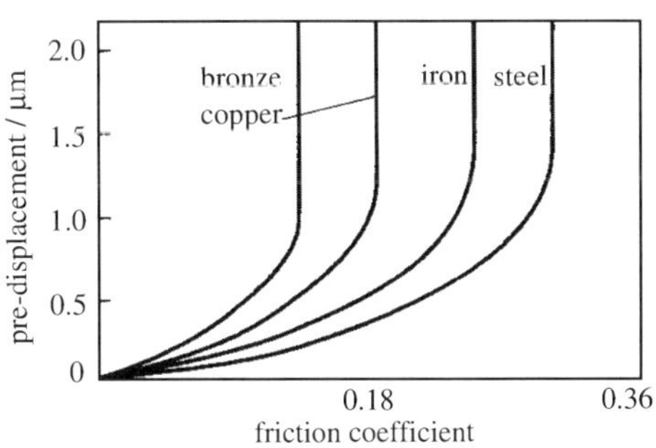

Figure 10.2 Pre-displacement curves via friction coefficient.

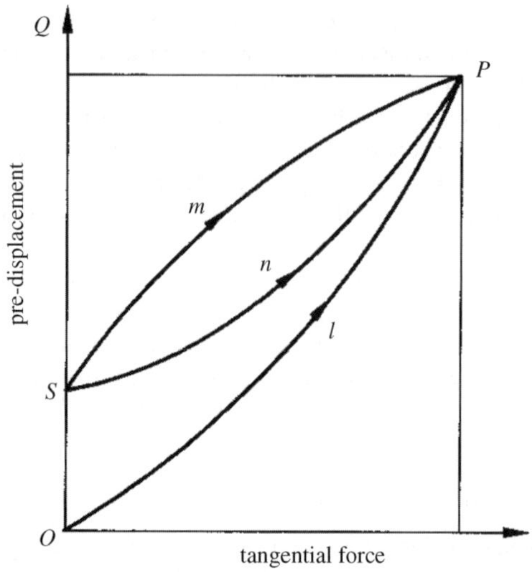

Figure 10.3 Partially reversible pre-displacement.

10.2 Macro-friction Theory

Friction is the resistance caused by the interaction between the two sliding contact surfaces with energy loss. Because the many factors influence on friction, different friction theories have been proposed. The main classical friction theories are as follows.

10.2.1 Mechanical Engagement Theory

The early friction theory thought that friction originates from surface roughness. The sliding friction energy is lost in roughness engagement, collision and plastical deformation, especially plowing through the hard asperities to embed in the soft surface to form furrows.

Figure 10.4 is the simplest friction model by Amontons (1699). He proposed that frictional force is equal to:

$$F = \sum \Delta F = \tan\varphi \sum \Delta W = fW, \tag{10.2}$$

where f is the friction coefficient, $f = \tan\varphi$, to be determined by the surface conditions.

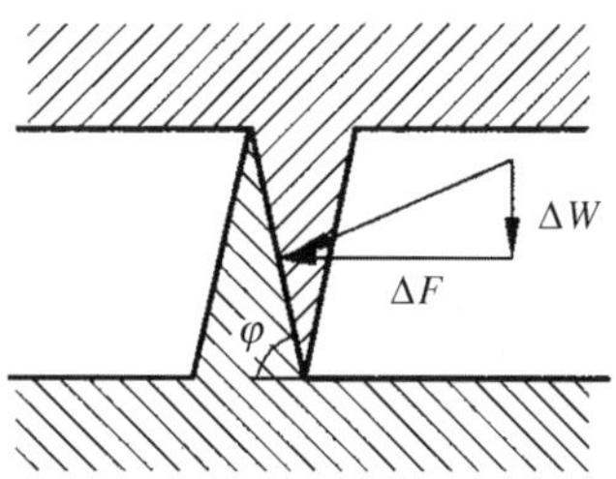

Figure 10.4 Mechanical engagement model.

Under the usual conditions, with decrease of the surface roughness, the friction coefficient decreases. However, the friction coefficient of the super-finished, surface increases sharply instead of decrease. Furthermore, when a surface is covered by a polar molecular adsorption layer, although its thickness is less than one-tenth of the polishing roughness, it can significantly reduce the frictional force. All these phenomena show that the mechanical engagement is not the only factor to generate the frictional force.

10.2.2 Molecular Action Theory

The action between molecules on the contact surfaces has been used to explain the sliding frictional force. Because the molecular mobility and the molecular force action can make the solid surfaces stick together, called the adhesion effect, it prevents the surfaces from sliding relatively.

Tomlinson (1929) first used the molecular action to explain the surface friction phenomena. He believed that the energy loss of the charge force between the molecules during sliding is the cause of friction. Thus, he derived the friction coefficient in the Amontons' formula.

Suppose, in the contact of two surfaces, some molecules produce the repulsion forces P_r, and the other molecules produce the attraction forces P_a, in which case the equilibrium condition is:

$$W + \sum P_a = \sum P_r. \tag{10.3}$$

Because P_a is very small, it can be omitted. If the contact number of molecules is n, the average repulsion of each molecule is P, then we have

$$W = \sum P_r = nP. \tag{10.4}$$

During sliding, the contact molecules continuously change, that is, contact and separate quickly, to form a new contact, but the balance condition is always satisfied. The energy loss caused by the conversion of the molecules should be equal to the friction work:

$$fWx = kQ, \tag{10.5}$$

where x is the sliding displacement; Q is the average power loss of the molecular conversion; k is the number of the conversion molecules, equal to

$$k = qn\frac{x}{l}, \tag{10.6}$$

where l is the average distance between molecules; q is the coefficient.

To solve Equations 10.5 and 10.6 simultaneously, the friction coefficient can be obtained as follows:

$$f = \frac{qQ}{Pl}. \tag{10.7}$$

Tomlinson's equation clearly points out that the molecular action has an influence on friction, but the formula cannot explain the friction phenomenon that the attraction force of the molecule sharply decreases with decrease of the distance between molecules (inversely proportional to 7 powers of the distance) such that the sliding resistance force generated by the contact surface molecules increases with increase of the actual contact area, but is not related to the normal load. Therefore, according to the molecular theory, the rougher the surface the smaller the actual contact area and the smaller the friction coefficient. Apparently, the equation is incorrect for most actual situations except for the heavy load condition.

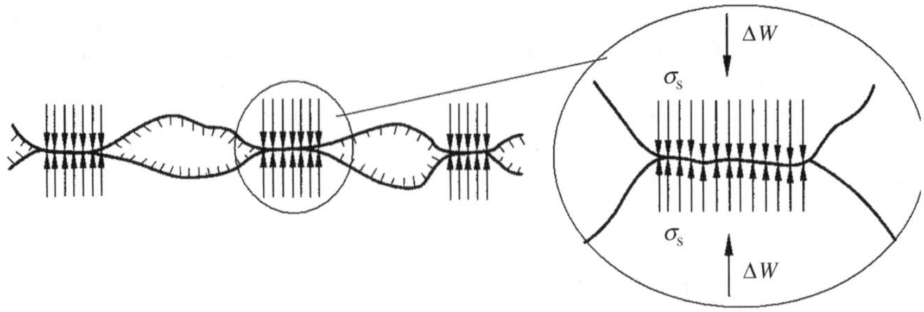

Figure 10.5 Plastic contact on the contact point.

10.2.3 Adhesive Friction Theory

As mentioned above, the classical friction theories, whether it is the mechanical or the molecular theory, are not perfect. The relationships between the friction coefficient, the roughness and the molecular actions are limited. At the end of the 1930s, the solid friction theory based on the mechanical-molecular combined action developed more quickly. Two theories, the adhesive friction theory and the binomial friction theory, had been established. These theories laid the theoretical foundation of the modern solid friction theory.

Bowden and Tabor, based on systematically experimental studies, established a more acceptable theory, the adhesive friction theory [1], which was important to friction and wear research.

10.2.3.1 Main Points of Adhesive Friction Theory

Bowden *et al.* (1945) put forward the simple adhesive friction theory which can be summarized as follows.

1. Friction surfaces in plastic contact
   Because the actual contact area is only a small fraction of the apparent contact area, the stress on the contact points reaches the yield stress σ_s to produce the plastic deformation. And, because the stress no longer increases, the contact area has to expand to support the load. Figure 10.5 shows the situation.
   As the stress on the contact point is equal to the soft material yield stress σ_s, if the actual contact area is A, then we have

$$W = A\sigma_s \tag{10.8}$$

or

$$A = \frac{W}{\sigma_s}. \tag{10.9}$$

2. Sliding friction changes alternatively between the adhesion and the jerking motion
   Because the contact point of the metal is in the plastic contact, it may transiently produce a high temperature to cause adhesion of the two metal surfaces. The adhesive node has a strong adhesion force. Subsequently, under the action of the frictional force, the adhesive points are sheared away to slide relatively. The sliding and the adhesion change alternatively to form the whole friction process.
   Figure 10.6 gives the measurement of the friction coefficient for steel to steel during the sliding friction process. The variation of the friction coefficient shows that the sliding friction is in the jerking motion. The experiments also show that with increase of the sliding velocity, the variation of the

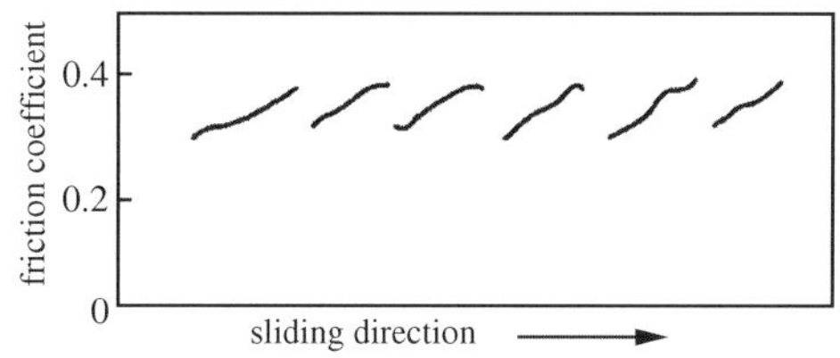

Figure 10.6 Jerking motions in sliding friction process.

friction coefficient and the adhesive time will reduce so that the friction coefficient and sliding process are smooth.

3. Frictional force is the sum of the resistances of the adhesion effect and the plowing effect
Figure 10.7 is the frictional force model composed of the adhesion effect and plowing effect. The rough peak on the hard surface embeds into the soft surface under the normal load W. Suppose that the peak shape is a semi-cylinder; the contact area consists of two parts. One is the cylindrical bottom surface, which is the adhesive effect area where the shear takes place during sliding. The other is the cylindrical front end, which is the plowing effect area. During sliding, the soft material is pushed forward and to the side by the hard peak (cylinder). Therefore, the frictional force F is equal to

$$F = T + P_e = A\tau_b + Sp_e, \tag{10.10}$$

where T is the shear force, $T = A\tau_b$; P_e is the plowing force, $P_e = Sp_e$; A is the adhesive area; τ_b is the shear strength; S is the plowing area; p_e is the force per unit plowing area.

The experimental results show that τ_b is related to the sliding velocity and the lubrication conditions and it is very close to the limit shear strength of the soft material. This shows that the shearing of the adhesive node usually occurs inside the soft material, causing the material to migrate.

p_e is determined by the nature of the soft material, with no relation to lubrication. Usually, p_e is proportional to the yield stress of the soft material. The depth of the hard peak embedding into the soft material decreases with increase of the yield stress. If a sphere embeds in a plate, the plowing force is inversely proportional to the square root of the yield strength of the soft material, that is, the harder the soft material, the smaller the plowing force.

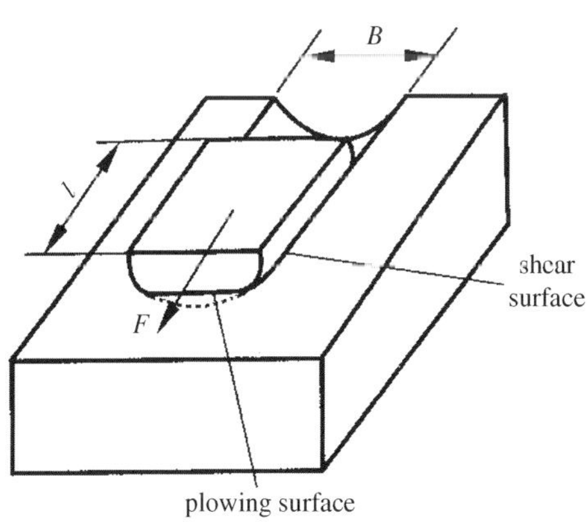

Figure 10.7 Frictional force of adhesion and plowing effects.

For the metal friction pairs, P_e is usually much smaller than T. The adhesion friction theory believes that the adhesion effect is the main reason to produce the frictional force. If we ignore the plowing effect, Equation 10.10 becomes

$$F = A\tau_b = \frac{W}{\sigma_s}\tau_b. \tag{10.11}$$

Therefore, the friction coefficient is equal to

$$f = \frac{F}{W} = \frac{\tau_b}{\sigma_s}. \tag{10.12}$$

10.2.3.2 Revised Adhesion Friction Theory

The friction coefficient obtained from Equation 10.12 is not consistent with the measured results. For example, for the most metallic materials, the relationship of the shear strength and the yield stress is $\tau_b = 0.2\sigma_s$. Then, the calculated friction coefficient $f = 0.2$. In fact, the friction coefficient of the most metal friction pairs is up to 0.5 in the air, and even higher in the vacuum. Therefore, Bowden *et al.* proposed some amendments to the simple adhesion friction theory.

Bowden believed that in the above adhesion friction theory, analysis of the actual contact area only considers the yield stress σ_s, but in the calculation of the frictional force it only considers the shear strength stress τ_b. This is reasonable for the static friction. However, for the sliding friction, because of the tangential force, the actual contact area and the contact point deformation depend on the combined effects of the compressive stress σ generated by the load and the shear stress τ by the tangential force.

Because the contact stress state on the peak point is complex and it is difficult to obtain the three-dimensional solution, then according to the general laws of the strength theory, the form of equivalent stress can be assumed as follows:

$$\sigma^2 + \alpha\tau^2 = k^2, \tag{10.13}$$

where α is the constant to be determined, $a > 1$; k is the equivalent stress.

α and k can be determined according to the extreme conditions. One extreme condition is $\tau = 0$, namely, the static friction state. At this point, the equivalent stress is equal to σ_s. Therefore, Equation 10.13 can be written as

$$\sigma^2 + \alpha\tau^2 = \sigma_s^2, \tag{10.14}$$

that is

$$\left(\frac{W}{A}\right)^2 + \alpha\left(\frac{F}{A}\right)^2 = \sigma_s^2, \tag{10.15}$$

or

$$A^2 = \left(\frac{W}{\sigma_s}\right)^2 + \alpha\left(\frac{F}{\sigma_s}\right)^2. \tag{10.16}$$

Another extreme condition is to make the tangential force F increase continuously. From Equation 10.16 we know that the actual contact area A increases correspondingly. Compared with F/A, W/A is very small and can be ignored. Therefore, Equation 10.15 becomes

$$\alpha\tau_b^2 = \sigma_s^2 \tag{10.17}$$

or

$$\alpha = \sigma_s^2/\tau_b^2. \tag{10.18}$$

Because most metals meet with the condition $\tau_b = 0.2\sigma_s$, by using Equation 10.17, we have $\alpha = 25$. Experiments show $\alpha < 25$, and Bowden *et al.* took $\alpha = 9$.

Equation 10.16 shows that W/σ_s represents the contact area of the static friction under the load W, while $\alpha(F/\sigma_s)^2$ reflects the increment of the contact area caused by the tangential force. Therefore, the contact area increases significantly so that the friction coefficient obtained by the revised theory is much larger than that of the simple adhesion theory. Thus, it is close to the reality.

As mentioned earlier, metal surfaces naturally form oxide film in the air or are polluted to form some other films. The films effectively reduce the friction coefficient. Sometimes, in order to reduce the friction coefficient, the surface of a hard metal is coated with a thin film of a soft material.

Such a phenomenon can be used to explain the revised adhesion friction theory as well. With a soft material film, the adhesive shear will occur in the film during sliding. Because the shear strength of the film is relatively low and its thickness is thin, the actual contact area is determined by the compression yield strength of the hard substrate material. Furthermore, the actual contact area is not large so the thin and soft surface film can reduce the friction coefficient.

Suppose the shear strength of the film $\tau_f = c\tau_b$, where c is the coefficient less than 1; τ_b is the shear strength of the material, by Equation 10.14 we have the start sliding condition of the friction pair as follows:

$$\sigma^2 + \alpha\tau_f^2 = \sigma_s^2. \tag{10.19}$$

According to Equation 10.17 we have

$$\sigma_s^2 = \alpha\tau_b^2 = \frac{\alpha}{c^2}\tau_f^2. \tag{10.20}$$

Thus, the revised friction coefficient is equal to

$$f = \frac{\tau_f}{\sigma} = \frac{c}{[\alpha(1-c^2)]^{\frac{1}{2}}}. \tag{10.21}$$

When c tends to 1, f tends to ∞. This shows that the pure metal surface in the vacuum is of an extremely high friction coefficient. When c continuously decreases, f decreases rapidly too. This indicates that the surface film of the soft material possesses a friction reduction function. When c is very small, Equation 10.21 becomes

$$f = \frac{\tau_f}{\sigma_s}. \tag{10.22}$$

Figure 10.8 shows the relationship between f and c. From Equation 10.22 it can be known that the revised adhesion friction theory is more accurate and can be used to explain the phenomenon which the simple adhesion theory cannot.

10.2.4 Plowing Effect

The plowing effect is that a hard metal asperity is embedded in the soft metal and during sliding it pushes the plastic flow of the soft metal away to form a furrow. The resistance in plowing is a part of the frictional force. In the abrasive wear, it is the major component of the frictional force.

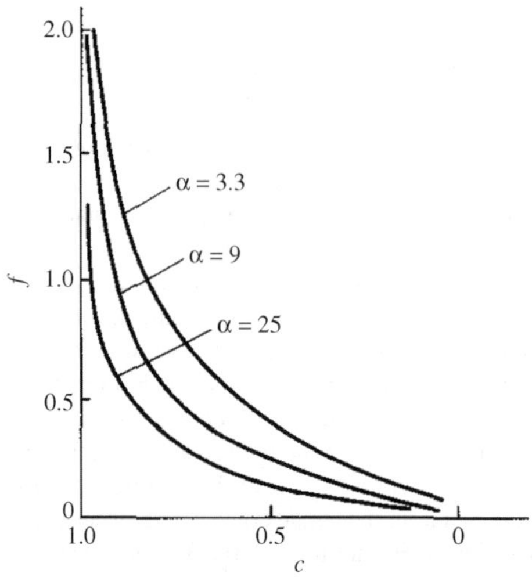

Figure 10.8 Relationship between f and c.

As shown in Figure 10.9, if we assume that the roughness of the hard metal surface is composed of a number of cones with a half-angle of θ. and under the normal load W, the embedded depth of the hard peak into the soft surface is h during sliding, the horizontal contact area $A = \pi d^2/8$. Furthermore, if we suppose that only the front surface of the cone contacts the soft metal, the contact area $S = dh/2$.

If the yield stress of the soft metal is σ_s, under the isotropic condition, the load W and the plowing force P_e are equal to

$$W = A\sigma_s = \frac{1}{8}\pi d^2 \sigma_s$$
$$P_e = S\sigma_s = \frac{1}{2} dh\sigma_s. \tag{10.23}$$

The friction coefficient of the plowing effect is

$$f = \frac{P_e}{W} = \frac{4h}{\pi d} = \frac{2}{\pi}\cot\theta. \tag{10.24}$$

When $\theta = 60°, f = 0.32$; and when $\theta = 30°, f = 1.1$. If the isotropic condition cannot be fully met we can introduce the coefficient k_p to correct Equation 10.24. It will enlarge f. Table 10.1 gives k_p of some metals.

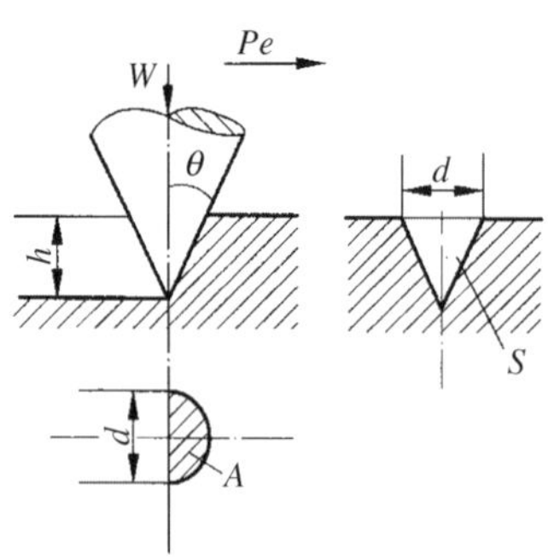

Figure 10.9 Cone peak plowing model.

Table 10.1 Correction coefficient k_p of some metals

Material	W	Steel	Iron	Copper	Tin	Lead
k_p	1.55	1.35–1.70	1.90	1.55	2.40	2.90

If we consider both the adhesion and plowing effects, the frictional force for a single asperity includes the shear and powing forces, that is,

$$F = A\tau_b + S\sigma_s. \tag{10.25}$$

The friction coefficient becomes

$$f = \frac{F}{W} = \frac{A\tau_b + S\sigma_s}{A\sigma_s} = \frac{\tau_b}{\sigma_s} + \frac{2}{\pi}\cot\theta. \tag{10.26}$$

For most cut surfaces, the average roughness peak angle θ is usually large so that the second item at the right-hand side of Equation 10.26 will be very small. Therefore, the plowing effect can be ignored so that Equation 10.26 will be transformed into Equation 10.12. However, if θ is small, the plowing item cannot be ignored.

It should be noted that the adhesion friction theory is an important development. Bowden *et al.* measured the actual contact area. It is only a very small fraction of the apparent area, They also revealed the effects of the plastic flow and the transient high temperature on the formation of the adhesive nodes on the contact point. At the same time, the adhesion friction theory has successfully explained a number of sliding friction phenomena such as: the anti-friction function of the surface film, the jerking motion, the adhesion wear mechanism and so on. The migration phenomenon in the adhesive wear of materials, derived from the theory, has been verified by the tracer technology radiology.

The adhesion friction theory simplifies the complex friction phenomenon although it is not perfect. For example, the actual surface is in the elastic-plastic deformation state, and the friction coefficient varies with the normal load. For another example, the transient high-temperature is not the inevitable phenomenon during sliding, nor is it necessary to form adhesion nodes. Although the adhesion nodes form because of plastic deformation on a very soft or very smooth surface, the adhesion phenomenon can be found under very light normal load. Besides, in the above analysis, the plowing resistance P_e is not related to the shear stress τ_b, but in fact, they are the indicators of the plastic flow of metal. In Equation 10.26, τ_b and σ_s are related to the stress state and the contact geometry of the surface layer, and therefore they are not constants.

10.2.5 Deformation Energy Friction Theory

The adhesion friction theory can only be applied to the cold welding of metal, but is not suitable for non-metallic materials. In addition, the theory obtains the frictional force in the force analysis. However, it is sometimes a more effective way to use the energy method to obtain a force.

Suppose that the two friction surfaces are of the same material. The elastic modulus is E, the shear modulus is G, the normal force is W and the frictional force is F. Then, the deformation energies generated by W and F are

$$E_N = \frac{1}{2}E\sigma^2 = \frac{1}{2}E\left(\frac{W}{A}\right)^2$$

$$E_F = \frac{1}{2}E\tau^2 = \frac{1}{2}E\left(\frac{F}{A}\right)^2, \tag{10.27}$$

where σ is the normal stress; τ is the shear stress; A is the contact area.

If the frictional force F is small, the shear deformation energy is less than the normal deformation energy generated by the load W. F is the static frictional force and the surfaces have no relative motion. With increase of F, if the shear deformation energy exceeds the normal deformation energy, the surfaces begin to slide relatively. This allows us to obtain the critical friction, or the maximum static frictional force. The sliding condition is

$$E_N = E_F. \tag{10.28}$$

Substitute Equation 10.28 into Equation 10.27, we have

$$\frac{1}{2}E\left(\frac{W}{A}\right)^2 = \frac{1}{2}E\left(\frac{F}{A}\right)^2. \tag{10.29}$$

So, the maximum static frictional force is:

$$F = N\sqrt{\frac{G}{E}}. \tag{10.30}$$

Therefore, the maximum static friction coefficient is equal to:

$$f = \frac{F}{N} = \sqrt{\frac{G}{E}}. \tag{10.31}$$

The friction coefficient obtained from the above analysis meets the main conclusion of the classical tribology. The first friction law is that the friction coefficient is proportional to the load, and the second law is that the friction coefficient is not related to the apparent contact area. It should be noted that: (1) although the above equations are obtained under the uniform stress condition, if the tensile or shear elastic modulus of the materials are the same, the equations can also be used in the non-uniform load; (2) if the materials are not the same, sliding will take place as long as the shear energy of any one material exceeds its normal energy.

Table 10.2 shows the tensile modulus, shear modulus and the corresponding maximum friction coefficient of some non-metallic materials. It can be seen from the table that without taking into

Table 10.2 Tensile modulus, shear modulus and the maximum static friction coefficient

Name	Tensile modulus E (GPa)	Shear modulus G (GPa)	Maximum static friction coefficient
Iron	110–160	45	0.53
Iron	151–160	61	0.63
Carbon steel, cast steel	200–220	81	0.61–0.64
Alloy	210	81	0.62
Bronze	105–115	42	0.53–0.632
Brass	91–110	40	0.603–0.62
Hard alloy	71	27	0.617–0.626
Rolled zinc	84	32	0.617
Lead	17	7	0.64
Glass	55	20–22	0.60–0.63
Concrete	14–23	4.9–15.7	0.59–0.83
Across grained wood	9.8–12	0.5	0.20–0.23
Longitudinal grain wood	0.5–0.7	0.44–0.64	0.94–0.97
Bakelite	1.96–2.94	0.69–2.06	0.59–0.83
Nylon	2.83	1.01	0.597

account the surface contamination, if a lubricant exists, the friction coefficient of the longitudinal grain wood is the smallest, that of the across grain wood is the largest, and those of the other materials are between 0.5–0.6.

10.2.6 Binomial Friction Theory

Kragelsky *et al.* believed that the sliding friction is a process to overcome both the surface roughness resistance and the molecular attraction [2]. Thus, the frictional force is the sum of the resistances of both the mechanical and molecular actions.

$$F = \tau_0 S_0 + \tau_m S_m, \tag{10.32}$$

where S_0 and S_m are the mechanical and molecular actions respectively; τ_0 and τ_m are the mechanical and the molecular shear stresses respectively.

According to their study, they proposed:

$$\tau_m = A_m + B_m p^a, \tag{10.33}$$

where p is the load per unit area; A_m is the tangential resistance of the mechanical action; B_m is the normal load coefficient; a is the index, less than, but tends to 1.

$$\tau_0 = A_0 + B_0 p^b, \tag{10.34}$$

where A_0 is the tangential resistance of the molecular attraction; B_0 is the roughness coefficient; b is the index.

Therefore

$$F = S_0\left(A_0 + B_0 p^b\right) + S_m\left(A_m + B_m p^b\right). \tag{10.35}$$

If $S_m = \gamma S_0$, where γ is the ratio constant, the actual contact area $A = S_0 + S_m$ and the normal load $W = pA$, we have

$$F = \frac{W}{\gamma + 1}(\gamma B_m + B_0) + \frac{A}{\gamma + 1}(\gamma A_m + A_0). \tag{10.36}$$

If set

$$\frac{\gamma B_m + B_0}{\gamma + 1} = \beta$$
$$\frac{\gamma A_m + A_0}{\gamma + 1} = \alpha. \tag{10.37}$$

Equation 10.36 can be written as

$$F = \alpha A + \beta W = \beta\left(\frac{\alpha}{\beta} A + W\right). \tag{10.38}$$

Equation 10.38 is called the friction binomial. β is the actual friction coefficient and it is a constant. α/β is the ratio from the molecular force transmitting to the normal load. α and β are determined by the physical and mechanical natures of the friction surfaces.

Comparing Equation 10.38 with Equation 10.1, the equivalent friction coefficient of Equation 10.38 is

$$f = \frac{\alpha A}{W} + \beta. \tag{10.39}$$

We can see that f is not a constant and is related to the ratio A/W. This equation is consistent with the experimental results.

Experiments show that if the friction pair consists of the plastic material, the surfaces are in the plastic contact. The actual contact area A and the normal load W have a linear relationship. Therefore, the friction coefficient f in Equation 10.39 becomes a constant. If the surfaces are in the elastic contact, the actual contact area is proportional to the 2/3 powers of the normal load. Therefore, the friction coefficient of Equation 10.39 decreases with increase of the load.

The experiments show that the binomial friction theory is suitable for the boundary lubrication, but it is also applicable to some actual contacts with the dry friction problems, such as problems of determining the sliding between the dam basis and the rock surface, to calculate the load-carrying capacity of an adhesive joint and so on.

10.3 Micro-friction Theory

On very smooth surfaces, such as the atomic-level flat crystal, the experiments show that friction does not completely disappear and is sometimes significant. This shows that in addition to the plastic deformation, the roughness engagement and the adhesion, there are more fundamental energy dissipation processes to generate the friction. Therefore, it is important to study the energy dissipation process from a micro-view perspective in order to find out the origin of friction and to control friction [3, 4].

A friction process is nonlinear and far from the equilibrium thermodynamic process. Essentially, the friction is the action that the objects resist the relative movement or the moving trend under an external force. It is an energy transferring phenomenon in the interface between the surfaces. When the two surfaces relatively move, the force caused by the movement does work, resulting in energy loss between the contact surfaces. It is known that 85–95% frictional work is converted to thermal energy, the remaining parts to surface energy, sound energy, light and so on.

10.3.1 "Cobblestone" Model

Adhesion friction theory and mechanical-molecular friction theory study the friction from the viewpoint of force. However, the relationships between the key parameters and the basic physical quantities of the surface and interface are unknown. From the microscopic scale, the energy dissipation has been studied and several friction models are established. Here, we will introduce the "cobblestone" model proposed by Israelachvili [5].

Israelachvili suggested considering the influences of the external load W and the internal molecular attraction separately after having studied the friction on a very smooth surface. Then, the frictional force F is superimposed in the following way:

$$F = S_c A + fW, \tag{10.40}$$

where S_c is the critical shear stress; A is the contact area; $S_c A$ is the part which contributed by the intermolecular force, fW is part of Coulomb's law; W is the external load, to be considered as a constant.

Comparing Equation 10.40 with Equation 10.38, we can see that they are quite similar, but the meanings of the parameters are different.

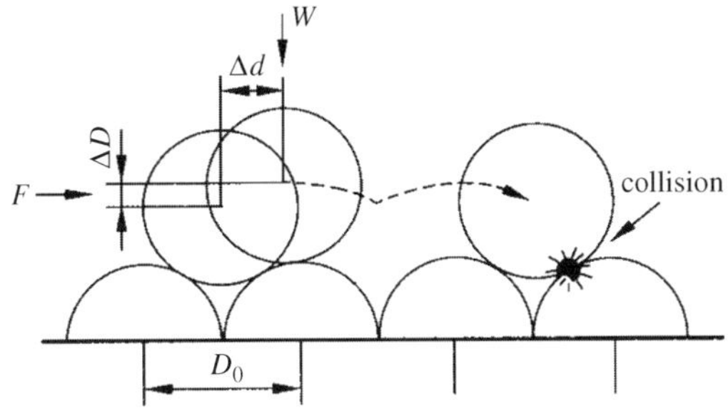

Figure 10.10 "Cobblestone" model.

In the "cobblestone" model, an atomic-scale smooth surface is considered and the relative sliding process is abstractly thought as a spherical molecule moving on a surface whose atoms are regularly arranged, as shown in Figure 10.10.

At first, assume that the spherical molecule is in the minimum potential state and is stable. When it moves forward to Δd in the horizontal direction, the spherical molecule must move up ΔD in the vertical direction. The work done by the frictional force in this process is equal to $F\Delta d$, which is equal to separating the two surfaces away from ΔD. the variation of the surface energy ΔE can be estimated by using the following formula:

$$\Delta E \approx 4\gamma A \frac{\Delta D}{D_0}, \tag{10.41}$$

where γ is the surface energy, D_0 is the equilibrium interface spacing.

During sliding, not all the energy has been dissipated or absorbed by the lattice vibration, but some is reflected back because of the molecular collision. Suppose the dissipated energy is equal to $\varepsilon\Delta E$, where $0 < \varepsilon < 1$ and is a constant, then according to the energy conservation law, we have:

$$F\Delta d = \varepsilon \Delta E. \tag{10.42}$$

Therefore, the critical shear stress S_c can be written as follows:

$$S_c = \frac{F}{A} = \frac{4\gamma\varepsilon\Delta D}{D_0 \Delta d}. \tag{10.43}$$

Israelachvili further assumed that the dissipation mechanisms of the friction energy and the adhesion energy are the same. When the two surfaces relatively slide with a characteristic length l, the friction and the critical shear stress can be written respectively as follows:

$$F = \frac{A\Delta\gamma}{l} = \frac{\pi r^2}{l}(\gamma_R - \gamma_A), \tag{10.44}$$

$$S_c = \frac{F}{A} = \frac{\gamma_R - \gamma_A}{l}, \tag{10.45}$$

where $\gamma_R - \gamma_A$ is the adhesion energy per area.

The model shows that the frictional force and the critical shear stress are proportional to the adhesion energy, but are not related to the adhesion force. This conclusion has partially been confirmed by experiments.

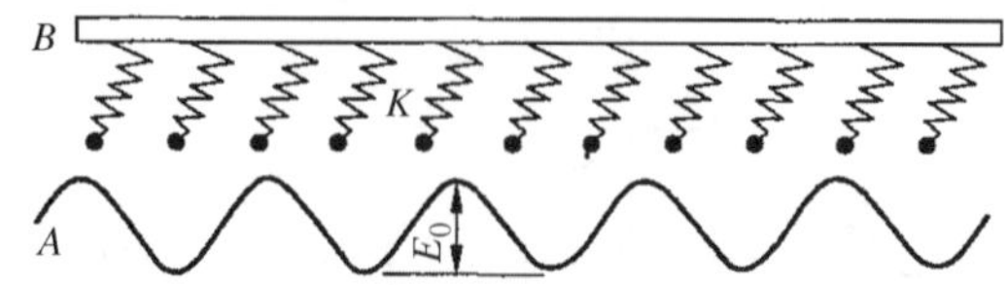

Figure 10.11 Independent oscillator model.

10.3.2 Oscillator Models

10.3.2.1 Independent Oscillator Model

Independent oscillator model (IO) was proposed by Tomlinson in 1929 [6]. He first used the IO model to explain the friction phenomena from the microscopic viewpoint and established the molecular friction theory. In the 1980s, people used the IO model to simulate and explain the experimental results [4].

The IO model is shown in Figure 10.11, where E_0 is the strength of the periodic potential, and K is the spring stiffness. Surface A is simply reduced to a single rigid surface with potential. The atoms on surface B do not interact with each other, but are acted upon by surface A and connected by the springs to the surface B. The springs transfer the energy to surface B so that the friction energy is dissipated.

Because the atoms of surface B do not interact with each other, we only need to study the movement of one atom B_0. The movement of B_0 depends on the integrated potential V_S of the periodic potential A. As shown in Figure 10.12, the black spot is B_0, the lower curve is V_S. In the beginning, B_0 is of minimum potential energy. At first, the movement is the quasi-static sliding (i.e. the sliding velocity is much smaller than the velocity of the solid deformation relaxation) such that B_0 maintains the minimum potential and V_S changes slowly. When B_0 suddenly jumps over the top to the next bottom of the potential to be stimulated by vibration, the periodic potential amplitude becomes large (Figure 10.12c). The energy is irreversibly dissipated away to send off the phonons in solid. The periodic potential has a tendency to transmit the moving energy into the vibrational energy. If the periodic potential is weak, B_0 slides smoothly without friction because V_S has no local minimum.

The IO model is widely used in micro-friction study, such as the influence of material parameters on the atomic scale stick-slip phenomena, influence of the elastic constant of the base material on the friction energy dissipation. Xu, based on the energy dissipation mechanism of the independent oscillator model, proposed a method to calculate the sliding frictional force and the friction coefficient in the elastic contact on a smooth interface [7]. The formulas are as follows:

$$
\begin{aligned}
F &= 4k\frac{\gamma A}{a_0}\left[1-\left(1+\frac{0.207a_0}{l}\right)e^{-\frac{0.207a_0}{l}}\right] \\
f &= 4k\frac{\gamma A}{a_0 N}\left[1-\left(1+\frac{0.207a_0}{l}\right)e^{-\frac{0.207a_0}{l}}\right],
\end{aligned}
\tag{10.46}
$$

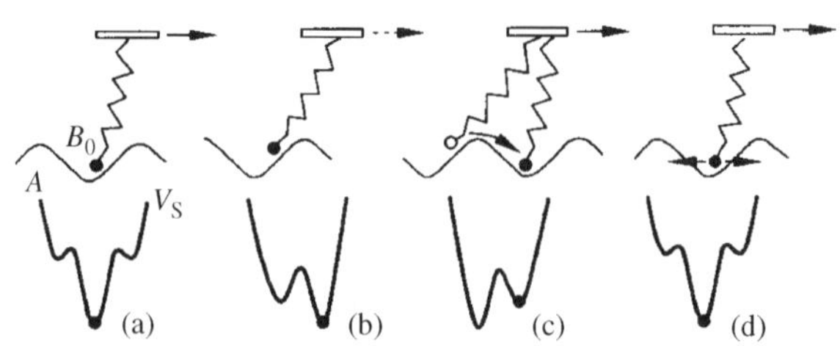

Figure 10.12 Energy dissipation mechanism of independent oscillator.

where F is the frictional force; f is the friction coefficient; k the coefficient related to the absolute temperature; γ is the interface free energy; A is the contact area; N is the normal load; a_0 is the lattice constant of the material; l is the proportional coefficient, determined by the following formula:

$$l \approx 2\left(\frac{2\gamma}{12\pi E r_{WS}}\right)^{0.5}, \tag{10.47}$$

where E is the bulk modulus; r_{WS} is the Wigner-Seitz radius of the metal crystal.

Xu's study shows that with increase of roughness the frictional force and friction coefficient also increase up to the critical value. At this time, the result obtained by his method is close to the result of Bowden's adhesion friction theory.

10.3.2.2 Composite Oscillator Model

In order to further study the friction and energy dissipation mechanism of a sliding process, Xu modified the independent oscillator model to the composite oscillator model, taking into consideration the friction on the smooth surface without wear [3, 8]. The composite oscillator model consists of the micro-elastic oscillator with the stiffnesses K_A and K_B respectively, and the multiple micro-independent oscillators on the interface with the stiffnesses, $K_{A,S}$ and $K_{B,S}$ respectively, as shown in Figure 10.13.

In the composite oscillator model, because the oscillators on the slow moving surface adsorbs the energy of the high moving surface and the energy cannot be returned to the high moving surface, the energy will be lost. Compared with the independent oscillator model, it can be found that the composite oscillator model does not simply take the friction interface as the periodic potential, but uses the same oscillators on both surfaces to express the periodic potential, where the contact stiffness is approximately expressed by the periodic potential. However, in the independent oscillator model, the energy of a tribo system cannot be transferred between the upper and lower surfaces, while for the

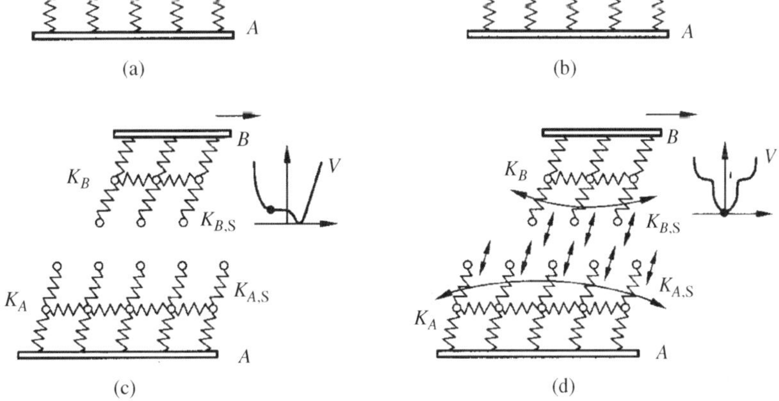

Figure 10.13 Composite oscillator model.

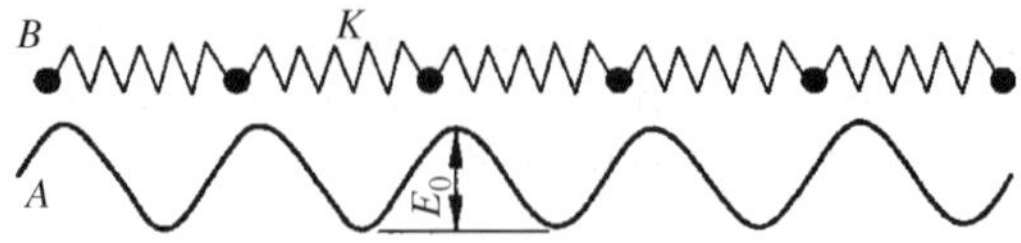

Figure 10.14 One-dimensional FK model.

composite oscillator model, the outside work will pass the energy to the other surface. This is obviously suitable for an actual tribo system.

10.3.2.3 FK Model

The FK model was proposed by Frenkel and Kontorova in 1938. After many years of development, it has become the basic model in low-dimensional and nonlinear physics [3, 4]. The FK model is composed of a linear one-dimensional atomic chain in the periodic potential, as shown in Figure 10.14. In moving, the interaction between the two atoms of the chain is simulated by the spring, with the spring stiffness of K. The interaction of the atomic potential energy surface is presented by the periodic potential field, where E_0 is the amplitude of the periodic potential.

The FK model can be used to study nonlinear friction phenomena. It has been successfully used in studying the static and dynamic characteristics of a quasi-kenamatic sliding friction, the mechanism of a micro-stick-slip phenomenon, phonon excitation and so on.

Weiss and Elmer, based on the Burridge-Knopoff model, proposed the FKT model (Frenkel-Kontorova-Tomlinson model), as shown in Figure 10.15. The FKT model includes two kinds of springs in the FK model and IO model, to take into account the supporting actions of the inter-surface atoms.

The FKT model takes into account both the interaction of surface atoms and the substrate influence. It is a suitable model for studying the interfacial friction and micro-mechanism of energy dissipation. For example, Gyalog, Thomas *et al.* used the two-dimensional FKT model to study the sliding friction phenomena on two infinite atomic-level smooth surfaces. They studied the mechanism of super-lubrication and believed that the common commensurability is the key factor for the super-lubrication. When the two surfaces changes from commensurability to incommensurability, friction disappears. However, the relationship of the frictional force and the staggered angle also needs more theoretical and experimental study.

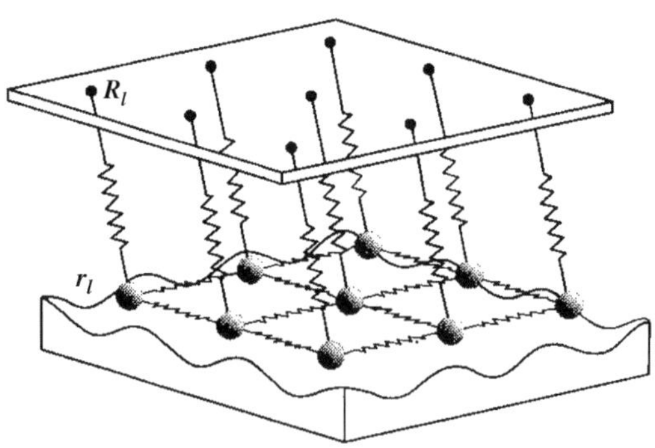

Figure 10.15 Two-dimensional FKT model.

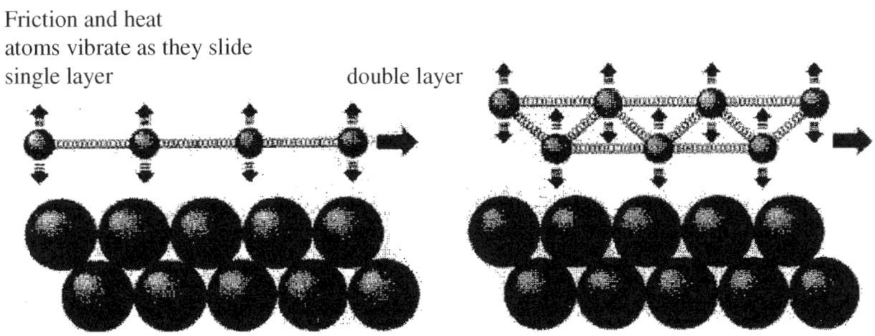

Figure 10.16 Schematic diagram of the phonon friction.

10.3.3 Phonon Friction Model

The concept of the phonon friction was first proposed by Tomlinson in 1929 [6]. In 1980s, Gary McClelland of IBM Almaden Research Center and Jeffrey Sokoloff of Northeastern University re-proposed the phonon friction. In 1991, Tabor proposed that the energy in the friction without wear will be dissipated in the form of the atomic (phonon) vibration on the basic theoretical research meeting funded by NATO [3].

In the study on the microscopic mechanisms of energy dissipation of the interfacial friction without wear, two major models were proposed: the phonon friction model and the electronic friction model. The phonon friction is believed to occur when the neighboring atoms in the surface relatively slide. It is related to the atomic vibration, which is mechanically activated by the sliding surface. The energy eventually dissipates in the form of heat, as shown in Figure 10.16. The electric friction is generated while the electrons of the metal interface are induced by sliding. The electronic friction model is related to the quantum theory, but the current research in this area is not fully carried out and its mechanism is not understood yet.

Krim and Widom experimentally confirmed the presence of phonons by a quartz crystal micro-balance (QCM) which has long been used to measure a very small weight and precise time [9]. In 1985, Krim, Widom *et al.* used the modified QCM to measure the sliding friction of an adsorbed film on the metal surface. They measured that the life of an existing phonon is no more than 0.1 ns.

The typical feature of the phonon friction is that the sliding surface is extremely sensitive to commensurability. Theoretically, if the two surfaces in contact change to be incommensurable, the sliding frictional force will be greatly reduced. Another typical feature of the phonon friction is that there is no static friction on a clean and elastic contact surface, that is, the frictional force is equal to the sliding velocity times the friction coefficient. However, at the micro-scale static friction is a common phenomenon. The force driving a stationary object to move is larger than the force keeping the object in motion. The static frictional force usually depends on the contact time so it is not a constant.

Till now, the energy dissipation mechanisms of the phonon and the electron frictions are still unclear. Furthermore, whether there is other mechanism for the frictional energy dissipation should be studied further.

10.4 Sliding Friction

It is an important subject to study the variation of the friction coefficient with the influencing factors in order to control a friction process and reduce the friction loss. The friction coefficient is an integrated nature of a tribo system, which is influenced by a variety of factors, such as: material matching, stationary contact time, load, friction pair stiffness, velocity, temperature, contact surface geometry, physical

Table 10.3 Influence of loading rate on the friction coefficient (bronze and steel)

	Dry friction			Lubrication		
Acceleration m/s^2	50	110	550	50	110	300
Friction coefficient	0.20	0.22	0.26	0.11	0.11	0.14

properties of surface layer and environmental media. Therefore, the friction coefficient varies significantly with the working conditions. Thus, it is very difficult to accurately pre-determine the working data and comprehensively estimate the influence factors of the friction coefficient.

10.4.1 Influence of Load

A load influences the frictional force by increasing the contact area and deformation. The friction always occurs on the peak points of the roughness because a surface processed by the common method is always rough. The number of contact points and the contact point area increases as the load increases. At first, the contact area increases, and then the number of points increases. The experimental results show that when the stress on a smooth surface is about half of the material hardness or the contact stress of a rough surface reaches 2–3 times the hardness, plastic deformation occurs.

When the surface is in plastic contact, the load has no influence on the friction coefficient anymore. Usually, the metal surface is in the elastic-plastic contact so the actual contact area is not linear to the load, and the friction coefficient decreases with increase of the load.

Because the friction surface in the elastic-plastic contacts, the friction coefficient will also change with the loading rate. When the load is very small, the influence of the loading rate is more significant. Table 10.3 shows the variation of the friction coefficient with load rate.

For the steel and cast iron friction pair, the friction coefficients at the different load rates are between 0.17 and 0.23.

10.4.2 Influence of Sliding Velocity

If the sliding velocity does not cause the nature of surface layer to change, the friction coefficient is almost independent of the sliding velocity. However, under common circumstances, the sliding velocity will cause the temperature to rise, resulting in deformation, inducing chemical reaction, wear and so on. Thus, it significantly affects the friction coefficient.

The experimental results of Figure 10.17 are obtained by Kragelsky *et al.* [2]. For the general friction pairs in the elastic-plastic contact, the friction coefficient increases to the maximum with the sliding velocity and then decreases. If the surface stiffness or load increases, the maximum coefficient will move toward the original point. However, if the load is extremely light, the friction coefficient increases continuously with the sliding velocity.

The relationship of the friction coefficient and the sliding velocity can be expressed by the following formula:

$$f = (a + bU)e^{-cU} + d, \tag{10.48}$$

where U is the sliding velocity; a, b, c and d are the constants to be determined by the material properties and the load as shown in Table 10.4.

The influence of the sliding velocity on the frictional force depends on the temperature condition. The sliding velocity induces the temperature to rise so as to change the nature of the surface layers, the interaction in the friction process, and the destruction conditions. Thereby, the friction coefficient is also

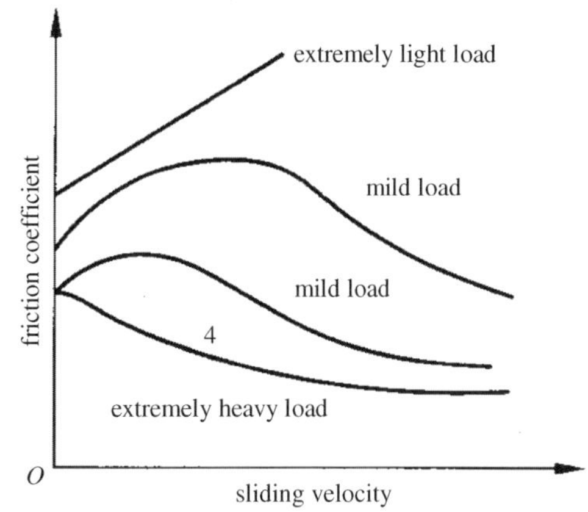

Figure 10.17 Friction coefficient via sliding velocity.

changed as well. If a material keeps the mechanical properties unchanged over a wide temperature range, such as graphite, its friction coefficient is almost not affected by the sliding velocity.

10.4.3 Influence of Temperature

In order to fully describe the surface temperature during a friction process, the instantaneous temperature, the average temperature, the bulk temperature, the temperature gradient, and the heat distribution are usually studied. Generally, the frictional heat influences the friction properties mainly in the following two aspects. One is to transform the lubrication state, such as from the hydrodynamic lubrication to the dry friction or the boundary lubrication. The other is to change the structure of the surface layer, namely, to change the friction surface and the surrounding medium, such as diffusion of the surface atoms or molecules, adsorption or desorption, surface structural variation and phase transformation.

The influence of temperature on the friction coefficient is closely related to the variation of the surface layer. Many experimental results show that as temperature increases, the friction coefficient increases. And when the surface temperature is high enough to soften the material, the friction coefficient will be reduced significantly.

10.4.4 Influence of Surface Film

The atoms on the metal surface are usually in an unbalanced state so they easily form a surface film with the surrounding media. The surface deformation and the temperature rise promote the formation of the

Table 10.4 Values of *a*, *b*, *c* and *d*.

Friction pair	Load per unit area N/mm^2	*a*	*b*	*c*	*d*
Iron-steel	1.9	0.006	0.114	0.94	0.226
	22	0.004	0.110	0.97	0.216
Iron-iron	8.3	0.022	0.054	0.55	0.125
	30.3	0.022	0.074	0.59	0.110

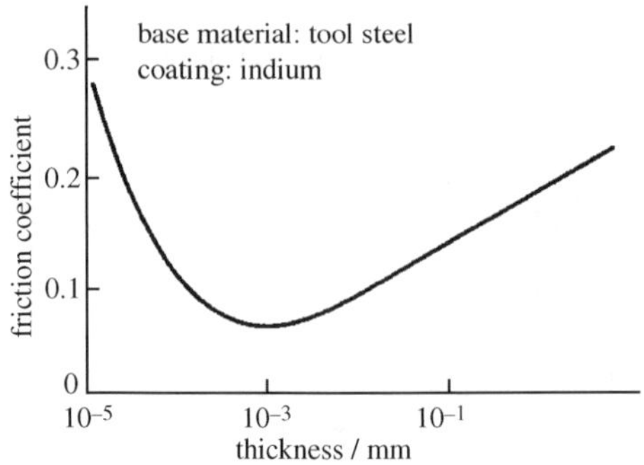

Figure 10.18 Influence of surface film thickness.

Table 10.5 Anti-friction abilities of oxide film and sulfide film

		Friction coefficient		
Friction condition	Friction pair	Pure surface	Oxide film	Sulfide film
Dry friction	Steel-Steel	0.78	0.27	0.39
	Copper-Copper	1.21	0.76	0.74
Stearate lubricant	Steel-Steel	0.11	0.19	0.16

surface film. Sometimes, in order to reduce friction, a thin film is artificially generated on the friction surface, such as the surface film of indium, cadmium, lead, soft metal sulfide, chloride, or phosphide.

A surface film similar to a lubricant film can reduce friction. It weakens the binding force between the atoms or substitutes the ion-binding force into the Van-der-Waals force so as to reduce the surface molecular force. Furthermore, the mechanical strength of a surface film is usually weaker than that of the bulk material so the sliding shear resistance is smaller.

The surface film thickness has a great influence on the friction coefficient. As shown in Figure 10.18, the experimental results obtained by Bowden gives the relationship of the film thickness and the friction coefficient while there is an indium film on the tool steel surface [1]. When the film thickness is 10^3 mm, the friction coefficient is the minimum. If the surface film is too thin or too thick, the friction coefficient is high.

Table 10.5 illustrates the anti-friction abilities of the surface films in the dry friction.

If the surface film is damaged, the friction coefficient will increase dramatically. The damage may be mechanical by a load. It depends on the hardness of the surface film and the connecting strength with the base material. For a lead, indium or other fusible metal surface film, the damage occurs when temperature rises to the melting point.

When a relatively rigid film, such as aluminum oxide, is formed, the connecting strength is usually low due to high brittleness. The cadmium film has a very good anti-friction effect and its connecting strength with the substrate is weak such that it is easily erased from the surface. A graphite film forming between metal and graphite is of a stable friction coefficient.

10.5 Rolling Friction

Although rolling was used much early, there has been little mechanism research or experimental data on rolling friction.

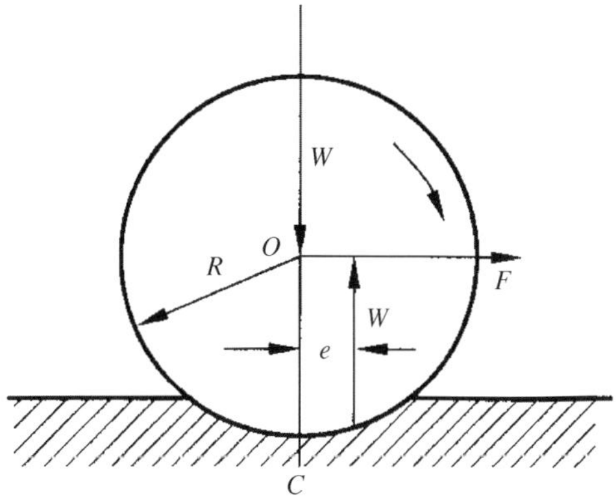

Figure 10.19 Rolling friction.

As shown in Figure 10.19, when a cylinder rolls on a plane, the deformation of the contact area is asymmetrical to the contact point C. As the contact pressure distribution is asymmetrical the counter-force of the surface should be offset. The moment of the force to the contact point is called the rolling friction moment.

The rolling friction coefficient k is defined as the ratio of the rolling friction moment and the load, that is,

$$k = \frac{FR}{W} = e, \tag{10.49}$$

where F is the driving force, W is the load; R is the radius of the cylinder; and e is the eccentricity.

It can be seen that the rolling friction coefficient, different from the sliding friction coefficient, is a dimensionless quantity.

In addition, the dimensionless quantity can also be used as we know the rolling resistance coefficient f_r which is the ratio of the driving force F to the load W, that is

$$f_r = \frac{F}{W} = \frac{k}{R}. \tag{10.50}$$

In 1785, Coulomb was the first to experimentally derive the rolling friction law: the product of the rolling resistance coefficient f_r and the cylinder radius R is a constant, that is, the rolling friction coefficient k or the eccentricity e is a constant. They depend on the friction pair material properties, but are not related to the load. Later, Dupuit proposed a revised formula, commonly referred to as the Dupuit law, which is

$$f_r = \frac{k}{\sqrt{D}}, \tag{10.51}$$

where D is the cylinder diameter.

The rolling friction coefficient k is the constant determined by the material and the surface conditions, and does not change with the velocity and the load. Clearly, the above-mentioned rolling friction law is the mechanism of rolling friction and can be approximately applied to engineering calculations.

10.5.1 Rolling Friction Mechanism

Usually, a rolling motion can be regarded as a combination of the following three basic rolling forms. The friction mechanisms and the surface actions of these three forms are different.

1. Free rolling
   That a cylinder or a sphere rolls in a flat and straight line without constraint is called free rolling, which is the simplest form of rolling.
2. Rolling with traction
   In the contact region there are both normal load and tangential traction, for example a friction wheel rolling.
3. Rolling with sliding
   When the geometries of the two rolling bodies cause the tangential velocities to be unequal, rolling is definitely companied by sliding. For example, the ball rolls on the raceway of a thrust ball bearing.

10.5.2 Resistances of Rolling Friction

The rolling friction mechanism is obviously different from that of sliding friction. Unless there is sliding, the rolling friction usually neither has a plowing effect, nor adhesion effect. The resistances of the rolling friction are mainly the following four types.

1. Micro-sliding
   Micro-sliding is a common phenomenon in the rolling process. When two objects with different elastic modulus are in the free rolling relatively, because of different tangential displacements, a micro-slip will occur. If the tangential traction exists on the rolling contact surface for transferring the mechanical work, it can produce a large amount of micro-sliding. When the geometry of the two surfaces makes the tangential velocities unequal, it will lead to even more micro-sliding.
   A large portion of the total rolling friction is the resistance produced by the micro-sliding and the friction mechanism is the same as the sliding friction.
2. Plastic deformation
   In a rolling process, when the surface contact stress reaches a certain value, the plastic deformation forms in the certain depth of the surface. The plastic deformation zone expands with increase in the load. Based on elastic-plastic calculation, the energy consumed by plastic deformation can be used to express rolling friction. For example, a ball freely rolls along a plane, the rolling frictional force F produced by the plastic deformation in the front of the ball movement can be expressed as follows:
   $$F \propto \frac{W^{\frac{2}{3}}}{R}, \tag{10.52}$$
   where W is the normal force; R is the ball radius.
3. Elastic hysteresis
   The elastic deformation produced in a rolling process needs some energy. Most elastic deformation recover after the contact finishes but a small part present elastic hysteresis. The elastic hysteresis energy of a viscoelastic material is much greater than that of a metal material and is often the major part of the rolling friction.
4. Adhesion effect
   The adhesion between the surfaces will be vertically separated during the rolling process. Because the separation is subject to a pulling force, the adhesive area does not expand in the process so the adhesion force is small. Usually, the adhesive resistance is only a small fraction of the rolling friction. It should

be pointed out that for railway transportation, the adhesive effect must exist between the wheel and the rail to prevent slipping and wear aggravatedly during rolling. Studies have shown that the adhesion between the wheel and rail is closely linked to material properties, contact conditions and environmental pollution [7].

To summarize, rolling friction is generally complex. Many factors affect rolling friction and the functions of these factors are different when the rolling types and working conditions change.

10.6 Special Friction and Friction Control

10.6.1 Special Friction

Many friction pairs of the modern machineries and equipment are often at high velocity, high or low temperature, in a vacuum or have other special operating conditions so the friction characteristics are different from those of the usual operating conditions.

10.6.1.1 High Velocity Friction

In aviation, with chemical or turbine machineries, the relative sliding velocity of the friction surface usually exceeds 50 m/s, even above 600 m/s. In such a case, a lot of friction heat is produced on the contact surfaces. With high sliding velocity and short contact time, a lot of friction heat is instantaneously produced and is hardly dissipated. Thus, in the interval, surface temperature is very high and temperature gradient is large so scuffing easily occurs.

The surface temperature of high-velocity friction can reach the melting point of the material, sometimes resulting in a thin melted layer on the contact area. The molten metal forms a liquid film to decrease the friction coefficient as velocity increases, such as in Table 10.6.

10.6.1.2 High Temperature Friction

High-temperature friction appears in engines, nuclear reactors and aerospace equipment. The friction materials working under high temperature are usually the metal compounds that have melted with difficulty, or ceramics such as metal compounds of steel, titanium, tungsten, and silicon carbide ceramics.

Studies have shown that under high temperature friction, the friction coefficient of various materials varies similarly with the temperature, that is, with increase of temperature, the friction coefficient first drops slowly and then increases quickly. In the process, there is minimum friction coefficient. For the typical high temperature friction materials, the minimum friction coefficient appears at about 600–700 °C.

10.6.1.3 Low Temperature Friction

Some friction pairs may work at low temperatures or in the cooling medium, where the ambient temperature is often below 0 °C. At this time, the influence of the frictional heat is very small, but the cold

Table 10.6 High-velocity friction coefficient

	Copper			Iron		Steel No. 3		
Sliding velocity m/s	135	250	350	140	330	150	250	350
Friction coefficient	0.056	0.040	0.035	0.063	0.027	0.052	0.024	0.023

Note: The friction piece is a steel ring composited with 0.7 carbon and with HB250 hardness; and the load per unit area is 8 MPa.

brittleness and organizational structure of the material will have a significant influence on the friction properties. Common low friction materials are the alloys of aluminum, nickel, lead, copper, zinc, and titanium, graphite and fluorine plastic.

10.6.1.4 Vacuum Friction

In the space or vacuum environments, the friction pairs have special features. For example, if the surrounding medium is very thin, and the surface adsorption film or oxide film ruptures, it is difficult to regenerate again. This results in the direct contact of metals to produce a severe adhesive friction and wear. The higher the vacuum, the larger the friction coefficient is. Furthermore, because there is no convection cooling in the vacuum condition so the frictional heat is difficult to be disputed away, the surface temperature is usually very high. Moreover, because of evaporation, a liquid lubricant cannot be used. Thus, solid lubricants or self-lubricating materials should be applied. In order to form a stable film to protect the surface, the friction pair can be made of self-lubricating materials of disulfide or selenide compounds or metal coating such as tin, silver, cadmium, gold, or lead.

10.6.2 Friction Control

The effective real-time friction control is a goal to be pursued in engineering. The common methods to reduce or increase the friction coefficient are through selecting lubricants or friction materials. However, because the friction coefficient depends on the load, velocity, temperature and other factors, it is difficult to predict it accurately. Therefore, it is more difficult to accurately adjust or control it to vary with variation of operating conditions and the run-time. In the following section, development in electric control friction technology is briefly described. The electric control friction is the method through an external electric field to change the friction coefficient.

10.6.2.1 Approach to Apply Voltage

The approach to applying an external voltage is an important method for electric friction control. There are three ways, as follows:

1. The direct method is to take the friction pair as two electrodes to connect both ends of the power to supply the voltage as shown in Figure 10.20a. However, this method has some limitations in that the friction pairs must be the conductors and the contact resistance must be small. Therefore, without a high-current source, only a low voltage can be applied. Usually, the voltage on the metal friction pair is only several milli voltages, which cannot bring about a significant electric friction effect.

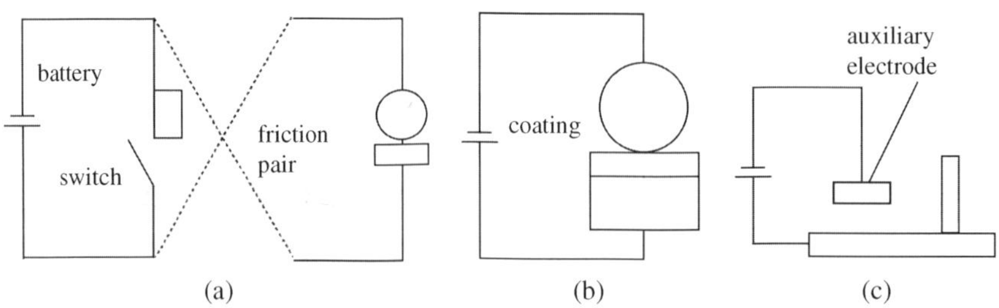

Figure 10.20 Approaches to applying the external voltage.

2. A coating method is to coat an insulated layer on the friction pair surfaces, as shown in Figure 10.20b. This method must ensure that the coating cannot be broken down under the friction and wear process. If a part of the friction pair is a conductor, the other is a semiconductor or conductive silicon rubber, the voltage applied can reach several voltages.
3. The indirect method does not take the friction pair as the electrodes, but introduces the auxiliary electrodes in the vicinity of the contact area, as shown in Figure 10.20c. One advantage of this method is that a higher voltage can be achieved whether the friction pair is conductor or insulator. And, if the auxiliary electrode serves as the anode, the friction pairs can be prevented from being electrochemically corroded. In addition, the auxiliary electrodes need not move together with the friction pair so it is easy to apply a voltage on the rotating friction pairs. The disadvantage of the method is that the direction, magnitude and distribution of the electric field in the contact region are more complex and it is very difficult to analyze the relationship between the applied voltage and the friction coefficient.

While applying a voltage, another important step is to select the lubrication condition. Whether there is a lubricant or not and what kind of lubricant is to be used have a decisive influence on friction control.

10.6.2.2 Effectiveness of Electronic Friction Control

Many researchers have tried to impose the external electric field to alter the friction coefficient of the metal friction pair in dry or lubricated condition. However, the results obtained are not remarkable enough because the friction coefficient varies generally only a small percentage to 30%. A recent study found that in water-based lubricant, to apply a voltage on the friction pair composed of metal and ceramic or silicon and ceramic can change the friction coefficient greatly, rapidly and reversibly [10, 11]. The results show that under some appropriate conditions, the electronically friction control can be realized successfully.

Figure 10.21 gives the experimental results of changing the friction coefficient by applying a voltage. In experiments, a silicon nitride ball slides on a stainless steel plate with a low velocity, known as a rotating

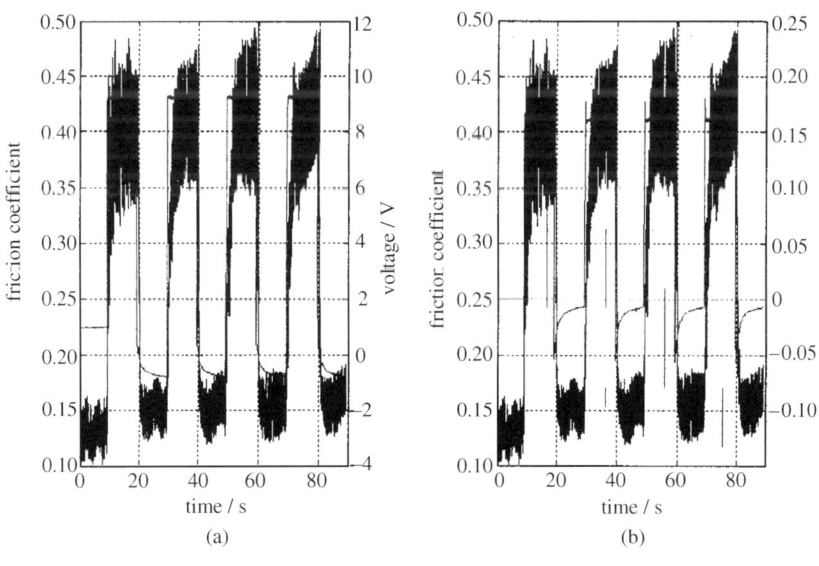

Figure 10.21 Electrically controlled friction coefficient.

pin disc test. The silicon nitride ball has a diameter of 4 mm, the load is 3.3 N, and the velocity is 0.3 m/s respectively. The lubricant is sodium dodecyl sulfate solution with a concentration of 0.01 mol/L. The voltage applied approach is shown in Figure 10.20c, and the voltage is a square-wave. Graphite auxiliary electrodes are used. During the experiment, the load and velocity remain unchanged, but magnitude and polarity of the external voltage vary.

From Figure 10.21, we can see that the first 10 seconds correspond to the state of nonapplied voltage. The current is zero. The voltage of the original battery, composed of the graphite-water solution-stainless steel, is about +1 V. At this time, the friction coefficient is about 0.13. In the subsequent 10 seconds, the external power supply is connected so the voltage rises up to +9.2 V and the current is about 0.16A. At the same time, the friction coefficient rises to about 0.4. And, after another subsequent 10 seconds, the polarity and magnitude of the external power supply are changed so that the voltage becomes −0.7 V, and the corresponding friction coefficient drops from 0.4 to about 0.15. The following changes have the same tendency as before. During experiments, the maximum variation of the friction coefficient is about 160%, the rising and falling time is less than 0.5 seconds. It can be seen that the friction coefficient can cause a rapid response in the voltage polarity switch during the experiments.

Studies have shown that the reason the applied external voltage leads to a higher friction coefficient is because the water electrolysis induces the decomposition of the adsorption film on the metal surface. If the voltage polarities are changed, the surfactant ions are re-adsorbed onto metal surfaces and decrease the friction coefficient [12].

10.6.2.3 Real-Time Friction Control

Figure 10.22 gives an example of using voltage to control the friction coefficient. The goal of the friction coefficient is the given curve (in this case a sine curve) [13]. In the experiment, the voltage is applied according to Figure 10.20c. The materials of the friction pair are ceramics and brass. The engineering ceramics contains α-Al_2O of 99.7 wt%, MgO of 0.25 wt% and impurities. The ceramic sample is sintered into a cylinder with ϕ 16 × 88 by grinding with roughness $R_a = 0.4$ μm. The grade of the brass block is H68 with 60 mm × 20 mm × 12 mm and roughness $R_a = 1.6$ μm. The testing machine is a reciprocating sliding pin disc friction tester; the lubricant is a zinc stearate solution.

In the experiment, the friction coefficient is firstly set to 0.3 according to the theoretical curve and then varies in the increment of 0.1/h, according to the sine curve. A cycle of the friction coefficient curve is divided into 12 steps so the length of each step is 5 min. Take the theoretical value as the goal of friction coefficient in each section. Using previous experimental results, the required voltage can be determined by the goal. The experimental results are shown in Figure 10.22. The histogram is designed to achieve the goal of the friction coefficient of the sine curve, and the dotted line is the actual experimental friction

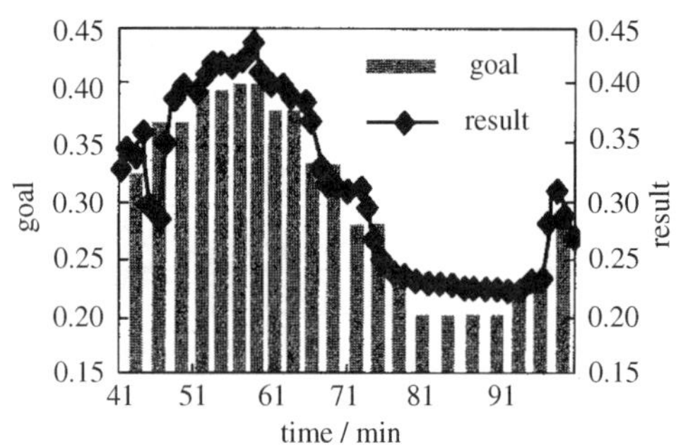

Figure 10.22 Electrical friction coefficient control test.

coefficient controlled by the given voltage. The results show that the electric controlled friction coefficient fits the pre-set curve well.

It should be pointed out that the electronic friction control technology is still at the laboratory research stage, and for engineering applications a lot of practical difficulties also need to be solved. However, with in-depth study, we believe that friction control can be realized in the near future.

References

1. Bowden, F.P. and Tabor, D. (1964) The Friction and Lubrication of Solid, Clarenden Press, Oxford.
2. Kragelsky, I.V., Dobychin, M.H. and Kombalov, V.S. (1977) Foundations of calculations for friction and wear, Mashinostroenie (in Russian), Moscow.
3. Xu, Z.M. (2006) Study on friction characteristics and computation based on barrier methods of contact interface. South China University of Technology PhD thesis.
4. Zhang, T., Wang, Hui, and Hu, Y.Z. (2001) Models of wearless friction at the atomic scale. *Journal of Tribology* **21** (5), 396–400.
5. Israelachvili, J.N. (2001) Microtribology and microrheology of molecularly thin liquid film, in Modern Tribology Handbook (ed. B. Bhushan), CRC Press LLC, New York, pp. 24–66.
6. Tomlinson, G.A. (1929) A molecular theory of friction. *Philosophical Magazine Series*, **7**, 905–939.
7. Youfu, Pei, Yuansheng, Jin, and Shizhu, Wen (1995) The impact factors and control measures of wheel-rail adhesion. *Foreign Rolling Stock*, **2**, 5–7.
8. Xu, Z.M. and Huang, P. (2006) Composite oscillator model for the energy dissipation mechanism of friction. *Acta Physica Sinica*, **55** (5), 2427–2433.
9. Krim, J. and Widom, A. (1988) Damping of a crystal oscillator by an adsorbed monolayer and its relation to interfacial viscosity. *Physical Review*, **B38**, 12184–12189.
10. Jiang, H.J., Meng, Y.G., and Wen, S.Z. (1998) Effect of external DC electric fields on friction and wear behavior of alumina/brass sliding pairs. *Science in China (Series E.)*, **41** (6), 617–625.
11. Meng, Y.G., Hu, B., and Chang, Q.Y. (2006) Control of local friction of metal/ceramic contacts in aqueous solutions with an electrochemical method. *Wear*, **260**, 305–309.
12. Chang, Q.Y., Meng, Y.G., and Wen, S.Z. (2002) Influence of interfacial potential on the tribological behavior of brass/silicon dioxide rubbing couple. *Applied Surface Science*, **202** (1–2), 120–125.
13. Meng, Y.G., Jiang, H.J., and Wong, P.L. (2001) An experimental study on voltage-controlled friction of alumina/brass couples in zinc stearate/water suspension. *Tribology Transactions*, **44** (4), 567–574.

11

Characteristics and Mechanisms of Wear

Wear is a continuous damage process of surfaces, which are in contact with a relative movement. Wear is the inevitable result of friction. The total damage caused by wear in the world is very great. According to statistics, the failure of a mechanical part mainly occurs in three ways: wear, fatigue and corrosion. Among them, wear is the largest factor, contributing about 60–80%. Therefore, study on wear mechanism and measures to improve wear-resistant ability will effectively save materials and energy, increase the performance and service life of machinery and equipment and reduce maintenance costs. This is of great significance.

With the rapid development of science and technology, study to improve wear of machinery and equipment has become crucial; especially in high-speed, heavy-load, precision, and special operation conditions, wear study has become urgent. At the same time, since the 1960s, materials science, surface physics and chemistry, as well as surface test technology have rapidly developed to promote the study of the wear mechanism.

The aim of wear study is to find the laws, influencing factors and characteristics of wear by analyzing the phenomena, in order to seek wear control methods and improve wear-resistant ability. Generally, the main contents of wear study include:

1. the conditions, characteristics and variation laws of occurrence of wear;
2. the influence factors of wear, including friction pair materials, surface topographs, lubrication conditions, environmental conditions, sliding speed, load, temperature and other working condition parameters;
3. the physical model of wear and wear calculation;
4. measures to improve wear resistant ability;
5. wear testing technology and experimental analysis.

11.1 Classification of Wear

The purpose of classification is to find the basic types of wear from a lot of phenomena so as to reasonably simplify the wear study and analyze the nature of wear. The wear classification expresses the recognition of the wear mechanisms. Although many different wear mechanisms have been put forward, there is no universally accepted one.

Principles of Tribology, First Edition. Wen Shizhu and Huang Ping.

11.1.1 Wear Categories

Initially, based on actions on the friction surface, wear is divided into the following three types.

11.1.1.1 Mechanical Wear

On a friction surface, the wear produced by mechanical action includes abrasive wear, surface plastic deformation, and brittle spalling. Among them, abrasive wear is the most common form in mechanical wear.

11.1.1.2 Molecular and Mechanical Wear

The molecular force will cause surface point adhesion, and then the mechanical force shears the point away and thus generate adhesive wear.

11.1.1.3 Corrosive and Mechanical Wear

The surface is first corroded by the chemical reaction caused and then the mechanical action in the friction process accelerates the corrosion process, which includes oxidation wear and chemical corrosion wear.

Clearly, although the above categories explain the causes of wear to a certain extent, it is too general to be used.

11.1.2 Wear Process

In 1962, Kragelsky proposed a more comprehensive classification of wear. According to his classification, he divided wear into three processes and used each process to illustrate the relationship of different wears, as shown in Figure 11.1 [1].

The three processes of wear shown in the figure are as follows.

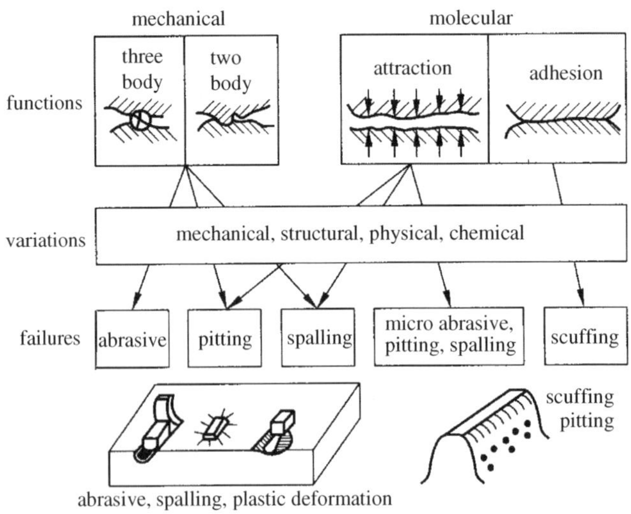

Figure 11.1 Wear classification of Kragelsky.

11.1.2.1 Surface Interaction

The interaction between two friction surfaces can be divided into two types: mechanical or molecular. The mechanical action includes elastic deformation, plastic deformation and the plowing effect, which can also be caused by direct engagement of the roughness of the two surfaces or by the external particles to produce the three-body friction and wear. The actions of the surface molecules include two kinds of effects, attraction and adhesion; the former force is smaller than the latter.

11.1.2.2 Variation of Surface

In the friction and wear process, the relationships between the different factors are complicated. Under the interaction, the mechanical, structural, physical and chemical properties of the surface will vary. These are caused by surface deformation, frictional force, temperature, environmental media and other factors, as shown in Figure 11.2.

Plastic deformation makes the metal surface harden and become brittle. If the surface withstands repeated elastic deformation, fatigue damage will occur. Friction heat will cause high temperature on the contact surfaces and annealing can make the contact surface metal soften and subsequently rapid cooling will result in re-crystallization or decomposition of the solid. The influence of the external environment is mainly on the dissipation of the medium in the surface, including oxidation and other chemical corrosions. Thus, it changes the organizational structure of the metal surface.

11.1.2.3 Forms of Surface Damage

In Figure 11.1 the following forms of wear are given:

1. **Abrasion:** the plowing effect on the frictional surface produces abrasive particles and grooves along the direction.
2. **Pitting:** the metal fatigue damage on the surface forms pits due to the repeated actions of the contact stresses.
3. **Peeling:** due to the deformation strengthening under the load, the metal surface becomes brittle, generating micro-cracks and causing some materials to peel off.
4. **Scuffing:** because of the adhesive effect, the surface forms adhesive points with high connection intensity such that the shear breaks the points, causing serious wear at a certain depth.
5. **Micro-wear:** all the above damage forms happen in the microscope.

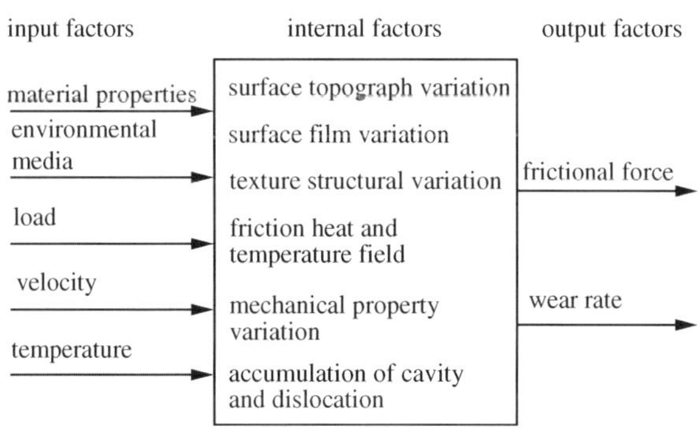

Figure 11.2 Friction and wear process.

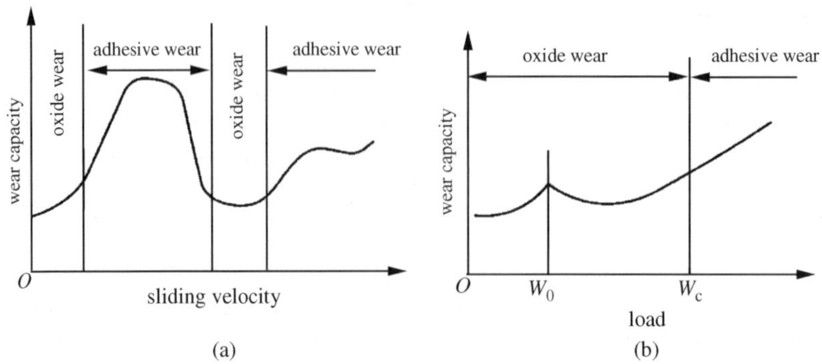

Figure 11.3 Wear conversion.

According to study, it is generally believed that classification of wear mechanisms is more appropriate and is usually of four basic types: abrasive wear, adhesive wear, surface fatigue wear and corrosion wear. Although the classification is not perfect, it outlines the common forms of wear. For examples, erosion wear formed by friction of the surfaces and solid particles contained in liquid can be classified as abrasive wear. Fretting is mainly due to oxidation of the contact surface so it can be classified as corrosive wear.

It should be pointed out that in actual wear phenomena, wear usually exists in several different forms. And, after the occurrence of one kind of wear, another may also appear. For example, fatigue wear debris can cause abrasive wear, and then abrasive wear may cause the clean surface to form corrosive or adhesive wear. The fretting wear is a typical complex wear. In fretting, adhesive wear, oxidation wear, abrasive wear and fatigue wear occur at the same time. With varying working conditions, primary wear also changes.

11.1.3 Conversion of Wear

The form of wear may also convert into variation of working conditions. Figure 11.3a gives the conversion of wear with variation of sliding velocity under a fixed load. When sliding velocity is low, wear occurs in the oxide film of the surface. This is oxidation wear and wear capacity is small. With increase in sliding velocity, wear debris size increases so the surface presents a metallic luster and becomes rough. Now, wear has been converted into adhesive wear, and wear capacity also increases. When the sliding velocity increases further, the surface oxide film will be regenerated because temperature rises. Therefore, wear turns into oxidation wear again and wear capacity becomes small. If the sliding velocity continue to increase, once again wear will be converted to adhesive wear, which will finally cause failure.

Figure 11.3b gives the experimental results that the sliding velocity is fixed while the load varies. If the load is small, it produces oxidation wear and wear debris is mainly Fe_2O_3. When the load reaches W_0, wear debris is the mixture of FeO, Fe_2O_3 and Fe_3O_4. If the load is larger than W_c, wear converts to hazardous adhesive wear.

11.2 Abrasive Wear

The phenomenon that external hard particles, hard bumps or rough peaks cause surface material to break or peel off is known as abrasive wear. The shovel teeth of an excavator, the rakes and backing block of a ball mill typically cause abrasive wear. A cutting chip can cause abrasive wear on the machine tool surface. The erosion of turbine blades and ship propellers working in water containing sediment also belongs to abrasive wear.

11.2.1 Types of Abrasive Wear

There are three types of abrasive wear:

1. An abrasive particle moves along a solid surface to produce surface wear which is called two-body abrasive wear. If an abrasive particle moves in the direction parallel to a solid surface, the contact stress on the surface is low such that scratches or minor furrows appear on the surface. If the abrasive particle moves in the direction vertical to the solid surface, the wear caused is referred to as impact wear. In such a situation, the particle collides on the surface in high stress such that a deep groove will be grinded on the surface and the large particular material is shed from the surface. Impact wear capacity is related to impact energy.
2. In a friction pair, where a hard surface roughness peak acts as an abrasive particle on the soft surface, this is also known as two-body abrasive wear and is usually a low-stress abrasive wear.
3. When the outside abrasive particles move between the two surfaces, similar to grinding, this is known as three-body abrasive wear. Usually, the three-body abrasive wear has high contact stress on the metal surface, often exceeding the crush strength of the abrasive particle. The compressive stress makes the friction surface of the ductile metal form the plastic deformation or fatigue, thus making the surface of the brittle metal form brittle fracture or spalling.

Abrasive wear is the most common form of wear. According to statistics, about half of the total loss of wear damage in production is caused by abrasive wear thus study of abrasive wear is very important. In general, abrasive wear mechanism is the plowing action of the abrasive particle so it is a micro-cutting process. Clearly, the relative hardness of the material to the abrasive particles, load and sliding velocity play important parts in abrasive wear.

11.2.2 Influence Factors on Abrasive Wear

In the laboratory, the study of abrasive wear is usually by rubbing a specimen material on an abrasive paper. Although this may omit the influence of impact, corrosion, temperature and other factors, the data obtained from the laboratory are different from what may occur in real situations. However, they also reflect the basic phenomenon and the law of abrasive wear so the conclusions are very useful.

First, the ratio of the abrasive hardness H_0 and the test piece material hardness H is called the relative hardness, which significantly influences abrasive wear characteristics as shown in Figure 11.4. When the hardness of the abrasive particles is lower than the specimen hardness, namely, $H_0 < (0.7-1)H$ no abrasive

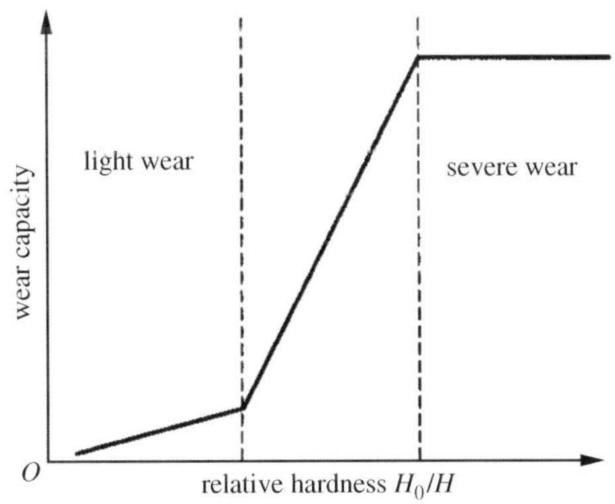

Figure 11.4 Relative hardness on wear.

wear or only mild abrasive wear occurs. If the abrasive particle is much harder than the specimen, the wear capacity increases with increase in the abrasive particle hardness. If the abrasive particle hardness is much higher, serious wear occurs, but the wear capacity no longer varies with change in hardness. Therefore, in order to prevent abrasive wear, the material hardness should be higher than the abrasive particle hardness. Generally, it is believed that if $H \geq 1.3H_0$, only minor abrasive wear appears.

The wear capacity can be expressed by the variation of the volume or the thickness. If the sliding displacement is s and the vertical worn thickness of the surface is h, the thickness per displacement dh/ds is known as the linear wear. The anti-abrasion ability E can be expressed as follows:

$$E = \frac{ds}{dh}. \tag{11.1}$$

Commonly, the relative wear-resistant ability R is used to illustrate abrasive wear capacity defined as:

$$R = \frac{E_s}{E_f}, \tag{11.2}$$

where E_s is the wear-resistant ability of the specimen material; E_f is the base wear-resistant ability, which is obtained when the adamantine spar serves as the wear with the hardness $H_0 = 2290\,\text{kgf/mm}^2$ and the specimen is composed of the antimony tin-lead alloy.

Khrushchev *et al.* studied abrasive wear systematically [2]. They pointed out that hardness is the main parameter to characterize the abrasive properties of a material and gave the following conclusions:

1. For the pure metal and steel without heat treatment, the wear-resistant ability is proportional to the hardness, as shown in Figure 11.5. It is generally believed that the hardness of annealing steel is proportional to carbon content. Therefore, it can be seen that the wear-resistant ability of steel under abrasive wear is linear to the carbon content.

   The straight line in Figure 11.5 can be expressed as follows:

$$R = 13.74 \times 10^{-2}H. \tag{11.3}$$

2. As shown in Figure 11.6, the hardness of steel can be enhanced by heat treatment. This can raise wear-resistant ability to slowly increase along a straight line, but the slope is smaller. Each straight line in the

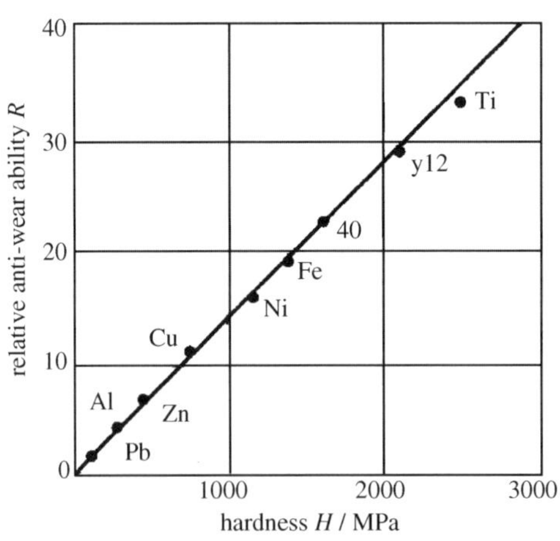

Figure 11.5 Relationship between relative wear-resistant ability and hardness.

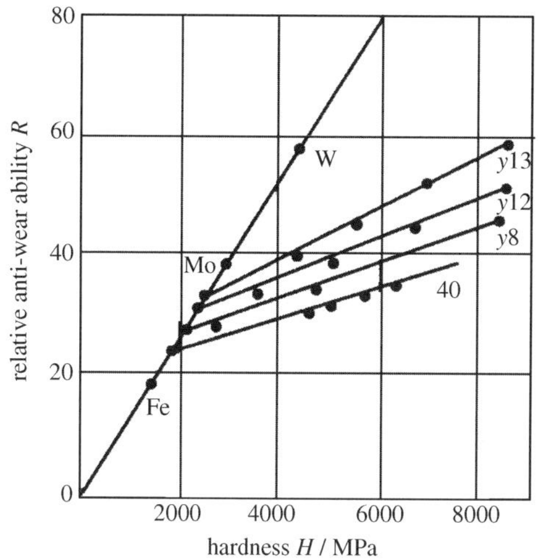

Figure 11.6 Influence of heat treatment on wear-resistant ability.

figure represents a kind of steel. The higher the carbon content, the greater the slope of the straight line. The intersection of the lines indicates the wear-resistant ability of the steel which is without heat treatment.

The influence of the heat treatment on wear-resistant ability of steel can be expressed as

$$E = E_p + C(H - H_p), \tag{11.4}$$

where H_p and E_p are the hardness and wear-resistant ability of the annealed steel; H and E are the hardness and wear-resistant ability after heat treatment; C is the coefficient of the heat treatment effect and increases with increase in carbon content.

3. Work hardening by plastic deformation of steel can raises the hardness of steel, but cannot improve abrasive ability.

   Хрущов *et al.*, based on the above analysis, concluded that abrasive wear-resistant ability is not related to work hardening because the furrows formed in wear have been strongly hardened. The hardening in wear is much severer than the original, but the wear-resistant ability of the metal material depends on the properties under the maximum hardening effect so the initial hardening has no effect on abrasive wear. Moreover, although heat treatment can raise a part of the hardness as a result of work hardening, this part is no use in improving wear-resistant ability so the effect of heat treatment is not very significant.

To summarize, there are three ways to improve hardness of steel: improve alloy compositions, use the heat treatment, or by work hardening. The wear-resistant ability of a material is related to the hardening method so we must use different ways to improve hardness after considering the relationship between them.

It should be noted that when surface hardness is greater than abrasive particle hardness, the surface may also be worn. This is because the particle can be pressed onto the metal, depending on the relative hardness and shape of the abrasive particle. For example, a solid plane can be pressed onto the metal by spherical, conical or other knife-edge particle with the same material to form an indentation.

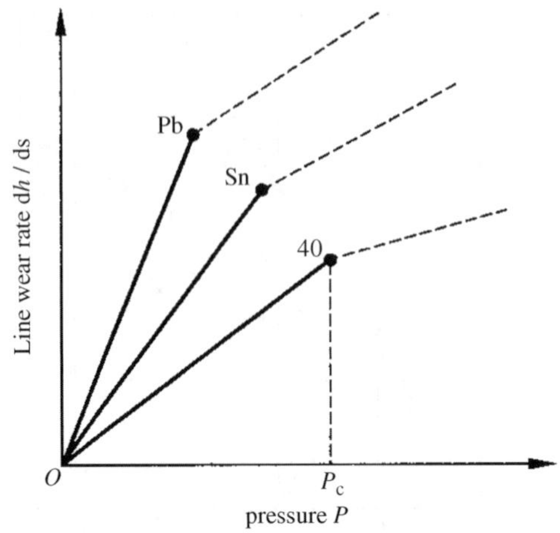

Figure 11.7 Relationship between surface pressure and line wear rate.

Therefore, in discussion of the abrasive wear, in addition to relative hardness, we should also consider the following factors.

1. Abrasive wear is related to wear particle hardness, strength, shape, sharpness, size, etc. Wear capacity is proportional to particle size, but when the particle is large to a certain value, the abrasive wear is no longer related to particle size.
2. The load significantly influences abrasive wear. Figure 11.7 shows that the line wear rate is proportional to surface pressure. When the pressure reached the turning point p_c, the line wear rate curve becomes flat. This is the result of transformation of the abrasive wear type. For different materials, the pressure turn points are different.
3. Figure 11.8 shows the relationship of the friction repeat number and the line wear rate. At the beginning, due to the running-in effect, the wear rate decreases with increase in the friction repeat number. At the same time, the surface roughness can be improved and the wear becomes slow.
4. If the sliding velocity is not too large, the tempering and annealing effects of the metal do not occur thus the line wear has nothing to do with the sliding velocity.

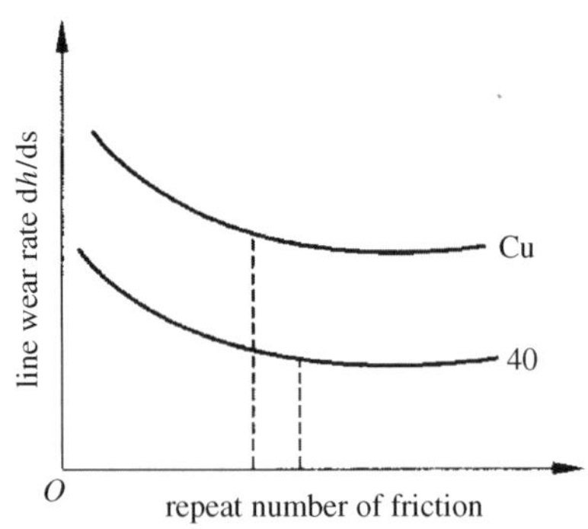

Figure 11.8 Relationship between the repeat number of friction and line wear rate.

11.2.3 Mechanism of Abrasive Wear

There are three kinds of abrasive wear mechanisms:

11.2.3.1 Micro-Cutting

The normal load will press the abrasive particles onto the friction surface. During sliding, the frictional force of the abrasive furrows shears, plows and cuts the groove-shape wear scars on the surface.

11.2.3.2 Squeezed Spalling

The abrasive particles under the action of the load are pressed onto the friction surface and leave indentations, and the plastic material is squeezed out to form flake-like spalling debris.

11.2.3.3 Fatigue Damage

The surface under the action of the cyclic contact stress of the abrasive particle becomes fatigued.

The simplest way to calculate abrasive wear is based on the mechanism of micro-cutting as shown in Figure 11.9.

Suppose the abrasive particle is a cone. If the half angle is θ and the indentation depth is h, then the projected area A pressed into the surface is equal to

$$A = \pi h^2 \tan^2 \theta. \tag{11.5}$$

If the yield stress of the surface is σ_s, the carrying load W of each abrasive particle is equal to

$$W = \sigma_s A = \sigma_s \pi h^2 \tan^2 \theta. \tag{11.6}$$

When the sliding distance is s, the volume of the material removed is equal to $V = sh^2 tg\theta$. If we define wear capacity rate as equal to dV/ds, it is

$$\frac{dV}{ds} = h^2 \tan \theta = \frac{W}{\sigma_s \pi \tan \theta}. \tag{11.7}$$

Because the yield stress σ_s is related to the hardness H, we have:

$$\frac{dV}{ds} = k_a \frac{W}{H}, \tag{11.8}$$

where k_a is the constant determined by the hardness, shape and number of abrasive particles in cutting.

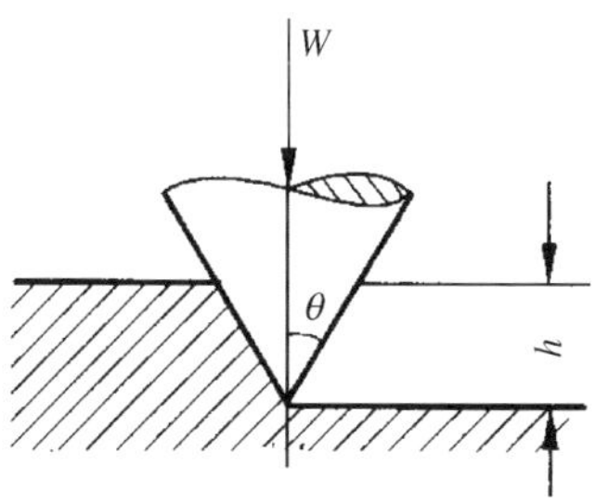

Figure 11.9 Cone abrasive wear model.

It should be noted that the above analysis has ignored a lot of practical factors, such as the distribution of abrasive particles, the elastic deformation and material accumulation in the front of the sliding so that Equation 11.6 can be approximately applied to the two-body abrasive wear. Because in three-body abrasive wear, a part of the abrasive particles roll along the surface, they do not produce the cutting effect. Thus, k_a in Equation 11.6 should be reduced.

Therefore, to improve wear-resistant ability, we must reduce micro-cutting. Various measures can be used: reducing abrasive force on the surface so as to make load distribution uniform; enhancing surface hardness; reducing surface roughness; increasing film thickness; filtering dusts to ensure that the friction surface is clean; and so on.

11.3 Adhesive Wear

When the surfaces slide relatively, the adhesive junctions of the friction pairs are sheared and the materials are cut off to form wear particles. Such a migration of the materials from one surface to another is referred to as adhesive wear.

Depending on the strength of the adhesive point and the damage position, adhesive wear is divided into light wear or heavy wear. Although wear forms, friction coefficients and wear rates may be different, material migration is the common feature and scratches are always along the sliding direction.

11.3.1 Types of Adhesive Wear

In accordance with severity, adhesive wear can be divided into four types.

11.3.1.1 Light Adhesive Wear

When the strength of the adhesive point is less than the strength of the two friction surfaces of metals, shear occurs in the junction. Although the friction coefficient may be large, wear is very small and material migration is not significant. Usually, such an adhesive wear takes place on the metal surface covered by oxide film, sulfide film or other coatings.

11.3.1.2 Common Adhesive Wear

If the strength of the adhesive point is higher than that of the soft friction surface, damage will occur on the soft metal surface not far away from the joint. Thus, the soft metal adheres on the hard metal surface. The friction coefficient is similar to the light adhesive wear, but the wear amount is significant.

11.3.1.3 Scratch

When the junction strength is higher than the two metal surfaces, scratches occur on the soft metal surface, but may also occur occasionally on the hard metal surface. The adhesive material migrates to the hard surface. This can also cause the soft surface to be scratched. Therefore, scratches mostly occur on the soft metal surface.

11.3.1.4 Scuffing

If the junction strength is much higher than the two surfaces, and the junction area is large, adhesive wear occurs in the depth of one metal surface. The two surfaces can be seriously worn, and even the friction surfaces are seized to stop sliding.

In the high-speed and heavy-load friction pair, because it is on the junction point, plastic deformation is large and surface temperature is high, the adhesive area of the adhesive point increases. This often results

Table 11.1 Critical pressure of scuffing

Friction materials	Critical pressure (N/mm^2)	Time of scuffing occurrence (min)
No. 3 steel-bronze	170	1.5
No. 3 steel-GCr15 Steel	180	2.0
No. 3 steel-cast iron	467	0.5

in scuffing. In the friction pair composed of the same metallic material, because it is near the adhesive point, material plastic deformation and work hardening are the same, the shear occurs in the depth so scuffing is very severe.

11.3.2 Influence Factors on Adhesive Wear

In addition to lubrication condition and friction material performance, the main outside factors affecting adhesive wear are the load and surface temperature. However, whether the load and temperature are the decisive factors is not well understood.

11.3.2.1 Load

Vinogradov systematically studied the influence of the load on scuffing (quoted in reference [13]). She believed that when the surface pressure reaches the critical value for a period of time, scuffing will occur. Therefore, the load is a decisive factor on scuffing. The critical pressures of several materials are presented in Table 11.1.

Through experimental study with a four-ball tester, it can be found that when the load reaches a certain value, the wear scar diameter suddenly increases. This load is called the scuffing load, as shown in Figure 11.10.

Experiments also show that if a specimen is immersed in heated oil and the load is below critical, scuffing cannot occur even at raised temperature. This shows that temperature is not the principal factor of scuffing.

However, the elastic-plastic deformation of the surface caused by the load must be accompanied by high temperature. Moreover, according to experiments, it has been found that critical load decreases

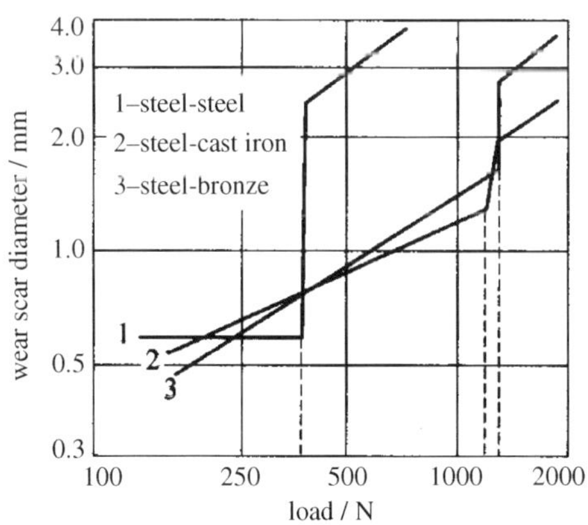

Figure 11.10 Four-ball tester experimental curves.

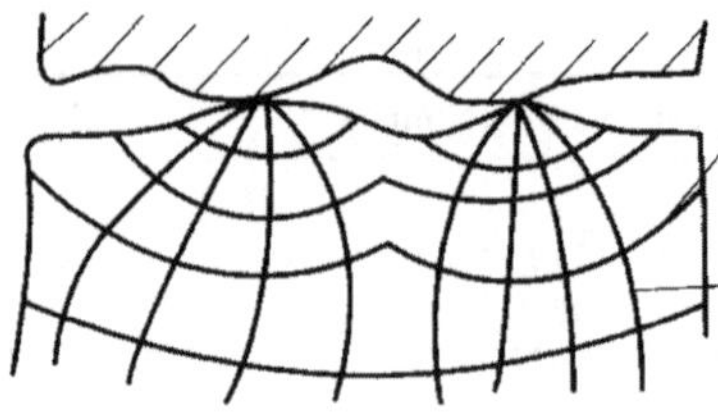

Figure 11.11 Temperature gradient line and contour.

with increase in sliding velocity. This shows that the occurrence of high temperature plays an important role in scuffing.

11.3.2.2 Surface Temperature

The friction heat generated in the friction process raises the surface temperature to form isothermal hemispherical contours near the contact point. In the depth of the surface, the contours of the contact points merge into the common isotherm contours, as shown in Figure 11.11.

Figure 11.12 gives the temperature distribution along the surface depth. The frictional heat is generated on top of the deformed zone, where the surface temperature T_s is the maximum. The thermal conductivity causes the temperature gradient in the deformation zone to be very large. However, the substrate temperature T_v in the body changes slowly.

The surface temperature characteristics have a significant influence on the interaction and damage of friction surfaces. The surface temperature can cause lubrication failure, while the temperature gradient causes the material properties and failure type to vary along the depth direction.

Figure 11.13 shows the experimental results by Rabinowicz (1965) [3]. He used the radioisotope method to measure the amount of metal migration. It can be seen that when the surface temperature reaches critical value (about 80 °C), the wear capacity and friction coefficient increase dramatically.

The surface pressure p and sliding velocity v are the two main factors affecting temperature characteristics. The velocity has a great influence. Therefore, to limit pv is an effective way to reduce adhesive wear and prevent scuffing. Based on the experimental and numerical analysis, the relationships of the surface temperature with velocity and pressure are shown in Table 11.2.

11.3.2.3 Materials

Adhesive wear-resistant ability of brittle material is higher than that of plastic material. The damage of the adhesive point of plastic material is in plastic flow. It mainly takes place at a certain depth (up to 0.2 mm) from the surface, and its debris is large, sometimes up to 3 mm. While damage of brittle material is mainly

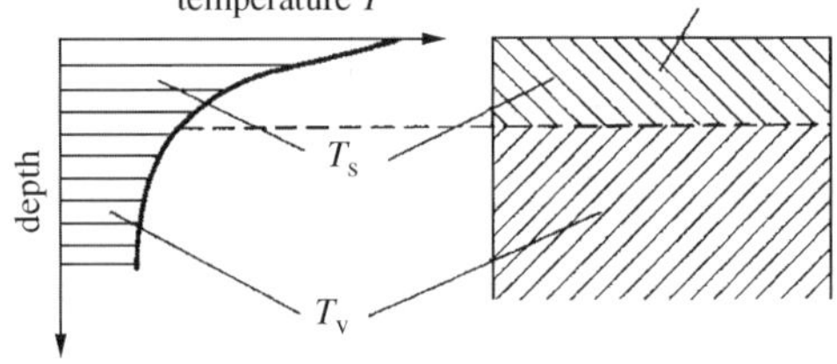

Figure 11.12 Temperature distribution under surface.

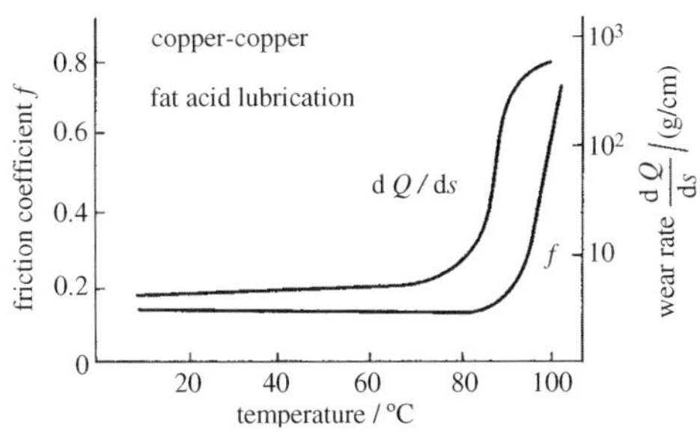

Figure 11.13 Influence of temperature on scuffing.

spalling, the damage position is near the surface, and its debris is easily sloughed and cannot be piled up on the surface. According to strength theory, the damage of brittle material is caused by normal stress, but the plastic material damage is determined by shear stress. The maximum normal stress is on the contact surface, but the maximum shear stress occurs beneath the surface. The higher the stress, the more severe the adhesive wear is.

The adhesion of the same metal or material with large inter-solubility is severe so adhesive wear is easily formed. Different metals or materials with small inter-solubility are of high wear-resistant ability. The friction pair composed of the metallic and non-metallic material is of higher anti-adhesive wear ability than that composed of the different metals. From the structure of the material, the multi-phase metal has a higher anti-adhesive wear ability than the single-phase metal.

Through surface treatment methods, the film of sulfides, phosphides, or chloride, and so on generated on metal surface will effectively reduce adhesion, while the surface film also limits the depth of the damage, thereby enhancing the ability of anti-adhesive wear.

In addition, it can improve anti-adhesive wear ability to improve lubrication conditions, such as adding the oil extreme pressure additives into oil or fat; using high thermal conductivity structure to enhance the cooling ability of the friction materials; lowering surface temperature; improving surface topography; or reducing contact pressure.

11.3.3 Adhesive Wear Mechanism

Usually, the actual contact area of the friction surface is only 0.1–0.01% of the apparent area. For the heavy-load and high-speed friction pair, pressure on the peak point can sometimes reach up to 5000 MPa, and produce an instantaneous temperature above 1000 °C. Because the volume of the friction pair is much larger than the contact peak point, once the contact points separate away, the peak point temperature will decrease rapidly; generally the local high temperature can only remain for a few milliseconds. If the friction surface is in such a state, the lubricating oil film, the adsorption film or other surface film will

Table 11.2 Relationships of surface temperature with velocity and pressure

Temperature	Contact state			
	Plastic contact		Elastic contact	
	Pressure p	Sliding velocity v	Pressure p	Sliding velocity v
Surface temperature θ_s		$\sqrt{v}$	p^n	$\sqrt{v}$
Temperature gradient		V	p^n	v
Substrate temperature θ_v	p	V	p	v

Note: $n < 1$.

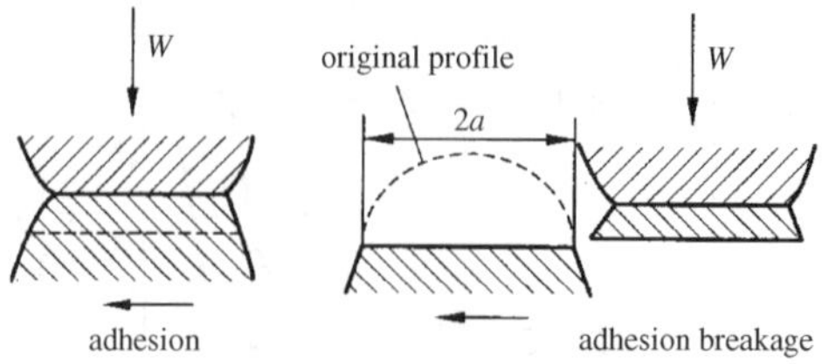

Figure 11.14 Adhesive wear model.

rupture so that the contact peak point will be adhered. With sliding, the adhesive junctions will be damaged. The adhesion, damage and re-adhesion turning alternatively form the adhesive wear process.

The reasons for the formation of the adhesive junctions are different. Bowdon believed that the plastic deformation of the adhesive point and the high transient temperature melt or soften the material to result in welding. He also proposed that with rise of temperature, the dissociated material similar to welding forms an adhesive point. Adhesive wear can also occur on non-metallic material surfaces, but the phenomenon cannot be explained from the high-temperature welding viewpoint.

Khrushchev *et al.* thought that adhesion is a kind of cold welding. It is unnecessary to reach the melting temperature before an adhesive point forms. Someone suggested that adhesion is due to the interaction of the friction surface molecules. Furthermore, others had tried to explain the adhesion phenomenon by using the movement of the valence electrons of metals or the movement and filling of the similar metal atoms in the crystal lattice planes. However, these viewpoints have not yet been supported by sufficient experimental data.

Although the adhesion mechanism is not yet clear, the adhesion phenomena must happen under a certain pressure and high temperature conditions. This understanding is very consistent.

The position of the adhesive point determines the severity of adhesive wear. The damage power is the frictional force which is not definitely connected with wear. Adhesive damage is very complex and is related to the relative strength of the friction material and the adhesive point.

The adhesive wear calculation is based on the model shown in Figure 11.14, which was proposed by Archard (1953) [4].

Suppose the adhesive area is a circle with a radius a, the contact adhesive area is πa^2 and the surface is in the plastic contact, the carrying load of each adhesive point is equal to

$$W = \pi a^2 \sigma_s, \tag{11.9}$$

where σ_s is the yield stress of the soft material.

Suppose the adhesive point is damaged in the spherical form, that is, the wear debris is a hemisphere and the sliding displacement of $2a$, the wear volume is $2/3\pi a^3$. Therefore, the wear rate can be written as

$$\frac{dV}{ds} = \frac{\frac{2}{3}\pi a^3}{2a} = \frac{W}{3\sigma_s}. \tag{11.10}$$

Consider that not all of the adhesive points form the hemispherical shape debris, the adhesive wear constant k_s should be introduced, where $k_s << 1$. Then, the Archard formula becomes

$$\frac{dV}{ds} = k_s \frac{W}{3\sigma_s}. \tag{11.11}$$

Equation 11.11 is similar to Equation 11.8.

Table 11.3 k_s of several lubricants (four-ball machine test, $W = 400$ N, $v = 0.5$ min/s)

Lubricant	Friction coefficient f	Wear constant k_s	Equivalent life of gear	
			Total revolution	Working time
Dry argon	0.5	10^{-2}	10^2	Second
Dry air	0.4	10^{-3}	10^3	Minute
Gasoline	0.3	10^{-5}	10^5	Hour
Lubricants	0.12	10^{-7}	10^7	Week
Lubricating oil with stearic acid (cooling)	0.08	10^{-9}	10^9	Year
Standard engine oil	0.07	10^{-10}	10^{10}	Year

Although the Archard model is approximate, it can be used to estimate the adhesive wear life. Fein [5] measured the anti-adhesive wear properties of several lubricants by a four-ball tester, as given in Table 11.3. Table 11.4 lists the results of Pooley and Tabor [6]. They measured k_s of several materials in the dry friction condition with a pin-plate wear tester.

In the tables, the adhesive wear constant k_s is much less than 1, indicating that only a very few of all the adhesive points are worn, and most of them do not produce wear debris. There is no satisfactory explanation for this phenomenon yet.

11.3.4 Criteria of Scuffing

Scuffing is the most harmful form of wear. The surface is not uniformly worn, sometimes the depth of the wear crack is up to 0.2 mm, and the worn material piles up so the friction coefficient is high and unstable. Scuffing often leads the friction pair to complete and rapid failure. Therefore, it should be avoided.

There is still no criteria to determine scuffing. The surface topography is sometimes used to determine scuffing, that is, when the roughness perpendicular to the sliding direction dramatically increases, scuffing easily occurs. It has also been proposed that friction temperature can be used as a criterion to determine the occurrence of scuffing. However, the commonly used criterion is the friction coefficient. When there is a sudden increase, scuffing occurs.

If the lubricating oil film has broken down, scuffing will depend on the chemical reaction film. Fan Yujin *et al.* discussed the scuffing criterion of GCr15 on 45# steel by measuring the friction surface temperature, the frictional force, the film variation, and surface reaction film formation under the oil lubrication condition [7]. Their experiments showed that the sliding velocity has a great influence on scuffing. At low speed, if the oil film ruptures, a chemical reaction film can be generated to prevent the surface from scuffing. Only when the surface temperature is too high, the reaction film will be a failure and scuffing occurs. At high speed, when the oil film has broken, it is difficult to form a reaction film so scuffing occurs immediately. Before scuffing, the temperature and friction are relatively low. At the medium speed, when the oil film has broken, a reaction film can form. Scuffing occurs if the reaction film is worn faster than it forms.

Scuffing usually appears in the worm drive, the rolling contact bearing or the journal bearing under the high speed and heavy load in poor lubrication condition. In order to prevent the occurrence of scuffing,

Table 11.4 k_s of several materials (pin-disc tester, dry friction in air, $W = 4000$N, $v = 1.8$ min/s)

Friction materials	Friction coefficient f	Wear content k_s
Mild steel–mild steel	0.6	10^{-2}
Cemented carbide–hardened steel	0.6	5×10^{-5}
Polyethylene–hardened steel	0.65	10^{-7}

scuffing is studied extensively, especially for gear scuffing. However, the present criteria are still semi-empirical formulas, with lack of sufficient and accurate data. Therefore they cannot be widely used yet.

In early studies, scuffing calculation was based on the static load with the aim of enhancing the surface hardness of the material as a measure to anti-seize. Subsequent studies showed that temperature plays an important part in scuffing and therefore it was proposed that scuffing criteria be based on thermal load.

Several commonly used criteria for calculation of scuffing are as follows:

11.3.4.1 $p_0 U_s \leq c$ Criterion

Almen and Boegehold [8], based on the statistics of the bevel gear scuffing failure of the rear axle, put forward the following criterion to prevent scuffing.

$$p_0 U_s \leq c, \tag{11.12}$$

where p_0 is the maximum Hertzian stress; U_s is the relative sliding velocity; c is the experimental constant, $c = 32 \times 10^2 - 15 \times 10^4$ MPa·m/s.

Equation 11.9 is an approximate result with the maximum dispersion up to 50%. However, it is simple. Therefore, it is commonly used as a preliminary calculation to select the anti-scuffing material.

Based on the experimental analysis, Blok [9] proposed to use $\sqrt{p_0^3 U_s}$ as a more realistic criterion for calculating the scuffing. Because $\sqrt{p_0^3 U_s}$ is proportional to the instantaneous contact temperature rise, the temperature factor has been actually considered in this scuffing criterion.

11.3.4.2 $WU_s^n \leq c$

Borsoff and Godet [10] obtained the results of Figure 11.15 based on study of gear scuffing. When scuffing occurs, the load W and sliding velocity U_s obey the exponential relationship. In recent years, the exponential criterion proposed can be expressed as the following formula:

$$WU_s^n \leq c. \tag{11.13}$$

If n can be determined by the experiments, this criterion can be satisfied with the results.

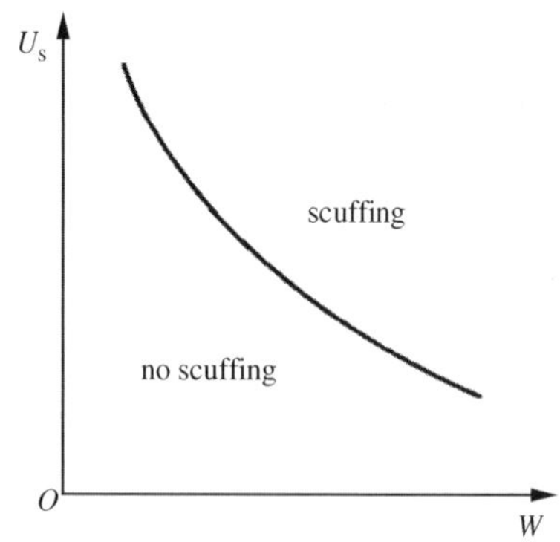

Figure 11.15 $WU_s^n \leq c$ criterion curve.

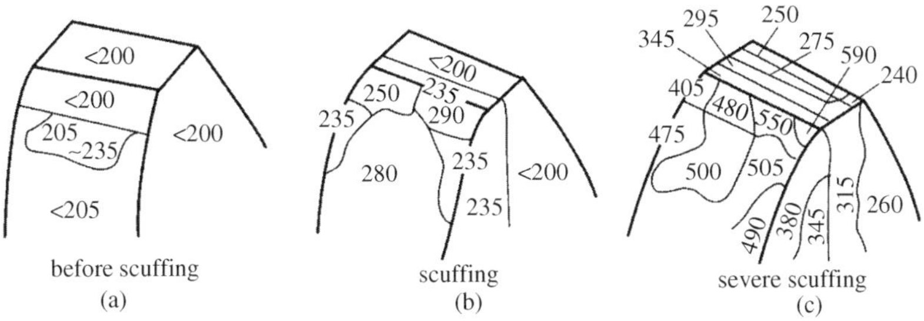

Figure 11.16 Scuffing temperature on tooth surface (°C).

11.3.4.3 Instantaneous Temperature Criterion

Figure 11.16 was given by Wilson [11], who presented the surface temperature distribution and variation on the gear teeth in the scuffing process. The figure shows that scuffing is closely related to temperature.

Blok [12] thought that scuffing is caused by the critical surface instantaneous temperature. The instantaneous temperature criterion he proposed is now widely used.

If the critical temperature is T_{sc}, the friction surface bulk temperature is T_b and the local instantaneous temperature rise is T_{fm}, the criterion without scuffing is

$$T_b + T_{fm} \leq T_{sc}. \tag{11.14}$$

The instantaneous temperature rise and the maximum temperature can be determined by the method in reference [13].

The scuffing critical temperature T_{sc} should be determined based on the friction pair materials, the lubricant and the lubrication state. For example, under the conditions of the general lubrication and the hardened gears, $T_{sc} = 150\,°\text{C} - 250\,°\text{C}$; while if the gears are made of nonhardened steel, $T_{sc} = 60\,°\text{C} - 150\,°\text{C}$. By using an ordinary mineral oil, the scuffing temperature is usually close to the evaporation temperature of oil.

It has been proven that the instantaneous temperature criterion is close to the exponential criterion. In the high-speed sliding condition, the instantaneous temperature criterion is equivalent to the exponential criterion of $n = 2/3$.

It should be pointed out that so far no criterion can accurately determine the occurrence of scuffing and the instantaneous temperature yet, whether the measuring or calculation methods are used.

11.3.4.4 Scuffing Factor Criterion

The scuffing factor t_f is defined as the time for the surface point to pass the contact area. Its unit is in seconds. If the half-width of the Hertzian contact area is b, then t_f is equal to

$$t_f = \frac{2b}{U_s}, \tag{11.15}$$

when scuffing occurs, the relationship of the critical load W_c and the scuffing factor t_f is

$$W_c = at_f + c, \tag{11.16}$$

where a and c are the experimental constants.

If $W_c >> c$, c can be omitted. Therefore, Equation 11.16 is close to Equation 11.13.

To summarize, because scuffing phenomena are complicated, the present criteria need to be improved further, and the mechanism of scuffing should be studied further.

11.4 Fatigue Wear

The pits forming due to fatigue on the two rolling or rolling and sliding friction surfaces and under the action of the cycle contact stress are known as fatigue wear or contact fatigue wear. In addition to the gear drive, the rolling bearings are the main mechanical elements appearing as failure wear. Micro-wear caused by variation of stress due to roughness also belongs to fatigue wear. However, surface micro-fatigue usually occurs only in running-in, and is nondevelopmental wear.

In general, surface fatigue wear is inevitable and even in a good oil film lubrication condition it will also occur. Slight fatigue wear will not develop in severe pits so as not to bring about failure during the normal working period.

11.4.1 Types of Fatigue Wear

11.4.1.1 Superficial Fatigue Wear and Surface Fatigue Wear

Superficial fatigue wear mainly occurs in the rolling friction pairs with the general quality of the steel. Under the action of cycle contact stress, fatigue wear begins from the crack inside the material because of stress concentration, such as the non-metallic inclusion or the hole. The initial crack is usually confined in a narrow area in the typical depth about 0.3 mm away from the surface, where the maximum shear stress is located. The crack first extends parallel to the rolling direction and then to the surface. After the material falls off from the surface, it forms the wear debris with a relatively smooth fracture. The initiation time of the fatigue crack is short, but the propagation of the crack is slow. The superficial fatigue wear is usually the main failure type of the rolling contact bearings.

In recent years, due to the development of vacuum melting technology, the internal quality of steel has improved remarkably in reducing the superficial crack such that the possibility of surface fatigue wear increases.

Surface fatigue wear occurs mainly in the sliding friction pairs with high-quality steel. The crack originates from the stress concentration source on the surface, such as the revolution mark, hit mark, staining or other wear mark. Then, the crack develops along the sliding direction from the surface into the internal with an angle of 20–40°. To a certain depth, the pit is formed because of the bifurcation of the crack. Its fracture is relatively rough. The formation time of the wear crack is very long, but the expansion rate is very rapid.

Because the edge of a superficial fatigue pit may be the source of the surface fatigue wear, these two kinds of fatigue wear usually exist at the same time.

11.4.1.2 Pitting and Peeling

According to the shape of the wear debris and the fatigue pit, the fatigue wear is usually divided into two kinds: pitting and peeling. The wear debris of the former is mostly the fan-shaped particle and there are many small and deep pits on the surface, while the wear debris of the latter is flake-shaped and there are large shallow pits. The shapes of the two wear pits are shown in Figure 11.17.

Fujita and Yoshida [14] carried out experiments of the steel samples with different heat treatments on a double-disc tester. They found that for the annealing steel and the quenched steel, the fatigue wear is in the form of pitting, while for the hardened steel, the form of fatigue wear is peeling.

Fujita and Yoshida proposed to use the ratio of the stress and hardness as the criterion. They believed that the crack is the maximum while the ratio is equal to $\sigma/\sqrt{3}H$ or τ/H. According to the measured

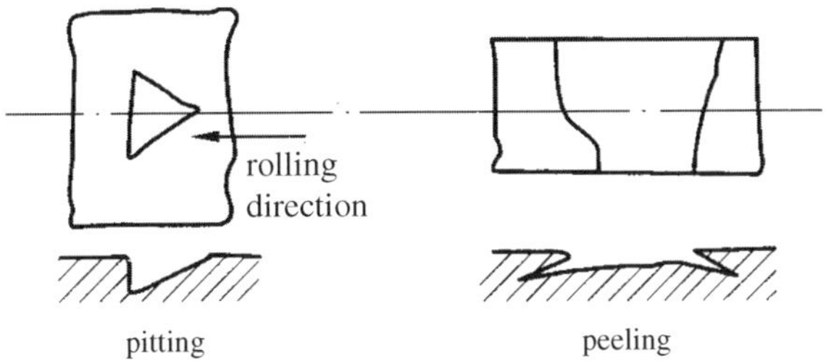

Figure 11.17 Pitting and peeling.

hardness and calculated stress in the depth, they made the following conclusion. For the soft materials, the maximum ratio is equal to $\sigma/\sqrt{3}H$ on the surface. Therefore, it can be used as the stress to determine the occurrence of pitting. For the hard material, the maximum ratio is equal to τ/H beneath the surface, which can be used to determine the occurrence of peeling.

Martin and Cameron [15] analyzed fatigue wear. They found that there are two kinds of wear debris, oval-shaped and fan-shaped. Oval-shaped wear debris is a flake, and its number is few. The crack of the fan-shaped wear debris radically expands from the surface to the superficial layer in an angle of 30–40°. Figure 11.18 gives the micro-hardness distribution along the depth. It can be seen that maximum hardness exists in the superficial layer, and its location is consistent with the depth of the maximum shear stress. This supports the Crook and Welsh conclusion that under the action of the cycle stress, the hardening layer is formed beneath the surface due to plastic deformation. The crack forming a deep pit is usually on the surface and extends with an angle of 40° downward to the hardened layer. Then, it develops in the parallel direction of the surface. The hardened layer constitutes a barrier to prevent the crack through downwards.

It is not sure whether any relationship exists between the pitting and the peeling. In practice, the different forms of wear debris exist simultaneously in the fatigue wear. In addition, although the fatigue wears are macroscopically different, the variations in the micro-structure of the material fatigues are the same.

11.4.2 *Influence Factors on Fatigue Wear*

In general, the factors that affect surface fatigue wear can be summarized into the following four aspects:

1. the macro-stress field in the dry friction or lubrication condition;
2. the mechanical properties and strength of the friction material;

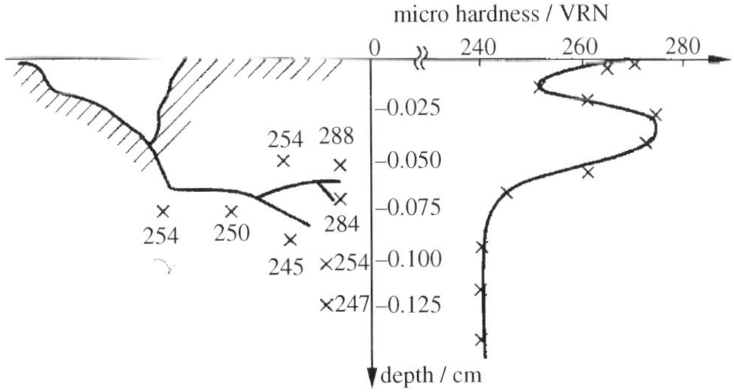

Figure 11.18 Micro-hardness distribution and crack propagation.

3. the geometry and the distribution of defaults inside the material;
4. the actions of lubricant or media on the surfaces of the friction pair.

Here, some key factors are introduced as follows.

11.4.2.1 Load Property

First, the load determines the macro-stress field in friction pair so as to directly influence the fatigue crack initiation and growth. It is generally believed that the load is the basic factor of the fatigue wear. In addition, the property of a load also has a tremendous influence. Pavlov, with an enclosed gear tester, has systematically studied the effect of a cyclical load on the contact fatigue. He first applied a constant contact stress 850 MPa on the unquenched gear until fatigue wear occurred. Then, by using the same specimen with 850 MPa as the basic load in the interval of 10×10^4 circles, he added the load to 950, 1050 and 1150 MPa resolutions respectively to continuously work in 2×10^4 circles and dropped to the basic load again, as shown in Figure 11.19a. His results showed that the total damage circles of the specimen increased after the loads were added, as shown in Figure 11.19b.

The experimental results show that the short-term peak load periodically added to the basic load does not reduce contact fatigue life but enhances it. Only when the applied time of the peak load is close to half the cycle period does the contact fatigue life begin to reduce.

The author has studied the effects of compound stress on contact fatigue wear, by using a ball and a cylindrical specimen with the extrusion in the maximum contact stress 2954 MPa [16]. The stress is less than 6% of the axial bending stress. Experimental results showed that added tensile bending stress significantly reduces contact fatigue life. However, the influence of compressive bending stress depends on magnitude. If added compressive stress is small, it increases fatigue life, but a large amount of compressive stress will reduce fatigue life. Therefore, a critical compressive bending stress exists and the corresponding fatigue life is the maximum, as shown in Figure 11.20.

The frictional force of the contact surface has a significant influence on fatigue wear. Figure 11.21 shows that a small amount of sliding will significantly reduce contact fatigue wear life. Usually, in pure rolling, the frictional force is only about 1–2% of the normal load. While sliding exists, the tangentially frictional force can increase up to 10% of the normal load. The frictional force increases contact fatigue wear because under the action of the frictional force maximum shear stress tends to move upwards to the surface. This increases the possibility of crack initiation. In addition, frictional force causes tensile stress to promote crack to grow quickly.

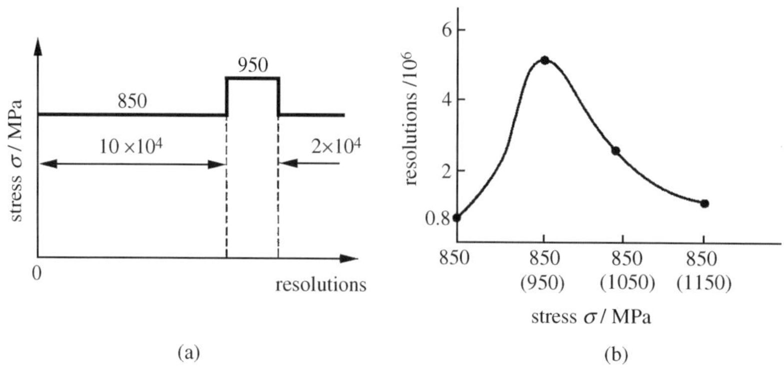

Figure 11.19 Influence of cyclical load on contact fatigue.

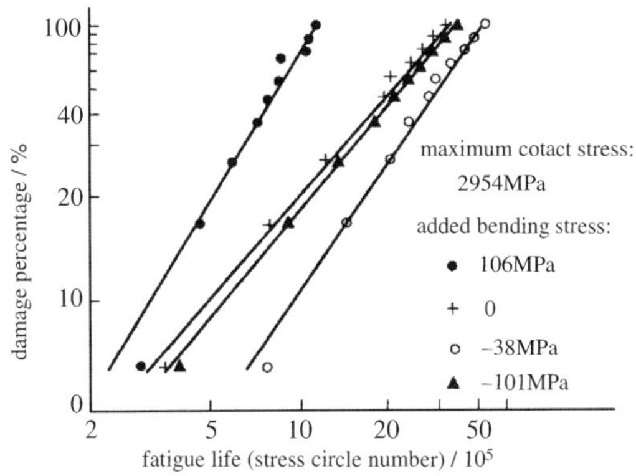

Figure 11.20 Fatigue life under compound stress.

The cyclic rate of the stress also affects the contact fatigue wear. Because the contact of the friction surfaces generates heat, the faster the stress circulates, the more the accumulation of heat and the higher the temperature on the surface. Therefore, the metal is softened and its mechanical properties are reduced so as to accelerate surface fatigue wear.

It should be noted that in the full film EHL, pressure distribution is different from Hertzian stress. This will changes the internal stress field of the surface. Particularly the secondary pressure peak and the necking will cause stress concentration which will affect fatigue wear. Contact fatigue wear research of the EHL is still not sufficient.

11.4.2.2 Material Property

The non-metallic dirt in steel breaks the continuity of the base body and this seriously reduces contact fatigue wear life. Particularly, under the action of the cyclic stress, the brittle dirt falls off from the substrate material to form cavities, becoming the sources of stress concentration to lead to early fatigue cracks.

The hardened layer of the case-hardened steel or other surface hardening steel affects the anti-fatigue wear ability. When the hardened layer is too thin, the fatigue crack will appear in the connections between the hardened layer and the substrate, easily forming spalling. A reasonable thickness of the hardened

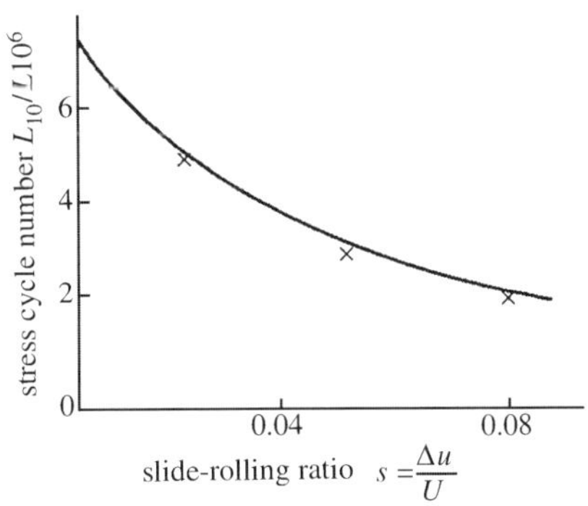

Figure 11.21 Slide rolling ratio on fatigue life.

Table 11.5 Influence of viscosity on gear contact fatigue wear

Oil	Oil temperature °C	Viscosity m²/s	Contact fatigue stress MPa	Transmission power kW
No.33 spindle oil	20	116×10^{-6}	450	4.9
Mechanical oil	20	757×10^{-6}	600	8.8
No.66 cylinder oil	82	84×10^{-6}	430	4.5
	57	303×10^{-6}	490	5.0
	45	757×10^{-6}	550	7.4

layer can make the fatigue crack appear in the hardened layer so as to increase the wear-resistant ability. In addition, reasonably increasing the substrate hardness can also improve fatigue wear life.

Although a high hardness can increase the anti-fatigue wear ability, if the hardness is too high, the brittleness will be increased as well. This may reduce the contact fatigue wear life.

The roughness of the friction surface is closely related to the fatigue life. The experimental data show that the contact fatigue wear life of the bearing with the roughness $R_a = 0.2$ is 2–3 times longer than that with $R_a = 0.4$; that with $R_a = 0.1$ is once longer than that with $R_a = 0.2$; and that with $R_a = 0.05$ is 0.4 times longer than that of $R_a = 0.1$. If roughness is less than $R_a = 0.05$, it hardly influences the contact fatigue wear life. In addition, under the partial EHL, the ratio of film thickness and surface roughness (the film ratio) is the important parameter to affect the surface fatigue life.

11.4.2.3 Physical and Chemical Effects of Lubricant

The experimental results show that with increase in viscosity, anti-contact fatigue wear ability increases. In Table 11.5, the fatigue lives under different experimental conditions are listed.

However, viewpoints differ on the mechanism of influence of viscosity on fatigue wear. Generally, it is considered that the increase of the lubricant viscosity can improve the fatigue life due to the formation of the EHL film, which can reduce the asperity interactions. However, this viewpoint cannot explain that the non-oil rolling does not bring about the fatigue wear, but if we add oil the contact fatigue wear will occur rapidly.

Way [17] proposed the hydraulic mechanism of the fatigue crack, as shown in Figure 11.22. In a friction process, the frictional force tries to force the surface metal to flow, and thus the fatigue crack tends to be directional, that is, with the same direction of the frictional force. As shown in the figure, the oil in the crack of the driving wheel is squeezed out during rolling, while the oil is sealed in the crack of the driven wheel so that the film pressure will promote the crack to grow. Because the oil is compressible and the metal is elastic, when the oil pressure reaches the crack tip, the pressure drops. The greater the viscosity of lubricating oil, the greater the pressure drops, that is, the lower the pressure in the crack tip, and the slower the crack grows.

Culp and Stover [18] experimentally compared the synthetic oil with the natural oil with the same viscosity. Their results showed that the contact fatigue wear life of the synthetic oil is higher. The reason is

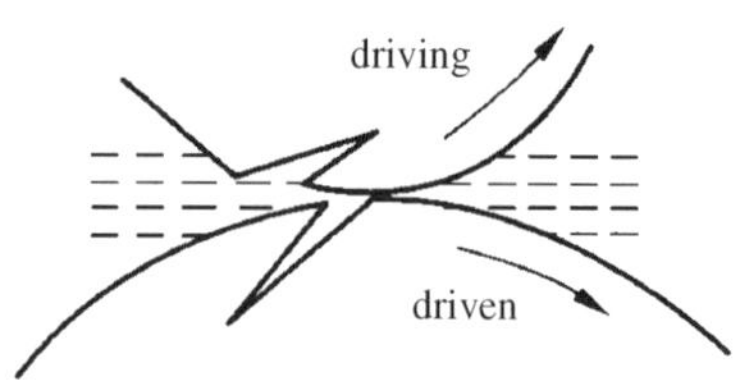

Figure 11.22 Hydraulic mechanism of fatigue crack.

that the viscosity-pressure coefficient of the synthetic is large. Therefore, it will produce a larger film thickness. This shows that the oil film thickness can prevent the formation of the crack.

The contact fatigue wear mechanism can be summarized as follows. At the beginning, the micro-crack forms. Whether oil exists or not, the cyclic stress plays an important role in fatigue wear. The crack initiates at the surface or is superficial at first, and soon extends to the surface. The viscosity of lubricating oil has also a major implication on the crack propagation.

In recent years, the chemical influence of the lubricant on the contact fatigue wear has been the focus of research. Studies have shown that the variation of the lubricant viscosity can change the contact fatigue wear life. Furthermore, the different chemical compositions of the lubricant can also influence the contact fatigue wear life significantly. In general, the oxygen and the water in the lubricant will drastically reduce the contact fatigue wear life. When the crack tip contains corrosive chemical compositions, they also significantly reduce contact fatigue wear life. If the additive can generate a strong surface film to reduce friction, it will improve anti-fatigue wear ability.

11.4.3 Criteria of Fatigue Strength and Fatigue Life

11.4.3.1 Contact Stress State

Strictly speaking, the applied conditions of Hertz contact theory should be without lubrication and in the static elastic deformation. However, the actual contact is of a relative motion and has lubricant as well. Therefore, the Hertz contact theory is only approximately suitable for the contact fatigue wear problem.

An elastic contact area is commonly an ellipse, and the contact stresses are shown in Figure 11.23. Here, the axis semi-lengths of the ellipse are a and b respectively, and the pressure distribution on the contact area is a half ellipsoid with maximum contact stress or Hertzian pressure p_H.

According to the analysis of the contact mechanics, the contact stress features can be summarized as follows.

1. The normal stress σ_x, σ_y and σ_z are the negative or the compressive stresses. They reach the maximum at the z axis. At the z axis, there is no shear stress so the only stresses are the normal stresses. Far away from the contact center (theoretical infinity), σ_x, σ_y and σ_z are equal to zero. It can be known that in the rolling process, the material is under normal stresses with varying pulsation.

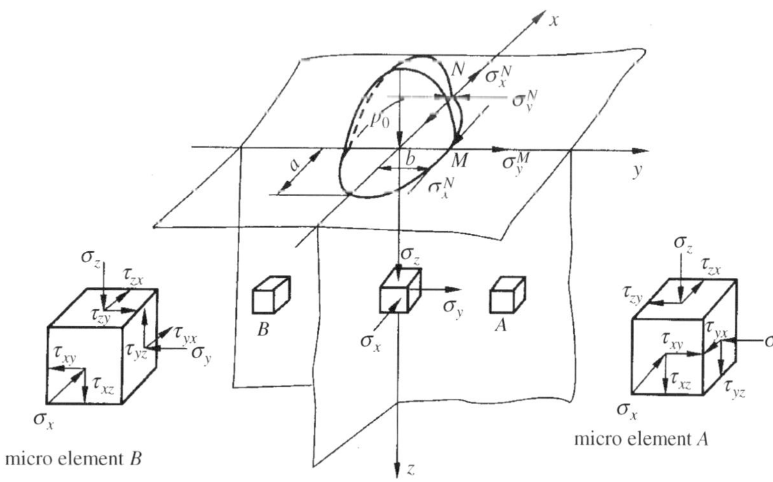

Figure 11.23 Contact stresses.

2. The sign of the shear stress τ_{xy} (or τ_{yx}) depends on the point position, which is equal to the sign of the product of the x and y coordinates. Far away from the contact center and $x=0$ (or $y=0$), $\tau_{xy}=0$. Therefore, in the rolling process, the two shear stresses are the alternate stresses.
3. The sign of the shear stress τ_{zx} (or τ_{xz}) depends on the point position, which is equal to the sign of the x coordinate. Far away from the contact center and $x=0$, $\tau_{zx}=0$. Similarly, the sign of the shear stress τ_{yz} (or τ_{zy}) is equal to the sign of the y coordinate. Far away from the contact center and $y=0$, $\tau_y=0$. Thus, in the rolling process, the four shear stresses are the alternate stresses.
4. The stress state on the contact surface is very complicated. Because of the possibility that the contact fatigue crack initiates on the surface, more attention should be paid to analysis of the surface stress state.

Here, we only introduce the stress states at the symmetrical axis of an ellipse. As shown in Figure 11.23, at the endpoints N and M, the radial stresses and the tangential stresses are equal and with the opposite sign, that is

$$\sigma_x^N = -\sigma_y^N$$

$$\sigma_x^M = -\sigma_y^M \tag{11.17}$$

So, in a pure-shear state, when $\sqrt{1-b^2/a^2} < 0.89$, the maximum surface shear stress reaches the ellipse endpoints N or M of the symmetrical axis.

This shows that in the rolling process, the contact stress components are different. Some are alternate stresses and some are pulsating stresses. Meanwhile, the normal stresses and shear stresses vary with positions at different phases. Therefore, it is very difficult to establish a relationship of the criteria of the contact fatigue strength with all the stress components. A variety of assumptions are put forward and some individual stress components are used as the contact fatigue criteria.

11.4.3.2 Contact Fatigue Strength Criteria

The commonly used criteria of the contact fatigue strength are as follows.

1. Maximum shear stress criterion
   According to the principal stresses at the z axis, the shear stress in the direction of 45° can be calculated. Analysis shows that the maximum shear stress of 45° is at a certain depth at the z axis. It is the maximum shear stress τ_{max} which is first used as the contact fatigue criterion. That is, when the maximum shear stress reaches a certain value, it will result in the contact fatigue wear. In the rolling process, the maximum shear stress is the pulsating stress, and its amplitude is τ_{max}.
2. Maximum orthogonal shear stress criterion
   Analysis shows that the maximum orthogonal shear stress τ_{yz} is at $x=0$ and at some y and z. Similarly, the maximum τ_{zx} is at $y=0$ and at some x and z. Thus, when the rolling surface coincides with one axis, the orthogonal shear stress will be the alternate stress. For example, if the rolling surface contains the short axis, the variation of the orthogonal shear stress τ_{yz} is from zero far away from the contact center to the maximum $+\tau_{yzmax}$ close to z, and then reduced to zero again at the z axis. Subsequently, the sign of the stress reverses and gradually reaches the negative maximum $-\tau_{yzmax}$, and then varies to zero. So, in each cycle, the maximum variation of the orthogonal shear stress τ_{yz} is $2\tau_{yzmax}$.

   It should be noted that although the orthogonal shear stress is usually less than the maximum shear stress, the variation of the orthogonal shear stress is larger than the maximum shear stress variation, namely, $2\tau_{yzmax} > \tau_{max}$. Because the fatigue phenomenon is directly related to the amplitude of the stress, ISO (International Organization for Standardization) and AFBMA (Anti-friction Bearing Manufacturer's Association) suggest using the maximum orthogonal shear stress as the criterion for the contact fatigue wear.

3. Maximum surface shear stress criterion
   Usually, the maximum shear stress is at the end point of the elliptical axis. For example, when rolling is in the same direction with the ellipse short axis, the maximum shear stress occurs at the endpoint of the long axis and is the pulsating stress.
   Although the surface shear stress is less than the maximum orthogonal shear stress, the defects on the surface and the rolling interaction greatly enhance the occurrence of the fatigue crack and the surface shear stresses.
4. Equivalent stress criterion
   The energy stored in material in the rolling process will change the volume and the shape of the contact body. The latter determines the fatigue damage. In accordance with the principle of the same deformation, the compound stress can be expressed by the equivalent stress instead:

$$\sigma_e^2 = \frac{1}{2}\left[\left(\sigma_x - \sigma_y\right)^2 + \left(\sigma_y - \sigma_z\right)^2 + \left(\sigma_z - \sigma_x\right)^2 + 3\left(\tau_{xy}^2 + \tau_{yz}^2 + \tau_{zx}^2\right)\right]. \tag{11.18}$$

The equivalent stress criterion considers the influence of all the stress components, but because of computational complexity and lack of data, it is not universally accepted yet.

Crook [19] found that in the disc rolling process, there is a plastic shear layer in the superficial. Because the plastic flow is confined to the thin layer, an elastic surface layer rolls relatively to the elastic core along the rolling direction. Hamilton [20] further experimentally proved that the plastic shear continues to accumulate as the stress cycles until the fatigue crack emerges. Johnson and Jefferies [21] analyzed the above phenomenon and proposed the plastic shear criterion of the contact fatigue under the condition with no continuous plastic shear flow:

$$p_H = 4k, \tag{11.19}$$

where p_H is the maximum Hertz stress; k is the yield shear stress. According to Tabor's empirical formula, $k = 6\mathrm{HV}$, where HV is the Vickers hardness.

As the maximum Hertz stress is larger than Equation 11.19, the orthogonal shear stress induces the plastic shear deformation parallel to the direction of the surface. When it is a rolling-sliding process, the frictional force is about ten times of the normal load. Then, $4k$ in Equation 11.16 should drop to $3.6k$.

The author has investigated a variety of contact fatigue criteria. The method is to add a small axial bending stress onto the contact stress field. Then, carry out the contact fatigue experiments for such a composite stress situation. The results are shown in Figure 11.20 [22]. In these experiments, the variation of the fatigue life is only caused by the axial stress. Thus, the results provide the basic guidelines to evaluate the contact fatigue. The calculation analysis show that the maximum shear stress criterion and orthogonal shear stress criterion cannot explain the experimental results, and the equivalent stress criterion can only partially explain the influence of the additional bending stress. The maximum surface shear stress position varies with the position and magnitude of the additional bending stress. Meanwhile, the contact fatigue life drops with increase of the maximum surface shear stress, that is, changing the additional bending stress, also changes the influence of the maximum surface shear stress on the fatigue life as well. Therefore, under experimental conditions, the maximum surface shear stress criterion is consistent with experimental results, while the fatigue crack initiates on the metal surface.

In addition, variable loads exist widespread in the engineering practice, and the contact fatigue wear of such a situation is more complex. Liu *et al.* studied the contact fatigue design criterion under a variable load [23].

11.4.3.3 Contact Fatigue Life

The contact fatigue phenomenon has a strong random feature. In the same condition, the fatigue lives of a number of specimens vary considerably. In order to ensure the reliability of the experimental data,

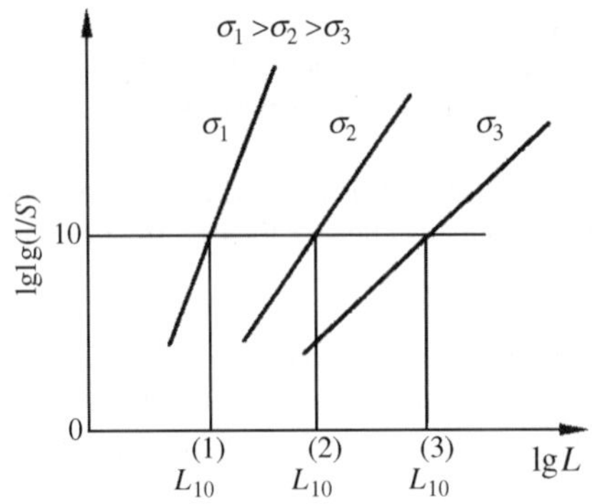

Figure 11.24 Weibull distribution.

the number of specimens should be more than 10 and the data should be treated according to the statistical method.

The contact fatigue life is usual in a Weibull distribution, that is

$$\log \log \left(\frac{1}{S} \right) = \beta \log L + \log A \tag{11.20}$$

where S is the probability of no damage; L is the actual life, usually expressed by the stress cycles N; A is a constant; β is the slope of the Weibull. For steel, $\beta = 1.1$–1.5, and for the pure steel, β is chosen the larger value; and for the rolling contact bearings: $\beta = 10/9$ for the ball bearing, and $\beta = 9/8$ for the roller bearing.

In the Weibull coordinates, Equation 11.17 will be a straight line, as shown in Figure 11.24.

When the experimental data have been obtained, the Weibull distribution curve can be drawn through the statistical calculation so as to obtain the slope β, the characteristic life L_{10} and L_{50}. L_{10} and L_{50} are the 10% and 50% of damage respectively. Thus the fatigue life, strictly speaking, is only between L_7 and L_{60} and is consistent with the Weibull distribution.

The slope β indicates the dispersion of the same group of the experimental data. As shown in Figure 11.25, when the load increases, the slope β also increases. Thus, the variation of the life expectancy reduces, that is, the dispersion reduces.

If the contact fatigue life of L_{10} or L_{50} is expressed by the stress cycles N, generally it is inversely proportional to the cube of the load. According to this approximate relationship, we can be obtained σ-N

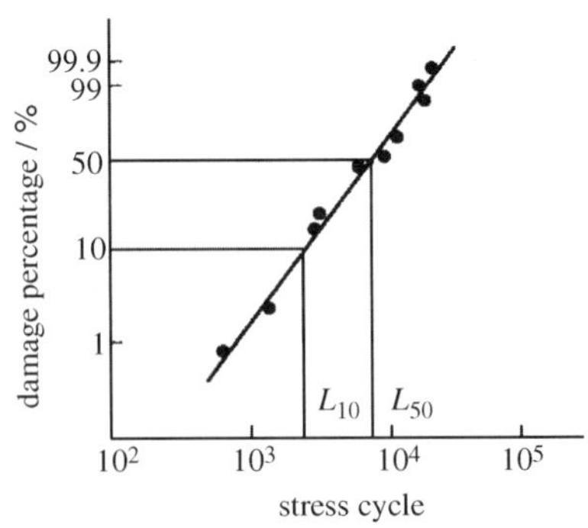

Figure 11.25 Distribution under different loads.

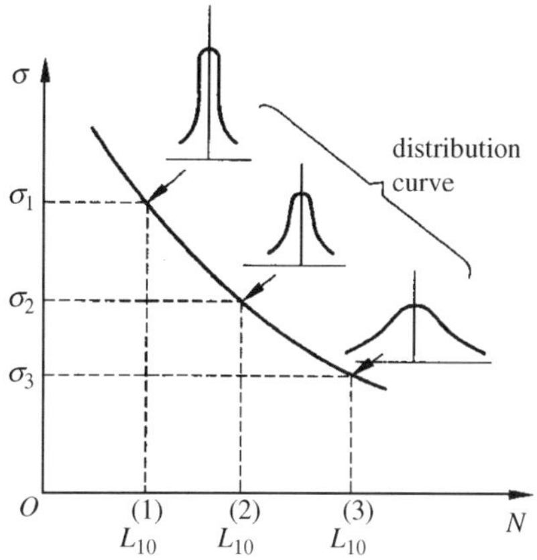

Figure 11.26 σ- N Curve.

curve, shown in Figure 11.26, where σ is the contact stress. Thus, from the curve we will be able to calculate the life under the condition of any stress.

11.5 Corrosive Wear

In the friction process, the surface damage caused by the chemical or electrochemical reaction of the friction surface metals and the surrounding medium is known as corrosive wear. The common corrosive wears are oxidation wear and special media corrosive wear.

11.5.1 Oxidation Wear

When the metal friction pair works in the oxidizing medium, the surface oxide film may be worn away and a new oxide film can quickly form. In the succession process the oxidation wear and the mechanical wear alternatively occur.

The oxidation wear depends on the strength of the oxide film and the oxidation rate. If the oxide film is brittle, its link to the substrate is weak and so is the shear strength. Or, if the oxide film formation rate is lower than the wear rate, the wear capacity is higher. While the oxide film is of a high toughness and the link to the substrate is high, or the oxidation rate is higher than the wear rate, the oxide film can play the role of anti-friction and wear-resistant. So, the oxidation wear is small.

For the steel friction pair, the oxidation reaction is related to the deformation of the surface contact. The plastic deformation promotes the oxygen in the air to diffuse into the deformed surface, and this also enhances the plastic deformation as well. First, the oxygen is saturated on the surface, and then it gradually spreads into the inner body. The oxygen concentration decreases from outside to inside. If the load, speed and temperature vary, different oxides are formed, such as the solid solution of iron and oxygen, granular oxide, or eutectic solid solution, for example, FeO, Fe_2O_3, Fe_3O_4 and so on. These oxides are hard and brittle.

The oxidation wear debris is a dark sheet or filamentary, but the flaky wear debris is reddish-brown, such as Fe_2O_3 or Fe_3O_4 is the gray and black filamentous debris. Sometimes, these features of the wear debris can be used to determine the oxidation process.

Factors affecting the oxidation wear include the contact load, the sliding friction pair speed, temperature, the hardness of oxide film, the oxygen media, the lubrication conditions, the material properties and so on. Usually, the oxidation wear rate is slower than the other wear rates. In Figure 11.27a,

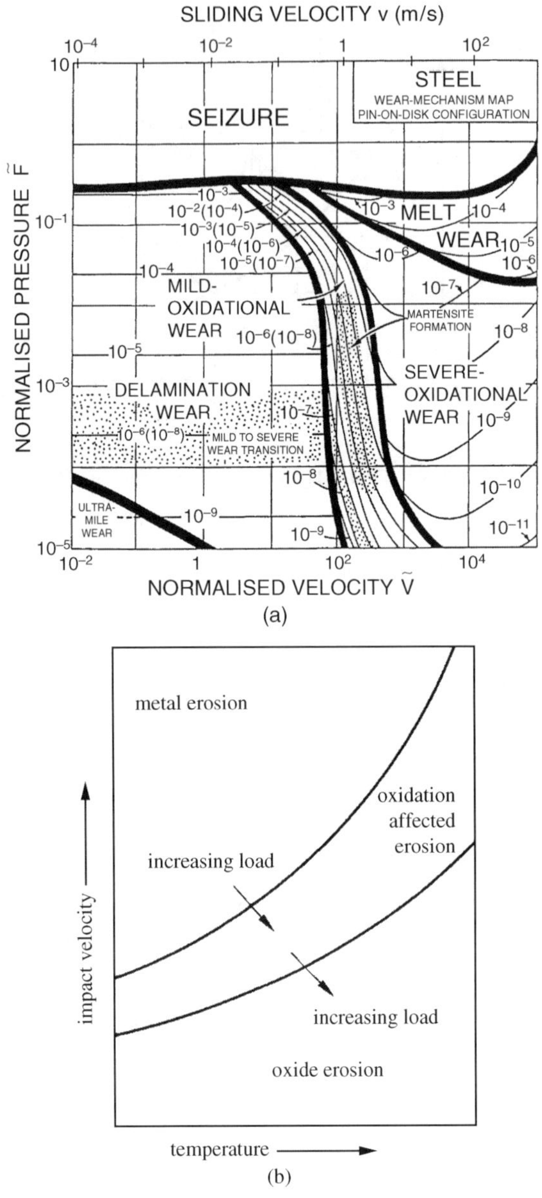

Figure 11.27 Influences of load, velocity and temperature on corrosion wear rate. (a) Wear map of steels for dry sliding; (b) Erosion-oxidation map for mild steel.

Lim *et al.* gave the experimental results with a pin-plate tester to study the influence of the velocity, load and temperature on the corrosion wear of steel [24].

It can be seen from the figure:

In the low velocity, the main components on the steel surface are the iron-oxide solid solution, granular oxide and eutectic solid solution. The wear capacity increases with increase of the sliding velocity. If the velocity is high; the main components are oxides. The wear capacity now is slightly low. When the sliding

Table 11.6 Corrosive capacities of commonly used bearing materials (g/h)

Bearing material	Tin Babbitt alloy	Lead-antimony alloy	Lead Babbitt alloy	Cu-Pb alloy	Ti-Al alloy
Corrosive capacity	0.001	0.002	0.004	0.453	1.724

velocity is much higher, the oxidation wear is transformed into adhesive wear due to the effect of frictional heat and the wear capacity increases sharply.

The influence of load on the oxidation wear is as follows. Under the load, the main components of the oxidation wear debris are Fe and FeO. While under heavy load, the main components mainly are Fe_2O_3 and Fe_3O_4, and seizure will appear.

Sundararajan *et al.* gave a map to show the temperature influence [25]. As shown in Figure 11.27b, the temperature enhances oxidation wear. Although the influence of the impacting velocity can increase metal erosion, it reduces oxidation wear.

11.5.2 Special Corrosive Wear

11.5.2.1 Factors Influencing the Corrosion Wear

In the chemical equipment, the metal surface of the friction pair reacts with the acid, alkali, salt or other media to induce corrosive wear. Corrosive wear is similar to oxidation wear but the wear trace is deeper and the wear capacity is more. The granular or filamentous corrosive wear debris is the medium compound of the metal surface and the surrounding media.

Because the lubricating oil contains some corrosive chemical compositions, corrosive wear can occur on the sliding bearing material. There are two kinds of corrosive wear: acid erosion and sulfidation corrosion. Beside choosing rationally a lubricant and limiting the acid and sulfur in oil, the bearing material is an important factor that influences corrosion wear. Table 11.6 shows the corrosive capacities of some commonly used bearing materials.

11.5.2.2 Chemical-Mechanical Polishing

Chemical-mechanical polishing is a comprehensive flattening technology applied in the ultra-large scale integrated circuit manufacturing process. This method can planarize a silicon wafer. It uses the corrosive wear and abrative wear as the basic methods to polish the wafer surface, the precision of which can be less than 1 Nm.

The chemical mechanical polishing is completed by chemical and mechanical actions. The steps are as follows. (1) The wafer forms chemical bonds with the oxygen and hydrogen on the polish solution particle surface. The wafer forms molecular bonds with the polish solution; (2) Under certain pressure, rotate the polishing pad to drive the slurry, and using mechanical action makes the slurry particles leave the surface so that the chemical or molecular bonds are broken to realize polishing. The chemical-mechanical polishing device and its working principle are shown in Figure 11.28.

11.5.3 Fretting

In 1937, serious damage was found on the smooth surface of some automobile products during transportation, which is called fretting. Fretting is caused as two surfaces are in relative motion with slight amplitude. It is a kind of corrosive wear so it is also called fretting corrosive wear.

Under a load, the peak of the contact surfaces forms adhesion. When the contact surfaces are suffered by external micro-vibration, they relatively slide to each other, usually no more than 0.25 mm. The adhesive points will be cut off and the cut surface forms oxidation wear, resulting in red-brown Fe_2O_3 wear debris

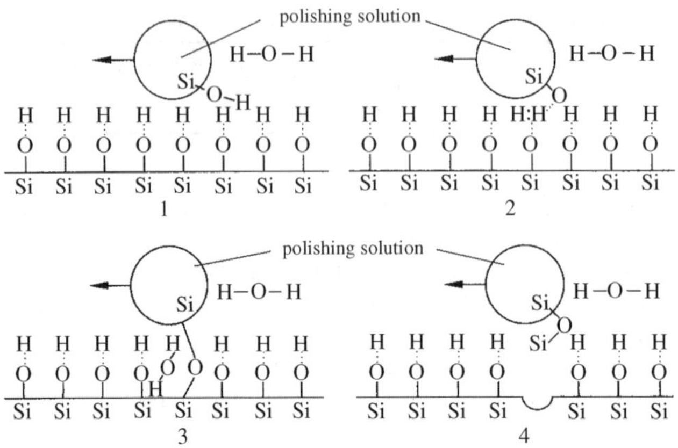

(a) chemical-mechanical polishing mechanism

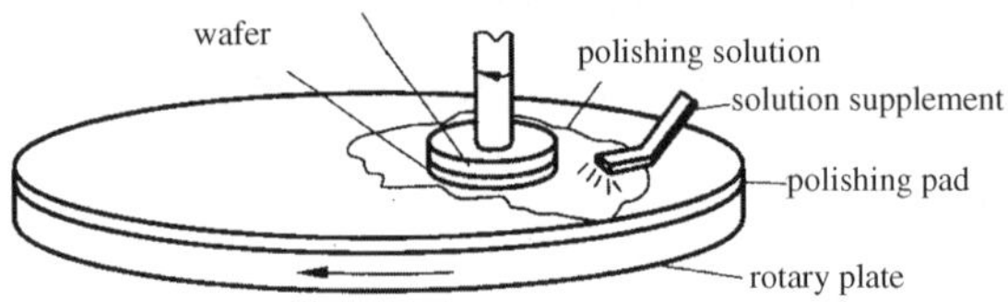

(b) chemical-mechanical polishing device

Figure 11.28 Chemical-mechanical polishing [26].

accumulating between the surfaces. Then, oxide wear debris plays the role of abrasive particle in the contact surface to cause abrasive wear.

Thus, minimal vibration and oxidation are major factors in fretting wear. Fretting wear belongs to the combination of adhesive wear, oxidation wear and abrasive wear.

The match of friction materials is an important way to avoid fretting wear. Generally, good anti-adhesive wear ability also has excellent anti-fretting wear ability. Improving the hardness can reduce fretting wear, but fretting wear is not related to surface roughness.

Proper lubrication can effectively improve anti-fretting ability because a lubricant film prevents the surface from oxidation. The usage of extreme pressure additive or coated molybdenum disulfide can also reduce fretting wear. Guo *et al.* studied the micro-damage problem of the steel cable of a bridge, and proposed to use polymer surface film to prevent fretting wear [27].

Fretting wear capacity increases with increase in the load, but as the load continues to increase, it will drop. Usually, the frequency of small amplitude vibration has no effect on the fretting of steel, but in large amplitude vibration, wear capacity increases with increase in vibration frequency.

11.5.4 Cavitation Erosion

Cavitation is surface damage caused by the relative motion of a solid surface and liquid. Usually, it occurs on the surface of the pump element, the turbine blade, the ship propeller and so on.

When pressure is below the evaporation pressure of liquid contacting with a solid surface, air bubbles will be formed near the surface or gas dissolved in the liquid may also form bubbles. When the bubble flows to the place where the liquid pressure exceeds the bubble pressure, the bubbles will collapse and instantaneously produce a severe impact and high temperature. If the solid surface is subjected to such an action repeatedly, fatigue occurs, such that some small pits form on the surface and then become sponge-like. Serious cavitation erosion can form a large pit on the surface, 20 mm deep.

The mechanism of cavitation erosion is due to the influence of the stress which causes surface fatigue, but chemical and electrochemical actions of the liquid accelerate cavitation erosion damage.

The effective measure to reduce cavitation erosion is to prevent the bubble from forming. First, the liquid movement should be streamlined to avoid the vortex, because in the low pressure vortex area, it is easy to form bubbles. Secondly, it should reduce gas concentration in the liquid and avoid disturbing liquid to limit the formation of air bubbles.

Appropriately selecting the material can improve anti-cavitation erosion ability. Usually, the metallic material with a high strength and toughness possesses a good anti-cavitation erosion ability and increasing corrosion resistance of the material will also reduce cavitation erosion damage.

It should be pointed out that the above-mentioned oxidation wear, special media corrosive wear, fretting and cavitation erosion are of a common phenomenon that the surface chemically reacts with the surrounding medium. Therefore, they can be collectively referred to as corrosive wear.

In most cases, the corrosive wear firstly has a chemical reaction and then under the mechanical action the chemical resultant falls off from the surface to form the wear debris. In the wear process, the lubricant additive reacts with the surface to form a chemical film. The chemical film may protect the surface from wearing, but the chemical reaction increases corrosion of the surface.

When the formation rate of the resultant surface chemistry and the worn rate are relatively balanced, their ratio may produce a different result. Here, take the extreme pressure additive to prevent from scuffing as an example to illustrate the different result.

A chemical reaction usually follows the Arrhenius principle so its reaction rate is

$$V = KCe^{E/RT}, \tag{11.21}$$

where V is the chemical reaction rate or the rate to generate the film; C is the concentration of the extreme pressure additive in the lubricant; E is the constant to show the activity of the extreme pressure additive; T is the absolute temperature, °K; R is the gas constant; K is the proportional constant.

Clearly, in a stable working condition, the corrosive wear rate depends on the chemistry reaction rate. By Equation 11.18, we can see that the corrosive wear rate is proportional to the concentration of the corrosive medium, while it is of an exponential relationship with temperature.

As pointed out before, when an extreme-pressure additive is used to reduce the adhesive wear, a suitable chemical activity, or additive compositions and concentration, should be carefully chosen. Figure 11.29

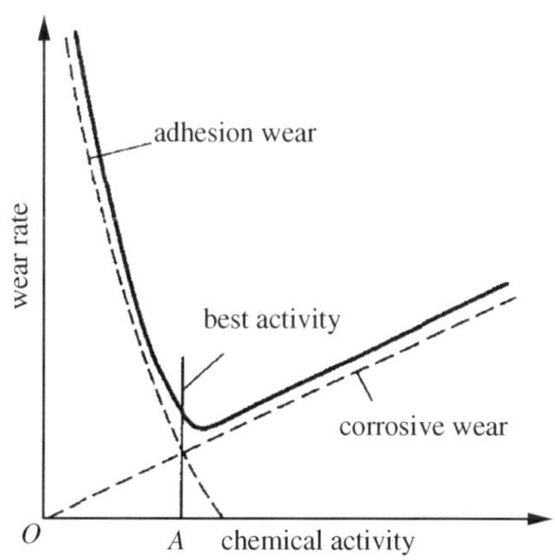

Figure 11.29 Best activity position.

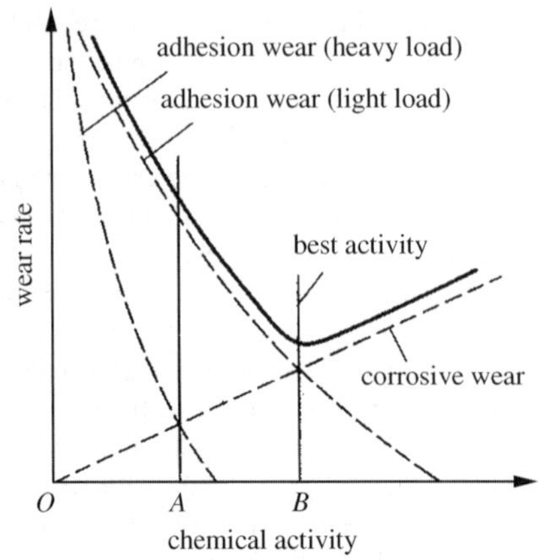

Figure 11.30 Best activity choice.

gives the relationship between adhesive wear and chemical activity of the extreme pressure additive which causes the corrosive wear. The adhesive wear rate increases as chemical activity decreases while corrosive wear rate increases linearly with chemical activity. Thus, Point A in the figure is the best activity choice, where total wear rate is the minimum.

In Figure 11.30 the adhesive wear curves change when the load is heavy. In this situation, Point B should be selected as the best chemical activity. In order to increase the chemical activity of the additive, we can increase the concentration of the additive, or choose a more active additive as well.

Therefore, it is shown that the anti-adhesive wear effect and corrosive wear effect are two aspects of the extreme pressure additive. Figure 11.31 gives the experimental results of one extreme pressure additive with the two wear testers. It is shown that although the anti-adhesive ability of the extreme pressure additive increases with its concentration, corrosive wear increases at the same time.

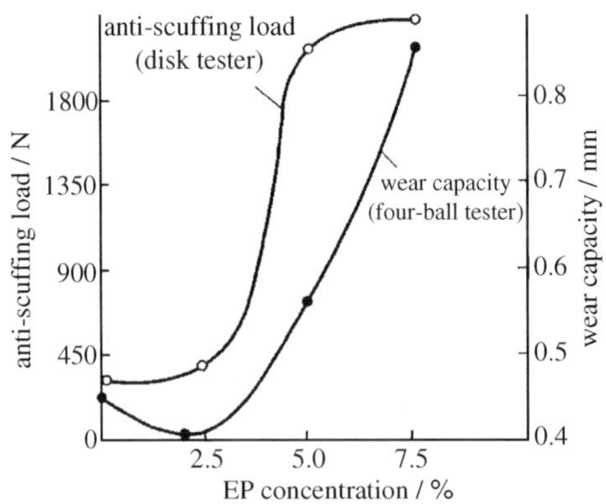

Figure 11.31 Influence of extreme pressure additive concentration.

References

1. Kragelsky, I.V., Dobychin M.H. and Kombalov V.S. (1977) Foundations of calculations for friction and wear, Mashinostroenie (in Russian), Moscow.
2. Khrushchev, M.M. and Babichev, M.A. (1960) A study of metal wear [in Russian], in *Izd. Akad. Nauk. SSSR*, Moscow.
3. Rabinowicza, E. and Mutisa, A. (1965) Effect of abrasive particle size on wear. *Wear*, **8** (5), 381–390.
4. Archard, J.F. (1953) Contact and rubbing of flat surface. *Journal of Applied Physics*, **24**, 981–988.
5. Fein, R.S. (1960) Transition temperatures with four ball machine. *ASLE Transactions*, **3**, 34–39.
6. Pooley, C.M. and Tabor, D. (1972) Friction and molecular structure: the behaviour of some thermoplastics. *Proceedings of the Royal Society of London*, **A329**, 251–274.
7. Fan, Y. J., Su, Z.W., and Wen, S.Z. (1988) Impact of oil-lubricated sliding velocity on surface scuffing. *Journal of Mechanical Engineering*, **24** (1), 18–87.
8. Almen, J.O. and Boegehold, A.C. (1935) Rear axle gears: factors which influence their life. *Proceedings of the American Society for Testing and Materials*, **35**, 99–146.
9. Blok, H. (1939) Seizure delay method for determining the protection against scuffing afforded by extreme pressure lubricants. *Journal of the Society of Automative Engineers*, **44** (5), 193–210.
10. Borsoff, V.N. and Godet, M. (1973) A scoring factor for gears. *ASLE Transactions*, **6**, 147–153.
11. Wilson, J.E., Stott, F.H., and Wood, G.C. (1980) The development of wear-protective oxides and their influence on sliding friction. *Proceedings of the Royal Society of London*, **A369**, 557–574.
12. Blok, H. (1963) The flash temperature concept. *Wear*, **6** (6), 483–494.
13. Wen, S.Z. (1990) *Principles of Tribology*, Tsinghua University Press, Beijing.
14. Fujita, K. and Yoshida, A. (1979) Surface failure of soft and surface-hardened steel rollers in rolling contact. *Wear*, **55**, 27–39.
15. Martin, J.B. and Cameron, A. (1961) Effect of oil on the pitting of rollers. *Journal of Mechanical Engineering Science*, **3**, 148–152.
16. Wen, S.Z. (1982) Effects of compound stress on contact fatigue. *Journal of Mechanical Engineering*, **18** (4), 1–7.
17. Way, S. (1935) Pitting due to rolling contact. *Journal of Applied Mechanics-Transactions of the ASME*, **57**, A49–A58.
18. Culp, D.V. and Stover, J.D. (1976) Bearing fatigue life tests in a synthetic traction lubricant. *Tribology Transactions*, **19** (3), 250–256.
19. Crook, A.W. (1957) Simulated gear tooth contacts: some experiments upon their lubrication and subsurface deformations. *Proceedings of the Institution of Mechanical Engineers*, **171**, 187–214.
20. Hamilton, G.M. (1963) Plastic flow in rollers loaded above the yield point. *Proceedings of the Institution of Mechanical Engineers*, **177**, 667–675.
21. Johnson, K.L. and Jefferies, J.A. (1963) Plastic flow and residual stresses in rolling and sliding contact. *Proceedings of the Institution of Mechanical Engineers*, **177**, 54–65.
22. Wen, S.Z. (1982) Evaluation of contact fatigue strength criterions. *Journal of Tsinghua University*, **22** (4), 9–18.
23. Liu, J.H., Wang, H., and Wen, S.Z. (1991) Experimental study on contact fatigue design criteria under varing load conditions. *Proceeding of the First National Conference of Tribological Design, Shenyang*, pp. 406–414.
24. Lim, S.G., Ashby, M.F., and Brunton, J.H. (1987) Wear-rate transitions and their relationship to wear mechanisms. *Acta Metallurgica*, **35**, 1343–1348.
25. Sundararajan, G., Roy, M., and Venkataraman, B. (1990) Erosion efficiency-a new parameter to characterize the dominant erosion micromechanism. *Wear*, **140** (2), 369–381.
26. Li, X. (1999) high-speed development of chemical-mechanical polishing technology. *Journal of Semiconductor*, **24** (3), 31–34.
27. Guo, Q., Wen, S.Z., and Luo, W.L. (1996) Fretting wear resistence mechanism of transferred film from organic high moleular materials. *Progress in Natural Science*, **6** (5), 593–601.

12

Macro-wear Theory

The wear problem is usually studied from the micro or the macro points of view. Micro-wear researches focus on the wear mechanism of the formation and variation from physics, chemistry, material science to establish a physical model in order to explore the essence and basic laws of wear. However, macro-wear researches regard wear as a common surface damage phenomenon by studying its topography variation, influence factors and measures to improve wear-resistant ability for engineering applications. The researches in these two aspects are all very important and it is effective to solve an actual wear problem by combining both aspects together.

In Chapter 11, we have discussed the different wear mechanisms. Abrasive wear is mainly caused by plowing and micro-cutting; adhesive wear is related to surface force and friction heat; fatigue wear is the result of initiation and propagation of the surface fatigue crack under the action of cyclic contact stress; while oxidation and corrosion wear are produced by the environmental medium and chemical reactions known as chemical wear.

An actual wear phenomenon is usually caused by more wear mechanisms so it is a comprehensive performance of different forms. For example, the wear of a plow harrow is mainly a form of abrasive wear, but because of the chemical action of certain substances in the water and soil, oxidation and corrosion wear also exist.

With variation in working conditions, the wear form of a mechanical part will change accordingly. In Figure 12.1, the gear failure types are given with the load and the velocity. In this chapter, wear will be regarded as a comprehensive phenomenon of the surface damage and the macro variations, the affecting factors and the measures of wear are discussed.

In order to design a mechanical part with adequate wear-resistant ability and possibly to estimate its wear life, we must establish a suitable calculation method. In recent years, due to the analysis of the wear states and the wear debris, some wear theories are put forward, which are the bases of the wear calculation.

A wear calculation method must consider the characteristics of the wear phenomenon. The characteristics are usually quite different from those of strength damage. For example, because the actual contact point of the friction pair is discrete, the bearing load area is very small and the wear process changes all the time. Thus the material damage form also changes. In addition, under the thermal effect and physical-chemical reaction, the establishment of wear theory is difficult.

Therefore, the dynamic characteristics and damage features of the wear process and the surrounding media are important for establishing the wear theory and calculation method. Because of these complexities, wear calculation is still not able to satisfy application requirements.

Principles of Tribology, First Edition. Wen Shizhu and Huang Ping.

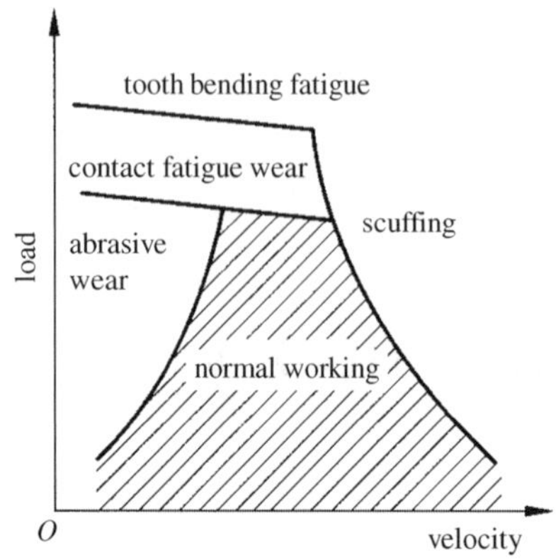

Figure 12.1 Gear failure types.

12.1 Friction Material

According to different requirements, friction materials can be divided into two kinds. One kind of friction material is used in the brake and clutch. The main requirements for this kind of material are high thermal stability and a high friction coefficient. The other kind of friction materials is also subdivided into the anti-friction materials and wear-resistant materials. The anti-friction and wear resistant properties of materials are usually concordant. Thus, a low friction coefficient usually corresponds to a good wear resistance. However, not all friction materials are of two good properties at the same time. Some are anti-friction but not wear resistant. Some are good wear resistant but of very high friction.

Selection of friction materials is mainly based on the pressure of the friction surface, the sliding velocity and working temperature. Because the sliding bearing is in the surface contact, the pressure is low so it mainly presents adhesive wear. Therefore, soft and hard materials can all be adopted. However, for the point or line contacts such as gears or rolling bearings, because the concentrated load will cause contact fatigue wear, hard material should be used.

12.1.1 Friction Material Properties

Usually, the main technical requirements for friction material are as follows.

12.1.1.1 Mechanical Properties

Because of the action of the load and movement of the friction surface, the friction material should have enough strength, malleability and especially compression resistance. In addition, fatigue strength is also very important. About 60% of failure in the sliding bearing is spalling due to surface fatigue.

The harder the metal material, the better its wear resistance is. A good plasticity can quickly have the friction surface running-in, but low plasticity will cause the material brittle and fractured under the impact load.

12.1.1.2 Anti-Friction and Wear-Resistance

A good wear-resistant material should also have a low coefficient friction. It should not only be wear-resistant but also should not cause excessive wear on the surface. Therefore, wear-resistant performance of friction material is essential for the compound performance of the material.

The running-in performance is an index to evaluate the material performances. A good running-in performance means that a high quality of the surface can be achieved in a short running-in time with a minor wear.

12.1.1.3 Thermal Property

In order to maintain a stable lubrication condition, particularly in the boundary lubrication, the friction material should have a good thermal conductivity to reduce the operating temperature of the friction surface. At the same time, the thermal expansion coefficient of the material should not be too large; otherwise the clearance may vary significantly and affect the lubrication performance.

12.1.1.4 Lubrication Ability

The friction material and the lubricant should be well oiled, that is, an adsorbed film should be firmly connected. In addition, the friction material and the lubricant should have better wettability so that it is easy for the lubricant to cover the whole friction surface.

12.1.2 Wear-Resistant Mechanism

It should be pointed out that in addition to being related to the compositions, the tribological properties of a material also depend on its structure. In order to develop a good friction material, various anti-friction and wear-resistant mechanisms have been proposed. The main mechanisms include the following.

12.1.2.1 Hard Phase Bearing Mechanism

It is generally believed that in the organization of a good anti-friction and wear-resistant material, some hard particles distribute heterogeneously in the soft plastic substrate. For example, the tin base Babbitt is the base solution of the plastic antimony and tin, in which many hard Sn-Sb cubic crystals and Cu-Sn needle-like crystals are distributed. Under the load, the hard phase mainly carries the load, while the soft phase supports the hard phase. Because the hard phase is in contact and has relative sliding, the friction coefficient and the wear rate are small. Also, because the hard phase is supported by the soft base, it is easy for the friction surface to deform such that the surface cannot be scratched easily. At the same time, the soft base can make the pressure uniform. When the load increases, the hard-phase particles under the larger pressure sink deeply into the soft base so that more hard particles attend to carry the load.

12.1.2.2 Soft Phase Bearing Mechanism

In contrast with the above view, it was believed that the anti-friction and wear-resistant mechanism is due to the soft phase bearing the load. In such a material, the thermal expansion coefficient of the soft phase is larger than that of the hard phase. During the friction process, friction heat causes expansion so that the soft phase is protruded up higher than the oil to bear the load. Because of high plasticity, the soft phase, thereby, reduces the friction.

12.1.2.3 Porous Saving Oil Mechanism

Powder metallurgy materials are widely used in modern machineries and equipment. Such a material is a mixture of the nonmetallic powder and metal powder with some solid lubricant, such as graphite, lead, sulfur, or sulfide to improve anti-friction ability. The can be made by molding and sintering.

The powder metallurgy material has pores of about 10–35% in volume. After the material is dipped into hot oil for a few hours, the pores will be filled with oil. When the friction pair is in relative sliding, friction heat causes material to expand so that the pores press the oil out to the surface. Therefore, the oil can be used for lubrication.

In the Babbitt alloy and lead the bronze bearing materials, because the various phases are of different thermal expansion coefficients, many small pores are formed after thermal expansion and contraction manufacturing processes. Therefore, these materials also have the same lubricating effect as powder metallurgy.

12.1.2.4 Plastic Coating Mechanism

In recent years, multi-layer materials have been increasingly widely used in the bearing and friction pairs. The hard base surface is covered with one or more layers of the soft metal coatings. The coating materials used are commonly lead, tin, indium and cadmium. Because the coatings are thin and plastic, they can be easily run in and reduce the friction coefficient.

12.2 Wear Process Curve

12.2.1 Types of Wear Process Curves

Figure 12.2 shows four typical curves, which indicate the relationship between the wear capacity Q and time t. A wear process curve is usually composed of three stages.

1. Running-in stage
   At this stage, the wear rate gradually decreases with increase of time. It appears in the initial running period of the friction pair.
2. Steady wear stage
   After running-in the friction surface will be at a steady state. At this stage, the wear rate keeps almost constant. This is the normal working period of the friction pair.
3. Severe wear stage
   At this stage, the wear rate increases rapidly with time such that the working conditions become worse drastically. This leads the friction pair to be quickly worn out.

Figure 12.2a is the typical curve of the wear process. In this working condition, the wear process curve consists of three stages.

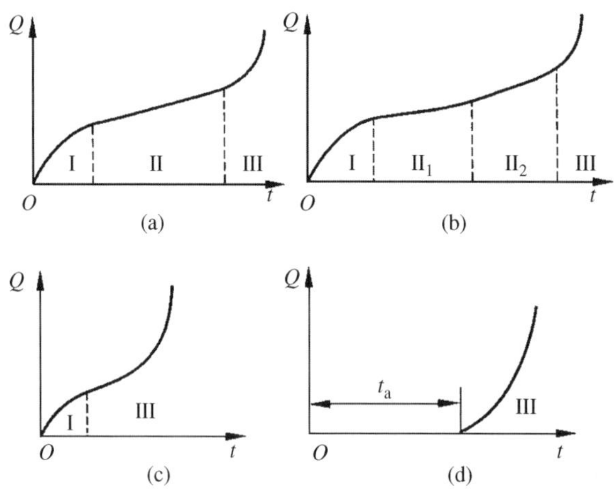

Figure 12.2 Typical wear process curves.

Figure 12.3 Surface morphologies before and after running-in.

The curve of Figure 12.2b indicates that after the running-in period the wear of friction pair went through two working conditions. Therefore, there are two steady wear stages. In the steady stages, the wear rates are different because the working condition changes.

Figure 12.2c is the wear curve under poor working condition. After the running-in, dramatic wear occurs so the steady wear stage cannot be established.

The curve of Figure 12.2d shows fatigue wear. When the element works over its contact fatigue life, it may exhibit the fatigue wear occasionally and rapidly develop into failure.

12.2.2 Running-In

After the friction pair is assembled, the micro and macro geometrical defects of the surface cause the actual contact pressure on the peak height thus wear is obvious. Therefore, before the normal operation of a new machine, it usually needs running-in according to the appropriate specification. During running-in, because of the wear and the plastic deformation of the peaks, the profile of the contact surface is gradually improved and the surface pressure, friction coefficient and wear rate decrease so that it enters a stable wear stage.

Because the surface topography has been dramatically changed in the running-in, the wear rate is usually 50–100 times higher than that of normal working conditions and the maximum asperity height h_{max} has been worn off about 65–75%.

The running-in not only let the friction pairs be adapted to each other in the geometry, but also allow the surface to obtain a stable structure suitable for the working condition as well.

Figure 12.3 indicates changes to surface topography before and after running-in. After running-in the contact area and peak radius have significantly increased. Figure 12.4 is the plasticity index curve. After running-in, the surface contact has transited from the plastic to the elastic-plastic contact, or even the elastic contact.

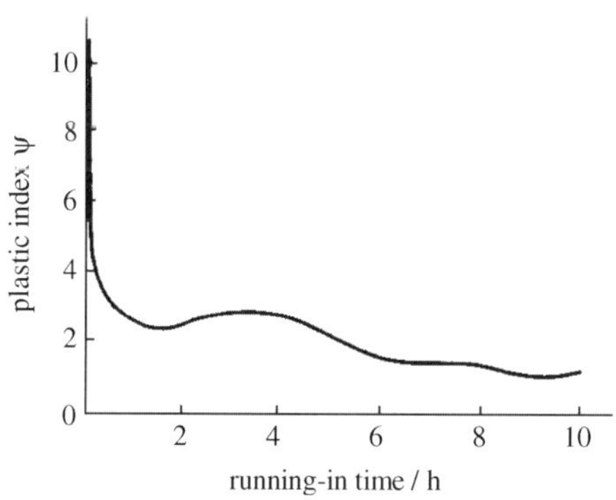

Figure 12.4 Plastic index via running-in time.

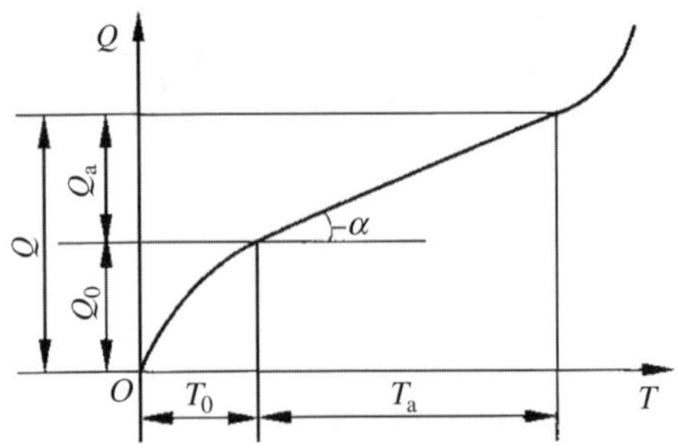

Figure 12.5 Running-in curve.

12.2.2.1 Working Life

The running-in time, the final wear volume and wear rate with different specifications are quite different. It has been practically proved that a good running-in can make the working life of the friction pair increase 1–2 times longer.

If the subscript 0 indicates the physical quantities of the running-in and *a* indicates those of the steady wear, the wear rate is equal to:

$$\gamma = \frac{dQ}{dT}. \tag{12.1}$$

The total wear capacity is $Q = Q_0 + Q_a$, and the steady wear rate $Q_a = \gamma_a T_a$, where $\gamma_a = \tan\alpha$ (Figure 12.5). Therefore, the normal life is equal to

$$T_a = \frac{1}{\gamma_a}(Q - Q_a). \tag{12.2}$$

It can be seen that the life T_a increases with decrease of Q_0 and Y_a.

Figure 12.6 gives three standard running-in curves obtained from three of the same type of engines. If the total wear amounts are the same as Q, their lives are quite different. We can see from the figure that No. 2 is better than No. 1. Although their surface qualities after running-in are the same, the running-in wear of No. 2 is less, that is, $Q_0 < Q_0'$. So, the life of No. 2 is longer than that of No.1, that is, $T_{a2} > T_{a1}$.

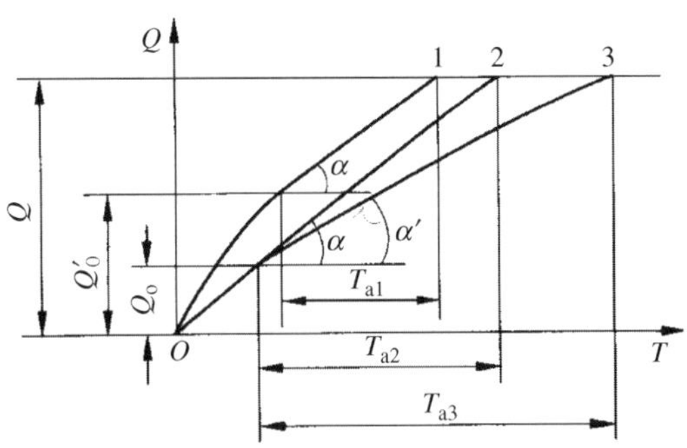

Figure 12.6 Three running-in curves of the same type engines.

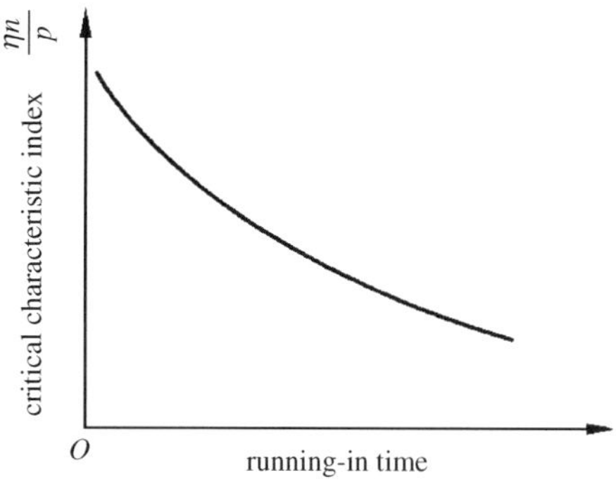

Figure 12.7 Running-in of the sliding bearing.

However, No. 3 is the best not only because the running-in wear is the smallest, but the steady wear rate is the lowest, that is, $\alpha' < \alpha$. Therefore, it has the longest life T_{a3}.

In addition, a good running-in can also effectively improve the other performances of the friction pair. As shown in Figure 12.7, the surface topography of a sliding bearing can be improved by running-in so that the critical bearing number can be reduced and a hydrodynamic lubricant film is easily established. For another example, a reasonable running-in engine can improve the surface quality of the cylinder piston ring so as to reduce the scratches, raise the surface match and save fuel consumption of the engine up to 50%.

12.2.2.2 Measures to Improve the Running-in Performance

A good running-in should be of a short running-in time, a small running-in wear volume and a high wear resistance. In order to improve the running-in performances, the following measures can be taken.

1. Choose a reasonable running-in specification
   When a new machine starts working, the load should not be too heavy, or the damage on the surface will be severe resulting in early wear failure. A reasonable running-in standard is to gradually increase the load and the friction velocity so that the surface quality can be improved accordingly. The working condition in the final running-in stage should be close to the working conditions of the usage.
2. Select an appropriate lubricant and additives
   The properties of a running-in oil have a significant influence on the surface. It can be found that the furrows of the surface before and after running-in are deeper and wider as the lubricating oil viscosity increases so that the surface wear resistance decreases. Because the thermal conductivity of the low viscosity oil is good and easy to form a surface adsorption film, the adhesive wear is slight in running-in so that the surface quality is good.

   If we add some appropriate additives into the lubricant oil, it can not only speed up the running-in process, but also strength the adsorption film to avoid serious adhesive wear. Thus, the surface quality can be improved.
3. Use appropriate matching materials
   The running-in performance of a friction pair is a combination of the properties of the matching materials. Material with a good running-in performance means that it is not only easy to run-in, but able to promote the running-in of the matched part.

   Take the sliding bearing material as an example. When the usual journal material is steel and the Babbitt alloy is used for the bearing material, the running-in performance is good. This is because the Babbitt alloy is of good plasticity and easy to run-in and its structure contains SnSb hard particles which promote the running-in of the journal surface. Although the soft texture of the lead bronze is easy for the bearing to run-in, it is difficult for the journal to run-in so that the running-in time is long. The Fe-Al_3 particles contained in FeAl bronze have a high hardness so they are hardly able to run-in without hurting the journal. Therefore, the matched journal must be quenched.

In order to improve the running-in performance of the material, a thin plastic coating can be plated on the surface. For example, the surface of the cast iron piston ring can be tin-plated. In order to speed up the running-in process of the matching surfaces, sometimes some suitable abrasive particles can be added between the friction surfaces, but they should be chosen carefully.

4. Control manufacturing precision and surface roughness

   Clearly, to raise the manufacturing and assembly precision of the friction pair will significantly reduce the wear capacity of the running-in. The choice of surface roughness should be determined based on working conditions.

   Khrushchev studied the running-in of the journal and the bearing surfaces and pointed out [1] that a different processing method gives a different roughness after the same running-in, and the running-in time is different. Many experiments proved that the final surface roughness after running is not related to the original roughness, but depends on the running-in working condition. The running-in roughness is best suited for the given operating condition. It ensures the lowest working wear rate. If the initial roughness is close to the best roughness, the running-in volume can be exponentially reduced.

12.3 Surface Quality and Wear

The friction surface processed by different methods has a different surface topography, such as roughness, waviness, macro-geometry deviation and directional processing traces as well as different physical qualities, such as work hardening, micro-hardness and residual stress. These will significantly influence the surface wear.

12.3.1 Influence of Geometric Quality

The geometric quality of a machined surface can be expressed by the surface topography parameters. If the peak height is H and the distance between the two peaks is L, the ratio L/H can be used to divide the profile parameters into roughness, waviness and macro deviation as shown in Figure 12.8.

Usually, surface waviness is cyclical, and its peak distance is long, generally 1–10 mm. The surface roughness is stochastic, its peak distance is about 2–800 μm, and the roughness peak height is about 0.03–400 μm.

The influence of roughness on wear has not yet been determined. In 1938, the engineers in the Chrysler automobile plant proposed that the smaller the surface roughness, that is, the smoother the surface, the smaller the wear capacity. Accordingly, their main component surfaces are designed and machined to be ultra-precise. This conclusion only reflects on the mechanical action of the surface roughness. However, in

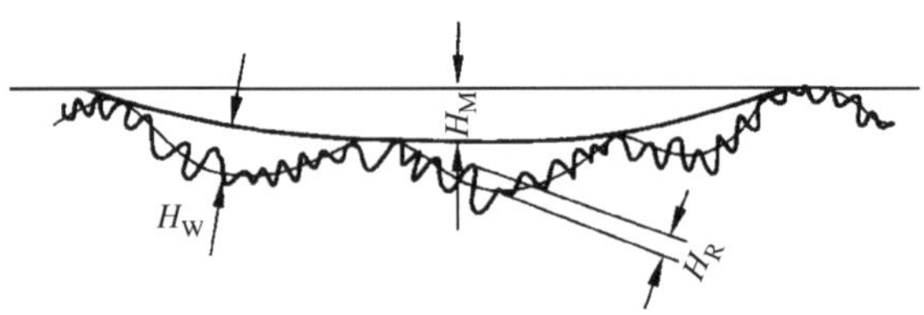

Figure 12.8 Surface profile divisions.

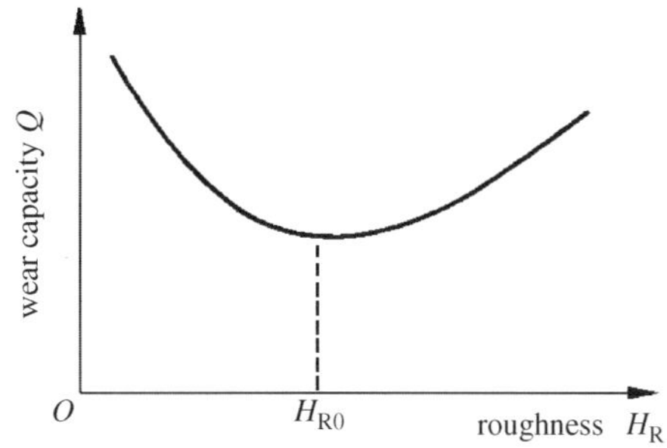

Figure 12.9 Optimal roughness.

1941, another factory, Buick, believed that the action of surface molecules was the main cause of friction and wear so that sufficient roughness could raise the wear-resistance. They suggested processing the element surface by corrosion. Khrushchev had studied the influence of the mechanical surface quality on wear systematically and gave the following conclusion. His study showed that under different working conditions, the surface roughness has an optimal value H_{R0}, where wear capacity is the smallest, as shown in Figure 12.9. This conclusion has been confirmed by many experiments.

The existence of optimal roughness shows that a wear process is a combination of mechanical and molecular actions. When the surface roughness is less than optimal, wear is mainly caused by molecular action. When the surface roughness is greater than optimal, wear is mainly produced by mechanical action.

Experiments also show that in different working conditions, optimal roughness is also different. In heavy working condition, the friction pair is worn seriously, and therefore optimal roughness will also increase accordingly. As shown in Figure 12.10, the working condition includes the load, sliding velocity, ambient temperature and lubrication condition.

Figure 12.11 shows that if surface roughness is different, the results are almost the same. Therefore, under certain working conditions, the optimal roughness after running-in is compatible with working conditions regardless of the initial roughness. Then, the surface will work stably under optimal roughness. The figure also shows that when $H_R > H_{R0}$, the intense mechanical wear occurs so that H_R tends to H_{R0}. While $H_R < H_{R0}$, the molecular wear is gradually H_R to H_{R0}. Only when the roughness is optimal, is wear at its minimum.

The influence of surface waviness on wear is similar to that of roughness. In addition, the surface waviness increases the wear capacity of running-in and the wear rate is stable after running-in.

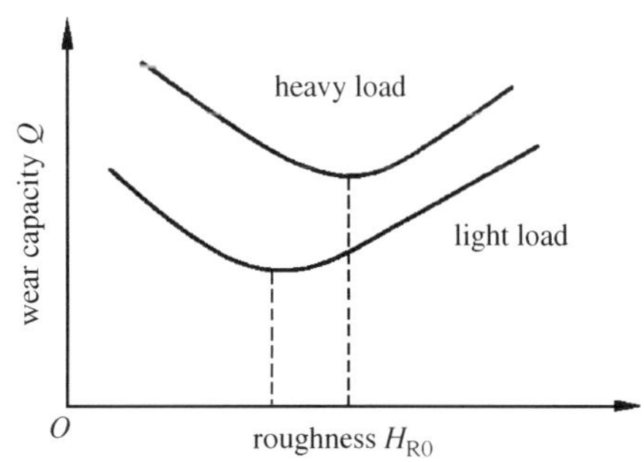

Figure 12.10 Optimal roughnesses of different working conditions.

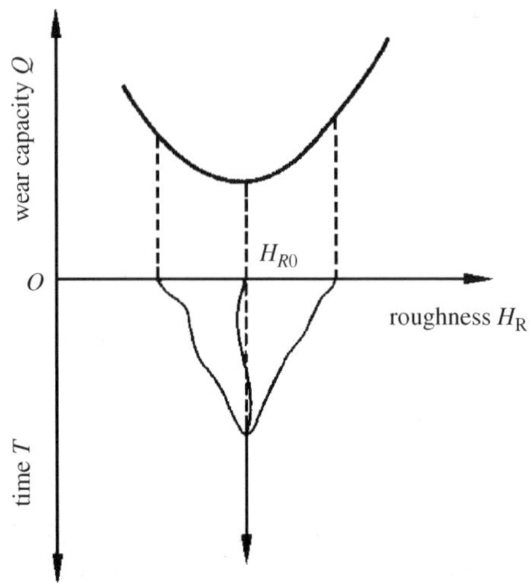

Figure 12.11 H_R varying in running-in process.

Figure 12.12 gives the running-in wear curves for a Babbitt specimen sliding on the steel surfaces with different waviness.

The direction of the trace on the friction surface influences the running-in time and wear capacity. The direction of the trace after running-in is always along the sliding direction and this is not related to the direction of the original trace.

Figures 12.13 and 12.14 show the influence of the direction of the trace on wear. The light load working condition refers to pressure $p = 14.2\,\text{MPa}$ and the heavy load working condition to $p = 66\,\text{MPa}$.

From the figures, we can see that in the light load working condition, the running-in is the smallest when the direction of the trace is parallel to the sliding direction. This is because under light load working conditions, the pressure is not high enough to easily form a lubricant film. Wear is mainly caused by mechanical action of the rough. However, under heavy load working conditions, the possibility of adhesive wear increases, the cross direction trace can avoid the large area contact so as to enhance wear-resistance. The machine rail belongs to the heavy working condition so the cross direction trace is usually adopted.

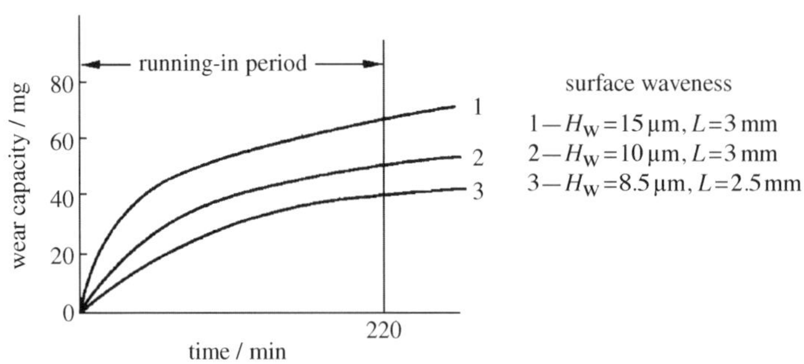

Figure 12.12 Running-in curves of Babbitt alloy and steel.

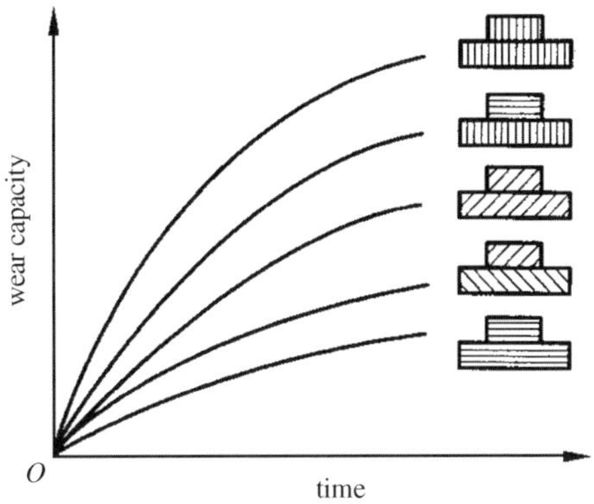

Figure 12.13 Wear under the heavy load condition.

12.3.2 Physical Quality

Due to dramatic change of heat during the cutting process and deformation, a processed surface possesses specific physical qualities, including work hardening, micro-hardness and residual stress distribution. These physical qualities significantly influence the wear performances of the surface. However, the influence is often ignored and researches on this are not adequate.

In the cold hardening process, plastic deformation of the surface promotes the diffusion of the oxygen in the metal to form a firmly connected oxide film which improves the ability of the anti-oxidation wear. After cold hardening, the plasticity of the surface reduces so that hardness is increased and adhesive wear is reduced to raise the anti-plowing ability. The initiation and development of the contact fatigue crack on the hardened layer of the surface must occur under high stress and after many stress cycles. Therefore, work hardening can increase the surface fatigue wear life. In general, after work hardening the surface wear resistance will be improved for any types of wear. Generally, the hardened layer of a roughing machined surface is about 0.3–1 mm deep; the hardened layer of a fine cutting or milling surface is about 0.1–0.2 mm deep; and that of the grinding is only 0.05–0.1 mm deep.

The stress state of the surface layer greatly affects the wear properties. In a cutting process, the deformation, the frictional force, the induced phase transition and the volume are changed by the cutting heat to induce surface residual stress. Residual stress distribution is affected by various factors and is

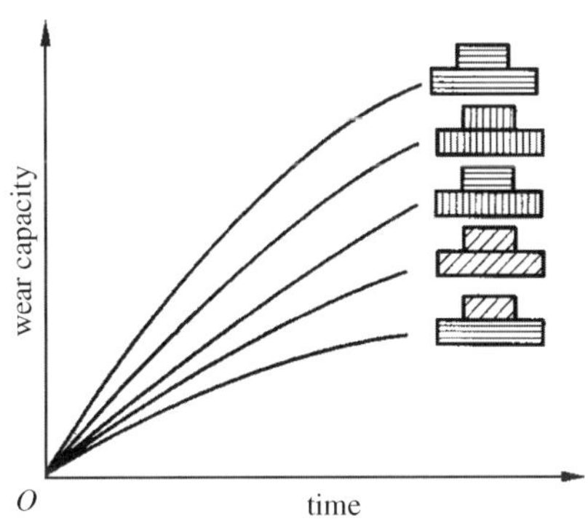

Figure 12.14 Wear under the light load condition.

very complicated. The influence of residual stress on wear is not very clear. Some people believe that both the tensile and compression residual stresses can improve abrasive wear resistance. The smaller the stress, the severer the wear is. This is because residual stress is body stress, which can reduce the activity of the metal atoms so as to slow down wear. Others think that only tensile residual stress can improve wear resistance. Because in the wear process, plastic deformation results in compressive stress, only when the compressive stress reaches a certain value, will the crack accelerate wear. Therefore, if the initial tensile stress is large, the time that the surface compressive stress reaches critical value is long so surface wear resistance is high. On the contrary, most experimental results show that the surface compression residual stress can improve the ability of the anti-contact fatigue wear, but tensile residual stress will reduce fatigue wear life. This conclusion can be obtained from the contact stress analysis, that is, surface compression residual stress can reduce maximum shear stress and the equivalent stress on the surface, and can also reduce the maximum shear stress under the surface.

In summary, the surface qualities including the geometric quality and the physical quality have important influences on wear. Because the surface qualities are determined by the processing and manufacturing conditions, the study of the relationship between surface quality and wear aims to discover suitable manufacturing method based on optimal surface qualities.

12.4 Theory of Adhesion Wear

In recent years, because the development of micro-analytical techniques of the surface wear has promoted the study of wear, many theories related to material wear have been proposed. Some important wear theories are as follows.

In earlier studies, Tonn attempted to set up the relationship between wear and the mechanical properties of materials, and he proposed an empirical formula of the abrasive wear [2]. According to action between the atoms, Holm derived the equation of the wear capacity per unit displacement [3]:

$$\frac{dV}{dS} = P\frac{W}{H}, \tag{12.3}$$

where V is the wear capacity; S is the sliding displacement; W is the load; H is the hardness; and P is the departure probability of the contact atoms.

Archard established his adhesive wear theory [4] and proposed the wear calculation formula similar to the Holm formula. Rowe modified the Archard equation. He considered the influence of the surface film as well as the tangential stress and the desorption of the boundary film to increase contact peak area, and derived the following wear formula [5]:

$$\frac{dV}{dS} = k_m(1 + \alpha f^2)^{\frac{1}{2}}\beta\frac{W}{\sigma_s}, \tag{12.4}$$

where k_m is the coefficient of the material; α is the constant; f is the friction coefficient; β is the factor relating to the surface film; and σ_s is the yield stress.

From the formulas of Holm and Achard, it can be known that the wear capacity is proportional to the sliding distance and the load, but for soft material, the wear capacity is inversely proportional to the yield stress or hardness.

Experimental studies have concluded that the wear capacity proportional to the sliding distance is basically suitable for most wear conditions. However, the conclusion that wear capacity is proportional to load is only suitable for a certain load range. For example, in the case of steel on steel, when the load is larger than $H/3$, the wear capacity will increase exponentially with the load. The conclusion that wear capacity is inversely proportional to hardness has also been confirmed by many experiments, particularly suitable for abrasive wear. The influences of other characteristics of materials on wear cannot be ignored as well.

Rabinowicz analyzed the wear debris formation of the adhesive wear from the energy viewpoint [6]. He pointed out that the formation condition of the debris is that the deformation energy before the separation of the wear debris must be greater than after separation. With the above suggestion and the Archard model, Rabinowicz analyzed the plastic deformation of the hemispherical abrasive debris and the energy stored in the adhesive nodes, and established that the stored energy per unit volume was equal to

$$e = \frac{p_s^2}{2E}, \tag{12.5}$$

where p_s is the surface compressive stress of the plastic deformation; and E is the elastic modulus.

If the contact circle radius of a wear debris is a and the surface energy per unit area is γ, the condition a wear debris forms from a plane surface is

$$\frac{2}{3}\pi a^3 \left(\frac{p_s^2}{2E}\right) > 2\pi a^2 \gamma. \tag{12.6}$$

By the elastic contact theory, it is known that $p_s = 1/3H$ for the metal material, where H is the hardness. Therefore, we have:

$$a > \frac{54E\gamma}{H^2} \quad \text{or} \quad a > \frac{KE\gamma}{H^2}, \tag{12.7}$$

where K is the coefficient to be determined according to the shape of wear debris.

In fact, in the friction process, there are other forms of energy. Therefore, the wear debris has disappeared before we encounter Equation 12.7 so a in Equation 12.7 should be the maximum size of a debris, that is

$$\alpha \leq \frac{KE\gamma}{H^2}. \tag{12.8}$$

12.5 Theory of Energy Wear

Fleisher proposed the energy wear theory, which is based on the energy consumption during the friction process [7]. The fundamental point of the energy wear theory is that the work done in the friction process is mostly dissipated in the form of the frictional heat, but about 9–16% stored in the friction material in the form of the potential energy. When the energy is accumulated to the critical value, the wear debris will be peeled off from the surface. Therefore, the wear is a process of energy transformation or consumption.

In Fleisher's analysis, the concept of energy density is introduced, which expresses energy absorbed or consumed in the unit volume. Suppose E_e is the absorbed energy density in each friction process, E_k is the energy density to form the wear debris in each friction process, then

$$E_k = \xi E_e, \tag{12.9}$$

where ξ is the coefficient taking into account the energy absorbed but not to form the wear debris.

If after n times friction, the wear debris is produced, all the energy to generate the wear debris in the first $(n-1)$ times is equal to $E_k(n-1)$. The energy absorbed by the last friction is E_e which is all consumed to peel the wear debris from the surface. Therefore, all energy density E_b' to form the debris is

$$E_b' = E_k(n-1) + E_e, \tag{12.10}$$

that is

$$E_b' = E_e[\xi(n-1)+1]. \tag{12.11}$$

The energy density of Equation 12.10 is obtained based on the condition that the absorbed energy is the same in each friction. Therefore, it is the average energy density. In fact, the actual absorbed energy is not the same in each friction.

According to Tross's study, the actual fracture energy density of the wear debris is K times the average energy density, where $K > 1$. Thus, the actual energy of the formation of a debris $E_b = KE_b'$. So, we have

$$E_e = \frac{E_b}{K[\xi(n-1)+1]}. \tag{12.12}$$

If we set E_R as the wear energy density, that is, the wear energy consumed per unit volume, it is equal to

$$E_R = \frac{\tau_y \Delta s}{\Delta h}. \tag{12.13}$$

The wear rate will be

$$\frac{dh}{ds} = \frac{\Delta h}{\Delta s} = \frac{\tau_y}{E_R}, \tag{12.14}$$

where τ_y is the frictional force per unit area; Δs is the sliding distance; Δh is the wear thickness; dh/ds is the line wear rate.

If E_e is the friction absorbed energy per unit volume, n is the number of the friction times to form the debris, the required wear energy per unit volume E_R is equal to

$$E_R = nE_e. \tag{12.15}$$

Taking into account that the deformation volume V_d storing the energy in the contact points is larger than the worn volume V_w, and setting their ratio:

$$\gamma = \frac{V_w}{V_d}, \tag{12.16}$$

we therefore have

$$E_R = \frac{nE_e}{\gamma}. \tag{12.17}$$

Substitute Equation 12.12 into Equation 12.17, then we have:

$$E_R = \frac{nE_b}{K[\xi(n-1)+1]\gamma}. \tag{12.18}$$

Because $n \gg 1$, Equation 12.18 can be rewritten as

$$E_R = \frac{nE_b}{K[\xi n+1]\gamma}. \tag{12.19}$$

Equation 12.19 has established the relationship of the number of the friction times, the wear energy density E_R and the energy density E_b to form a wear debris. In order to calculate the line wear rate, substitute Equation 12.17 into Equation 12.14, Thus, we have

$$\frac{dh}{ds} = \frac{\tau_y \gamma}{nE_e}. \tag{12.20}$$

Or substitute Equation 12.19 into Equation 12.14, it will be

$$\frac{dh}{ds} = \frac{\tau_y K(\xi n + 1)\gamma}{nE_b}. \tag{12.21}$$

The above coefficients K, ξ and γ are related to the physical properties and the structure of the friction material and the critical friction number n is influenced by the load, the energy absorption and storage capacity of the material. Furthermore, the energy accumulation in the friction process also depends on the storage volume, which is related to the micro-geometry of the contact peak point.

12.6 Delamination Wear Theory and Fatigue Wear Theory

12.6.1 Delamination Wear Theory

It is generally believed that the mechanisms of the abrasive wear and the corrosion wear are relatively understood. However, although the adhesive wear, fretting and fatigue wear have many common features, there is no theory to explain these three kinds of wear yet. The delamination wear theory was proposed by Suh in 1973 [8]. This theory is based on the analysis and experiments of the elastic-plastic mechanics and summarizes the extensive research results. It can well explain a lot of wear phenomena. It has been proven that the delamination wear theory promotes the in-depth study on the common nature of wear.

Analysis with a scanning electron microscope showed that the shape of wear debris is of a long and thin layer structure, which is produced from the surface crack. The delamination wear theory is based on dislocation theory, the fracture and plastic deformation near the surface explain the formation of the wear debris.

The basic arguments are: during the relative sliding friction, the roughness of the soft surface is easily deformed. Therefore, under the cyclic load, the soft asperity is first fractured to be smoothed. In this way, the contact state is no longer between the asperities, but the hard roughness to the relatively smooth soft surface. When a hard asperity slides on the soft surface, the soft surface is subjected to the cyclic load and plastic shear deformation in the surface layer continually accumulates to bring about cyclic dislocation beneath the surface. Because of the action of the image force, dislocation beneath the surface of tens of microns thick disappears. So, the dislocation density close to the surface is less than that inside the body, that is, the maximum shear deformation occurs at a certain depth. With the continuous accumulation of the shear deformation, cracks or holes finally form. When a crack has formed, according to stress analysis, it can only extend parallel to the surface because the normal stress prevents it from developing in the depth direction. When the crack extends to the critical length, it will be peeled off the surface to form the debris.

The delamination wear theory can well describe many wear phenomena. For example, the surface layer deformation, the formation and expansion of the crack, the formation of the Babbitt layer as well as the influences of the lubricant, the sliding speed and the composite load on wear.

According to the delamination wear theory a simple formula can be derived to calculate wear capacity. When a hard surface slides on a soft surface, total wear capacity Q can be expressed as follows.

$$Q = k_0 Ws, \tag{12.22}$$

where k_0 is the wear coefficient; W is the load; and s is the sliding distance.

The thickness h of a flake debris, according to the thickness of low dislocation density area, can be determined as follows:

$$h = \frac{Gb}{4\pi(1 - v)\tau_j}, \tag{12.23}$$

where G is the shear modulus of elasticity; v is the Poisson's ratio; τ_j is the surface shear stress; and b is the Burger vector.

The critical sliding distance s_0 refers to the sliding distance through which a hole or crack just forms the debris. If the wear volume is V, and the sliding distance is s, the relationship between them is given by the following formula:

$$V = \left(\frac{s}{s_0}\right)Ah, \tag{12.24}$$

where A is the flake debris area, which is related to the load and the material yield limit, $A = W/\sigma_s$.

Substitute A and h into Equation 12.24, we have

$$V = \frac{WsGb}{4\pi s_0\sigma_s(1 - v)\tau_j}. \tag{12.25}$$

If set

$$K = \frac{Gb}{4\pi s_0(1 - v)\tau_j}. \tag{12.26}$$

Finally we have:

$$\frac{dV}{ds} = \frac{V}{s} = K\frac{W}{\sigma_s}. \tag{12.27}$$

It can be seen from Equation 12.27 that the wear capacity of the delamination wear theory is proportional to the load and inversely to the sliding distance, but not directly related to the hardness of the material. This is different from the adhesive wear result.

12.6.2 Fatigue Wear Theory

From study of the fatigue wear process, Kragelsky [9] had put forward the fatigue wear theory to which attention was paid widely. The basic views of fatigue wear theory are that: (1) due to surface roughness and waviness, the contact surface is not continuous so the friction surface is subjected to the cyclic load; (2) material wear is the mechanical damage process due to local deformation and stress of the contact point; (3) fatigue wear depends on the stress state of the contact point.

In a wear process, the contact point is subjected to large cyclic stress. When the stress cycles reach a certain number, the fatigue crack expands to form a wear debris. Fatigue wear on the contact point is related to the contact state. In the elastic contact, the destroyed stress cycles are usually 1000 times or more, while in the plastic contact, only about a dozen times or more, that is, low cycle fatigue damage.

Fatigue wear belongs to the material fatigue damage and it occurs under repeat friction actions. The friction number that cause the wear can be determined in accordance with the damage state of the contact point and is related to the damage form and the stress state. Therefore, according to the load, movement condition, surface topography and material properties, the stress state of the contact point can be determined and the wear calculation formula can be established.

The fatigue wear theory established by Kragelsky *et al.* has been proven by the experiments of metal and non-metallic materials, including rubber, polymer plastic and self-lubricating materials. According to the theory, a number of the wear calculation methods for the mechanical elements have been developed. However, because fatigue wear is quite complex, many parameters lack accurate data so their applications are limited.

12.7 Wear Calculation

12.7.1 IBM Wear Calculation Method

Bayer *et al.* proposed for a wear calculation model [10]. They experimentally obtained the data and directly developed the calculation method to predict the wear life of the machine elements.

First, wear can be divided into two types: zero wear and measurable wear. The zero wear amount does not exceed the roughness height significantly while measurable wear means that wear thickness apparently exceeds surface roughness.

A large number of experiments show that in order to ensure the friction pair works in the zero wear state for a certain period of time, the following condition must be met with:

$$\tau_{\max} \leq \gamma \tau_s, \tag{12.28}$$

where $\tau_{\max}$ is the maximum shear stress of the mechanical element; τ_s is the shear yield stress; and γ is the coefficient, related to the material, lubrication state and the work duration.

In the IBM calculation method, wear life is expressed by the number N of repeat trips. A trip is equal to the contact length along the sliding direction. Usually, $N = 2000$ is used to determine the zero wear coefficient. Then, γ is presented by γ_0 because within this period, the wear characteristic can be stably displayed.

Experiments show that when $N = 2000$, $\gamma_0 = 1$ for the lubrication state; $\gamma_0 = 0.2$ for the dry friction; $\gamma_0 = 0.2$ for the boundary lubrication; if the boundary lubricant contains the active additives, $\gamma_0 = 0.54$.

Refer to the fatigue curve of the metallic material, the relationship between the zero wear trip number and the maximum shear stress can be expressed as follows:

$$\tau_{\max}^9 N = (\gamma_0 \tau_s)^9 \times 2000$$

$$\tau_{\max} = \left(\frac{2000}{N}\right)^{\frac{1}{9}} \gamma_0 \tau_s. \tag{12.29}$$

By using Equation 12.29 to predict the working life of the zero wear, the working hours should be converted into the number of trips. The shear yield stress τ_s in the above equation can be determined experimentally by the curve of Figure 12.15.

For measurable wear, the proposed calculation model is that the wear capacity is the function of the trip number and the wear energy consumed for each trip. The relationship between the variables can be expressed by the following differential equation:

$$dQ = \left(\frac{\partial Q}{\partial E}\right)_N dE + \left(\frac{\partial Q}{\partial N}\right)_E dN \tag{12.30}$$

where Q is the measurable wear capacity; E is the wear energy consumed for each trip; N is the trip number.

The measurable wear can be calculated in accordance with the following two types.

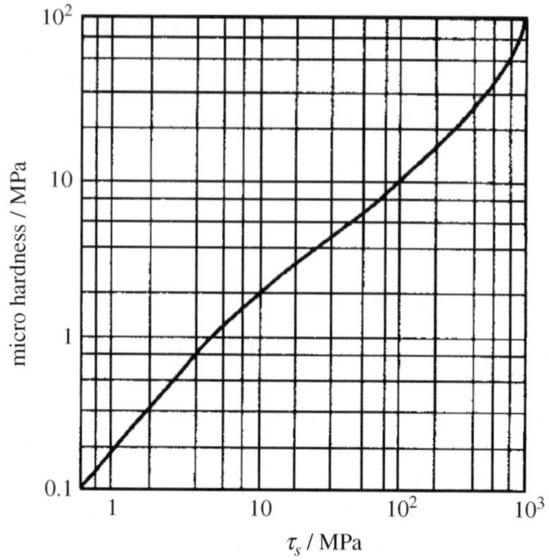

Figure 12.15 Micro-hardness via shear yield stress.

12.7.1.1 Type A

The energy consumption of this type wear will keep constant. It mainly appears in the dry friction or under heavy load with serious material transformation and abrasive wear.

For Type A wear, Equation 12.28 can be simplified as:

$$dQ = cdN, \tag{12.31}$$

where c is the wear constant of the system, which can be determined by experiments.

12.7.1.2 Type B

The energy consumption of this type of wear changes for each trip it appears in the lubricated or light-load conditions. The wear usually belongs to the fatigue wear.

For Type B wear, Equation 12.21 can be written as

$$d\left[\frac{Q}{(\tau_{\max}s)^{\frac{9}{2}}}\right] = cdN, \tag{12.32}$$

where s is the sliding distances of each trip.

Integrating Equation 12.31 or 12.32, the relationship between the wear capacity and the trip number can be obtained.

12.7.2 *Calculation Method of Combined Wear*

Pronikov proposed a wear calculation method for the relative sliding friction [11]. He divided the wear into surface wear and combined wear. The surface wear occurs in the normal direction of the surface and the worn thick is usually non-unform. The combined wear is the wear of the matching friction surfaces, causing the variation of the relative positions of the surfaces in the friction process. Clearly, the combined wear will change the nature of the friction pair and affect the work performances of the mechanical elements.

The basic principle of the combined wear calculation is to determine the allowed variations of the positions of the matching surfaces, that is, the combined wear capacity, and then calculate the life of the mechanical elements.

The main steps to calculate the combined wear calculation are briefly introduced as follows.

1. First, based on the actual working condition, determine the wear curve and the corresponding wear rate.
   The usual wear calculation only considers the two cases in Figure 12.2a and b. For normal working condition, the steady wear is the longest so it will take the steady wear time as the actual life of the mechanical element.
   Because the steady wear rate is unchanged, if the wear thickness h is used as the wear capacity and t is the wear time, the line wear rate is equal to

$$\gamma = \frac{dh}{dt} = \tan\alpha = \text{const.} \tag{12.33}$$

   Experimental results show that Equation 12.33 is mainly suitable for abrasive wear, but it can be approximately adopted for other types of wear except for fatigue wear.
2. Experimentally determine the line wear rate and the relationship of the working parameters.
   Generally, the wear rate depends primarily on surface pressure p and the sliding velocity v, that is

$$\gamma = Kp^m v^n, \tag{12.34}$$

   where K is the coefficient for the working conditions, related to the material, surface quality, lubrication conditions and other factors. For example, in the usual lubrication condition, $K = 3.35$ for bronze, and $K = 0.92$ for steel; m and n are the influence indicators respectively to the surface pressure and the sliding velocity. Their values are between 0.6 and 1.2 based on different working conditions.
   Experiments prove that the line wear rate is proportional to the surface pressure, but has nothing to do with sliding velocity, that is

$$\frac{dh}{ds} = Kp. \tag{12.35}$$

   So

$$\frac{dh}{dt} = \frac{dh}{ds} \cdot \frac{ds}{dt} = Kpv. \tag{12.36}$$

   Therefore, the abrasive wear indicators $m = n = 1$ and the line wear rate can be simply written as

$$\gamma = Kpv. \tag{12.37}$$

   Determine the combined wear and the relationship between the wear capacities of the two surfaces. Because the wear capacity of a friction surface is usually expressed by the vertical thickness, but the combined wear capacity is measured according to the positions of the two matching surfaces, it must be based on the geometric structures of the mechanical parts to determine the combined wear capacity of the two surfaces.
   As shown in Figure 12.16, the surface wear capacities of the conical thrust journal 1 and the bearing 2 are respectively expressed by the thicknesses of h_1 and h_2. After the wear, the relative position variations are the axial displacements of them so the combined wear capacity is h.

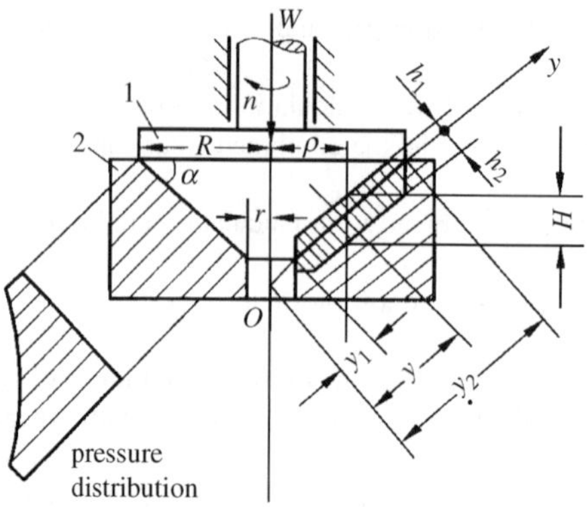

Figure 12.16 Combined wear of cone thrust bearing.

According to the geometry, it can be derived that the combined wear capacity is equal to

$$h = \frac{h_1 + h_2}{\cos \alpha}. \tag{12.38}$$

It should be noted that for the journal surface or the bearing surface the wear capacity may be different at the corresponding point, that is, surface wear distribution is not uniform. However, according to the condition that the two surfaces must keep contact with each other, the combined wear capacity h at each point must be equal.

Figure 12.17 is a block brake. After the wear, the two surfaces will displace radically. This may result in the blocks and the ring becoming loose and affect the braking torque. For such a situation, the relationship between the combined capacity h and the surface wear capacities h_1 and h_2 is the same as Equation 12.38, but α varies with the position of the contact point.

3. Select the combined wear limit according to the working performances and the usage requirements of the mechanical parts.
   For example, the maximum combined wear capacity of the cam and the tappet should be the maximum allowable motion error; the combined wear limit of the screw and nut depends on the transmission accuracy or the magnitude of the empty rotation displacement; and the combined wear capacity of the

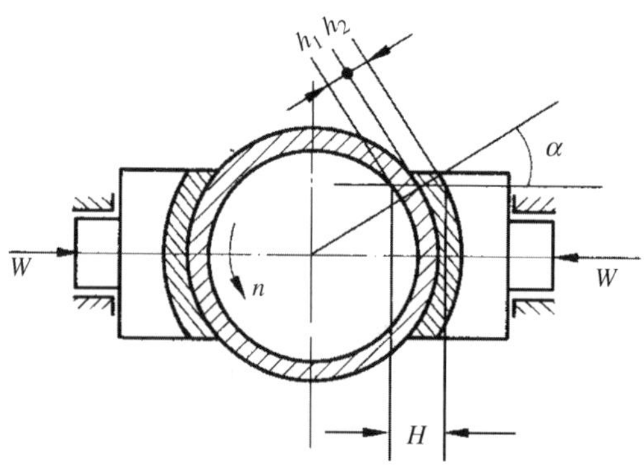

Figure 12.17 Combined wear of block brake.

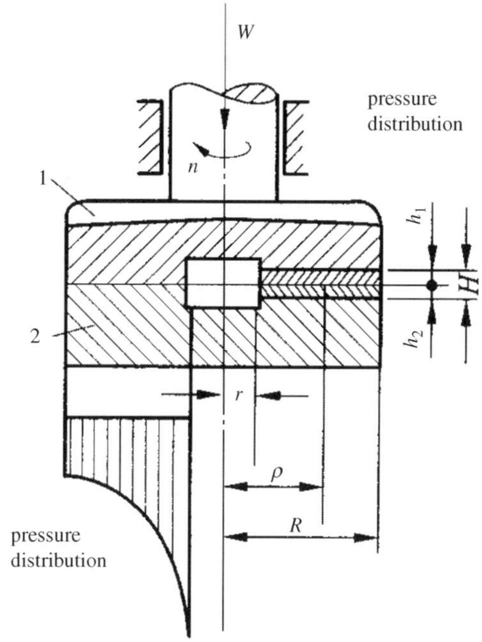

Figure 12.18 Combined wear calculation of flat thrust bearing.

gear drive is determined by the reference accuracy, the limited impact load, the smoothness of the tooth and so on

4. Calculate wear life

   Here, we take the plane thrust bearing of Figure 12.18 as an example to show the wear life calculation.

Under the action of the axial load *W*, the journal rotates with a rotational speed *n*. If the abrasive wear is the main wear type, from Equation 12.37 the line wear rate at the point with the radius ρ is equal to

$$\gamma_1 = K_1 p \times 2\pi n\rho$$
$$\gamma_2 = K_2 p \times 2\pi n\rho. \tag{12.39}$$

Because the combined wear capacity $H = h_1 + h_2$, the combined wear rate γ is equal to

$$\gamma = \gamma_1 + \gamma_2 = 2\pi n\rho(K_1 + K_2)p \tag{12.40}$$

or

$$p = \frac{\gamma}{2\pi n(K_1 + K_2)} \cdot \frac{1}{\rho}. \tag{12.41}$$

Equation 12.41 shows that when the journal rotates, the pressure along the radius direction has a hyperbolic distribution.

Then, the total load carrying capacity W is

$$W = \int_r^R 2\pi\rho p d\rho = \frac{\gamma}{n(K_1 + K_2)} \int_r^R d\rho = \frac{\gamma(R - r)}{n(K_1 + K_2)}. \tag{12.42}$$

So

$$\gamma = \frac{Wn(K_1 + K_2)}{(R - r)}. \tag{12.43}$$

The combined wear capacity h can be expressed as follows:

$$h = \gamma T = \frac{Wn(K_1 + K_2)}{R - r} T, \tag{12.44}$$

where T is the wear life.

It is easy to obtain that the wear thicknesses of the plate and the bearing surfaces are respectively equal to

$$\begin{aligned} h_1 &= \gamma_1 T = \frac{WnK_1}{R - r} T \\ h_2 &= \gamma_2 T = \frac{WnK_2}{R - r} T. \end{aligned} \tag{12.45}$$

From the above equations, it is known that h_1 and h_2 are not related to ρ. Therefore, the wear of the plate or the bearing surfaces is uniform. Substitute the maximum wear capacity h into Equation 12.44, we can obtain the thrust bearing wear life T.

For the cone thrust bearing of Figure 12.16, a similar approach can be used to analyze. As α is a constant, by Equation 12.38, we have

$$\gamma = \frac{\gamma_1 + \gamma_2}{\cos\alpha}. \tag{12.46}$$

Select the coordinate axis Oy as Figure 12.16, the sliding velocity at the point of ρ on the friction surface is

$$v = 2\pi\rho n = 2\pi n y \cos\alpha. \tag{12.47}$$

The line wear rates of the journal and the bearing surfaces are

$$\begin{aligned} \gamma_1 &= 2\pi K_1 n p y \cos\alpha \\ \gamma_2 &= 2\pi K_2 n p y \cos\alpha. \end{aligned} \tag{12.48}$$

So, the combined wear rate is

$$\gamma = \frac{\gamma_1 + \gamma_2}{\cos\alpha} = 2\pi n p y (K_1 + K_2). \tag{12.49}$$

And, the pressure is equal to

$$p = \frac{\gamma}{2\pi n(K_1 + K_2)} \cdot \frac{1}{y}. \tag{12.50}$$

We can see that the pressure distribution on the cone thrust bearing surface along the generating line is hyperbolic. In order to determine γ, the relationship of the load W and the surface pressure p should be used.

$$W = \int_{y_1}^{y_2} 2\pi p\rho \cos \alpha dy = 2\pi \cos^2 \alpha \int_{y_1}^{y_2} pydy, \tag{12.51}$$

where $y_1 = r/\cos \alpha$, $y_2 = R/\cos \alpha$, $\rho = y \cos \alpha$.

Substituting Equation 12.50 into Equation 12.51 and integrating it, we can obtain

$$\gamma = \frac{Wn(K_1 + K_2)}{(R - r)\cos \alpha}. \tag{12.52}$$

Therefore, the relationship of the combined wear capacity and the wear life is

$$h = \gamma T = \frac{Wn(K_1 + K_2)}{(R - r)\cos \alpha} T. \tag{12.53}$$

Finally, it should be pointed out that the wear phenomenon is a micro-dynamic process of the surface. The wear performances are not only related to the inherent properties of the material, but are the comprehensive performance of the tribological system so the influence factors are very complex. Therefore, the wear problem in the tribology theory and practice is not yet well-known. It is reported [12] that in recent decades, hundreds of wear formulas have been released. The parameters that are related to materials, mechanics, thermal physics, chemistry and others are more than 100 so as to limit the applicability of these formulas. Obviously, it is very difficult to create a unified quantitative wear formula.

References

1. Khrushchev, M.M. (1946) Study of running-bearing alloys and pins. Moscow-Leningrad: Izd-vo AN, USSR.
2. Tonn, W. (1937) Beitrag Zur Kenntnis des Verschleissvorganges beim Kurzversuck. *Z. Meta Ukd.*, **29**, 196–198.
3. Holm, R. (1938) The friction force over the real area of contact. *Wiss. Vëroff Siemens Werken*, **17**, 38.
4. Archard, J.F. (1953) Contact and rubbing of flat surface. *Journal of Applied Physics*, **24**, 981–988.
5. Rowe, C.N. (1966) Some aspects of the heat of adsorption in the function of a boundary lubricant. *ASLE Transactions*, **9**, 101–111.
6. Rabinowicza, E., and Mutisa, A., (1965) Effect of abrasive particle size on wear. *Wear*, **8** (5), 381–390.
7. Fleisher, G. (1973) Energische methode der bestimung der ver-schleibes schmierungs- technik. **9** (4), 269–274.
8. Suh, N.P. (1973) The delamination theory of wear. *Wear*, **25** (1), 111–124.
9. Kragelsky, I.V., Dobychin, M.H., and Kombalov, V.S. (1977) Foundations of calculations for friction and wear, Mashinostroenie (in Russian), Moscow.
10. Bayer, R.G., Clinton, W.C., Nelson, C.W., and Schumacher, R.A. (1962) Engineering model for wear. *Wear*, **5** (5), 378–391.
11. Pronikov, A.S. (1957) Wear and durability of machines. Moscow: Mashgiz.
12. Ludema, K.C. (1996) Mechanism-based modeling of friction and wear. *Wear*, **200**, 1–7.

13

Anti-Wear Design and Surface Coating

With development of industrial technologies, demand for the precision of the surfaces of machineries and equipment have greatly increased. Under high speed, high temperature or corrosive working conditions, the local surface damage of the element often leads to the failure of the whole equipment. The anti-wear design helps us increase the wear resistance of a component so as to extend the service life. Therefore, more attention has been paid to this aspect in engineering.

The most effective way to design a wear-resistant mechanical part is to establish a fluid lubricating film, a surface adsorption film or a chemical reaction film between the friction surfaces. This must be based on the working condition of the friction pair to correctly select the lubricant oil or grease, or to appropriately use additives to create a lubricant film with the special properties. Another important aspect of anti-wear design is to correctly match the material friction pairs and to reasonably choose the surface enhanced measure. In addition, the filtration and sealing of the lubricating oil supply system are also important aspects of anti-wear design [1].

Surface coating is a newly important technology which can effectively improve the service life of mechanical parts. By bead welding, thermal spraying, brush plating or other physical-chemical methods, the technology has the surface coated with the wear-resistant, pyroceram, corrosion-resistant and other special performances to gain significant economic benefits.

13.1 Selection of Lubricant and Additive

An adequate thick lubricant film will protect or reduce surface wear. This practice has proved that in most cases, the film can also effectively lubricate even without being thick enough to completely cover the asperity of the surface. If the film is too thick, it may bring about some adverse effects. For example, the rigidness will be small. Usually, the film thickness ratio $\lambda = h_{\min}/\sigma$ is used as the parameter to determine the lubrication state. It is generally believed that if $\lambda \geq 1.5$, wear can be controlled within a slight range to obtain a reasonable life.

Further divisions of the ratio λ are as follows. For low speed or surface roughness, $\lambda = 0.5$–1, while for high speed or roughness, $\lambda \geq 2$. For running-in surface, $\lambda \geq 0.5$–1, while for non-running-in surface,

Principles of Tribology, First Edition. Wen Shizhu and Huang Ping.

λ should be larger than 2. For flat or cylindrical contact surface, $\lambda = 2$–5 or even greater in order to compensate for surface waviness or shape error. Under unstable load, λ must also be increased.

13.1.1 Lubricant Selection

The lubricant selection should be based on working conditions. The main characteristics of some base oils are listed in Table 13.1.

The general requirements of the lubricant are as follows.

13.1.1.1 Viscosity, Viscosity Index and Viscosity-Pressure Coefficient

The appropriate viscosity can form a film thick enough, but if viscosity is too high friction will increase that will raise temperature. Viscosity is affected by temperature. When the working temperature and the ambient temperature change significantly, we also need to choose a suitable viscosity index, which is the measuring parameter of the thermal stability. The higher the viscosity index the lower the influence of temperature. If viscosity is low or the viscosity-temperature characteristic is not good enough, a tackifier can be added to improve them. The tackifiers commonly used are: polyethylene n-butyl ether, poly methyl acrylate ester, polyisobutylene and so on. These high polymers not only increase the oil viscosity but also change their molecular chain shapes with temperature. At low temperature, the chains curl into small balls with less influence on the viscosity, while at high temperature the chains stretch to form lines, and increase the influence on viscosity so as to improve the viscosity-temperature characteristic. As mentioned above, the viscosity-pressure coefficient of a lubricant is of significant impact on the film thickness of EHL.

13.1.1.2 Stability

In lubrication, the oxidation causes the lubricant to deteriorate and significantly reduce the working life. Therefore, good stability of a lubricant is needed. Figure 13.1 shows the temperature limits of several commonly used synthetic oils. If the temperature exceeds the maximum limit allowed, oxidation will speed up.

The commonly used anti-oxidant additives include: dialkyl disulfide phosphate parahydroxydiphenylamine and 2,6-di-tert-butyl-p-methylphenol diphenylamine. They can not only form a protective film on the metal surface to prevent corrosion, but also prevent the metal from being catalyzed and oxidize the oil thereby reducing the oxidation rate.

13.1.1.3 Other Requirements

Other requirements for a lubricant include the abilities of cooling, sealing, anti-corrosion, chip removal, fire safety, compatibility with the environment and so on. Table 13.2 lists some basic principles of the lubricant selection.

Other notes are:

1. If a lubricant is easily oxidized or there is a circulation lubrication system, the lubricant should not be mixed with animal fat or vegetable oil.
2. The gasoline and diesel should not be used in a damp place.
3. Transformation oil is unsuitable for a lubricant.
4. The steam turbine oil and hydraulic oil should not be used in high-temperature machinery, such as the internal combustion engine.
5. When the working temperature is low or low viscosity is required, a lubricant can be mixed with some kerosene, but the amount of kerosene cannot be more than 50%.
6. The kerosene should not be used in precision machinery.

Table 13.1 Characteristics of some base oils

Features of base oil	Dibasic acid ester	Compound ester	Phosphate	Polymethyl silicone oil	Phenyl methyl silicone oil	Chlorinated phenyl methyl silicone oil	Polyethylene glycol	Polyphenylene oxide	Mineral oil
T_{max} ° with no oxygen °C	250	300	120	220	320	305	200	450	200
T_{max} with oxygen °C	210	240	120	180	250	230	200	320	150
T_{min} °C	−35	−65	−55	−50	−30	−65	−20	0	−50–0
Density g/cm^3	0.91	1.01	1.12	0.97	1.06	1.04	1.02	1.19	0.88
Viscosity Index	145	140	0	200	175	195	160	-60	0–140
Flash point °C	230	250	200	310	290	270	180	275	150–200
Self-ignition point	low	middle	very high	high	high	very high	middle	high	low
Boundary lubrication	well	well	very good	fair	fair	well	very good	fair	well
Toxicity	slight	slight	little	non	non	non	low	low	slight
Relative prices	5	10	10	25	50	60	5	250	1

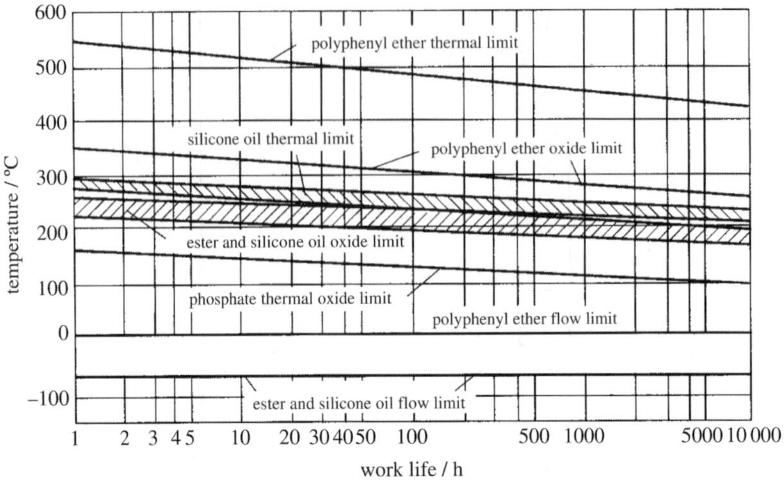

Figure 13.1 Temperature limits for some synthetic oils.

Table 13.2 Basic principles of lubricant selection

Working condition	Characteristics of lubricant
Heavy load	High viscosity lubricant
High speed	Low viscosity lubricant with the cyclic supply system
Variable speed or load	To raise the viscosity about 25%
Precision machine tools or hydraulic machinery	Low viscosity to avoid heat
High temperature rise	High viscosity and anti-oxidation mineral oil or synthetic oil
Large temperature variation	High viscosity index lubricant
Low temperature	Lubricant with a solidifying point lower than 50 °C
Severe wear	High viscosity lubricant to add the anti-wear or oily additive
More debris	To increase the amount of lubricant and add the filter to the cyclic supply system. If necessary, use the clean or dispersed additive
Long service life	High viscosity and oxidation-resistance lubricant
Large tolerance or roughness	High viscosity lubricant
Fire protection	Fire resistant or water lubricant with anti-wear or extreme pressure additives

13.1.2 Grease Selection

13.1.2.1 The Composition of Grease

Grease is made up of lubricating oil with some densifier added at high temperature. In the lubricating grease, oil is the main component, occupying about 75–85% of the total weight, the densifier is about 10–20%, and other additives about 0.5–5%. Therefore, the lubricating oil determines the lubrication performance, low-temperature performance and anti-oxidation ability of the grease. Grease to be used under high-speed and light-load conditions should be mineral oil with low viscosity, and under low temperature, the grease should be mineral oil with low pour point.

13.1.2.2 Function of Densifier

The function of a densifier is to reduce the fluidity of the oil, and also to enhance the abilities of sealing, pressure-resistance, bufferness and so on. The temperature resistance, the water resistance and the hardness of grease depend mainly on the variety and contents of the densifier. For example, grease with calcium soap as a densifier is water-resistant, but not temperature resistant, while sodium soap densifier is not water and temperature resistant. Grease is classified into calcium-based grease, sodium-based grease, and so on.

13.1.2.3 Grease Additives

The function of additives in grease is similar to that in the lubricating oil. By adding some graphite or MoS_2, the anti-wear and pressure-resistance abilities of grease can be improved. By adding amine compounds, the anti-oxidation ability can be enhanced.

13.1.3 Solid Lubricants

Solid lubricants are low-shear strength solids, such as soft metal, soft metal compounds, inorganic, organic and self-lubricating composite materials. The characteristics of solid lubricant are good thermal resistance, high chemical stability, and high pressure resistance. It is also non-volatile and non-polluting. Particularly, they can be suitable for situations without sealing or with no lubricant supply system. Solid lubricant is especially useful for situations in which conventional lubrication is difficult to be used, such as in the atomic energy industry, plastics industry, rockets, satellites and other particular fields. The disadvantage of the solid lubricant is that the surface wear is usually higher than that of the oil lubrication. Because it cannot effectively remove heat, spalling may result. The characteristics of commonly used solid lubricants are listed in Table 13.3.

Finally, Figure 13.2 shows the approximate ranges of some lubricants for design reference.

Table 13.3 Characteristics of common solid lubricants

Lubricant type	Temperature limits °C	Typical coefficient of friction	Usage
1. Layered solid			
Molybdenum disulfide	350 (in air)	0.1	Powder, adhesive film, or cathodic vacuum coating
Graphite	500 (in air)	0.2	Powder
Tungsten disulfide	440 (in air)	0.1	Powder
Calcium fluoride	1000		Melt coating
Graphite fluoride		0.1	Brush or cathodic vacuum coating
Talc		0.1	Powder
2. Febrile material			
PTFE (unfilled)	280	0.1	Powder, solid blocks, or adhesive film
Nylon 66	100	0.25	Solid block
Polyimide	260	0.5	Solid block
Acetal	175	0.2	Solid block
Polyphenylene sulfide (filled)	230	0.1	Solid block or coating
Polyurethane	100	0.2	Solid block
Polytetrafluoroethylene (Fill)	300	0.1	Solid block
Nylon tip 66 (Fill)	200	0.25	Solid block
3. Others			
Al_2O_3	800		Powder
Phthalocyanin	380		Powder
Lead	200		Brush or cathodic vacuum coating

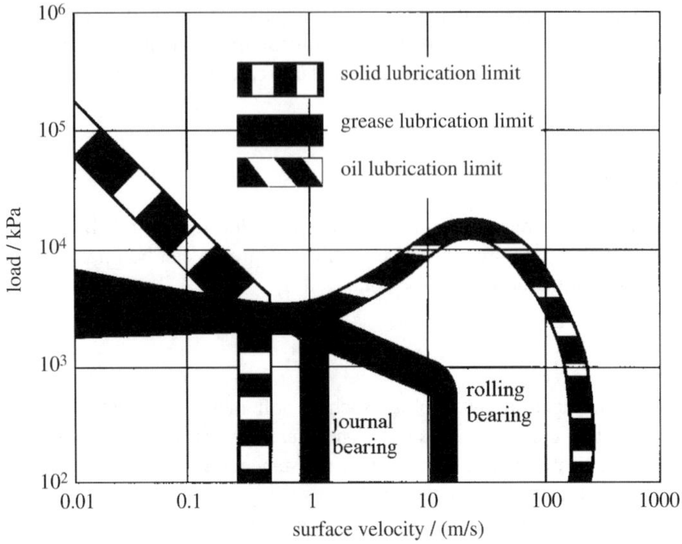

Figure 13.2 Use ranges of lubricants.

13.1.4 Seal and Filter

Relatively moving parts in a mechanical system need lubricating and cooling. The lubricant or coolant is often contaminated by the outside dust that increase the surface wear. It is shown that whether the lubricant is clean or not it can have an effect on the friction life up to 10 times of difference. Therefore, we must take sealing and filtering measures for the lubricant and the coolant to remove harmful particles from the friction surface. These are the important measures of anti-wear design.

The particles in the lubricant are mainly of two types: the hard and the soft particles. The hard particles can easily lead to premature wear and obstruct the oil groove.

The mechanisms that the particles cause surface wear can be divided into three types.

1. The hard particles embedded in the friction surface cut the friction surfaces, causing abrasive wear. The severity of the wear is proportional to the number and hardness of the particles.
2. The hard particles between the friction surfaces continuously roll, draw, and squeeze so as to bring about the local plastic deformation and atomic dislocation of the friction surface. Eventually, these lead to surface fatigue wear.
3. The hard particles roll, draw, and squeeze to produce high ridges on the surface. These high ridges will lead to direct contact of the metals and develop adhesive wear.

The different types of particles in the lubricant and their sources are given in Table 13.4.

13.2 Matching Principles of Friction Materials

The wear resistance of the material is an important characteristic for matching. The wear resistance covers the hardness, toughness, solubility, heat resistance, corrosion resistance and other properties of the material. Because the wear mechanism is different, one or two aspects of the above properties are particularly demanded for the different wears. In addition, the material mating must be paid attention to.

Table 13.4 Types and sources of particles in lubricant

Types	Sources
Metal particle	Wear product of processing, assembly, casting and so on, especially in the new assembled components
Metal oxides (such as aluminum oxide), metal salts (such as chloride or sulfide)	Corroded metal particles suspending on the friction surface
Oil sludge deposit	Combustion product; heat and aging oil, deposit of oil mixed with water and salt
Rubber particle	Worn products of seals, flexible pipes, gaskets and others
Fiber	Abscission products of cotton or filter
Inorganic particle (e.g., sand)	Enter while in operation or maintenance from the surrounding environment.

Sometimes, hard to hard is good, such as rolling contact bearing, sometimes hard to soft is good, such as sliding bearing, and sometimes wear is specially limited to a particular part. For example, in the cylinder, wear is limited on the piston ring rather than on the cylinder. The following mating types of materials are used for the different wear pairs.

13.2.1 Material Mating for Abrasive Wear

As mentioned before, for abrasive wear, wear resistance of the pure metal or steel without heat treatment is proportional to natural hardness. Even by using heat treatment to improve the hardness of annealed steel, the wear resistance does not increase as high as the hardness. However, the higher the carbon contents, the higher the wear resistance for the hardened steel with the same hardness.

The wear resistance is related to the metal micro-structure. The wear resistance of martensite is better than that of pearlite and ferrite. For pearlite, the wear-resistance of the flake form is better than that of the ball form, and the thin flake form better than the thick flake form. The wear resistance of tempered martensite is often better than that of the non-tempered because the micro-structure of the non-tempered martensite is hard and brittle.

For steel with the same hardness, the wear resistance of the steel containing alloy carbide cementite is better than that of steel containing common ordinary cementite. The more carbon atom there are, the better is the wear resistance. The wear resistance of some alloy elements, such as Ti, Zr, Hf, V, Nb, Ta, W, and Mo are better than Cr and Mn because the carbide can be easily formed so as to improve wear resistance.

For abrasive wear caused by the impact of solid particles, proper matings of the hardness and the toughness are needed. For a small impact angle, that is, the impact velocity is closely parallel to the surface, such as the plow share or the slot board for transporting ore as shown in Figure 13.3, the mating surface material should be of high hardness. The hardened steel, ceramics, cast stone, or tungsten carbide can be used to prevent cutting wear. For a large impact angle, the mating surface material should be of adequate toughness so that the rubber, high manganese steel with austenite, plastic and so on, can be used.

Figure 13.3 Abrasive wear with small impacting angle.

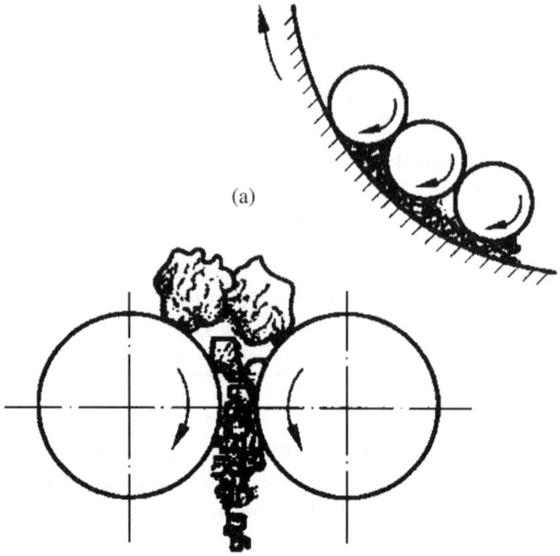

Figure 13.4 Abrasive wear under high impacting stress.

Otherwise, the kinetic energy of the collision can cause the material to crack and peel off the surface easily. For high stress impact, such as the crusher roller, ball roller, rail, and so on, as shown in Figure 13.4, high manganese steel with austenite can be used because under high impact stress it can be hardened after deformation.

For three-body abrasion, the general measure is to enhance the hardness. As the hardness is about 1.4 times larger than that of the particle, the wear resistance of the friction surface is the best and it is not useful to raise the hardness further. The size of the particle for the three-body wear also has an impact on the wear rate. Experiments show that when the particle size $<100\,\mu m$, the smaller the particle, the lower the wear rate is. If the particle size is $>100\,\mu m$, the wear rate is not affected by the particle size any more.

13.2.2 Material Mating for Adhesive Wear

As described earlier, adhesion often occurs because friction heat may cause the material to recrystallize, accelarate the diffusion or soften the surface material. The local high pressure and high temperature in the contact area may cause surface material melting. Therefore, adhesive wear is closely related to material mating. Material mating has the following laws.

Two materials with low solid solubility are not easy to be adhered each other. In general, if the lattice types and lattice constants are close, the solubility of the materials is high so they may easily adhere.

If the two materials can form intermetallic compound, they cannot easily adhere because the intermetallic compound has a weak covalent bond. Plastic material is often more adhesive than brittle material. Furthermore, because the strength of the connect point of the plastic material is often larger than that of the base metal, wear often occurs in the sub-surface, resulting in severe abrasion.

The higher the melting point, recrystallization temperature and critical tempering temperature of the material, the lower is the surface energy, and the lesser is the adhesion. The adhesion of the multi-phase metallographic structure is lower than that of the single-phase. For example, the adhesive effect of pearlite is weaker than that of ferrite or austenite. The metal compound has a lower adhesion possibility than the single-phase solid solution, and the hexagonal structure has a lower adhesion possibility than the cubic crystal structure. Metal mating with non-metal, such as carbide, ceramics, polymers and so on, has a higher adhesive resistance than mating with metal. Polytetrafluoroethylene (PTFE) mating with steel has a high adhesion resistance, low friction coefficient, and low surface temperature. The heat-resistant thermosetting plastic has better adhesion resistance than the thermoplastics.

In similar conditions, to improve surface hardness it is not easy to produce the plastic deformation as it is difficult to be adhered. For example, if the hardness of the steel is above 700HV or HRC 70, adhesive wear can be avoided.

13.2.3 Material Mating for Contact Fatigue Wear

Contact fatigue wear is the process when the cyclic stress of the surface causes the crack to initiate and grow. Because the anti-fatigue wear ability is positively related to hardness, to raise the hardness is beneficial to the anti-contact fatigue wear ability.

When the surface hardness is too high, the material will be too brittle so the anti-contact fatigue wear ability will fall. As shown in Figure 13.5, when the hardness of the bearing steel is 62HRC, the contact fatigue wear resistance is highest. If the hardness is further enhanced, the average life expectancy will decrease.

The mating materials of the high pair should have a difference in hardness from 50 to 70 so that the surfaces are easy for running-in to be beneficial in raising contact fatigue wear resistance.

In order to eliminate the initial crack and non-metallic inclusion, the smelting and rolling processes of the material must be strictly controlled. Therefore, the electrical furnace smelting, the vacuum or the electroslag remelting technologies are regularly used in the bearing steel.

Although the hardness of the gray cast iron is lower than that of carbon steel, because the graphite sheet is not directional, the friction coefficient of the gray cast iron is low. Therefore, it has a good contact fatigue wear resistance. The contact fatigue wear resistances of the alloy iron and chilled cast iron are better. The ceramic material usually has a high hardness and a good contact fatigue resistance with good high-temperature performances, but is mostly brittle and not impact-proof.

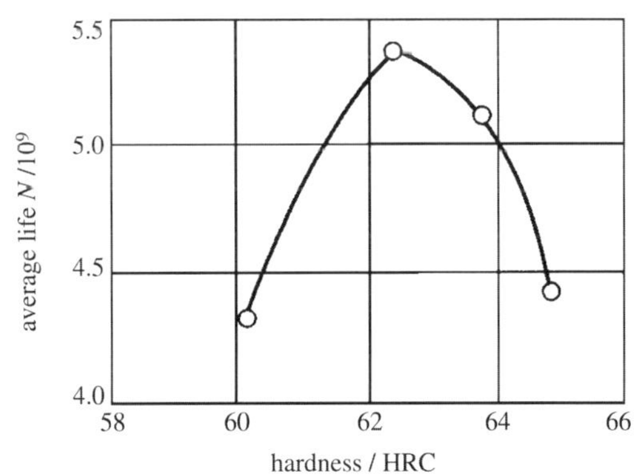

Figure 13.5 Relationship of fatigue wear life and hardness.

13.2.4 Material Mating for Fretting Wear

Fretting wear is the composite form of adhesive wear, oxidation wear and abrasive wear. Generally, the material mating suitable for adhesive wear resistance is also suitable for fretting wear resistance. In fact, in the whole fretting process, any mating having a limit to fretting is useful. For example, material good in anti-oxidation wear or anti-abrasive wear can improve fretting wear resistance.

13.2.5 Material Mating for Corrosion Wear

The material chosen should have good corrosion resistance, especially if its oxide film is firmly connected with the surface and has a good toughness. The more compact the material, the better is the corrosion wear resistance.

13.2.6 Surface Hardening

Surface hardening improve wear resistance by modifying the material surface. There are three types of commonly used surface hardening methods. They are mechanical processing, diffusion treatment and surface coating.

1. Mechanical processing does not change the chemical composition of the surface. It changes the structure, mechanical properties or geometric shape of the surface by processing so as to strengthen the surface.
2. Diffusion treatment depends on infiltration or injection of some elements or heat treatment to alter the surface chemical composition so that the surface strength can be enhanced. For example, most chemical and chemical-heat treatment belong to diffusion treatment.
3. Surface coating involves directly plating or brushing a reinforced surface layer on the material surface with physical or chemical method. The layers are divided into two kinds: hard coating and soft coating. The hard coatings are usually aluminum, bead welding, spray carbide or ceramics. Soft coating is often used for adhesive wear resistance, and its purpose is to reduce the friction coefficient or raise temperature resistance. Soft coatings include the coatings of copper, indium, gold, silver and other soft metals, and solid lubricant coatings, such as PTFE and MoS_2.

Evaluations of surface hardening can be presented by the parameters: f/f_0, k/k_0 and F/F_0. Here, f_0 and f are the friction coefficient before and after hardening; k_0 and k are the wear indicators before and after hardening (such as fatigue load); F_0 and F are the scuffing load before and after hardening. Table 13.5 shows the evaluation data of the common chemical-heat treatment for surface hardening.

Table 13.5 Strengthening effect of common chemical-heat treatment

Chemical heat treatment	Recommended Material	f/f_0	k/k_0	F/F_0
Carburizing	Carbon steel and alloy steel	0.8–1.0	2–3	1.0–1.5
Nitride	Alloy steel	0.8–1.0	2–4	1.0–1.5
Carbonitriding	Quenched and tempered carbon steel and alloy steel	0.7–0.8	2–5	1.5–2.0
Cyanide	Quenched and tempered carbon steel and alloy steel	0.7–0.8	2–5	1.5–2.0
Boronizing	Medium carbon steel and alloy steel	—	2–5	—
Thiocyanate permeation	Carbon steel, alloy steel and stainless steel	0.5–0.6	2–5	4–5
Sulfurizing	Carbon steel and cast iron	0.4–0.5	1.5–3	5–10
Iodine-Cd bath treatment	Titanium alloy	0.5–0.6	—	5–10

13.3 Surface Coating

Surface coating is one or more thin layers of different materials coated on the solid surface to strengthen the surface or to enable the surface to possess a special function. Different manufacturing technologies of the coating will obtain different coating properties. A coating is only used in specific situation so as to obtain a good result. Therefore, coating can be effectively used in engineering applications only to fully understand the types, performances and design criteria of coatings. The plating methods, types, characteristics and main usages of the commonly used surface coatings are introduced as follows [2].

13.3.1 Common Plating Methods

13.3.1.1 Bead Welding

Bead welding is a method of using welding to cover the surface with a wear-resistant, heat-resistant or corrosion-resistant coating of a certain metal. The metallurgical process and thermo-physical process of bead welding are basically the same as the common welding process, but its purpose is to obtain the special properties of the surface. Therefore, it is not exactly the same as welding.

The commonly used bead welding methods include general bead welding, arc bead welding, submerged arc bead welding, plasma bead welding, automatic bead welding protected by carbon dioxide gas and so on, as shown in Figure 13.6.

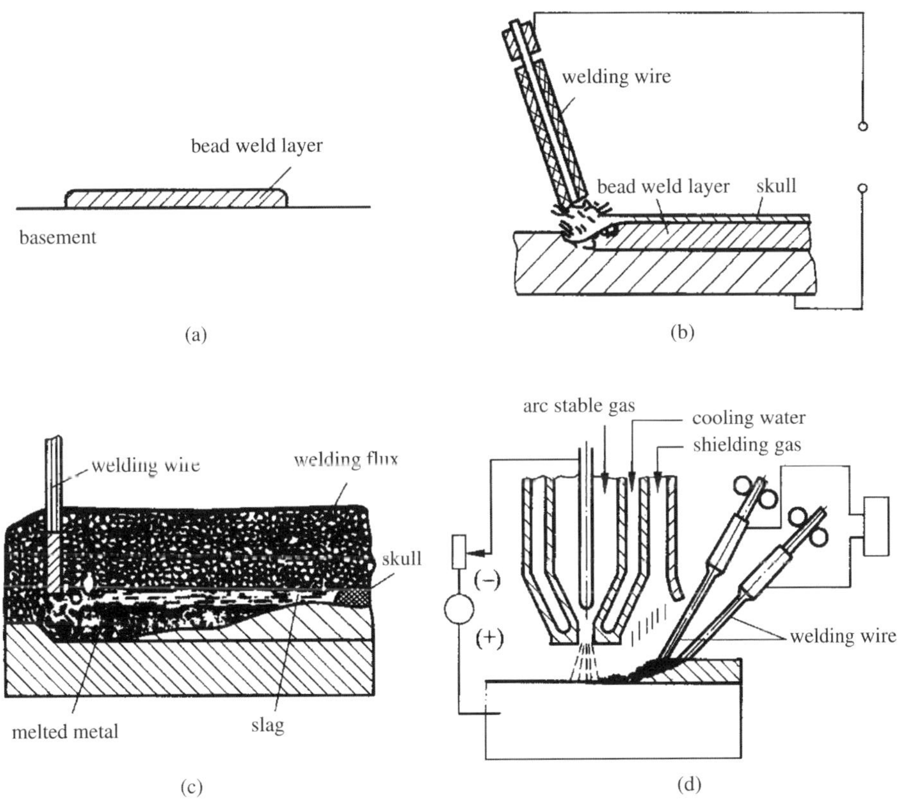

Figure 13.6 Bead weld methods: (a) common bead welding; (b) arc bead welding; (c) submerged arc bead welding; (d) plasma bead welding.

In common bead welding, oxygen-acetylene is used as the heat source. Because its flame temperature is low, generally, a uniform layer less than 1 mm thick can be obtained, which is suitable for smaller part surface protection.

The arc bead welding is of high production efficiency. However, because the protective effect of the arc zone is poor, sometimes pores or cracks can easily form on the surface. Spraying water vapor and carbon dioxide on the protected area may improve the quality of the bead welding coating.

Because the plasma arc bead welding is of a high temperature, bead welding material is refractory. In addition, it has a very high speed and high bead speed, but a low dilution rate so it has been widely used.

13.3.1.2 Thermal Spraying

Thermal spraying is to spray the molten or semi-molten material particles or powder with high speed to the surface to obtain the necessary surface coating. Thermal spraying has many advantages: the shape and size of the base material and parts are generally unrestricted; the coatings are various; the base material does not change after the spraying process; the coating thickness may vary in a large range and so on.

The materials of the thermal spray can be metals, alloys, metal compounds, ceramics, plastics, glass, composite materials and so on. Some thermal spray materials and their characteristics are given in Table 13.6.

The thermal spraying methods are as follows:

1. Flame spraying uses the energy of gas and oxidation combustion to melt the coating material in order to spray the molten particles onto the surface with compressed gas.
2. Arc spraying uses arc heat and discharge energy to melt the coating material in order to spray molten particles onto the surface with compressed gas through deposition to form the coating.

Table 13.6 Commonly used thermal spray materials and their characteristics

Thermal spray material type	Material	Characteristics
Metal wire	Zn, Al, Zn-Al alloys; Cu and its alloys; Ni and its alloys; Pb and its alloys; Mo and its alloys; and carbon steel and stainless steel	Widely used, with characteristics of wear-resistance, corrosion-resistance, heat-resistance
Alloy powder, self-fluxing powder and composite powder	Ni-based, Fe-based and Cu-based alloy powder and Ni-B-Si, Ni-Cr-B-Si, Ni-Cr-B-Si-Mo and Co-Cr-W self-fluxing alloy and the composite powder	Good spraying and spray fusing properties as well as characteristics of wear-resistance and corrosion-resistance
Heat-resistant alloy	Ni/Cr, Ni/Al, Ni-Cr/Al + McrAlX (where M can be Ni, Co, Fe, Ti, V, Zr, Ta, Ni-Co, Ni-Fe, and X can be Y, Hf, Sc, Ce, La, Th, Si, Ti, Zr, Ta, Pt, Rh, C, Y_2O_3, Al_2O_3, ThO_2)	Good creep strength at high temperature, high ductility, corrosion-resistance, wear-resistance, fatigue-resistance, impact-resistance
Ceramics	Cr_2O_3, Al_2O_3, $ZrSiO_4$, ZrO_2, Al_2O_3-TiO_2	Refractory oxide, insulation, thermal insulation
Plastics	Polyethylene, nylon, EVA resin, epoxy resin + TiO_2, $CaCO_3$, SiO_2	Corrosion-resistant

3. Plasma spraying uses the plasma flame to heat the coating powder to the molten or semi-molten state in order to spray them onto the surface to form a coating.

In addition, there are gas explosion spraying, high-energy density gas spraying, laser spraying, and water-stable plasma spraying methods.

13.3.1.3 Slurry Coating

Slurry coating brushes the solid–liquid mixture onto the solid surface, and then solidifies it under certain conditions to form slurry coating. The advantage of slurry coating is that the temperature to form the coating is low. Usually, the ceramic or metal particles in the slurry will not metallurgically conjugate at a low temperature, but conjugate themselves or with the agglomerant forming *in situ* by chemical reaction. The formation and performances of the agglomerant are critically important to the properties of the coating. Therefore, the additive selection and adding amount in the slurry must be strictly controlled or be determined according to the base material and the required performances of the coating.

According to coating forming methods, the slurry coatings can be divided as follows.

1. Slurry coating
   The steps of slurry coating are as follows. The first step is to brush or spray the slurry containing the agglomerant, solid particles and liquid carrier on the surface. Then, the coating is dried at a lower temperature. During the drying process, partial or complete agglomerant evaporated. The next step is to sinter the coating at high temperature to form the required surface coating. The sintering process is sometimes under atmospheric pressure, but in the most cases, in a vacuum or inert gas environment. Such a coating is generally for high temperature usage, and its base material is usually a heat resistant alloy.
2. Glue coating
   Glue coating (or cold coating) can be divided into two types, organic and inorganic. The former mixes the adhesive resin, curing agent, fillers and other solid powders in a certain proportion to form the slurry with some viscosity, and then the slurry is painted on the surface to form a coating after solidification. Depending on the glue, solidification can occur at room temperature or at a certain temperature. By adding different solid fillers, such as wear-resistant, anti-friction, and anti-corrosion fillers, different performances of the coating can be obtained. The combination of the coating and the substrate is usually a simple mechanical bond. At present, such a coating is mostly adapted to low stress or at a slightly high temperature, such as in the sliding rail of the machine tool without abrasive wear, erosion wear, cavitation wear and corrosive wear. Based on research data, as long as the combination of the coating and the substrate is firm, the coating life is generally 7–10 times than the ordinary metal material.
3. Thermochemical reaction slurry coating
   The thermochemical reaction slurry coating is obtained from the *in situ* conversion of the compound contained in the slurry. The combination between the coating and substrate is chemical bond. The main advantage of the thermochemical reaction slurry coating is that at a relatively low temperature and relatively simple conditions, a good coating can be obtained with temperature of 316–538 °C.
4. Thermal chemical reaction slurry coating based on the valence conversion of Cr
   The thermal chemical reaction slurry coating based on the valence conversion of Cr is obtained by repeatly painting the slurry containing the soluble chromium compounds, solid particles, and other added substances onto the metal surface, and carrying out the heat treatment at a certain temperature to change the valence of the chromium compounds. In the heating process and through chemical reactions, some compositions in the slurry are transmitted to the required substances.

Shao *et al.* [3] prepared wear-resistant and corrosion-resistant Cr_2O_3 ceramic coating by using the method at low temperature (190–200 °C) with three rounds of painting on the surfaces of aluminum, aluminum alloy and iron alloy. Their experimental results showed that the micro-hardness of the coating is up to HV400–800. To select an appropriate formulation, the obtained slurry coating is not only of the good tribological properties at room temperature, but also works well at high temperature of 400 °C above the ambient temperature.

13.3.1.4 Electric Brush Plating

An electric brush plating system is composed of the brush connecting with the anode of DC power and the workpiece connecting with the cathode. The electrical brush is often made of high purity and fine graphite block, covered with a cotton or durable polyester cowl. Electric brush plating is carried out by having the electrical brush soak in the plating solution and the workpiece surface move relatively under a suitable pressure. Due to the existence of an electric field, the metal ions of the solution diffuse to the surface of the workpiece through contact of the brush with the workpiece and acquire the electrons from the surface to reduce to metal atoms. These metal atoms deposit to form a crystalline coating.

It is generally believed that electric brush plating coating and the basement form the mechanical, physical and electrochemical combinations. The mechanical combination is to use the mosaic effect to form a coating on the surface; the physical combination is an electron exchange process caused by the contact; the electrochemical combination is to reduce the numerous metal ions to the metal atoms to form a firm metal coating on the surface.

Because the chemical composition of the coating metal is generally different from that of the base metal, each coating metal atom in the interface forms a certain lattice with the substrate atom. The atoms are not simply piling up together, but there is a strong interaction due to acquiring or losing electrons. The strength of the metallic bond is determined by the nature of the crystal structure and the crystal surface of the two interfaces, while the coating combination strength depends on the bond strength. Thus, the main combination strength of the coating and substrate is the electrochemical combination, as well as the mechanical and physical.

The fundamental principle of electric brush plating is similar to general electric plating, but also has its own characteristics. Electric brush plating equipment is usually portable with small size, light weight, and has low consumption of water and electricity. Electric brush plating solutions are mostly the water solutions of organic complex metal compounds. The complex in the water is of high solubility and good stability. The higher the level of the metal ions, the more stable the performances are. They can be used in wide current density and temperature ranges, being non-flammable, non-noxious and non-corrosive.

Table 13.7 shows the common types of electric plating solutions.

Electric brush plating can be used to improve the refractory, corrosion-resistant and wear-resistant performances of mechanical parts, repair the worn size and geometry of mechanical parts, fill the scratches and grooves of the surface and repair the out-of-tolerance parts.

13.3.1.5 Plating

Here, the vapor deposition plating technology is mainly introduced, which includes two categories: the physical and the chemical. The physical vapor deposition is also called PVD, mainly includes vacuum evaporation plating, sputter plating and ion plating. Chemical vapor deposition includes the chemical vapor deposition (CVD) and the plasma enhanced chemical vapor deposition (PECVD).

1. Vacuum evaporation plating
   Vacuum evaporation plating involves heating and melting the material in a vacuum environment until it reaches evaporation (or sublimation) so that a large number of atoms, molecules, or atomic

Table 13.7 Commonly electric plating solutions and their components

Name	Type	Main ingredients
Electrical net fluid	Surface treating solution	Sodium hydroxide, tertiary sodium phosphate, gallium sodium, sodium chloride, and so on
Activated liquid		Activated complex liquid, argental activated liquid
Purification liquid		Zinc purification liquid, silver purification liquid
Stripping solution		Nickel, copper, zinc, steel, chromium, copper-nickel-chromium, ferrocobalt, soldering tin, lead-tin
Ni type	Monometallic solution	Special nickel, fast nickel, semi-bright nickel, dense fast nickel, acid nickel, neutral nickel, alkaline nickel, low-stress nickel, high temperature nickel, high pile-up nickel, high-level semi-bright nickel, shaft nickel, black nickel
Cu type		fast copper, acid copper, alkaline copper, copper alloy, high pile-up copper, semi-bright steel
Fe type		Semi-bright neutral iron, semi-bright alkaline iron, acid iron
Co type		Alkaline cobalt, semi-bright and neutral cobalt, acid cobalt
Sn type		Alkaline tin, neutral tin, acidic tin
Pb type		Alkaline lead, acid lead, lead alloys
Cd type		Low hydrogen brittleness cadmium, alkaline cadmium, acid cadmium, weak acid cadmium
Zn type		Alkaline zinc, acid zinc
Cr type		Neutral chromium, acid chrome
Au type		Neutral gold, gold 518, gold 529
Ag type		Low-hydrogen silver, neutral silver, thick silver
Others		Alkaline steel, arsenic, antimony, gallium, platinum, rhodium, aluminum
Binary alloy	Alloy plating liquid	Ni-Co, Ni-W, Ni-W (D), Ni-Fe, Ni-P Co-W, Co-Ag, Sn-Zn, Sn-In, Sn-Sb, Sn-Pb, Au-Sb, Au-Co, Au-Ni
Ternary alloy		Ni-Fe-Co, Ni-Fe-W, Ni-Co-P, Ni-Pb-Sb, babbitt

groups leave the molten surface and condense on the substrate to form a surface coating, as shown in Figure 13.7. The evaporated material can be made of metal, alloy or compound so that the coating can be a metal, an alloy or a compound film respectively. The coating made by vacuum evaporation plating possesses the features of purity, various species and high quality. It is mainly applied in optics, micro-electronics, magnetics and decoration for anti-corrosion, anti-friction and wear-resistance.

2. Sputtering plating

   Sputtering plating uses energetic particles as the coating material, including positive ions produced by glow discharge or an ion source, to bombard the target material and, with the momentum of the particles, knock out the atoms of the target material or other particles, thereby depositing a thin coating on the substrate surface, as shown in Figure 13.8.

   In accordance with the sputtering methods, the sputtering plating is divided into DC sputtering, RF sputtering, magnetron sputtering and ion beam sputtering. Sputtering plating has many unique advantages. For example, it can realize high speed and large area deposition; almost all metals, compounds, media can be used as target materials and form a coating on various substrate materials. Therefore, sputter plating technology is widely used in industrial applications.

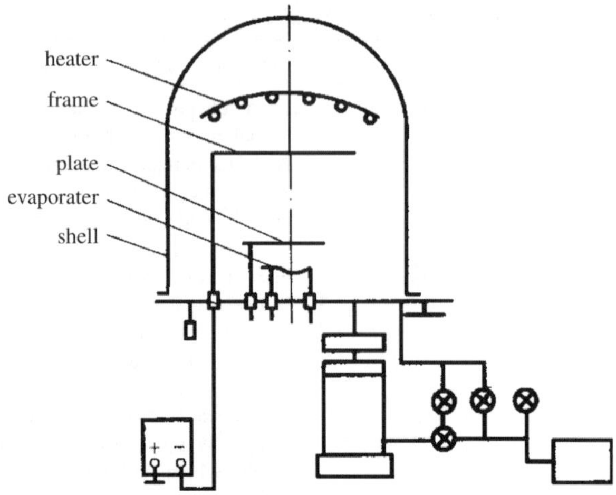

Figure 13.7 Vacuum plating.

3. Ion plating

   The evaporated particles from the evaporation source are atoms or molecules with an average kinetic energy of about 0.2 eV. Sputtering particles are composed of target atoms with an average energy between 5 and 10 eV. In ion plating, before coagulation, parts per thousand or percent of the particles are ionized as positive ions with energy from a few to hundreds of eV. During the process of coagulation and growth, accompanied by the bombardment of energetic ions, they can become coating material ions or working gas ions. Both ions simultaneously exist during multi-technology processing.

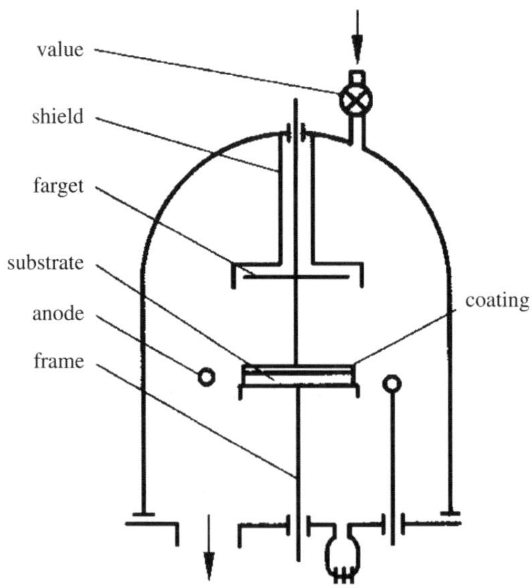

Figure 13.8 Sputtering plating.

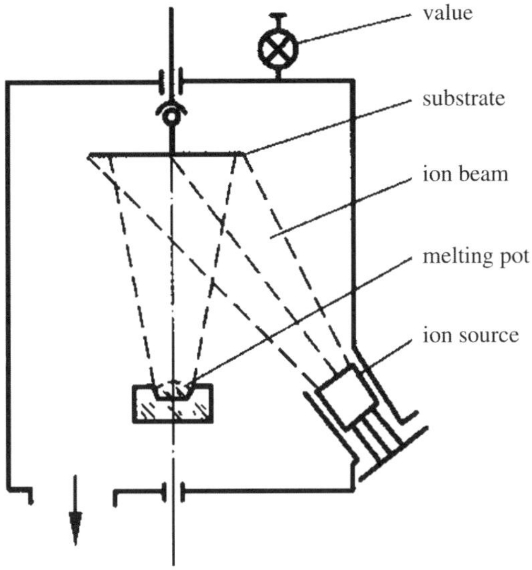

Figure 13.9 Ion plating.

The coating formed by ion plating has the advantages of firm connection with the substrate and dense structure. Ion plating is another powerful PVD plating method other than evaporation plating and sputter plating as shown in Figure 13.9.

With ion plating method, the coating can be plated on a metal, a non-metal, alloy, ceramics and compound substrates [4]. It can be used for anti-corrosion, anti-wear, lubrication, decoration and so on.

4. Chemical vapor deposition

Chemical vapor deposition film, also known as CVD film, is a solid film formed by using one or more gas compounds or elementary gas to form a chemical reaction on the substrate surface. Figure 13.10 is the schematic diagram of the chemical vapor deposition film. With CVD technology, not only can a glassy film be deposited, but also high purity and high integrity thin film with crystalline structure can be deposited. Compared with other thin-film preparation techniques, CVD technology can easily and accurately control film chemical compositions and structure.

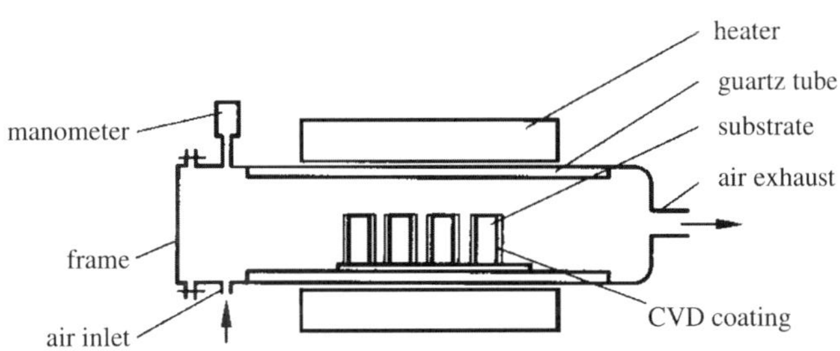

Figure 13.10 Chemical vapor deposition plating.

With CVD technology, we can prepare a pure metal film, an alloy film as well as an intermetallic compound film, such as boron, carbon, silicon, germanium, boride, silicide, carbide, amide, oxide, sulfide, diamond, or diamond-like carbon, which can be used as decorative, optical, electrical or other functional film for anti-wear or anti-corrosion.

The main disadvantage of CVD is that the temperature of the substrate is high during deposition and that limits its application. For example, to deposit nitride or boride as a hard film, the substrate must be heated up to 900 °C or above.

13.3.2 Design of Surface Coating

The surface coating technology uses metallurgy or plating coating to enhance surface strength. In order to obtain a high quality surface coating, we must first understand the working conditions of the pre-coated parts and the possible failure types and then design and select coating materials and coating performances. Furthermore, we should select the appropriate plating process according to the features and applications of the plating method. Surface coating design is an important task.

13.3.2.1 General Principles of Coating Design

1. Satisfy the requirements of working conditions
   The type of coating is chosen according to the force and working conditions. For example, in the atmosphere or corrosive media, thermal spray coating can be used and ceramics, plastics or other non-metallics can be selected as coating materials. In order to improve surface wear resistance, ceramics or alloys should be considered as coating material. If the coating works at a high temperature or the temperature varies significantly, heat-resistant steel, alloy and ceramics coatings should be used.
2. Appropriate structure and properties
   The thickness, bond strength and size tolerance of the coating should be designed based on working conditions as well as determining whether to allow any holes inside the coating, whether there is a need for mechanical processing and how much surface roughness there is after processing.
3. Adaptive to the material and performances of the substrate
   The coating should be adaptive to the material, size, shape, physical and chemical properties, thermal expansion coefficient and surface heat treatment of the substrate.
4. Technically feasible
   In order to realize the designed performances of a surface coating, we should analyze the feasibility of the selected coating method. If the performances cannot be met by single coating, compound coating should be considered.

13.3.2.2 Selection of Surface Plating Method

In order to select a plating method, the following are usually considered.

1. Melting point of coating material
   For example, the melting point of the ceramic coating material is higher than that of the metallic material so the plasma spray method can be used.
2. Coating thickness
   Usually, the optimum coating thickness is not the same. Under normal circumstances, the thickness range of bead welding is about 2–5 mm; the thermal spray coating thickness is around 0.2–0.6 mm; the spray coating thickness is around 0.2–1.2 mm; the electric brush plating coating thickness is below 0.5 mm; while the common plating coating thickness is below 0.05 mm.

3. Bond strength of coating and substrate
   The bead welding and thermal spray coatings are of high bond strength. For example, bond strength of spray coating of Ni-base self-melting alloy powder with the substrate is above 3.5 MPa; and that of thermal spray coating is generally 0.3–0.5 MPa. The bond strength of electric brush plating coating and the substrate is generally higher than that of electric brush plating coating, but lower than that of spray coating.
4. Heat-resistant temperature of substrate
   Bead welding can melt the substrate surface; the molten spraying may let the substrate surface reach 1000 °C; and common spraying to 300 °C or less. However, electric brush plating or slurry coating can be carried out at room temperature. The plating substrate temperature of the coating process is also low, usually at room temperature or a little higher.

13.4 Coating Performance Testing

With the wide applications of the coatings and surface modification technology, the coating quality and its performance testing are given more and more attention. Because a coating is obtained through different processes and methods, the coating performance testing methods are also different. For some coatings, only qualitative or semi-quantitative testing can be used for evaluation. Common coating performance testing methods in practical applications are introduced as follows.

13.4.1 Appearance and Structure

13.4.1.1 Coating Appearance

The coating surface should be smooth, dense, with no bubbles, no peeling, no fall off and has uniform color. If the coating is not smooth or has some small pinholes, it should be polished.

13.4.1.2 Measurement of Coating Thickness

Commonly used method to measure the coating thickness is by using a microscope, a micro-meter or sensors as follows.

1. Place the cross-section of the coating specimen under a microscope and measure the thicknesses in at least two view fields and more than five measured points for each view field. The intervals of the points should be equal. Take the average or minimum value as the thickness of the coating. For thermal spray coating, magnification of the microscope is 20 times and for brush plating coating, magnification is 200–500 times.
2. Measure the thickness of the substrate with the micro-meter. After plating or brushing, measure the thickness again at the same points. The measured points are at least three. Compared with the measured results of the substrate to obtain the average coating thickness or the minimum thickness.
3. The eddy current, magnetic or contact scanning sensors can be used to measure the coating thickness. The eddy current measurement uses the amplitude and the phase of the eddy current generated on the coating surface to measure the thickness. If the coating thickness is different, the magnitude of the amplitude and phase of the eddy current varies. Therefore, the variation of the amplitude and phase can be used to determine the thickness of the coating.

13.4.1.3 Determination of Coating Porosity

Pores always exist in the coating. They can store the lubricant and abrasives so that the coating is more anti-wear. However, corrosive medium may pass through the holes into the substrate surface and weaken the bond strength of the coating, resulting in stripping. The porosity refers to the number of pores per unit area of the coating, usually 5–15%. Some commonly used methods to determine porosity are as follows.

1. Strip the coating from the substrate, place it in the air of 105–120 °C to dry for about 2 hours, and weigh out the quality m_1. Then, immerse the coating into distilled water at room temperature to exhaust the air. Weigh out the quality containing water m_2. Dry the surface water of the sample and weigh out the quality m_3. Then, the surface porosity ε can be calculated as follows:

$$\varepsilon = \frac{m_3 - m_1}{m_3 - m_2} \times 100\,\%. \tag{13.1}$$

2. Spray the coating on the concave surface of the determined cylindrical blank and then precisely grind it to the standard cylinder. From the original size of the blank, the volume of the coating can be known. And, after accurately weighing out the grinding cylinder, we can calculate the quality and density of the coating, and then calculate the porosity of the coating with the following formula.

$$\varepsilon = \left(1 - \frac{\rho_a}{\rho}\right) \times 100\%, \tag{13.2}$$

where ρ is the true density of the coating; ρ_a is the apparent density of the coating.

13.4.2 Bond Strength Test

The bond strength of the coating includes the bond strength between coating the substrate and the bond strength between the coating particles. However, for spraying and brush coatings, it is only necessary to detect the bond strength between the coating and substrate.

13.4.2.1 Drop Hammer Impact Test

The hammer weighs 500 g. Drop it from a height of 100 mm and it impacts the same point of the coating repeatedly until the coating is peeled off. The number of dropping times is used as the testing standard, as shown in Figure 13.11.

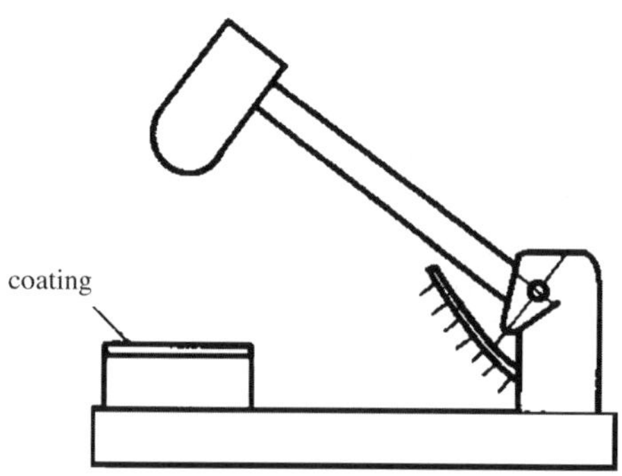

Figure 13.11 Drop hammer impact test.

13.4.2.2 Vibrator Impact Test

Use a vibrator to impact the coating until the coating cracks or peels off. The impact energy is used to characterize the bond strength of the coating and substrate. This method is suitable for thermal spray ceramic coating.

13.4.2.3 Scratch Test

Use a needle tool to scratch perpendicularly the coating thoroughly and determine the bond strength by the specification according to coating type. This method is suitable for soft metal spray coating, plastic coating and brush coating of aluminum, zinc, lead and so on.

13.4.2.4 Broken Test

Plate a coating 0.1 mm thick on a low-carbon steel plate of 1 mm thick. Clamp the sample plate with a vice and bend it repeatedly until it fractures. If there is no coordinate for the coating, the bond strength is good.

13.4.2.5 Tensile Bond Strength Test

Usually, the test is carried out under the condition of the tensile test. The tested specimen can be divided into two groups: with and without agglomerant.

Figure 13.12 presents the tensile bond test with no agglomerant. Drill a center hole in the base. Use the sliding fit between the pins and the center hole. Have the pin face and base surface in the same plane. Spray a coating on the plane and then apply a load to test the bond strength as shown in the figure.

The disadvantage of the test with no agglomerant is that even if the mating between the pin and base is of high precision, a bridgelike coating may form between them to cause the stress concentration which is a fracture source. Therefore, the measurement result of this testing method is generally lower than the actual. In addition, the coating is required to be of a certain thickness; otherwise the coating may generate a shear fracture and the tensile bond strength will not be obtained. However, the thicker the coating, the lower is the bond strength.

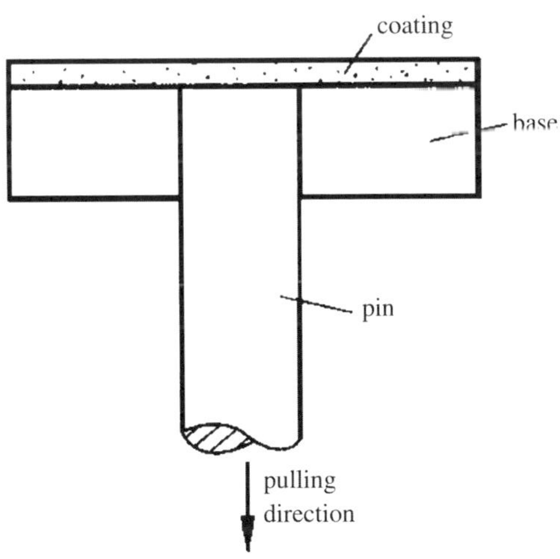

Figure 13.12 Bond strength test with no agglomerant.

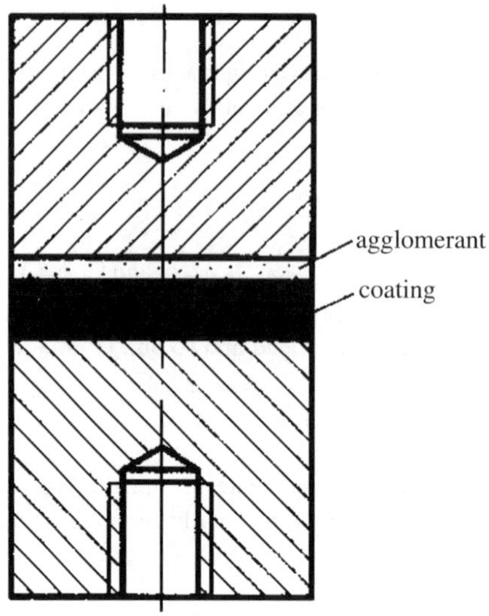

Figure 13.13 Bond strength test with agglomerant.

Figure 13.13 gives the tensile test method with the agglomerant. Spray a coating on the face of the dual specimen and bond another dual to the coating face with the agglomerant. Grind off the spill agglomerant and the spray coating for testing.

Because the agglomerant inevitably penetrates into the pores of the coating in the bonding process, it becomes difficult to directly compare the present test result with the result with no agglomerant. The intensity obtained from the test with the agglomerant is higher. It can usually obtain a more satisfactory result for plastic coating, but is not suitable for spray coating.

13.4.2.6 Shear Bond Strength Test

The shear bond strength test of a coating can also be divided into two types: with and without agglomerant as shown in Figure 13.14. Figure 13.14a uses an agglomerant to bond the two coated pieces. To pull the pieces to break the coating, we can obtain the shear bond strength of the coating. Figure 13.14b uses two parallel plates as shear specimens to test the bond strength of the coating. Other shear tests can be found in reference [2].

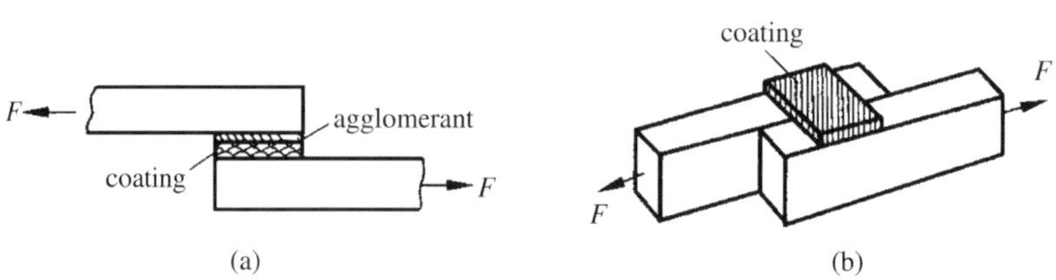

Figure 13.14 Shear bond strength test of coating.

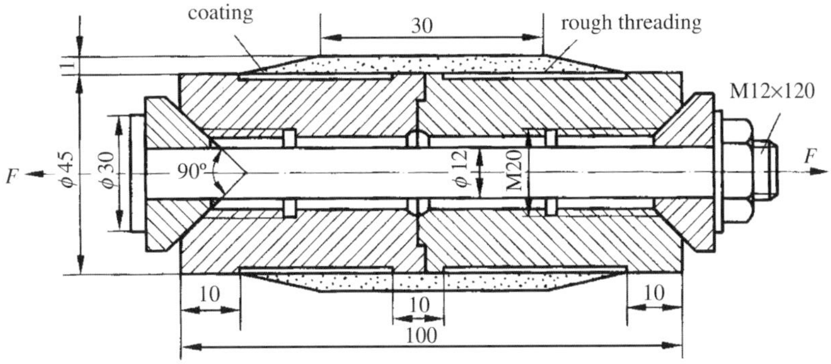

Figure 13.15 Parallel coating strength test.

13.4.2.7 Measurement of Internal Bond Strength of Coating

The internal bond strength of the coating is the adhesion between the coating particles, which reflects the cohesion between particles, also known as the coating strength. The coating strengths parallel to and perpendicular to the surface are significantly different.

The fixture and specimen of the parallel strength test are showed in Figure 13.15. The primary load applying rate is 9807 N/min and then we can easily obtain the strength.

The perpendicular coating strength test is shown in Figure 13.16. First, paint a low melting point solder film on the one side face of the blank sample, blow the surface roughness by the shot blasting method, and then spray the surface coating to be tested. Then, melt the solder to take off the coating. As shown in Figure 13.16b, adhere the end faces of the two rods and carry out the tensile test. The agglomerant strength used should be larger than the coating strength.

The high-speed particle beam is used to impact the coating to measure its bond strength. As shown in Figure 13.17, high-speed particles are ejected from the nozzle which can precisely control the particle velocity, the flow rate as well as the impact point. Under the impact, part of the coating particles fall off. According to the amount of coating particles falling off, we can determine the bond strength between the coating particles. It is a high-precision testing method to measure the bond strength between the coating particles.

Similar to perpendicular test, this test is not suitable for spray coating. It is mainly used for hard metal coating, such as ceramic coating.

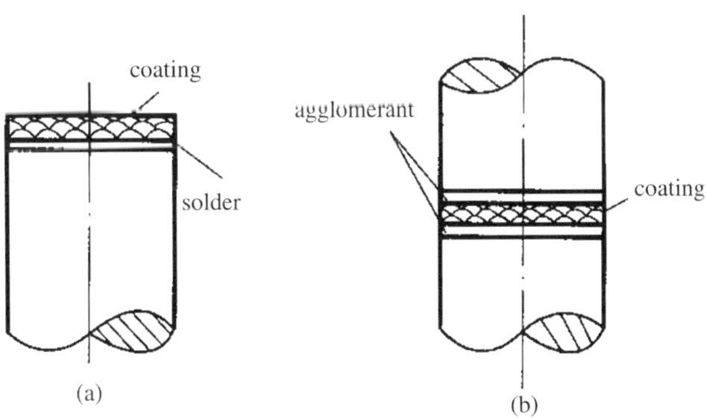

Figure 13.16 Vertical coating strength test.

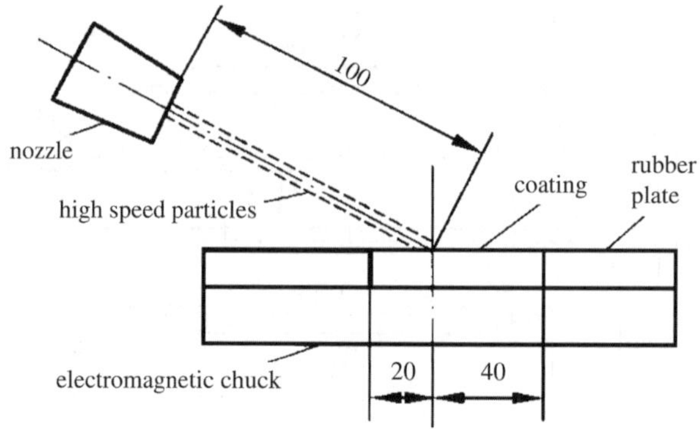

Figure 13.17 High-speed particle beam impact test.

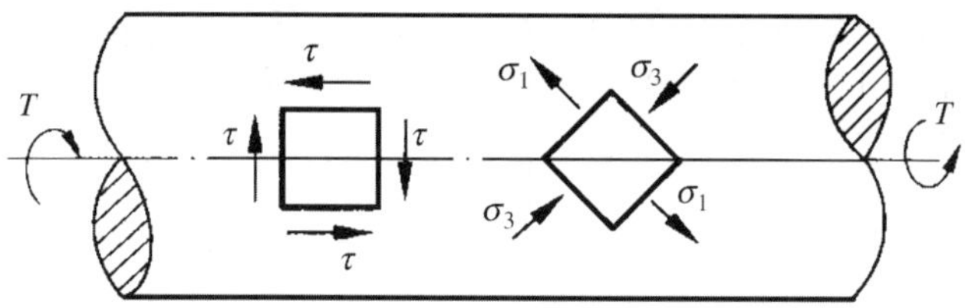

Figure 13.18 Brush coating strength torsion test.

The torsion test method of brush coating is shown in Figure 13.18. The specimen is under pure distortion. The cracking of the coating is taken as the testing standard. The shear strength of the coating is measured. If the brush coating falls off, the shear stress is approximately equal to the bond strength between the coating and the substrate.

The torsion test is of high precision and has good repeatability. However, because the brush coating thickness has a great impact on the measured data, the uniform coating thickness should be maintained.

It should be pointed out that among the variety of testing methods above, if the coating formation process and the standard are different, the measured results will be different. When the coating is made under different conditions, the bond strengths are also not the same.

13.4.3 Hardness Test

The hardness test of a coating can be in several ways, but the different test methods may obtain different hardness. Common methods to measure hardness are usually divided into two kinds: static (indentation) and dynamic. Static hardnesses include Brinell hardness, Rockwell hardness, Vickers hardness, and so on, while dynamic hardnesses include the hammer, Shore hardness, and so on. In addition, surface scratch hardness is also used in the coating test. Because conventional hardness testing methods can be found in common textbooks, only two kinds of special hardness tests will be introduced here.

13.4.3.1 Micro-Hardness (Hm) Testing

The characteristic of the micro-hardness test is to reduce the test sample within the microscopic scale, which is commonly used to determine the hardness of the compositions of the material or a group of phases. In addition, using the method to measure the micro-hardness of a single spray particle is also effective. The working principle of a micro-hardness tester is the same as that for a Vickers hardness tester, but the applied load is smaller. The commonly used loads are 2, 5, 10, 20, 50, 100 or 200 g. It is suitable for all coatings, particularly for brush coating less than 0.3 mm thick, except for plastic coating.

Micro-hardness has been widely used in the phase structure study of spray coating.

13.4.3.2 Hoffman Scratch Hardness Testing

The Hoffman scratch hardness test indirectly measures coating hardness and wear resistance. It is suitable for soft-metal coatings and plastic coatings, but it requires the minimum thickness of 0.89 mm. By using a 6 mm taper head under the load of 19.60 N, the coating is scratched. The width of the scratch can be used to present hardness and the wear resistance of the coating. The wider the scratch, the lower the hardness is, so the weaker the coating bond strength is. Hoffman scratch hardness H_N can be calculated as follows:

$$H_N = \frac{b}{5} \times 10^{-3}, \tag{13.3}$$

where b is the scratch width measured in inches.

13.4.4 Wear Test

A coating is the most widely used wear resistance method. The wear resistance of the spray coating depends on match of the micro-structure and the hardness of its metallic phase. Because there are pores in the spray coating and oxide as a thin film between the coating particles or as other forms (e.g. granular) in the coating, the wear resistance may be high although the macro hardness is not so high. Particularly if there is a lubricant, the coating pores can store lubricant so as to have a certain lubricity. The wear resistance of the brush coating is not only related to coating hardness, but also directly related to the coating structure, the composition of the plating solution and the parameters of the plating process. In addition, the coating friction and wear properties are not the inherent properties of the material, but are determined by the many factors of the tribo system. If the working conditions change, the wear test data will change significantly. Hence, only under certain specified conditions can the wear resistance of a coating be evaluated. The wear test methods will be introduced in Chapter 14.

13.4.5 Tests of Other Performances

13.4.5.1 Fatigue Test

The commonly used fatigue test is the four-point bending test. The number of cycles as the crack appears is used to represent the performance of the anti-fatigue damage ability of the coating. In addition, the distortion, rotation or other bending methods can be applied to test the coating fatigue.

13.4.5.2 Measurement of Residual Stress

The shrink of solidification or the unbalanced crystallization can cause residual stress in the coating. The larger the difference of the expansion coefficients of the substrate material and the coating material, the larger the residual stress is in the coating. If the plating solution compositions, plating temperature, substrate temperature and operation parameters of the coating process were changed, the residual stress also changes. Generally, the thicker the coating, the larger is the residual stress. In addition, the residual stress is related to the coating material melting point and other factors. The different thermal spraying methods and coating materials will produce different residual stresses. When the coating material is metal, the residual stress of a plasma flame spray coating is higher than that of oxyacetylene flame coating. If the coating material is ceramic, the residual stress of oxyacetylene flame spray coating is higher than that of plasma flame spray coating.

1. X-ray diffraction method
   Before the test, polish the coating specimen surface with a piece of sandpaper to make the specimen smooth with the maximum roughness $R_a = 40–70\,\mu m$. Then, use the X-ray diffraction device to measure the stress in the coating from room temperature to 600 °C. The main advantage of the X-ray diffraction method is that it is a non-destructive determination of surface stress, but the disadvantage is that it is difficult to accurately determine the stress distribution and stress level of the interface if the coating is too thick. If the coating thickness is not thicker than 0.15 mm, the measuring result will be fairly accurate.
2. Bending curvature method of ring specimen
   First, machine the sample blank to a ring, and spray or brush a coating on its surface. The residual stress in the coating will cause the curvature of the sample to change. Therefore, according to the measured curvature, we can calculate the residual stress using the following formula:

$$\sigma_r = \frac{1}{1-\mu^2}\left[\frac{h_1^3E_1 + h_2^3E_2}{6\rho h_1(h_1+h_2)} + \frac{E_1(h_1^3E_1 + h_2^3E_2)}{12\rho^2(h_1E_1+h_2E_2)} + \frac{E_1E_2h_2(h_1+h_2)}{2\rho(h_1E_1+h_2E_2)}\right], \tag{13.4}$$

   where E_1, and E_2 are the elastic modulus of the coating and the substrate respectively; h_1 and h_2 are the thicknesses of the coating and the substrate respectively; ρ is the curvature radius of the surface; μ is Poisson's ratio; σ_r is the residual stress.

In addition to the above methods, there are some other methods, such as the strain gage, the projection and the trapezoidal groove methods, which can also be used to measure the residual stress, but are not as good, convenient and accurate as the X-ray diffraction method.

References

1. GE, Z.M., Hou, Y.K. and Wen, S.Z. (1991) *Wear Resistance Design*, Mechanical Industry Press, Beijing.
2. Chen, X.D. and Han, W.Z. (1994) *Surface Coating Technology*, Mechanical Industry Press, Beijing.
3. Shao, T.M. and Jin, Y.S. (1996) Slurry coatings study progress, progress in tribology, State Key Laboratory of Tribology, Tsinghua University, 3, 31–35.
4. Wang, Y.L., Jin, Y.S., and Wen, S.Z. (1988) The analysis of the friction and wear mechanisms of plasma-sprayed ceramic coating at 450 °C. *Wear*, **128**, 265–276.

14

Tribological Experiments

14.1 Tribological Experimental Method and Device

Friction and wear experiments are designed to examine the tribological characteristics of a tribo system and their variations under working conditions, in order to reveal the influence factors on friction and wear properties so as to reasonably determine design parameters.

Because the phenomena of friction and wear are complicated, experimental methods and devices are various and experimental data obtained are conditional and often difficult to compare. In recent years, the standardization of the experimental methods has been paid more and more attention in order to make the test methods uniform.

Friction and wear performances are the synthesis performance of a variety of affecting factors. Therefore, only by strictly controlling experimental conditions can we obtain a reliable conclusion.

14.1.1 Experimental Methods

Current experimental methods can be divided into the following three categories.

14.1.1.1 Laboratory Specimen Test

A laboratory specimen test uses the universal testing machine to carry out the experiment with a specimen according to given working conditions. The advantage of the laboratory specimen test is that because the environmental and working parameters can be easily controlled, the reproducibility of experimental data is high and the experimental period is short, so many systematic data can be obtained in a short time. However, because experimental conditions are not fully met with the actual working conditions, the results are often less practical. A laboratory specimen test is mainly used to study friction and wear mechanisms and influence factors such as friction pair materials and to evaluate lubricant performances.

14.1.1.2 Simulation Test

After the laboratory specimen test, a simulation test can be carried out further with the actual component designed according to selected parameters. Because test conditions are close to actual working conditions, it enhances the reliability of the experimental results. At the same time, through intensifyed and strictly controlled experimental conditions, a series of experimental data can also be obtained within a short

Principles of Tribology, First Edition. Wen Shizhu and Huang Ping.

period and the factors influencing wear performance can also be studied individually. The main purpose of the simulation test is to verify the reliability of the data and the rationality of the wear design of the part.

14.1.1.3 Actual Test

Based on the above two tests, an actual test can finally be used, which is much more actual and reliable, but its test cycle is long and costly. Furthermore, experimental results are influenced by various factors so it is difficult to analyze the results in-depth. This test is usually used as a means to test the first two experimental data.

The above three tests can be selected based on the needs of the experimental study. It should be noted that the friction and wear performances of a tribo systems under given conditions are the synthesis performance so the experimental results are not universal. Therefore, in the laboratory test, it should possibly simulate actual working conditions, such as sliding velocity, surface pressure, temperature, lubrication state, environmental media conditions, surface contact form, and so on. For the high-speed friction and wear test, temperature is the main effecting factor. Therefore, thermal conditions and temperature distribution of the specimen should be close to the actual situation. In the low-speed friction experiment, because the running-in time is long, in order to eliminate the impact of running-in on the experimental results, it can run-in the friction surface of the specimen in advance so as to create conditions compatible with the quality of the used surface. For the specimen without running-in, the first several measured data are usually not adopted because they are unstable.

Generally, the most common friction and wear testing machines are mainly used to evaluate the performances of different materials and lubricants under different velocity, load and temperature conditions. They can also be used to study wear mechanisms.

Figure 14.1 shows the contact and movement types of common friction and wear testing machines. The relative motion between the samples can be pure sliding, pure rolling or sliding-rolling. The motion of a tester may be a reciprocating movement or a rotary movement. The contact forms of a specimen are of three kinds: surface contact, line contact and point contact. Usually, the pressure of the surface contact is only about 50–100 MPa, commonly used in the abrasive wear test. The pressure of the line contact can reach 1000–1500 MPa, suitable for the contact fatigue wear and adhesive wear tests. The pressure of the point contact can be higher than 5000 MPa. It is used for the spalling test or contact fatigue wear test of the high strength material with very high contact pressure.

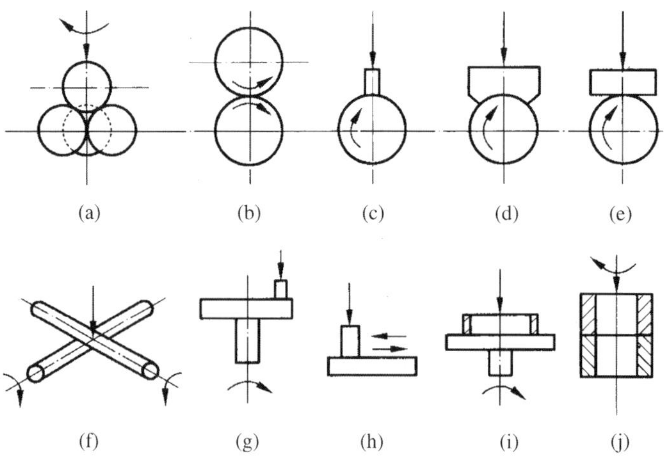

Figure 14.1 Types of contact and movement of friction and wear testers.

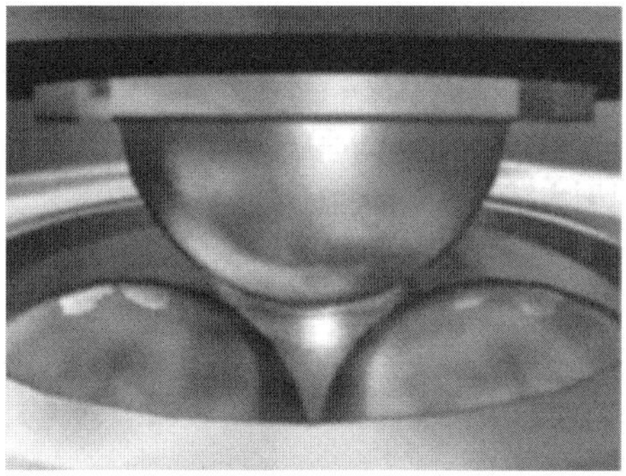

Figure 14.2 Four-ball machine.

14.1.2 Commonly Used Friction and Wear Testing Machines

The four-ball testing machine is made by the principle as shown in Figure 14.1a. The three balls of the four are clamped into the retainer cup with a raceway, see Figure 14.2. The other ball is placed at the top of the other three balls to maintain the point contact under the vertical load. While the ball is driven by the rotational shaft, it relatively slides to the three balls below. The four-ball testing machine is often used to evaluate the performances of the lubricant additives. According to the diameter of the wear scar and friction coefficient, we can analyze the testing results.

The Timken testing machine is also called the ring-block testing machine, which is designed based on the form of Figure 14.1e, belonging to the line contact test. The rotating ring serves as the base, which is generally the outer ring of a tapered roller bearing or a standard ring. It is pressed on a rectangular block serving as the testing sample. As rotation time increases, a rectangular wear scar will appear on the surface of the block. By measuring the width of the wear scar, we can evaluate the friction and wear properties of the tested lubricants or the material of the block.

The pin disc testing machine is designed according to Figure 14.1g and h, belonging to the surface contact. The pin is pressed on a cylinder or a disc, which continuously slides or reciprocates. The specimen can easily be made by any material.

At present, there are many new types of multi-function friction and wear testing machines, which have a variety of contact and movement forms. As long as we replace the specimen, we can complete many different types of friction and wear tests or combination tests.

Although the usual friction and wear testing machines have the function to measure the frictional force, the accuracy is usually low and cannot meet the needs of studying the friction performances yet.

14.1.3 EHL and Thin Film Lubrication Test

14.1.3.1 EHL and Thin Film Lubrication Test Machine

Figure 14.3a shows the EHL tester, the author designed according to the principle of the optical interference [1]. The load is added through the lever 1. In the relative movement, an elastohydrodynamic film forms between the ball 2 and the glass disc 3. The outer light 4 is projected through the semi-anti and semi-translucent film 5 of the glass disc. The reflected lights on the semi-anti and semi-translucent film and on the steel ball surface interfere with each other. Through the microscope 6, the spectral monochromator 7 and CCD, the interfered images are finally sent to the computer to be displayed (see Figure 14.3b). By using the relative light intensity principle, we can obtain the EHL film thickness

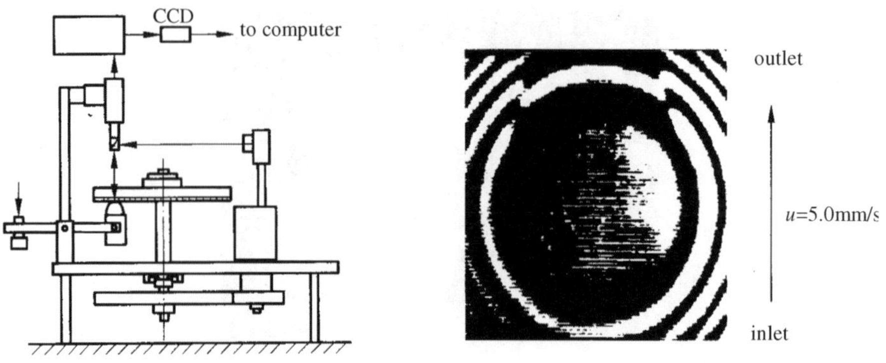

(a) (b)

Figure 14.3 EHL film thickness tester.

curves [2]. The instrument can be used to measure the EHL film thickness and shape in the point contact, the relationship of the EHL film thickness with the load, the velocity and temperature or to study the film-forming ability of the special media (such as the high-water-based medium, the micro-polar fluid) or additives. It can also be used to study lubrication failure, the transition from EHL to the boundary lubrication or mixed lubrication.

14.1.3.2 Principle of Relative Light Intensity

By using the optical method to measure film thickness, the measurement results are usually associated with absolute intensity of the light. Therefore, the variation of the external light source will change the measurement results significantly. The problem is how to accurately calibrate the light intensity at the point to obtain the corresponding film thickness. Here, the principle of the relative light intensity to measure the film thickness is introduced.

As shown in Figure 14.4, a lubricant film exists between the ball and the glass disc. When the incident light beam through the glass disc semi-anti and semi-translucent film and the steel ball is reflected to produce light ① and light ②, they interfere because their optical distances are different. According to the

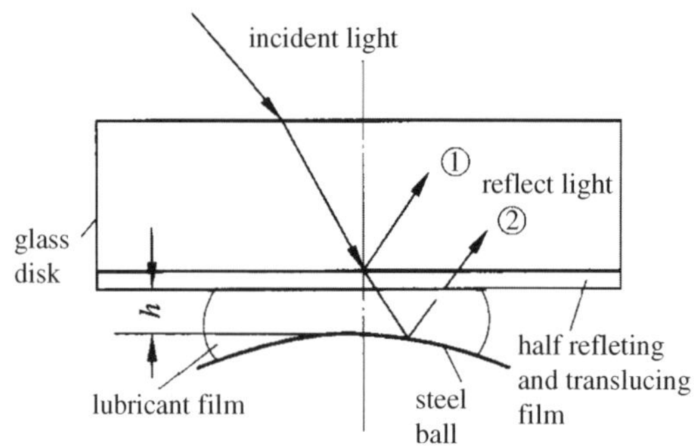

Figure 14.4 Principle of relative light intensity.

optical interference principle, when the incidence is vertical, the relationship of the film thickness and the intensities of the reflected lights and interfered light at any point is as follows:

$$I = I_1 + I_2 + 2\sqrt{I_1 I_2}\cos\left(\frac{4\pi nh}{\lambda} + \varphi\right), \quad (14.1)$$

where, I is the intensity of the interfered light; I_1 and I_2 are respectively the reflected light ① and light ②; λ is the optical wavelength; n is the refractive index of the lubricant; φ is the phase difference caused by the coating: and h is the film thickness to be measured.

Because the film thickness between the steel ball and glass plate varies, the interfered light intensity changes.

The maximum and minimum intensities have the following relationships with the reflected intensities:

$$\begin{aligned} I_1 &= (I_{\max} + I_{\min})/4 + \sqrt{I_{\max} I_{\min}/2} \\ I_1 &= (I_{\max} + I_{\min})/4 - \sqrt{I_{\max} I_{\min}/2} \end{aligned}. \quad (14.2)$$

If we set the relative light intensity as follows:

$$\bar{I} = 2(I - I_a)/I_d, \quad (14.3)$$

where $I_a = (I_{\max} + I_{\min})/2$ is the average light intensity; $I_d = I_{\max} - I_{\min}$ is the intensity difference, which shows the relative position of the light intensity between the maximum and the minimum of the interferogram.

Substitute Equation 14.2 into Equation 14.1 and use Equation 14.3, we can get the formulas of the relationship between film thickness, optical wavelength, refractive index and relative light intensity as follows:

$$h = \frac{\lambda}{4\pi n}[\arccos(\bar{I}) - \varphi]. \quad (14.4)$$

The relative intensity I_0 can be obtained by calibrating when the film thickness is 0. Substitute it into the above equation to determine φ, we can obtain the film thickness formulas in the first interference fringe as follows:

$$h = \frac{\lambda}{4\pi n}[\arccos(\bar{I}) - \arccos(\bar{I}_0)]. \quad (14.5)$$

The film thickness formula by the principle of the relative intensity is derived from optical interference theory. It omits high-order reflections and some other factors, such as: lightwave loss in each interface, metal absorption, coating thickness and so on.

The relative intensity interference principle uses the maximum and minimum intensities as the upper and lower limits to normalize the interference light. Because the difference in the optical path in the same interference level is the same, the adjacent film thickness difference can also be determined by the thickness. By using the image processing technique, we can obtain the high-resolution thickness. The resolution across the film thickness is

$$\text{Resolution} = \frac{\lambda}{4nI_d}. \quad (14.6)$$

For example, if the corresponding optical wavelength is 600 nm, the lubricant refractive index is 1.5 and the maximum intensity difference $I_a > 100$, the relative light intensity measurement resolution can be less than 1 nm.

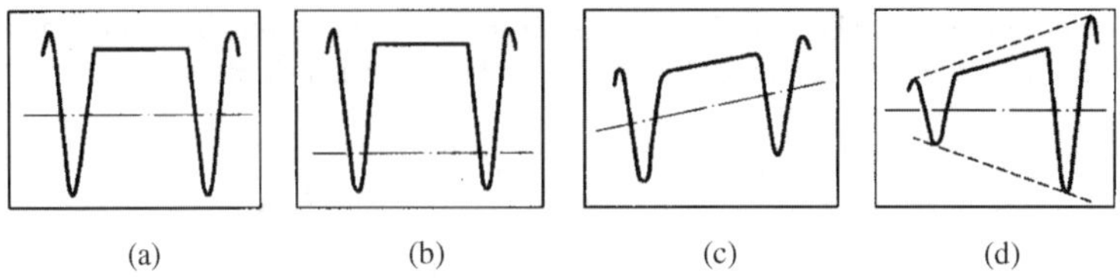

Figure 14.5 Modification of outside interference.

The relative intensity principle has a high anti-interference ability of the outside light. While the outside light is uniform, the actual interference image is equal to add a relatively constant to the original light intensity, as shown in Figure 14.5b. When an oblique outside light field exists, the curve becomes inclined (Figure 14.5c) or compressed (Figure 14.5d). The extreme points can be used by a linear transformation to correct the relative light intensity curve as the standard (Figure 14.5a). Therefore, the external influence can be greatly reduced.

14.2 Measurement of Wear Capacity

The wear capacity of a mechanical element can be represented in the weight, volume or thickness loss of the material. The wear weight or volume is the sum of the worn part, while the wear thickness reflects the wear distribution along the friction surface.

The commonly used wear measurement methods are as follows.

14.2.1 Weighing Method

By weighing the specimen weight before and after the experiment, the wear capacity can be determined, commonly with a sophisticated analytical balance with the measurement accuracy of 0.1 mg. Because the measuring range is limited, the weighing method is only applied to the small specimen. For the micro-wear, it will take a very long time to obtain the measurable weight. If the specimen surface has a significantly plastic deformation during the wear test, the weighing method is not suitable because the weight loss is not much, but the shape of the specimen changes a lot.

14.2.2 Length Measurement Method

With a precision measuring instrument, such as the length measuring instrument, the universal tool microscope or the other noncontact micro-meter, we can measure the size or surface distance variation corresponding to the base level before and after the experiment.

The method can be used to measure wear distribution. However, it has some significant defects, such as the measurement data contain the size change caused by deformation and the measured results of the contact measurement instrument are influenced by the contact situation, the temperature and so on.

14.2.3 Profile Method

With the surface profiler 1 as shown in Figure 14.6, we can directly measure the worn surface profile changes before and after the experiment and can even determine the zero-wear, that is, the worn thickness is no more than the surface roughness. In order to ensure that profile positions are accurately the same before and after wear, microscope 2 can be used to determine the baseline 3 on the position of specimen 4.

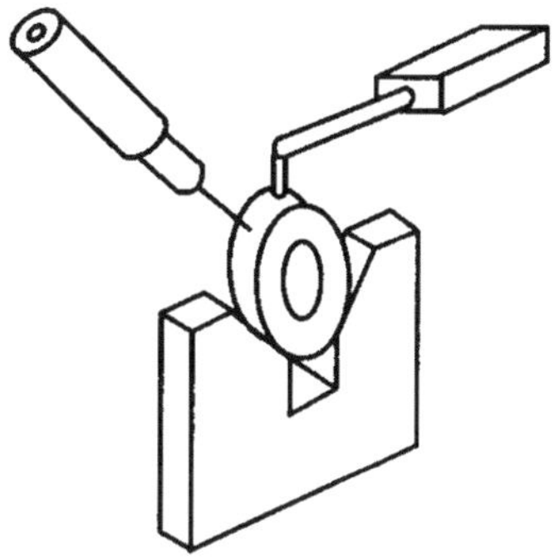

Figure 14.6 Measurement of surface profile.

When the profile method is used to measure surface wear, the worn thickness must be larger than the surface roughness. Figure 14.7 shows two kinds of wear. In Figure 14.7a, the worn surface is used as a data. In Figure 14.7b, there is a wedge-shaped groove on the surface with the width b and B before and after wear. Therefore, both wear capacities can be calculated with the wear thickness h.

The profile method can record the profile variation and the wear distribution during the wear process. However, its measurement procedures are complicated so its applications are restricted by the shape and size of a part and the measurement range.

14.2.4 Indentation Method

The method is to make an indentation or groove artificially as the data and to measure the depth of the data to obtain the wear capacity. If we arrange the data at the different points, we can also measure the wear distribution along the surface. The indentation method is commonly used by the Vickers hardness tester. A square pyramidal indentation made by the pressure head on the surface is as shown in Figure 14.8.

If the cone angle is α (typically $\alpha = 136°$) and the diagonal length is d, then the height h is equal to

$$h = \frac{d}{2\sqrt{2}\tan\frac{\alpha}{2}} = \frac{d}{m}, \tag{14.7}$$

where $m = 2\sqrt{2}\tan \alpha/2 \approx 7$.

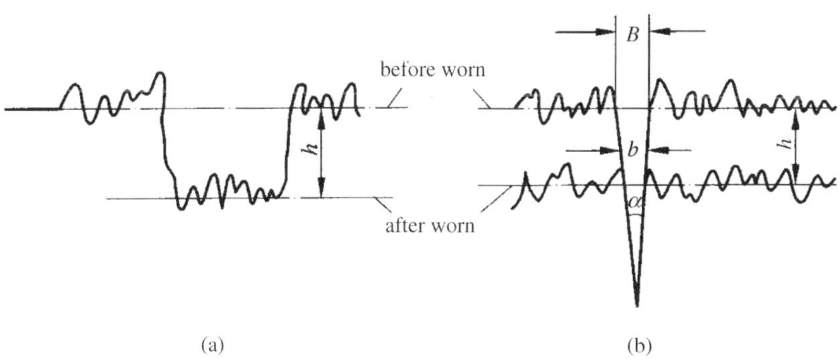

Figure 14.7 Measurement of two kinds of measurable wear profiles.

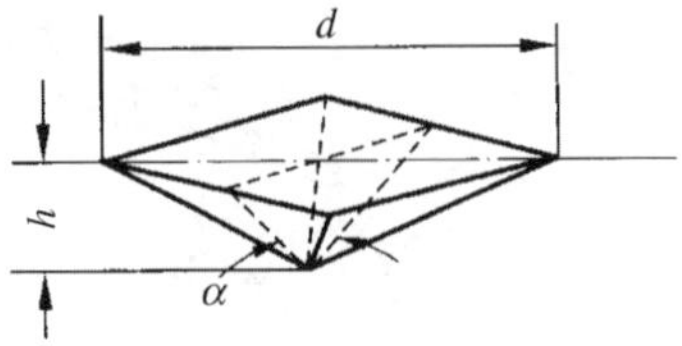

Figure 14.8 Square pyramid indentation.

After the surface is worn, calculate the height variation by measuring the diagonal increment according to the following formulas. If the diagonal changes from d to d_1 after wear, the wear thickness δ is equal to

$$\delta = h - h_1 = \frac{1}{m}(d - d_1). \tag{14.8}$$

The indentation method can also be used to measure the wear capacity of the cylindrical surface. Figure 14.9 shows an internal face indentation. If the inner circle radius before wear is R, d and d_1 are the diagonals before and after wear, the wear thickness δ of the internal face will be

$$\delta = \frac{1}{m}(d - d_1) - \frac{1}{8R}(d^2 - d_1^2). \tag{14.9}$$

An external face indentation is shown in Figure 14.10. The wear formulas to calculate the thickness is:

$$\delta = \frac{1}{m}(d - d_1) + \frac{1}{8R}(d^2 - d_1^2). \tag{14.10}$$

It should be noted that in accordance with the above formula to calculate the wear thickness, some errors exist. Because the deformation in the indentation process is not entirely plastic, the shapes of the indentation and the pressure head are not exactly the same. Therefore, m should be increased to consider the influence of the elastic deformation. When the cone angle $\alpha = 136°$, the following m can be selected: for good plastic material such as lead, $m = 7$; for cast iron, $m = 7.6$–8.2, with an average of 7.9; and for bearing steel, $m = 7.7$–8.4, with an average of 8.

Another factor to bring about errors is that heave exists surrounding the indentation caused by pressure. This makes the surface shape change and influence the performances of friction and wear, as well as the accuracy of the wear measurement. Figure 14.11 shows the surrounding heave of the indentation, where a and a_0 are determined according to material nature.

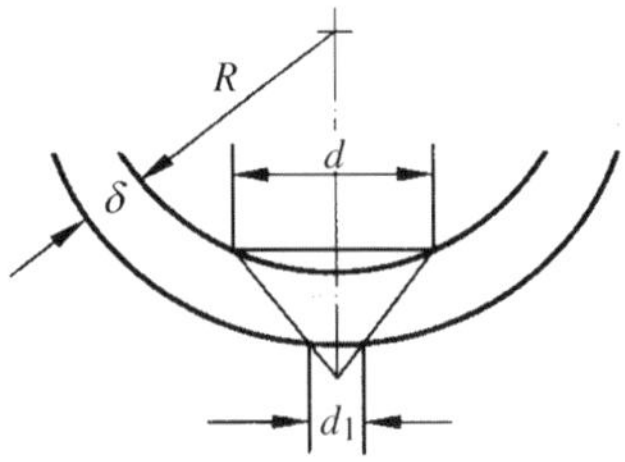

Figure 14.9 Internal face indentation.

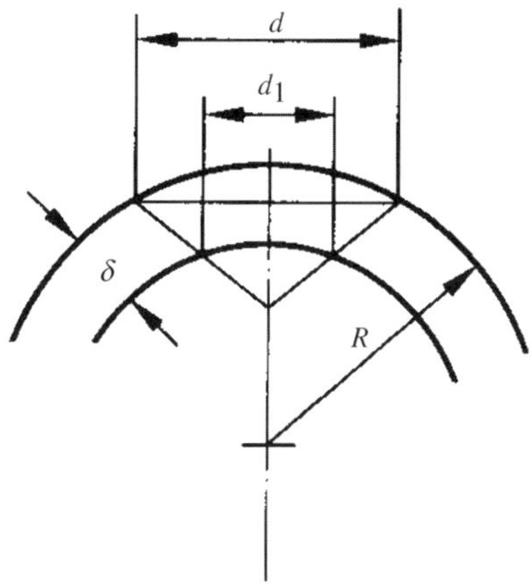

Figure 14.10 External face indentation.

If the diagonal before and after wear is reduced from d_1 to d_2, the actual wear thickness h is equal to

$$h = \frac{1}{m}(d_1 - d_2). \tag{14.11}$$

However, if we consider the influence of the heave, the wear thickness h' is equal to

$$h' = \frac{1}{m}(d_0 - d_2). \tag{14.12}$$

The measurement error E is equal to

$$E = \frac{h' - h}{h} = \frac{d_0 - d_1}{d_1 - d_2}. \tag{14.13}$$

Usually, E is up to 60% or even higher. Therefore, to reduce measurement error, we must use a special tool to repair the surface, carry out a fully running-in before the experiment or eliminate the heave around the indentation before the end of measurement.

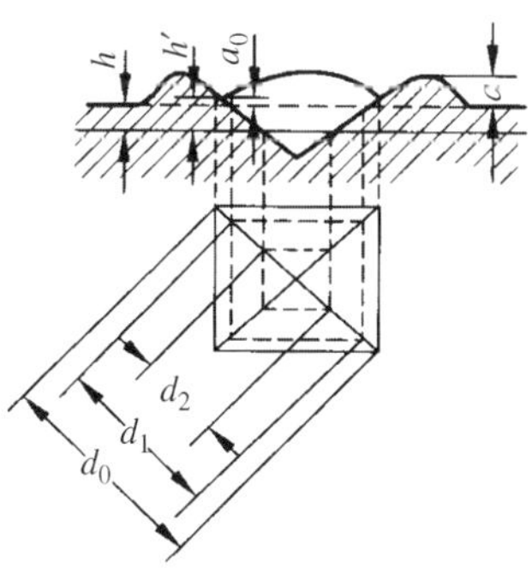

Figure 14.11 Indentation deformation.

If we wish to measure the distribution of wear, a series of indentation can be arranged on the friction surface. In order to ensure the measurement accuracy, it should have one diagonal vertical to the sliding direction, and the other along the same direction of the sliding. The size of each indentation should be the same as possible. In order to make each indentation the same size, the special load limiter is often used in the indentation. The force applied to press the head into the surface can be calculated as follows:

$$F = 54\frac{d^2}{H_v} \tag{14.14}$$

where F is the force to press the head into the surface, N; d is the diagonal length, mm. usually $d = 1$ mm; H_v is the Vickers hardness of the material, MPa.

14.2.5 Grooving Method

The grooving method is very similar to the indentation method, but the grooving method eliminates the influence of the elastic deformation rebound and the surrounding heave. Although elastic deformation and cutting heat are the factors to cause the groove geometric error, it is generally no more than 5%. Therefore, the accuracy of the groove to be used in this method is higher than that of the indentation method. Figure 14.12 shows the sizes of the goove.

According to the geometry relationship, we have

$$h = r - \sqrt{r^2 - \frac{l^2}{4}} \tag{14.15}$$

or

$$h \approx \frac{l^2}{8r}. \tag{14.16}$$

The maximum calculated error of Equation 14.16 is less than 1%.

The wear thickness δ calculated according to the variation of the groove width l is equal to

$$\delta = \frac{l^2 - l_1^2}{8r}. \tag{14.17}$$

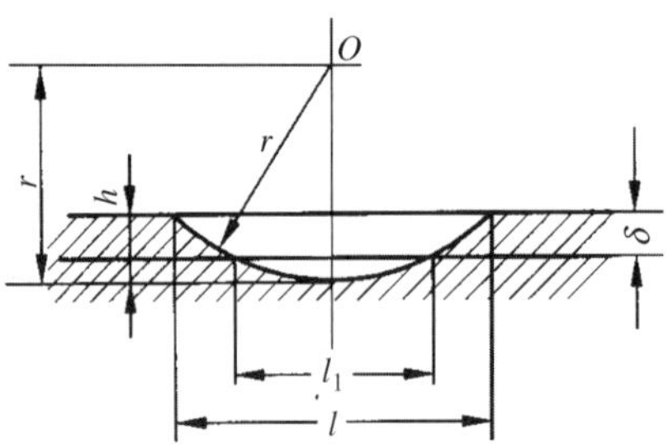

Figure 14.12 Sizes of groove.

If we measure the wear capacity of a cylinder surface with the radius R after wear, the wear thickness δ of the inner circle will be

$$\delta = \frac{1}{8}(l^2 - l_1^2)\left(\frac{1}{r} - \frac{1}{R}\right). \tag{14.18}$$

The wear thickness δ of the external circle will be

$$\delta = \frac{1}{8}(l^2 - l_1^2)\left(\frac{1}{r} + \frac{1}{R}\right). \tag{14.19}$$

In order to avoid debris clogging the groove and affecting the accuracy of the measurement, the longitudinal direction of the groove should be perpendicular to the sliding direction. The groove length is often chosen about 1.5 mm while the groove depth should exceed surface roughness and wear thickness.

The indentation method and grooving method are suitable for measuring the small wear capacity of the specimen with a smooth surface. Because the two methods must partially damage the surface of the specimen, they cannot be used for studying the wear process with the structural variation on the surface.

It should be pointed out that the wear measurement methods described above have the common shortcoming that before measuring, the machine must be disassembled so the measurement procedure is complicated. In addition, the wear capacity varies with time. Therefore, wear working conditions will be changed for each disassembly. The following two methods are real-time measurements of the wear capacity and thus avoid the above mentioned shortcomings.

14.2.6 Precipitation Method and Chemical Analysis Method

The precipitation method is to separate debris contained in the lubricating oil by filtration or sedimentation and to weigh out the weight of the wear debris.

Furthermore, the quantitative analysis chemistry method can be used to determine the compositions and weight of the wear debris contained in the lubricant. It can also be used to determine the wear position as well.

If we remove the oil sample from a lubrication system in a regular interval to measure by the above method, it can determine the variation of the wear with time. However, the results are the total wear capacity of the entire system so it is not possible to determine the wear distribution on the surface. In addition, reasonable sampling from the lubricant is the key to ensuring measurement accuracy.

14.2.7 Radioactive Method

The radioactive method is to activate the friction surface to let the debris possess radioactivity. When it falls into the lubricating oil during the wear process, we can regularly collect and measure the radioactive intensity of the lubricating oil to know that the wear capacity varies with time.

Figure 14.13 is the device for the radioactive method for the journal bearing 1. When the radioactive debris circulating with the lubricant passes through the Geiger counter 8, the calibrating device records the number of radioactive pulses so that the wear capacity can be continuously measured.

The radioactive method has a high sensitivity up to 10^{-7}–10^{-8} g. Furthermore, it can simultaneously measure the wear of several locations or surfaces.

The activation methods of the specimen surface include the coating method, casting method, embedding method, exposure method and proliferation method. The protective measure must be adopted according to the sample radioactivity.

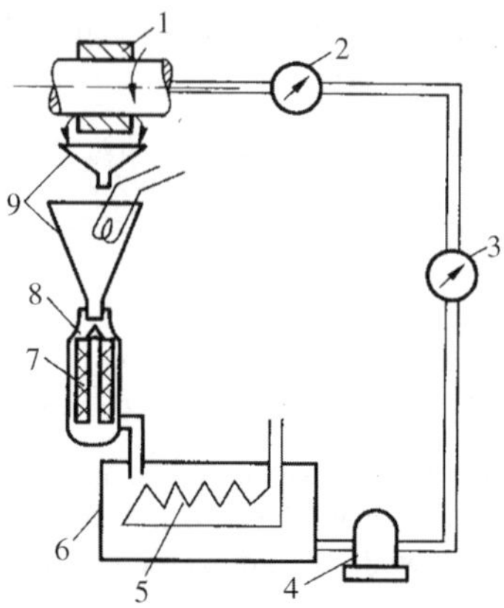

Figure 14.13 Radioactive measuring wear method: 1. bearing; 2. flux meter; 3. manograph; 4. pump; 5. cooler; 6. oil tank; 7. filter; 8. Geiger counter; 9. hopper.

14.3 Analysis of Friction Surface Morphology

Because the tribological phenomenon occurs on the surface, the surface structure variation is the key to studying the mechanism of friction and wear. Modern surface testing techniques have been successively used to study friction and wear phenomena.

14.3.1 Analysis of Surface Topography

The surface profiler or electron microscopy can be used to analyze the variation of the topography of the friction surface.

A surface profiler measures the surface profile by detecting the vertical motion of a probe while it uniformly moves along the surface. After amplifying the measured signal, the surface profile curve can be drawn out. With a computer, we can also quickly obtain the parameters of the surface topography.

The transmission electronic microscope (TEM) and the reflecting electronic microscope (REM) can be used to study the surface topography, the damage properties of the subsurface and topography of the surface oxide film. Because they are only used in the replication detection and are of large measuring errors and inconvenient operation, they have gradually been replaced by the scanning electron microscope (SEM).

The scanning electron microscope is used to directly observe the topography of the surface and variation in the friction process. The SEM image is clear and has a three-dimensional effect with a wide magnification range and the large detection range. It can be directly used to measure the friction surface of the small part.

It should be noted that the surface analysis mentioned above is limited before and after wear, and they cannot be used to examine the variation during the wear process. Wang and Wen [3] used the in-situ observation technique to study the variation of the dry friction surface during the wear process.

14.3.2 Atomic Force Microscope (AFM)

As shown in Figure 14.14, when two atoms are very close to each other, the repulsion force of their electron clouds is larger than the attraction forces of the nucleus and the electron clouds so the resultant force is the repulsion force. Otherwise, if two atoms are separated by a certain distance, the repulsion force

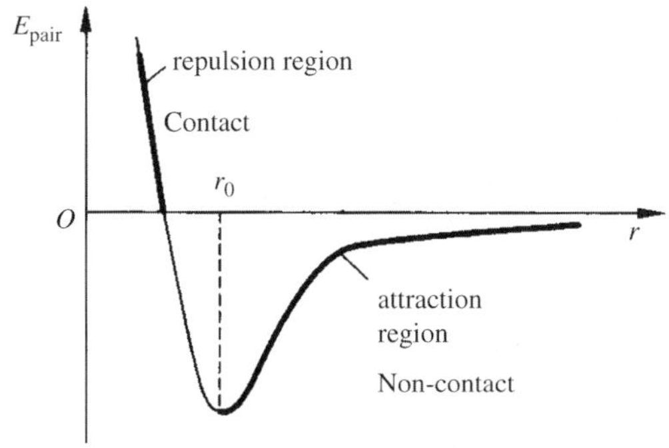

Figure 14.14 Contact energy vs. distance.

of their electron clouds is smaller than the attraction forces of the nucleus and the electron clouds such that the resultant force is the attraction force. This is the working principle of the AFM. Figure 14.15 shows the schematic diagram of AFM.

From the energy viewpoint, the relationship of the energy and distance between atoms can be verified by the Lennard-Jones formula as follows.

$$E_{pair}(r) = 4\varepsilon\left[\left(\frac{\sigma}{r}\right)^{12} - \left(\frac{\sigma}{r}\right)^{6}\right], \qquad (14.20)$$

where σ is the diameter of the atom; and r is the distance between the atoms.

With Equation 14.20, we can obtain that r_0 in Figure 14.15 is equal to $2^{1/6}\sigma$. And, by Equation 14.20, we can see that the relationship of the distance r and the energy E is determined.

The atomic force microscope works as shown in Figure 14.15. We can change the distance to measure the energy or keep the energy fixed to measure the distance between probe and the surface. With

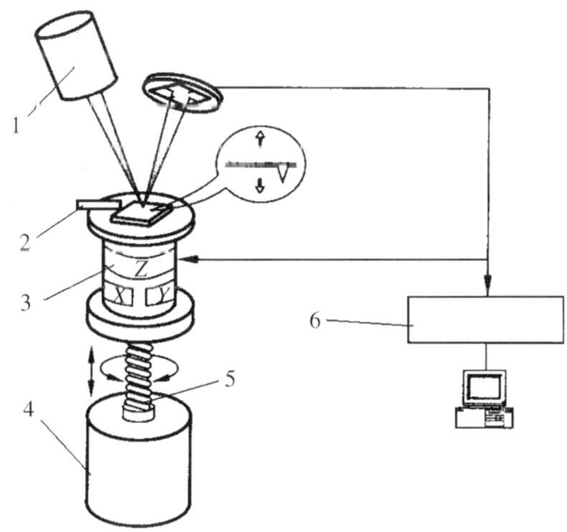

Figure 14.15 Schematic diagram of AFM: 1. laser; 2. probe lever; 3. piezoelectric ceramic tube; 4. stepping motor; 5. screw drive; 6. SPM controller.

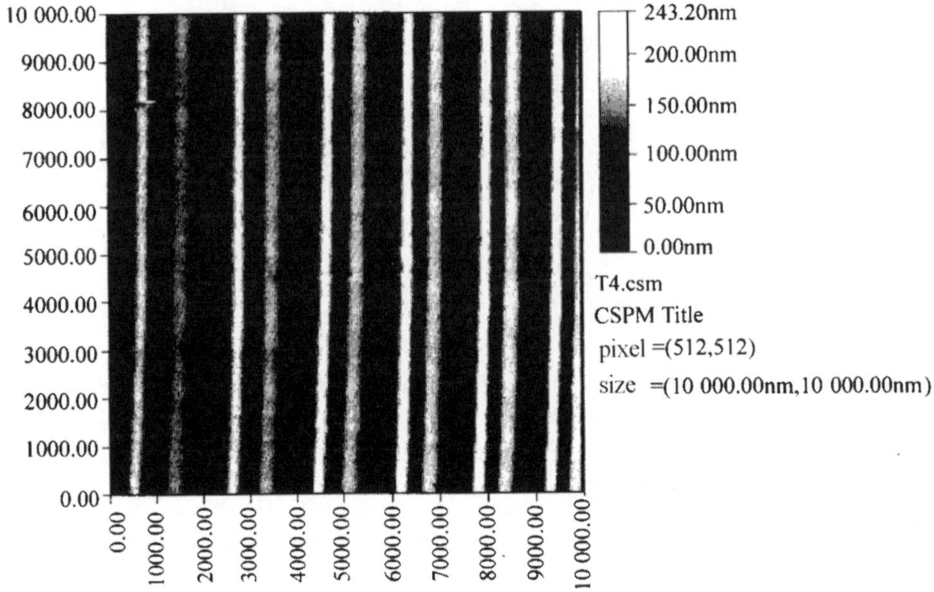

Figure 14.16 Optical disk surface topography measured with AFM.

the stepper motor to adjust the vertical distance which is the surface shape, we can obtain the surface profile. Figure 14.16 shows an image of the optical disc surface topography measured by the atomic force microscope.

14.3.3 Surface Structure Analysis

The variation of the metal surface structure in the wear process is commonly analyzed with the surface diffraction technique, which is to project an electron beam onto the wear surface. Because the metal atoms are arranged orderly in the crystal, the electron scatter in the specific direction form the diffraction spot. The distribution of the diffraction spots for the different atom and crystal is different so we can use this feature to analyze the surface structure and its variation.

The penetration ability of the electron diffraction is weak so the scattering thickness is only about 10^{-9}–10^{-10} m. The electron diffraction can be used to analyze the thin friction surfaces, such as metal adhesive wear and migration phenomena of the friction material.

The X-ray can produce the same diffraction spots as the electron beam and its penetration ability is strong such that its scattering thickness reaches up to 10^{-6}–10^{-4} m. The X-ray diffraction is often used to study the thick friction surface structure, or the lubrication mechanism of the additive in the metal wear.

As indicated above, if we study the material wear with an instrument such as optical microscope, scanning electron microscope, X-ray diffractometer and so on, the worn surface topography and wear particle can be observed, but such an instrument cannot be used to study the state of the subsurface of material. In a wear process, the sub-surface has undergone significant change. If we use these surface instruments to obtain the subsurface information, we must cut the sample to carry out the cross-section analysis. If so, the wear characteristics of the original sample may be significantly changed.

The acoustic microscope can reveal the subsurface structure of the material without cutting the sample so as to preserve the original characteristics of the subsurface. The basic principles and application of the acoustic microscope are introduced as follows [4, 5].

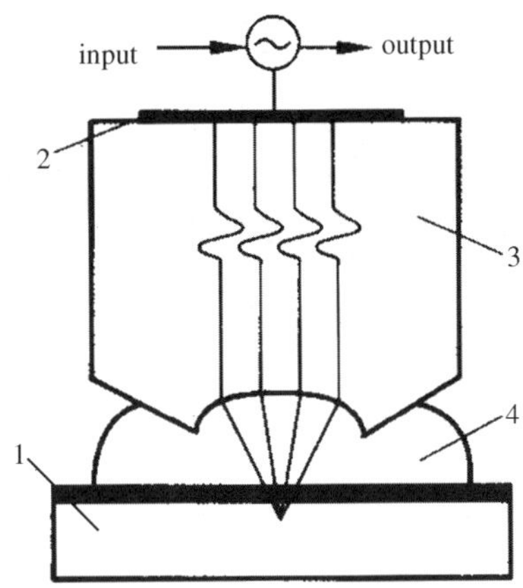

Figure 14.17 Schematic diagram of acoustic microscope: 1. sample; 2. energy transducer; 3. damping rod; 4. coupling liquid.

Figure 14.17 shows the basic principle of the acoustic microscope. It consists of four parts: the energy transducer-lens, signal detection circuit, mechanical scanning system and image display system. The energy transducer-lens is a key part, including a piezoelectric transducer, quartz damping rod and spherical acoustic lens. Its function is to motivate and focus the ultrasound. One end of the damping rod is polished and plated by a metal coating on which there is the piezoelectric transducer. The other end is grounded into a ball to form an acoustic lens. The high-frequency electric signal motivates the transducer to send off the ultrasonic wave in the longitudinal form in the damping rod. Then, it is focused on the acoustic lens and finally reaches the sample surface and interior through the coupling liquid (usually water). When the ultrasonic encounters the discontinuous medium or defects such as hollow, cracks, and inclusions, it is reflected and received by the transducer to be converted to the electrical signal by the detector. After amplification, the grayscale signal can be used to study the subsurface structure. By scanning the sample point by point and collecting the signal of each point, the scanning acoustic microscope image can be constituted.

There are two kinds of imaging modes for the acoustic scanning microscope: surface/subsurface imaging and internal imaging. Figure 14.18a shows the surface/subsurface imaging mode. When the incident angle of the sound wave is larger than the critical angle of the Rayleigh wave, a Rayleigh wave will be excited on the sample surface. Because the Rayleigh wave is a surface wave, its transmission depth is only one wavelength. In the transmission process, the energy will be sent off in the Rayleigh angle, also known as the leaked Rayleigh wave. In this way, the information of the subsurface of one

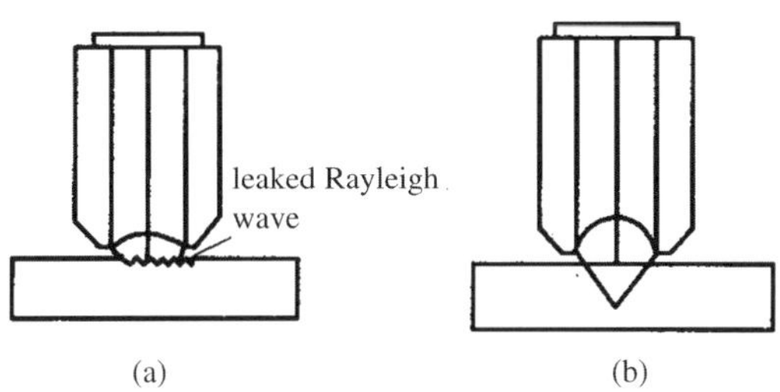

Figure 14.18 Acoustic microscope imaging mode.

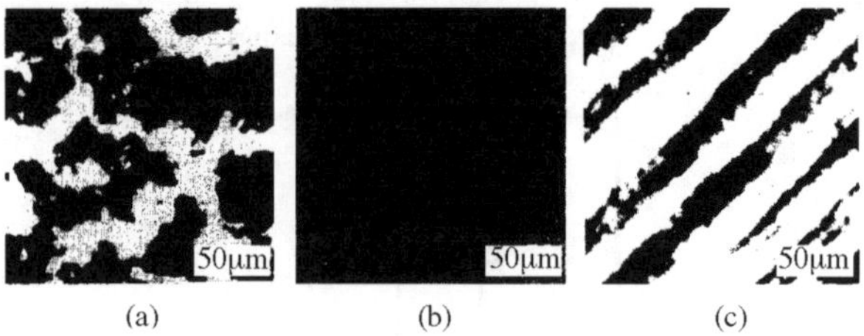

Figure 14.19 Three acoustic microscope images ($f = 150$ MHz, $z = -25$ μm, $A = 2.5 \times 2.5$ μm). (a) CoCrMoSi + Al_2O_3 + TiO_2; (b) CoCrMoSi + Fe_3O_4; (c) carbon steel.

Rayleigh wavelength deep beneath the surface has been obtained from the leaked Rayleigh wave. Figure 14.18b shows the schematic diagram of the internal imaging mode. In the internal imaging, a low frequency and large curvature lens are used. Without exciting a Rayleigh wave, the longitudinal wave is focused through the lens into a certain depth of the sample, and then the returning sound wave from the depth is received to form the image. Because frequency is limited, the resolution of the image is generally not very high.

Figure 14.19 shows the three acoustic microscope images in the subsurface coating obtained by Peng *et al.* [5]. Two kinds of areas, white and black, exist in the images. The white area is the metal alloy and the black is the mixture of the oxide and the alloy.

It should be pointed out that the acoustic microscope can observe the material subsurface structure with no destruction, but also some other means are needed for verification because it is still at an early stage. It can be expected that with the improvement, the acoustic microscopy will be an important tool for wear research.

14.3.4 Surface Chemical Composition Analysis

The chemical composition analysis is very important to the study of the mechanisms of friction and wear, because the chemical composition and distribution characteristics of the surfaces reflect the chemical reaction and the element migration between the surfaces. Common analysis methods are as follows.

14.3.4.1 Energy Spectrum Analysis

The electron beam, X-ray or vacuum ultraviolet light beam projects on the specimen surface to excite the Auger electron or photoelectron. By measuring and analyzing the energy, the surface chemical composition can be determined.

An Auger electron spectrometer projects an electron beam onto the surface to excite the secondary Auger electron. The Auger electron energy of one kind of element is different from those of the others so collecting the electrons and analyzing the energy spectrum will help us study composition variation in two or three atoms deep beneath the friction surface. If the Auger electron spectroscopy configures a scanning device, we will be able to quantitatively analyze a large range of chemical composition and element distribution of the surface.

An X-ray photoelectron spectroscopy is to use the photoelectron stimulated on the surface by X-ray to carry out the energy spectrum analysis. It is suitable for studying the formation of surface film as well as the influence of the additives.

14.3.4.2 Electron Probe Micro-Analysis (EPMA)

Under the action of the electron beam on the specimen surface, any element emits the corresponding ray spectrum. According to the different wavelength and intensity of the ray, we can determine the composition and the content of the surface. Usually, an electron micro-probe and scanning electron microscopy can be combined into a scanning electron micro-probe, which is not only used to determine the composition within the micro-meter range to the fixed point, but can also scan the surface to analyze the surface element distribution.

It should be noted that with development of the electronic technology and the ultrahigh vacuum technology, more multi-functional measuring instruments can be developed by combining a variety of surface analysis techniques. To analyze simultaneously the surface topography, structure and chemical composition will play a very important part in studying the friction and wear mechanisms.

14.4 Wear State Detection

In the large electricity generating set or mechanical system, the detection of the operation of the key friction and wear parts is often required to predict their working conditions in time and take effective maintenance measures to prevent sudden damage or accident [6, 7].

Usually, a physical or chemical detection method can be used periodically or continuously to show the wear state of the machinery and equipment. Commonly used detection methods are as follows.

14.4.1 Ferrography Analysis

The ferrography analysis is the separation and analysis technology of the ferro wear debris contained in the lubricating oil. The ferro wear debris forms in the most mechanical abrasion process. According to the size, shape and composition of the debris, we can determine the wear form, wear stage and wear location.

As shown in Figure 14.20, a small amount of lubricating oil is extracted from the lubricating oil circulatory system. The oil sample flows through the spectrometer at a low and stable velocity and is limited to a narrow strap in the slope glass center. Because the distance from the magnetic pole to the inlet is slightly shorter than the outlet, the magnetic field gradient gradually decreases downwards. Therefore, under the actions of the viscous force and the magnetic field, the larger ferro debrises deposit first at the front of the glass and the smaller ones deposit at the rear. When a certain amount of oil has flowed through the glass, cleaning to remove the excess oil and fixing the debris, we can produce a ferrogram for detection.

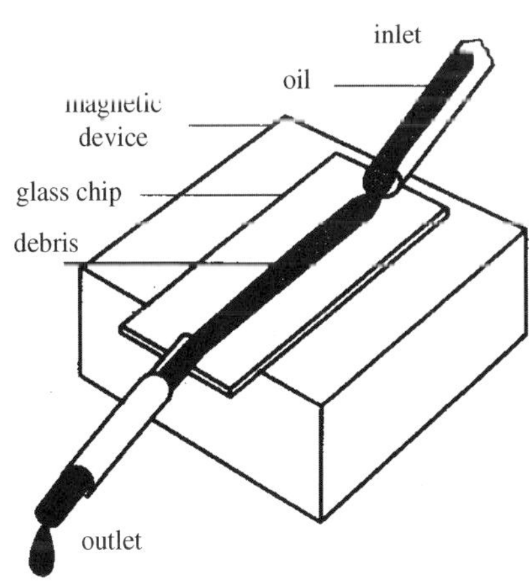

Figure 14.20 Ferrography measuring principle.

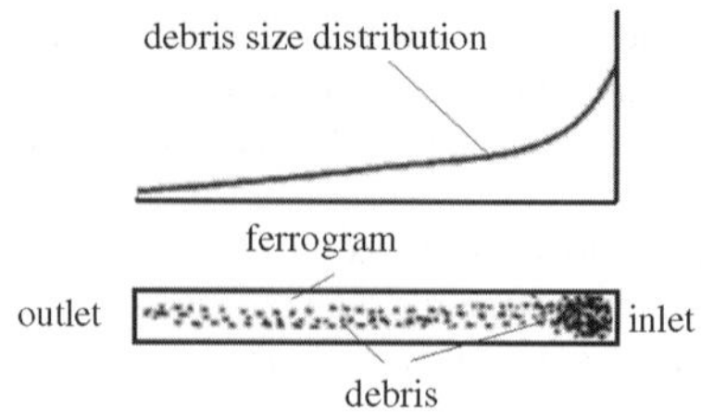

Figure 14.21 A typical ferrogram.

Figure 14.21 is a typical ferrogram, on which the size of the debris deposited decreases gradually along the flow direction. In a typical case, the total weight of the deposited debris is about 10 μg. To detect the density of the wear debris with the ferro spectrometer, we can estimate the distribution of the debris.

Usually, a large number of testing results can be compiled as the standard wear debris image database to be used as reference during ferro spectrum measurement. Using an optical microscope to observe the ferrogram obtained from the ferro spectrometer and comparing it with the images in the database, we can determine the wear state according to the shape and distribution density of the debris. For example, the normal wear debris is flaky; the abrasive or furrow wear forms the spiral or curl debris while the spherical wear debris is often generated by the surface contact fatigue wear. For oxidation wear or corrosion wear, the wear debris is the compound, which presents a different color under a colorful light. In addition, the concentration of wear debris shows the severity of the wear. Figure 14.22 gives some different shapes of the debris observed under optical microscopy.

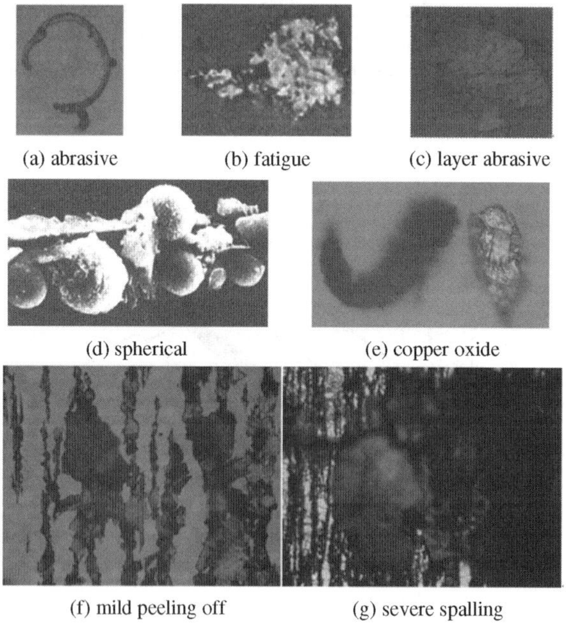

Figure 14.22 Ferrograms of wear debris.

14.4.2 Spectral Analysis

The spectral analysis uses the property that an atom of any substance emits the typical spectra under certain conditions to analyze metallic compositions in a lubricant.

Usually, the atom is in a stable state. When energy increases to a certain level, it is excited. The excited atom is in an unstable state for a very short period. After excess energy has been released in the form of the optical spectrum, the atom returns to the stable state. Because the atom of one substance sends off a spectral signal different from that of another, by spectral analysis we can determine the contents of the metal in the lubricating oil so as to predict the wear state. The spectral analysis method is usually used to detect the wear of the locomotive, marine diesel and airplane engine to prevent them from failure.

14.4.3 Lubricant Composition Analysis

The lubricant composition analysis is to use the physical or chemical analysis equipment to detect oil acidity, additive concentration, and concentration and composition of insoluble material. It is a useful wear state analysis method.

14.4.4 Mechanical Vibration and Noise Analysis

The vibration or noise measurement is an important way to determine the wear state of the mechanical equipment because it can continuously work during operation. With help of the spectrum analysis on the measured vibration or noise signals, we can predict the emergence of serious wear.

This indirect detection method is effective for the low-noise component, such as the journal bearing. However, for the gear drive because the vibration and noise are large, it is often difficult to distinguish normal vibration or vibration produced by severe wear that affects the reliability of the detection.

14.4.5 Lubrication State Analysis

For full-film lubrication, sensors can be used to measure film thickness between friction surfaces, friction coefficient, contact condition, surface temperature and other parameters to determine the lubrication state.

If we use electrical measurement techniques, they can usually measure and even automatically control the lubrication system continuously. For example, when the film thickness or temperature reaches a critical value, the automatic control device will adjust the working parameters to ensure normal operation or shut down the machine to prevent an accident.

14.5 Wear Failure Analysis

Wear failure analysis is very important to practical production. In order to accurately determine the cause of wear failure and make a quick response decision, a wide range of expertise and experience are needed. Here, the general ways of the wear failure analysis are briefly introduced.

In order to determine the cause of wear failure, the following steps can be followed.

14.5.1 Site Investigation

Collect information on site as much as possible to understand the failure process and the situation. The investigation mainly includes the following four areas.

1. Find the worn part and its drawing. If possible, collect the same new part so as to easily carry out analysis and comparison.

2. Find out the working conditions of the wear failure component, including the load, velocity, working temperature and other parameters as well as the destructive process and location.
3. Check the lubricant supply system and its technical performances.
4. Collect the maintenance and operation specifications.

14.5.2 Lubricant and its Supply System

It is important to inspect the following aspects.

1. Check the type and performances of the lubricant; the oil viscosity and the viscosity-temperature property; check the oil ash content after combustion, and the contents of Zn, Ba and Ca in the used additives. If necessary, use spectral analysis to determine the chemical compositions of the lubricant and additives.
2. Find the deterioration and pollution levels of the lubricant. Check the acidity or alkalinity of the lubricant, as well as the contents of the insoluble material, water, ethylene glycol and pollutants. Check whether the lubricant contains the excessive amount of the large or unusual wear debris. Check the lubricating oil replacement period. If necessary, carry out ferrography analysis.
3. Find out the working condition of the lubricating oil supply system, including the work performances of the pump, filter and system rated flow.

14.5.3 Worn Part Analysis

1. Analyze the initial damage location and the failure developing process of the worn part.
2. Identify the main wear form. With the optical microscope, observe the friction surface. According to the wear feature, determine the wear form. By cutting a cross section, observe the fatigue crack and surface structure variation. Because a large part is inconveniently analyzed in the laboratory, use the special polymer to copy the worn surface of the damage part for observation and analysis in the laboratory.
3. Perform reasonable analysis of material selection. With energy spectrum technology, analyze the chemical composition and contents of the material. Compare the mechanical properties, such as the topography and hardness before and after failure. Use the cut cross section to observe the size and distribution of the nonmetallic inclusion.

14.5.4 Design and Operation

Analyze whether the design of the failure component is reasonable. For full-film lubrication, check film thickness, film thickness ratio, lubricant flow rate, surface pressure and temperature. Find the variations of the load, speed, temperature and vibration, noise before and after failure. The above observation and analysis will help us determine the failure cause and the main wear type. In general, the most common causes of wear failure are as follows:

1. poor choice of material or lubricant;
2. the lubricating oil supply system does not work properly, or the lubricant has deteriorated and is contaminated;
3. the working conditions are over the extent of the parameters permitted by the design;
4. the manufacturing and installation errors cause bad working conditions.

Tables 14.1 and 14.2 are statistics tables provided by the Shell company after years of investigation and study, and sum up the wear failure modes of the parts.

Table 14.1 Statistics of failure modes

Failure mode	Amount
Corrosion (electric corrosion, weak organic acid corrosion, low temperature corrosion)	408
Sedimentation	266
Wear	146
High temperature corrosion	136
Fatigue fracture	155
Surface contact fatigue wear	149
Adhesive wear, abrasive wear	83
Fracture	102
Erosion	70
Cavitation	41
Abrasion	40
Spalling	26
Melting and softening	24
Fretting	18
Others	55
Total	1719

Table 14.2 Statistics of failure parts

Parts	Amount
Journal bearings, rolling bearings	320
Cylinder and piston ring	241
Combustion device components (such as an exhaust valve)	150
Marine device	110
Fuel oil system parts	73
Boiler, heater and pipe	72
Turbine blade	64
Gear drive	55
Fuel haul and storage equipment	41
Framework and basement	38
Lubrication system Parts	31
Compressor and turbocharger	30
Cooling system parts	15
Others	138
Total	1378

References

1. Huang, P., Luo, J.B., Zou, Q., and Wen, S.Z. (1994) NGY-2 Interferometer for nanometer film thickness measurement. *Tribology*, **14** (2), 175–179.
2. Huang, P., Luo, J.B., Zou, Q., and Wen, S.Z. (1995) Investigation into measure nanometer lubrication film thickness by relative light intensity principle. *Lubrication Engineering*, **1**, 32–34.
3. Wang, W.Q. and Wen, S.Z. (1993) *In situ* observation and study of the unlubricated wear process. *Wear*, **171**, 19–23.
4. Wang, Y.L., Jin, Y.S., and Wen, S.Z. (1989) The inspection of the sliding surface and subsurface of plasma-sprayed ceramic coating using scanning acoustic microscopy. *Wear*, **134**, 399–411.

5. Peng, H.T., Jin, Y.S., and Yang, Y.Y. (1999) Study of wear mechanism of materials by SAM. *Tribology*, **19** (4), 294–298.
6. Fan, J.C., Zhou, H.S., He, P. *et al.* (1997) Tribology application in mechanical condition monitoring. *Lubrication Engineering*, **44**, 13–14.
7. Huo, Y.X., Chen, D.R., and Wen, S.Z. (1997) Monitoring of the wear condition and research on the wear process for running equipment. *Tribology Transactions*, **40** (1), 87–90.

Part III

Applied Tribology

15

Micro-tribology

Micro-tribology, also known as nano tribology or molecular tribology, studies the behavior and damage of the friction interface at the molecular scale [1].

The basic disciplines, research methods and test equipment of micro-tribology are quite different from those of macro-tribology. The main experimental instrument of micro-tribology is the scanning probe microscope, which includes the atomic force microscopy (AFM) and the friction force microscope (FFM). In theoretical research, macro-tribology is usually based on the response of bulk properties to the friction interface to characterize tribological behaviors. It analyzes the tribological problem with the continuum mechanics, including fracture mechanics and fatigue theory. However, micro-tribology starts with the atom or molecular structure to study the tribological behaviors of the nano scale surface and interface molecular layer. Its theoretical basis is surface physics and surface chemistry; its theoretical analysis method mainly depends on the molecular dynamics simulation, and its experimental test instrument is the scanning probe microscopy. In this chapter, micro-friction, micro-contact, micro-adhesion and micro-wear phenomena are introduced. In addition, molecular film lubrication will also be discussed.

15.1 Micro-friction

15.1.1 Macro-friction and Micro-friction

Bhushan and Koinkar [2] used the ball-plate friction tester and the FFM to study macro- and micro-friction coefficients experimentally. The results are shown in Table 15.1. The macro-friction coefficient was measured with a 3 mm diameter aluminum ball and plate specimen under a relative sliding velocity of 0.8 mm/s and the load of 0.1 N corresponding the Hertz stress of 0.3 GPa. The micro-friction coefficient is determined with the FFM on the sliding sample. The probe material is Si_3N_4, tip radius is 50 nm, sliding velocity is 5 μm/s, scanning area is 1 μm × 1 μm, and the load is between 10 nN and 150 nN corresponding to the Hertzian stress of 2.5 GPa–6.1 GPa.

The results of Table 15.1 show that the micro-friction coefficient is far lower than the macro-friction coefficient. Bhushan *et al.* believed that because the hardness and elastic modulus in the micro-scale are higher under very light load conditions than those of the macro, in the micro-friction process material wear is very little so the friction coefficient is low. Furthermore, in micro-friction, less particles are embedded in the surface. This also reduces the frictional force of the furrow. Their experiments also demonstrated that when the load increases, the micro-friction coefficient and the micro-wear significantly increase. This shows that the conclusion that the friction coefficient is independent of the contact area in the classic

Principles of Tribology, First Edition. Wen Shizhu and Huang Ping.

Table 15.1 Macro- and micro-friction coefficients

Specimen material	Roughness Ra/nm	Macro-friction coefficient	Micro-friction coefficient
Si(111)	0.11	0.18	0.03
C^+ – into Si	0.33	0.18	0.02

Amontons formula is not suitable for micro-tribology. They also pointed out that under a light load, a very low friction and zero wear may be possibly achieved on the small sliding surface.

15.1.2 Micro-friction and Surface Topography

With the FFM, Ruan and Bhushan [3] carried out the sliding friction experiments on the new split surface of the highly oriented pyrolytic graphite (HOPG) with the purity of 99.99%. Their experiments showed that when the probe slides over the HOPG surface, the variation of the atomic scale frictional force corresponds with the surface topography and has the same periodical variation. However, the variation of the peak location of the frictional force is of some displacement relative to the roughness peak location, as shown in Figure 15.1.

In the left picture of Figure 15.1a, the grayscale image of the surface topography of the new split HOPG area is given in the 1 nm × 1 nm area. The right one is the grayscale image of the frictional force variation within the same area. Figure 15.1b is the grayscale image overlapping the surface topography and the frictional force, based on Figure 15.1a. The figure shows that the location of the frictional force peak has a displacement away from the location of the corresponding the roughness peak. Analysis showed that the displacement is caused by the slope of the roughness peak.

The surface topography also makes the micro-frictional force significantly anisotropic, that is, the sliding frictional force to the different direction is different. The experimental results are listed in Figure 15.2, illustrating the directional properties of the frictional force. Figure 15.2a is plotted based on the distribution of the grayscale images HOPG frictional force, while Figure 15.2b and c are the variation of the average frictional force along *AA* and *BB*, respectively. Obviously, the frictional force along *AA* is larger than that along *BB*. Sometimes, the directionality of the frictional force can also be observed in the macro-sliding.

Ruan and Bhushan, according to that the frictional force and the topography shown in Figure 15.1 have the same period and the similar characteristics, proposed the "ratchet" model of micro-friction. They believed that the probe slides along the substrate surface similar to the movement of a pawl ratchet tooth edge. Therefore, the asperity slope is a key factor of the friction coefficient.

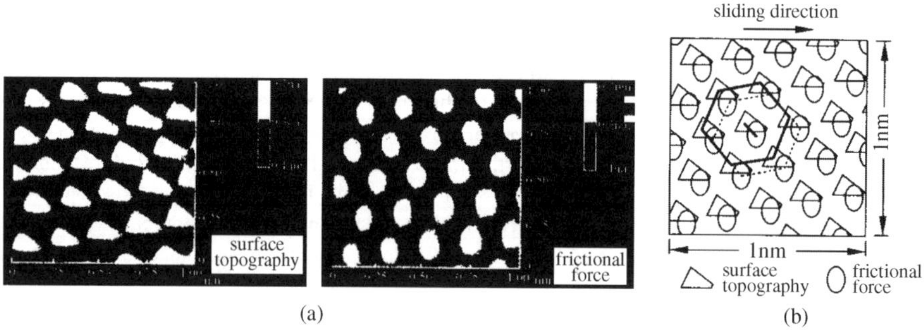

Figure 15.1 Micro-friction images of HOPG [3].

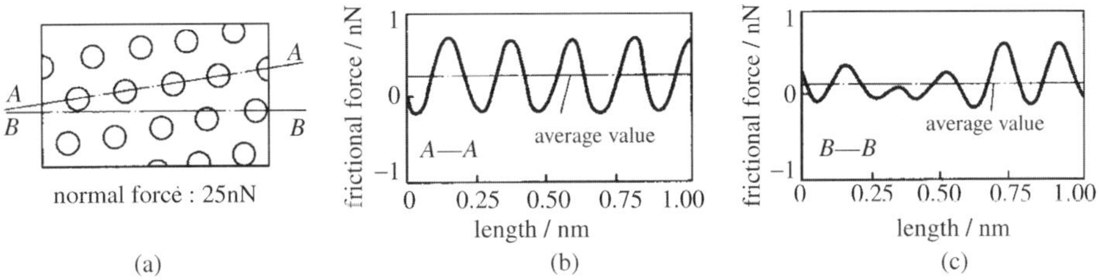

Figure 15.2 Directionality of frictional force [3].

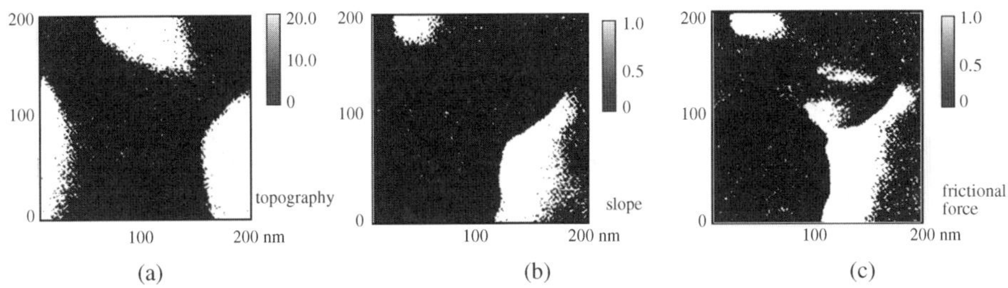

Figure 15.3 Micro-friction on single-crystal diamond [3].

It has been evidenced that the micro-frictional force is related to the roughness peak slope by the FFM friction experiments using a Si_3N_4 probe sliding on the HOPG substrate or on the single crystal diamond substrate. Figure 15.3 shows the grayscale maps of the roughness peak height, the slope and the frictional force distribution on the single crystal diamond surface in the 200 nm × 200 nm area. The diamond surface is polished and the load in the experiments is 50 nN. From the figure we can see that the frictional force distribution and slope distribution of the roughness are almost the same, but they are not related to the peak height.

Ruan and Bhushan [4] using the same method studied the relationship between the frictional force and topography on the pyrolytic graphite HOPG with the Si_3N_4 probe. The probe, under the load of 42 nN, slides on the graphite surface at the velocity of 1 μm/s. Figure 15.4 shows the relationship between the distributions of the frictional force and the surface roughness height distribution within the 1 μm × 1 μm area. It can be seen from the figure that the main part of HOPG substrate is atomic-scale smooth and the

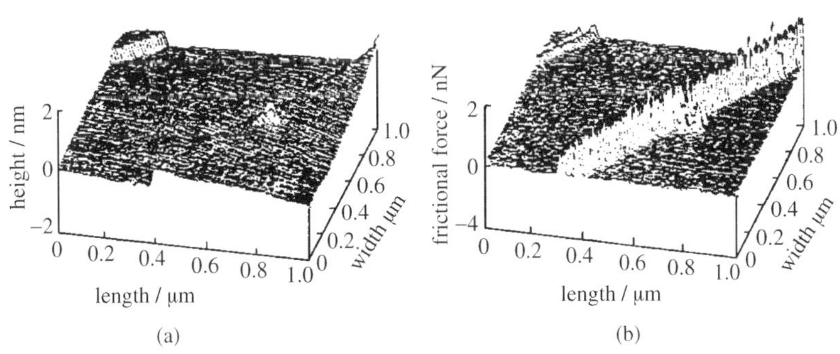

Figure 15.4 Micro-topography and frictional force of HOPG [4].

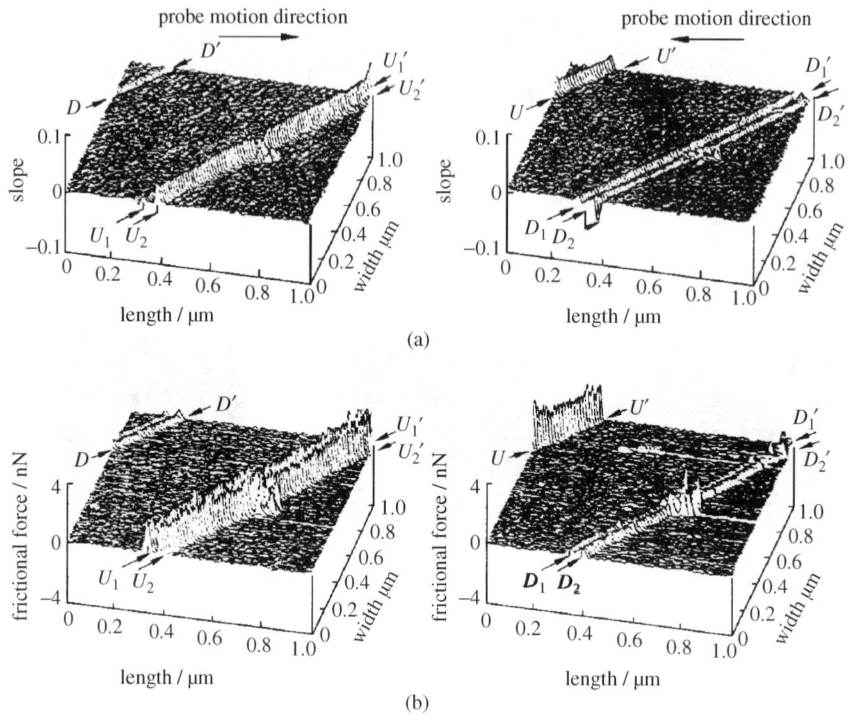

Figure 15.5 Roughness slope and frictional force in strip region [4].

friction coefficient is very low. However, the friction coefficient in the region with the strip surface topography is very high.

In order to investigate the influence of the strip region on the frictional force as well as the applicability of the ratchet model, Ruan and Bhushan further studied the relationship of the roughness slope and the frictional force with the probe sliding reciprocally, as shown in Figure 15.5.

Figure 15.5a gives the roughness peak slope distribution as the probe moves forwards and backwords. Figure 15.5b gives the frictional force distribution. Comparing the left two figures, we can see that in the negative slope of the strip region DD', the frictional force is higher than that in the smooth region, while in the positive slope region U_1U_1' and U_2U_2', the frictional force is even higher. It can also be seen that the slope along U_1U_1' and U_2U_2' varies significantly, but the frictional force is almost the same. Similar phenomena exist in the right two figures while the probe slides in the opposite direction. In the positive slope stripe UU' area, the frictional force is large, but in the negative slope stripe D_1D_1' and D_2D_2', the frictional force is slightly higher than that of the smooth region. In addition, although in D_1D_1' and D_2D_2', the difference of the slopes is significant, the frictional forces are nearly equal.

The experimental results of Figure 15.7 show that it cannot simply use the slope to characterize the frictional force for the roughening treated and complex topography, particularly for the strip topography. By the electron microscopy analysis on the strip region, it is shown that there are two reasons for the increase in the frictional force. One is that with increase in roughness height, most of the directions of the surface lattice change, not only the direction of the (0001) plane. Another is that the amorphous carbon appears in the strip region after the roughening treatment. Therefore, the change of the roughening structure causes the increase in the frictional force.

In summary, the frictional force sliding on the graphite material is closely related to the crystal lattice direction. Along the (0001) plane, the frictional force is the lowest. Studies have shown that other materials also have a similar feature, that is, have obvious anisotropic characteristic in micro-friction.

15.1.3 Plowing Effect and Adhesion Effect

15.1.3.1 Plowing Effect

Micro-friction is the friction between the molecular smooth surfaces under a very light load. In such a condition, some new problems existing in adhesion theory need to be studied, such as plowing effect, material transformation behavior, zero or negative load friction mechanism, origination of interfacial friction without adhesion and so on.

In the friction model of the adhesion theory proposed by Bowdon and Tabor as mentioned in Chapter 10, when a hard asperity slides in a soft surface, the frictional force includes the force to push the front material away and the force to separate adhesion on the contact area. If we consider the plowing effect, the adhesive force is usually excluded in measuring the friction coefficient μ_p which shows the strength of the plowing effect. Analysis shows that the plowing friction coefficient produced by a spherical asperity depends on the spherical radius and pressured depth. The friction coefficient of a cone asperity is only related to the cone vertex.

Guo *et al.* observed that the plowing force fluctuates over time by using a cone probe sliding on a sodium chloride substrate [5]. The material in front of the probe moves non-uniformly and this is also related to the fluctuating plastic deformation of the material. This is the important characteristic of the plowing of the ductile material and is one factor to cause sliding friction instability.

Guo *et al.* also carried out the plowing experiments with a diamond cone probe sliding on a hard and diamond-like carbon coating. They found that when the load is low, the coating is in the plastic deformation. When the load is high, the surface appears to have many intermittent small fractures and the frictional force drops suddenly when the fracture occurs. Figure 15.6 shows the variation of the frictional force during sliding. The symbol *A* in the figure indicates that the micro-fracture occurs. They are corresponding to the sudden drops in frictional force.

To sum up, in the plowing the mechanical behaviors of different materials are different. The ductile material produces plastic fluctuation deformation, while the brittle material produces intermittent micro-fracture. All these result in the variation of the plowing effect.

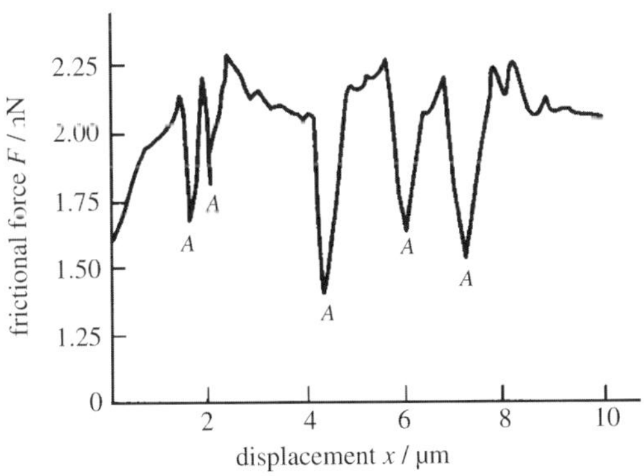

Figure 15.6 Ceramic brittle fracture [5].

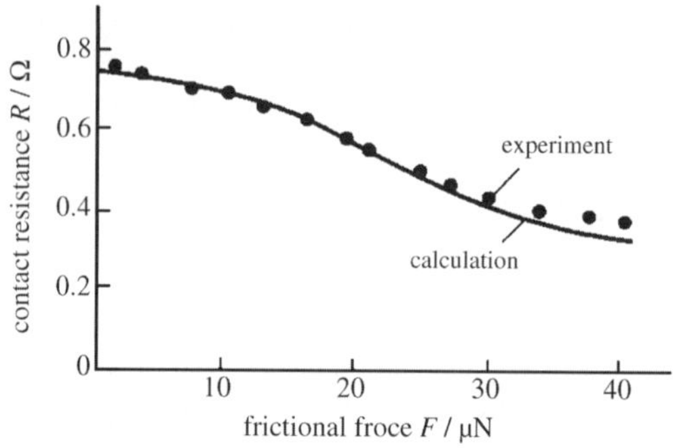

Figure 15.7 Friction under zero load [5].

15.1.3.2 Adhesion Effect

Guo *et al.* found the toughness behavior and stick-slip phenomenon when they used the tungsten probe and the gold substrate to study the micro-friction under high-vacuum conditions with a friction force microscope [5]. They analyzed the relationship of the frictional force and the contact resistance according to the revised adhesion friction model proposed by Bowdon and Tabor. Figure 15.7 shows the calculated and measured results which are well in agreement. In the figure, the frictional force is inversely proportional to contact resistance. Therefore, with increase in frictional force, the contact area increases. Because the load is equal to zero during sliding, there is adhesive resistance on the contact area so the frictional force is generated by the adhesive effect.

In the static contact, the adhesion force between the surfaces is often more important than the external load. Figure 15.7 shows that the adhesive force is greater or no less than the external load during sliding. Experiments show that even the external load is negative, for example, $P = -0.6$ μN, the sliding frictional force also exists. This is a very important characteristic of micro-friction.

In order to reveal the friction mechanism under zero or even a negative load, the friction pair materials should be selected to be of an appropriate adhesive strength and not easy to produce the surface damage or material transformation as well as the slip occurring on the interface. Pivin *et al.* carried out some friction experiments under the load $W = -0.9$ μN in the high vacuum environment by using an iridium (Ir) probe and the common-valence compound Ni_3B substrate to compose the friction pair [6]. Ir is a hard metal material. Ni_3B is a single crystal and its surface is very smooth with the electrical conductivity. Therefore, contact resistance can be measured to determine the contact area. Figure 15.8 shows the variations of

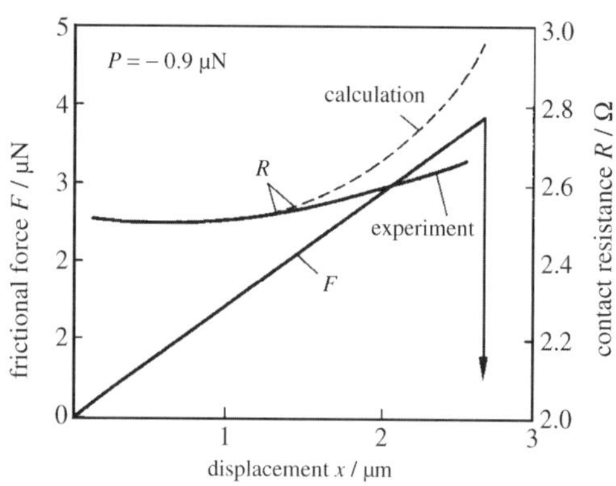

Figure 15.8 Friction under negative load [6].

frictional force and contact resistance with the sliding displacement in the adhesion process, where contact resistance is inversely proportional to the contact area.

The experiments show that the stick-slip phenomenon is significant when the load is negative. Figure 15.8 is the variations of the frictional force and the contact area in the adhesion process. The frictional force increases, but the contact area reduces. However, the experimental results of Figure 15.7 and Figure 15.6 are contrary, which show that with increase in frictional force, the contact area increases.

15.2 Micro-contact and Micro-adhesion

15.2.1 Solid Micro-contact

15.2.1.1 Zero Load Contact

In 1981, Pollock found that some combinations of materials may contact and deform under zero load condition according to the micro-contact experiments in the high vacuum with a probe and substrate [7]. Figure 15.9 shows the relationship of the contact area and the load of different combinations of materials. The abscissa in the figure is the external load *P* and the vertical coordinate is the contact resistance *R* which is inversely proportional to the radius of the contact point.

In Figure 15.9a, the combination of materials is the tungsten probe and the gold substrate. The materials are purified by the ion erosion and are annealed near the melting temperature. *A* and *B* in the figure are the results under a very small or the zero load conditions, where contact resistance rapidly decreases and the contact area increases quickly. This is because the surface adhesion causes the surface contact and plastic deformation. In Figure 15.9b, the combination materials are also the tungsten probe and gold substrate. Before the experiment, they are pre-exposed to the oxygen so that the surfaces are contaminated. Therefore, the adhesion contact at the zero load is not obvious. In Figure 15.9c, the combination materials are the tungsten probe and the Ti_4O_7 substrate. The experiments show that there is no contact under the zero load and the load to separate the contact is very small, indicating less adhesion.

The above experiments show that the interfacial adhesion energy and the surface force significantly influence contact and deformation. The adhesion energy of a material combination gives the corresponding contact state. Therefore, when studying surface contact and deformation, these influences must be considered, especially in nano-tribology.

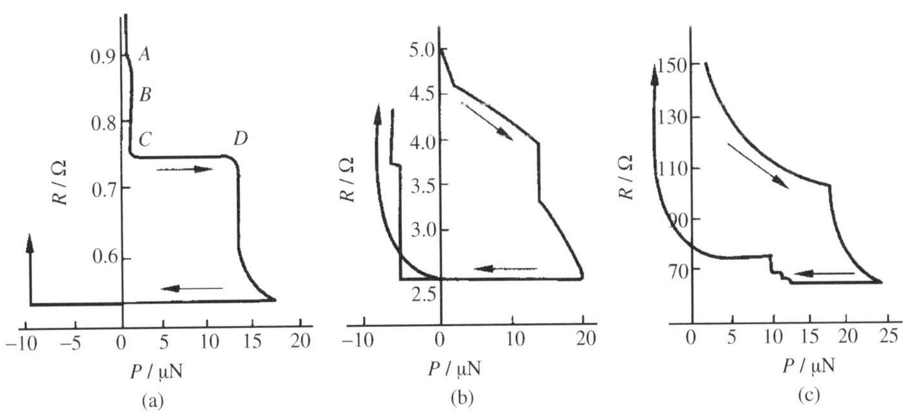

Figure 15.9 The relationship of contact area and load [7].

15.2.1.2 Elastic, Elastic-Plastic and Plastic Contacts

Pollock *et al.* applied the concepts in macro-mechanics, such as the hardness, elastic-plastic, and toughness, to micro-contact analysis to derive the approximate formula [5].

The surface force S can be equivalent to a load. According to geometric simulation and elastic simulation, the contact of the two rough surfaces can be equivalent to the contact of a rough and elastic hemisphere with a smooth and rigid plane. In convenience, usually a single asperity contact is analyzed. If an elastic hemisphere with the radius r contacts with the rigid plane under the load P, the total load $(P + 2S)$ will produce a contact circle with the radius a. According to the Hertz elastic contact theory, the formula for the elastic contact can be modified as follows:

$$P+2S=K\frac{a^3}{r}. \tag{15.1}$$

It should be pointed out that Equation 15.1 is similar to the elastic contact formula proposed by Johnson, Kendall, and Roberts in 1971 which is commonly referred to as the JKR formulas to consider the surface energy. Equation 15.1 shows that surface adhesion influences the contact area. It is larger than that in classic Hertz elastic contact theory. This means that all the friction and wear properties relevant to the contact area will be increased accordingly.

Subsequently, according to the contact of the hemisphere and the plane, the stress field can be calculated when the contact center axis begins to appear in the plastic deformation. The approximate relation is

$$P+1.5S=1.1\pi a^2\sigma_e, \tag{15.2}$$

where σ_e is the elastic yield stress.

When the contact force P satisfies with Equation 15.1, the contact state enters the elastic-plastic.

If we have not considered the influence of adhesion energy, it is derived from the classical contact theory that when the plastic deformation zone extends to the whole contact surface, that is, in fully plastic contact, the contact radius a_p is approximately equal to

$$a_p \approx 60r\frac{\sigma_e}{E}. \tag{15.3}$$

When we consider the influence of the adhesion energy, the plastic contact formulas according to the classic plastic theory will be

$$P+2\pi wr=\pi a^2 H, \tag{15.4}$$

where H is the hardness; and w the adhesive energy.

When Equations 15.3 and 15.4 are satisfied at the same time, the contact state is fully plastic. It can be seen from the above formula that even though $P = 0$, that is, the zero load condition, the existence of the adhesion energy w can also produce the plastic deformation on the contact surface.

15.2.2 Solid Adhesion and Surface Force

15.2.2.1 Solid Adhesion Phenomena

One of the most influential studies on the micro-mechanism of the solid adhesion was carried out by Landman, Luedtke *et al.* in 1990 [8]. They used the molecular dynamics simulation to study the normal approach and separation processes of the hard nickel probe and the soft gold substrate. The results showed that when the probe moves slowly down to 0.4 nm above the substrate surface in the quasi-steady manner, the probe begins to appear unstable. At the same time, the gold substrate gradually blows up under the action of surface force. Subsequently, the gold atom in the crystal suddenly jumps up to the nickel-probe in a very short time, in 10–12 s, and the jumping distance of the gold atom is about 0.2 nm. Then the two surfaces contact adhesively, that is, a single gold molecular film forms on the probe surface. The phenomenon is because the probe and substrate surface have different surface energies similar to liquid wets on a solid surface.

When the probe continues to move downwards to the gold substrate, the adhesive gold atoms on the probe surface gradually increase. Because the lattice of the gold substrate generates more slip and defect, the elastic deformation of the gold substrate transforms to the plastic deformation.

When the probe moves upwards, the substrate material connected to the probe begins to become tensile and necking occurs. It significantly appears as a plastic flow and there is material transformation. Thus, a gold wire is formed to connect the probe and substrate in the atomic-scale. The connecting wire remains in the crystalline state. Finally the connecting wire breaks such that the probe and substrate are completely separated. After separation, the gold substrate appears to have surface damage, while the nickel-gold probe has some gold material.

15.2.2.2 Adhesion and Surface Force

As shown in Figure 15.10, when the surface a of Solid A and the surface b of Solid B adhesively contact, they constitute the interface ab. If an external force is exerted to separate the adhesive interface and move them to infinity, the required power per area is defined as adhesive work w or Dupré adhesion energy.

According to surface physics theory, we have

$$w = \gamma_a + \gamma_b - \gamma_{ab}, \tag{15.5}$$

where γ_a and γ_b are the free energies of the two solid surfaces a and b respectively.

Free energy is defined as: the work or energy needed to increase the unit area of the surface. It can also be understood to be the work or energy to move the molecule of the unit area from the inside body to the surface. Thus, it can be known that the energy of the molecule at the surface is higher than that in the body. γ_a or γ_b is the interfacial energy, which can be defined as the work or energy required to increase each unit interface. It can also be understood as the energy or work required to force the molecules of the unit area on the surface a across the interface to the surface b. Therefore, γ_a and γ_b are equal.

In fact, the Dupré adhesion energy expresses the work to overcome the attraction of the two surfaces during separating the adhesive surfaces. If the total adhesion energy of the contact surface is E, the pulling force is P and the displacement is δ, the surface force S is equal to

$$S = -\left(\frac{\partial E}{\partial \delta}\right)_P. \tag{15.6}$$

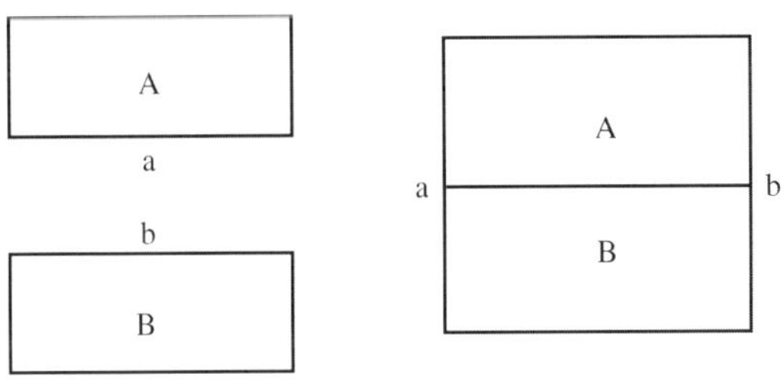

Figure 15.10 Solids contact and separation.

The engineering surface contact is generally considered as the roughness contact. By using the geometric simulation, the two contact surfaces can be equivalent to the contact of an elastic sphere with a rigid plane. The contact area is a circle, which is the axisymmetric problem. In the nano-tribology, the atomic-scale contact problem is also axisymmetric. The area of a single contact point is usually 0.01 μm^2. Therefore, in the scanning probe microscopy experiment, the metal or the ceramic probe tip radius should be 0.3 μm, and a load of 100 nN is applied.

If the contact radius is a, the adhesion energy $E = \pi a^2 w$. According to the energy balance condition, for an axisymmetric contact problem, Equation 15.5 can be used to calculate the surface force. Therefore, we have

$$S = \left(\frac{3}{2}\pi w K a^3\right)^{1/2}, \tag{15.7}$$

where K is the equivalent elastic constant.

In Equation 15.7, the adhesion energy w can be obtained according to the potential function at the interface of the two solid-phases A and B. It is equal to the work required to move the molecules on the surfaces a and b individually from the contact surface to the infinity. The interface potential function $\varepsilon_{ab}(r)$ of the two phases at the interface satisfies the geometric law, that is

$$\varepsilon_{ab}(r) = [\varepsilon_a(r) \cdot \varepsilon_b(r)]^{1/2}, \tag{15.8}$$

where r is the coordinate; $\varepsilon_a(r)$ and $\varepsilon_b(r)$ are the potential functions of the two solid surfaces a and b, respectively.

Then, the relationship between the adhesion energy and surface free energy can be obtained as follows:

$$w = 2\varphi(r_a r_b)^{1/2}, \tag{15.9}$$

where φ is a constant.

If the adhesion energy is obtained, the surface force can be determined from Equation 15.6.

15.3 Micro-Wear

15.3.1 Micro-Wear Experiment

The micro-wear is the surface damage process under a very light load on the molecular layer. The wear depth is usually in the nanometer scale so sometimes it is also called the nano-wear. The main micro-wear experiments use the specially developed nano-wear testers, such as the atomic force microscope (AFM), the friction force microscope (FFM) and so on. The formation of the micro-wear is because when the cone-shaped probe slides, a normal load is applied on the substrate surface of the testing material. The relative sliding manner is the combination of the vertical and horizontal scanning steps, which is composed of a two-dimensional wear plane. That is, after the probe moves a certain length along the longitude on the surface, it moves a small step along the lateral direction to repeat the vertical sliding.

Usually, the measurement and characterization methods of the nano wear are different from the macro-wear. For example, the weighing method cannot be used in the nano wear. In the micro-wear study, according to the quality of the sample surface, the wear depth or number of wear times is often used to present the material wear resistance or coating wear life. For the smooth surface, the variation of the worn surface height can be used to determine the wear depth and to characterize the material anti-wear ability. However, for the rough surface, usually the number of the times to wear a certain thickness is used to evaluate the anti-wear ability of the material.

In order to reasonably choose the parameters of the working condition in the nano wear test, Jiang *et al.* carried out a study on the influences of the probe load, wear number, vertical and horizontal sliding

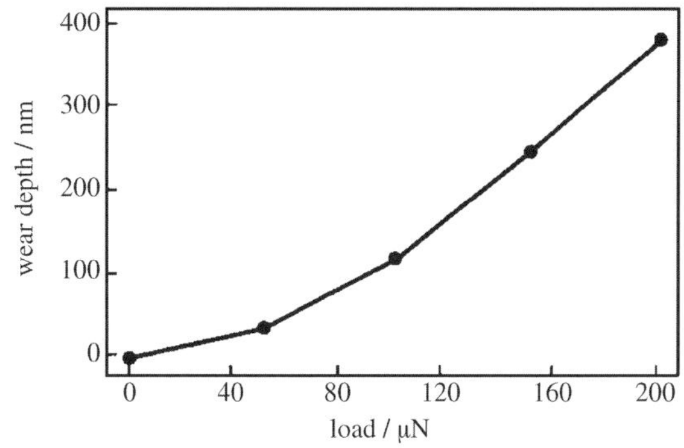

Figure 15.11 Relationship of wear depth and load [9].

velocities, and moving step length on the wear depth [9]. In their experiments, they used the diamond probe slingding on the silicon substrate with a gold coating with the thickness of 800 nm plated by the CVD method. Because the sample surface is very smooth, the wear depth is used to present the anti-wear ability. Figure 15.11 shows the relationship of the wear depth and the load with the sliding velocity of 3.06 μm/s, and the step of 30 nm for three times. Figure 15.12 indicates that under a different load, the wear depth of the gold film surface varies with the wear number under the same experimental conditions as in Figure 15.11.

From the figure we can see that the wear depth increases linearly with wear number. The same thickness is worn for each trip, which means that the gold film is homogeneously worn across the thickness.

Figures 15.13 and 15.14 give the relationship of the two kinds of loads with the next wear depth and the relationship of the velocity and the step distance, respectively. In Figure 15.13, the step distance is 30 nm, while in Figure 15.14 the sliding velocity is 3.06 μm/s.

Figure 15.13 shows that the sliding velocity has little influence on the nano-wear. This is because the sliding velocity in the experiment is low and the gold film is soft.

It should be noted that according to reports so far, the micro-wear experiment or molecular dynamics simulation are carried out on an ideal surface, which is quite different from what happens in practice. Even if the actual friction material is simple crystal material, its strength is only 10^{-4}–10^{-5} of the ideal crystal strength. This is because there are many deficiencies, such as dislocations and micro-cracks in the real crystal. In addition, a lot of materials are polycrystal or noncrystal and their surfaces are often

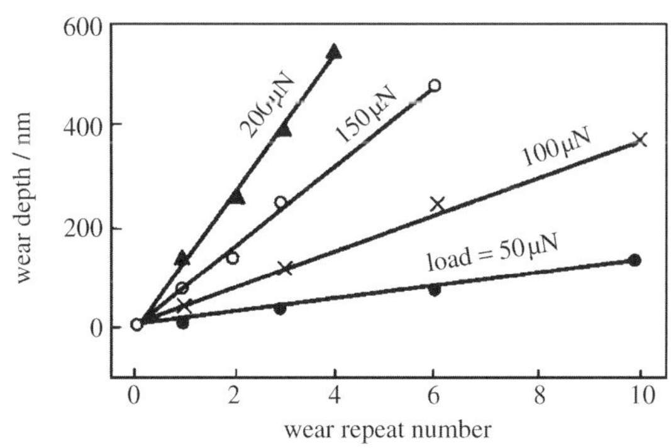

Figure 15.12 Relationship of wear depth and wear number [9].

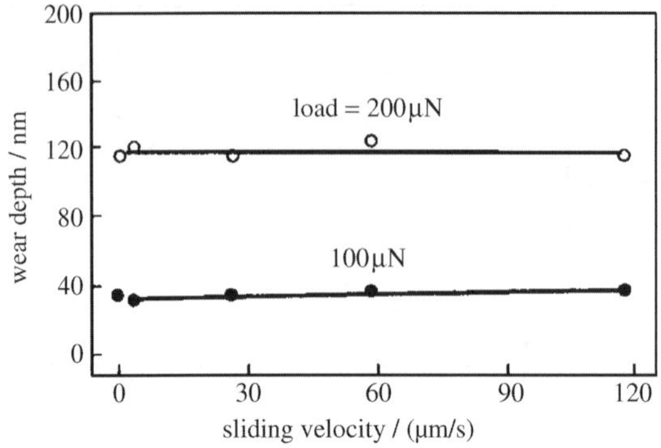

Figure 15.13 Relationship of wear depth and sliding velocity [9].

contaminated so the quality is not homogeneous. Therefore, it is often difficult to directly apply results obtained from the micro-wear experiment to engineering or to carry out a quantitative analysis. However, it is still very important to use them to carry out qualitative analysis.

15.3.2 Micro-Wear of Magnetic Head and Disk

Magnetic recording devices, such as magnetic heads and disks, work in the clearance with the average height of 25 nm–76 nm, while the relative velocity of the two surfaces is 3 m/s–30 m/s. Clearly, if the contact brings about friction and wear, it may seriously affect the accuracy and reliability of the magnetic recording device. Therefore, one of the key research fields of the micro-tribology is in the high-density magnetic recording device.

For the magnetic head and disk system, Bhushan and Koinkar experimentally studied the friction and wear properties of a variety of silicon materials [2]. Their micro-experiment is carried out by using FFM and the experimental conditions and experimental methods can be found in Reference [1,2]. The large numbers of experimental results are summarized in Table 15.2 to show the friction and wear properties of the silicon materials.

Figure 15.15 gives the surface topography under the different loads after ten wear times. The figure shows four kinds of sample materials: (a) Si(111) surface, (b) PECVD oxide Si(111) surface, (c) dry

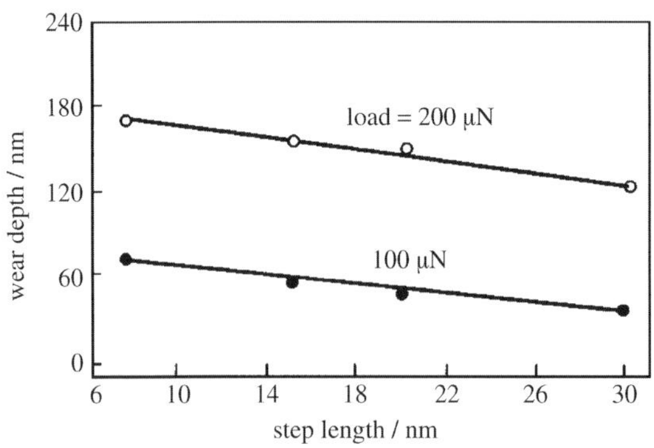

Figure 15.14 Relationship of wear depth and step distance.

Table 15.2 Friction and wear properties of silicon materials [1,2]

No.	Material	Roughness $^{a}R_a$	Friction coefficient	Depicts depth^{b}/nm	Wear depth^{b}/nm	Micro-hardness^{c}/GPa
1	Monocrystalline Si(111)	0.11	0.03	20	27	11.7
2	Monocrystalline Si(110)	0.09	0.04	20		
3	Monocrystalline Si(100)	0.12	0.03	25		
4	Polysilicon	1.07	0.04	18		
5	Polysilicon (polished)	0.16	0.05	18	25	12.5
6	PECVD silicon oxide Si(111)	1.50	0.01	8	5	18.0
7	Dry thermal oxidation of silicon Si(111)	0.11	0.04	16	14	17.0
8	Wet thermal oxidation of silicon Si(111)	0.25	0.04	17	18	14.4
9	C^+ ion implantation into the silicon, such as Si(111)	0.33	0.02	20	23	18.6

[a] Measuring area: $A = 500\,\text{nm} \times 500\,\text{nm}$.
[b] Probe load: $W = 40\,\mu\text{N}$.
[c] Probe load: $W = 150\,\mu\text{N}$.

and thermal oxide Si(111) surface and (d) C+ ion implantation Si(111) surface, respectively. It can be seen that the PECVD oxide Si(111) surface is rough and is of a high anti-wear ability.

Figure 15.16 gives the four wear scar images with the silicon surface topography under the same working conditions. Figure 15.15a–d are the Si(111) surfaces corresponding to the untreated, intensified by PECVD oxide, the dry environment and the thermal oxide, and C+ ion implantation. The experimental results show that the surface of the PECVD oxide has the highest anti-wear ability, while the wear resistance and anti-carved ability of the four silicon surfaces are of the same order. In addition, it can be

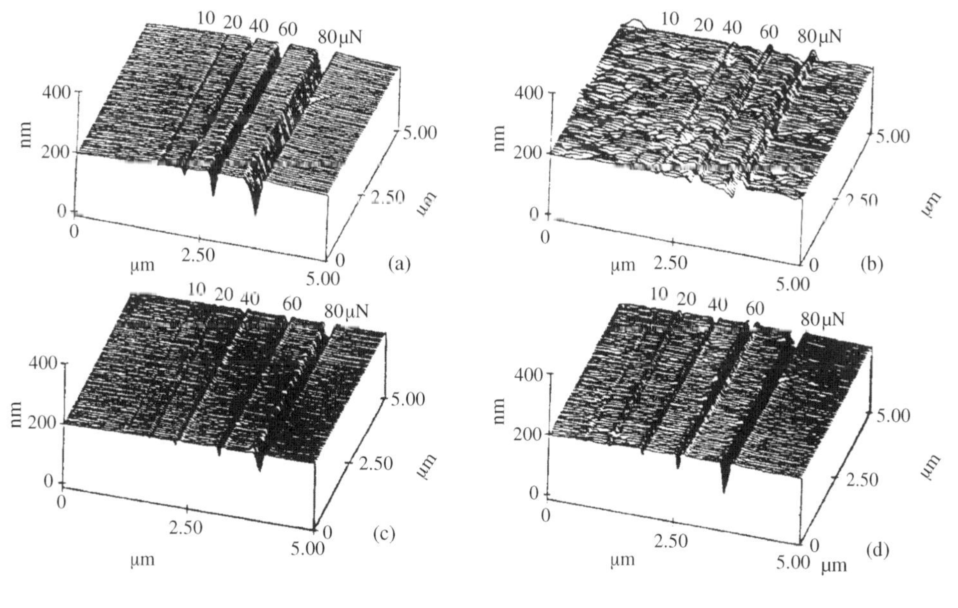

Figure 15.15 Surface topography after micro-wear [2].

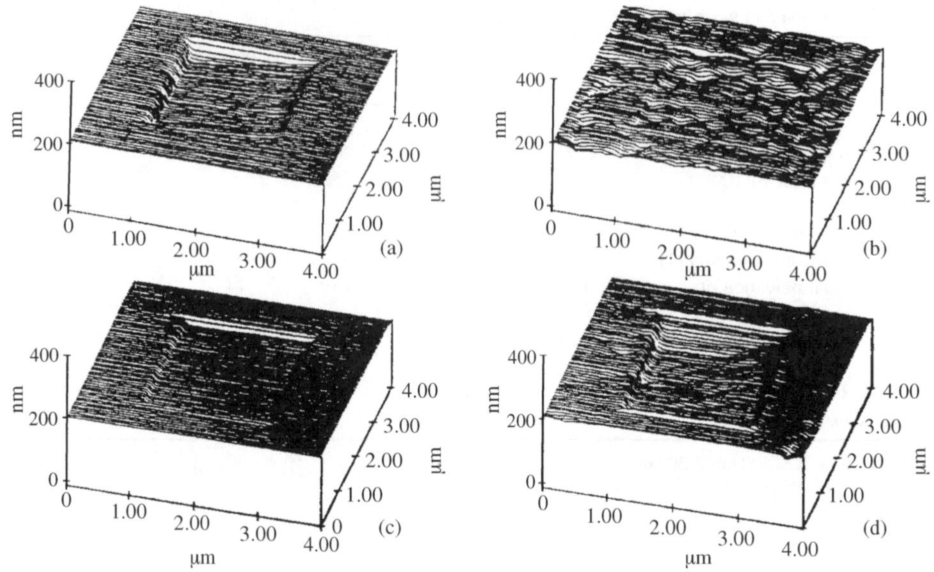

Figure 15.16 Images of wear surface topography [2].

observed that because the nano-wear depth is very small, the wear debris may easily be removed from the surface by the probe during scanning. Therefore, abrasive debris can fall off automatically.

Bhushan and Koinkar used the AFM to carry out the wear experiment on the aluminum surface coated by DLC. Under the load of 20 µN and after different times of wear, the surface topographies are shown in Figure 15.17. From the figure, it can be seen that the initial defect of micro-abrasion on the

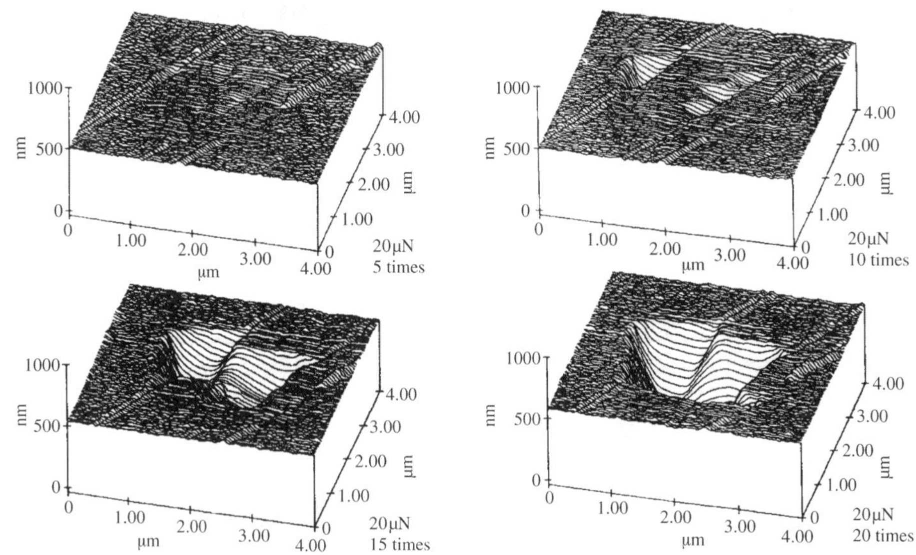

Figure 15.17 Non-uniform micro-wear.

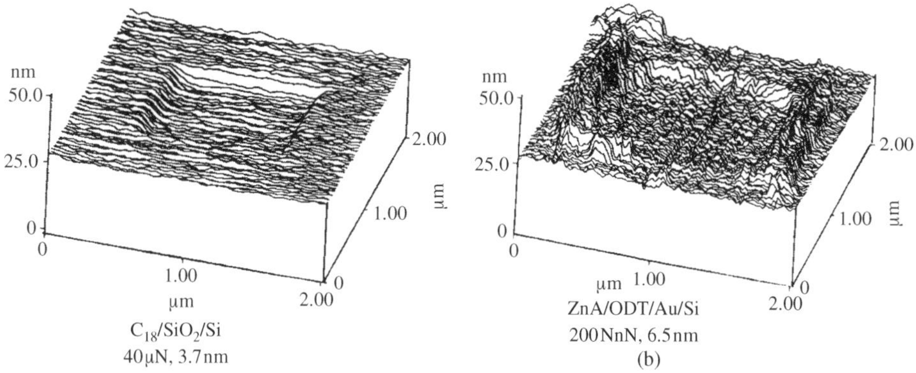

Figure 15.18 SAM and LB film lubricated and wear topography [10].

surface is the scratch. Then, because the surface energy at the scratch is high, it becomes weak and expands gradually. The part without scratches has a relatively high anti-wear ability and micro-abrasion is non-uniform.

The hard disk is made of the magnetic media covered by a corrosion-resistant protective coating and with a thin lubricant to improve the anti-friction and anti-wear properties. Recently, the ordered monolayer film, such as the self-assembled monolayer (SAM) or LB film, has been applied as a lubricant in the hard disk-head. Bhushan *et al.* experimentally studied the influence of lubricant film on micro-friction and wear [10]. The results are shown in Figure 15.18. In the figure, (a) is the SAM lubricated and worn surface topography measured. The substrate is the silicon by oxidation treatment, and then implant of the C_{18} monolayer on the surface with the amino-silane through chemical reaction. The surface structure is multi-layer, that is, $C_{18}/SiO_2/Si$. (b) is the LB film lubricated and worn surface topography. The substrate structure is silicon covered by gold coating and then plated with octadecylthiol (ODT). Then, the zine arachidate (ZnA) is adsorbed. The structure is ZnA/ODT/Au/Si.

The experimental results show that the lubrication performances of the SAM film with C_{18} is better than those of ZnA of the LB film. The friction coefficient is low and the wear life is long. With FFM, the friction coefficient of SAM film lubrication is measured to be 0.018 under the load of 40 μN and with the wear depth of 3.7 nm. The worn surface is smooth. The friction coefficient of the LB film is 0.03 under the load of 200 nN with the wear depth up to 6.5 nm. The edges on the surface are very rough. This is because the connection of LB film and substrate relies on the Van der Waals attraction force so the combination is weak. However, the connection of the SAM film and the substrate is the chemical combination so it is strong. The appropriate choice of molecular length and polar group can further improve the boundary lubrication ability of the SAM film. Qian [11] and Jiang [12] fully studied the lubrication performances of the self-assembled membrane by using an atomic force microscopy with a micro-sphere-plate tester.

15.4 Molecular Film and Boundary Lubrication

In the past, the boundary lubrication researches mostly focused on revealing the boundary film formation and failure mechanism in the physical-chemistry or the chemistry aspect, but there are few researches in the physics or the mechanics aspect. Furthermore, due to limitation of the surface testing equipment, the previous research seldom studied the relationship between the microstructure and properties in the atomic or molecular scale. Therefore, the establishment of boundary lubrication physical model is difficult.

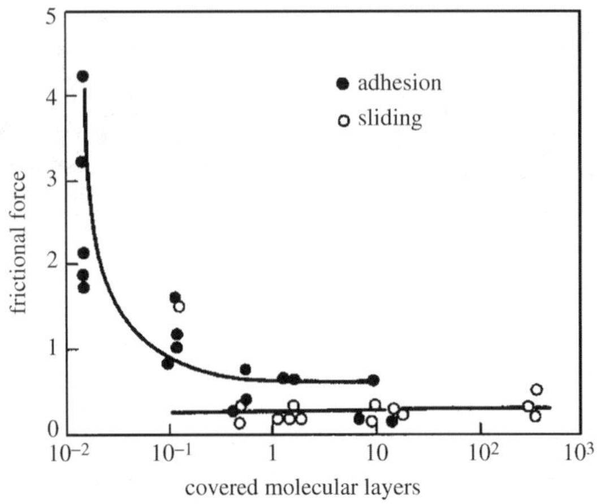

Figure 15.19 Relationship of friction coefficient and coverage of ethanol C_2H_6O molecules [13].

15.4.1 Static Shear Property of Molecular Layer

The liquid between two surfaces with a small gap is called the confined liquid. The solid surface can affect the molecular structure of the liquid near it. If the two surfaces affect together, the changes in molecular structure are great. Therefore, the molecular structure and the properties of the confined liquid are quite different from those of the bulk phase.

Figure 15.19 is the relationship between the friction coefficient and the layer number of the ethanol C_2H_6O molecules given by Ko and Gellman. Experiments were carried out on the Ni(100) surface at the velocity of 10 μm/s and the temperature of 120 K. When the surface is not entirely covered by the molecular layer, the friction coefficient is high, adhesion is severe, and wear is serious. When the surface is completely covered by one or several molecular layers, the friction coefficient is stable, about 0.2 and is not related to the layer number further. Therefore, the decisive influence factor on the friction is the monolayer originally forming.

Vinet described the shear behaviors of a series of lubricants [14]. The relationship of the contact pressure and the shear elastic modulus G_c of the polystyrene obtained under the condition of a constant Poisson's ratio is as shown in Figure 15.20.

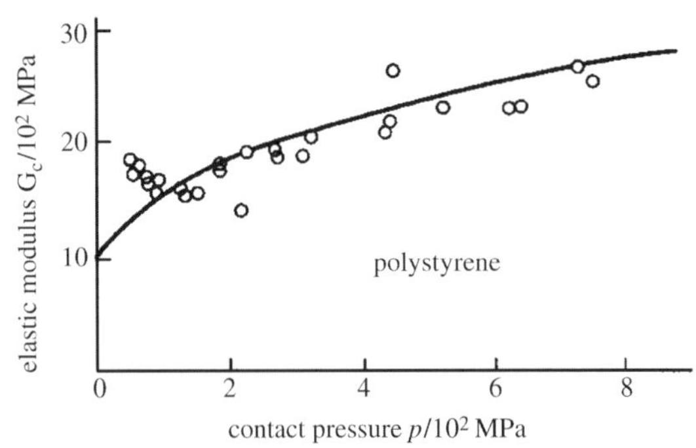

Figure 15.20 Relationship of shear elastic modulus G_c and contact pressure of polystyrene [14].

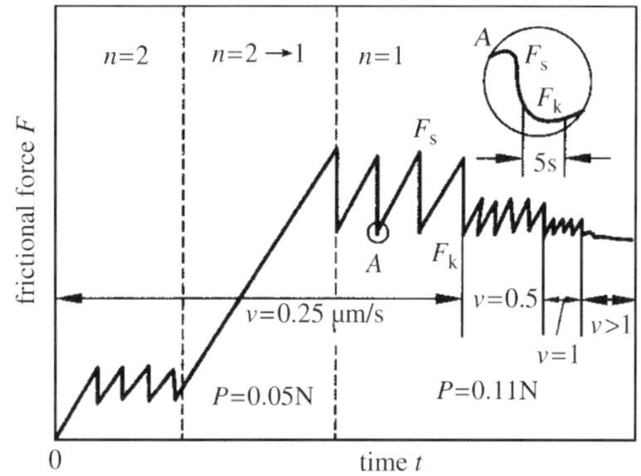

Figure 15.21 Typical stick-slip curve [15].

15.4.2 Dynamic Shear Property of Monolayer and Stick-Slip Phenomenon

Israelachvili *et al.* carried out a series of studies on the stick-slip on the interface of molecular layer [15,17]. They used mica as the friction surface and octamethylcyclotetrasiloxan (OMCTS) as a lubricant. The surface force apparatus was used to experimentally study the stick-slip. OMCTS is a nonpolar silicon fluid which can form clear molecular layer on the mica surface. Figure 15.21 is a typical stick-slip curve. In the figure, n is the molecular layer number; P is the normal load; and v is the sliding velocity.

From the figure we can see that during sliding the frictional force fluctuates. From adhesive contact to sliding, the frictional force increases steadily to the maximum, which is the static frictional force F_s, which is the static limit of the shear stress. Then, the surface suddenly slides to a new adhesive contact state. The frictional force reduces to the minimum, which is the dynamic friction force F_k, which is the dynamic limit of the shear stress. Also, adhesion and sliding alternatively change but the surface is not worn. In addition, after the static frictional force arrives, the sliding movement is very fast, but the change from dynamic friction to static friction is a gradual process. As shown in the figure, such a transformation needs five seconds.

The static and dynamic frictional forces are related to the molecular layers and sliding velocity. As the molecular layers n decreases, F_s and F_k increase, and the increment amplitude $\Delta F = F_s - F_k$ also increases, but the varying frequency f decreases. As the sliding velocity v increases, the amplitude decreases but the frequency increases. Until the sliding velocity reaches the critical v_c, the stick-slip phenomenon disappears. Then, the sliding is smooth and the frictional force is F_k. Obviously, this is the ideal friction condition.

From the friction formula $F = \tau_c A$ of the adhesion theory, we have

$$\begin{aligned} \Delta F &= A(\tau_{cs} - \tau_{ck}) = A\Delta\tau_c \\ \Delta\tau_c &= \tau_{cs} - \tau_{ck}, \end{aligned} \tag{15.10}$$

where A is the contact area; τ_{cs} and τ_{ck} are the static and the dynamic shear stress limits, respectively.

Figure 15.22 gives the experimental results that the increment ΔF of the frictional force varies with the contact area A with different sliding velocities. As shown in the figure, ΔF and A have a linear relationship. This means $\Delta\tau_c$ is a constant and the conclusion that the frictional force F is proportional to the contact area A in the adhesion theory is correct.

With Figure 15.22, the relationship of $\Delta\tau_c$ and v can be established as shown in Figure 15.23. To extend the line segment until outwards to the intersection point of the abscissa, we can obtain the critical sliding velocity when the stick-slip disappears. At this time, F_s and F_k are equal. In the figure, as $n = 1$, the critical

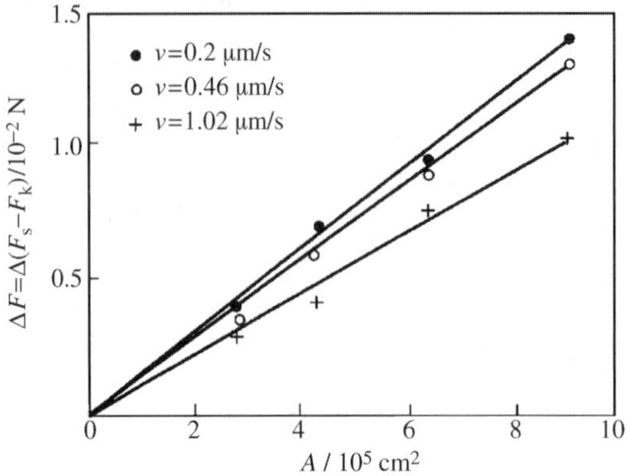

Figure 15.22 Relationship of F and A.

sliding velocity of OMCTS is about 3 μm/s. As $n=2$, v_c is between 3 μm/s and 2 μm/s, and $\Delta\tau_c$ and v_c are lower than those as $n=1$.

For the usual boundary lubrication, the typical friction curve is shown in Figure 15.23. It can be divided into the two regions, the stick-slip zone $v < v_c$ and the smooth zone $v \geq v_c$. This is because the molecular film has a periodical phase variation, that is, from solid-like adhesion (condensed state) to liquid-like sliding (molten). When the velocity exceeds the critical velocity, the molecular film cannot condense in time so as to keep the liquid-like state to smoothly slide.

For some boundary lubrication systems, such as the mica surface implanted dihexadecyldimethylammonium acetate (DHDAA) monolayer, the superkinetic friction appears.

Figure 15.24 gives the friction curves the two mica surfaces are implanted the DHDAA monolayer in the dry friction condition at 25 °C. In the low velocity zone, the stick-slip phenomenon occurs and the frictional force varies between F_s and F_k. After the sliding velocity increases to the critical velocity $v_c=0.1$ μm/s, the stick-slip disappears and the sliding is smooth with the dynamic frictional force F_k. However, when the velocity is higher, to arrive at the supercritical velocity $v_c^* \approx 0.4$ μm/s, it appears as superkinetic friction. The frictional force varies between F_k and F_{sk}. F_{sk} is the superkinetic frictional force, and $F_{sk} \ll F_k$.

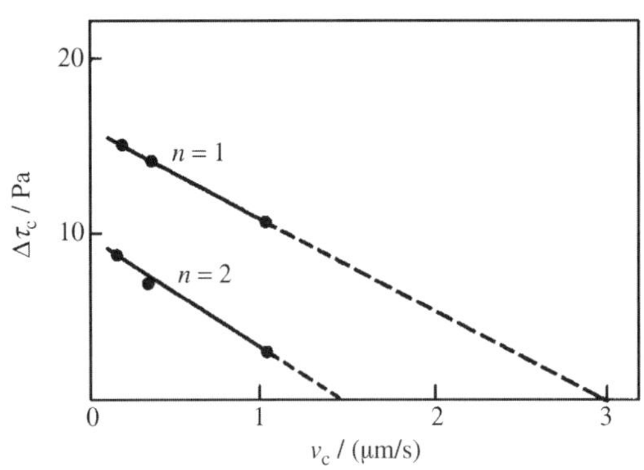

Figure 15.23 Relationship between $\Delta\tau_c$ and v [15].

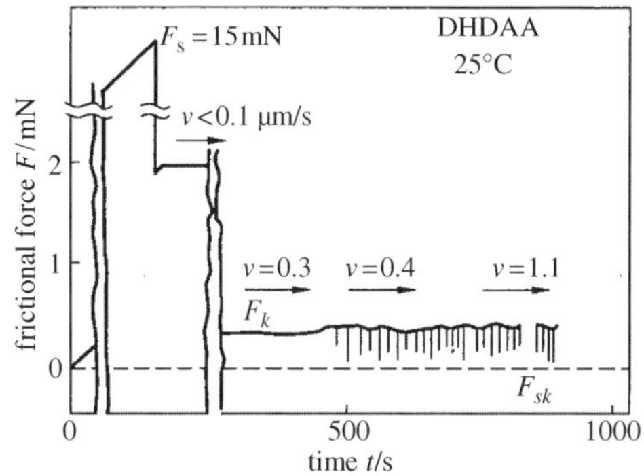

Figure 15.24 Superkinetic friction [16].

Superkinetic friction exists under specific conditions. Because one end of the DHDAA molecule firmly connects the surface, the entire molecule can swing freely. At super critical velocity, the molecule tilts along the sliding direction, thereby being of a very low friction.

The experiments show that the increase of the normal load will drop the super critical velocity, that is, the load will speed up sliding into the superkinetic friction state. This is because the increase of load reduces the gap between the molecules so as to tilt the lubricating molecules. In the conventional stick-slip friction, the load also promotes the disappearance of the stick-slip because the load increases the relaxation time of the lubricating molecules so as to increase the time of the structure variation of the molecular film. Thus, this prevents the phase transition of the molecular film to maintain the liquid-like state friction.

15.4.3 Physical State and Phase Change

The friction state of the molecular film in the interface is very special. This confined liquid in a small gap is not only subjected to structural force and normal force, but also subjected to tangential shear. Therefore, its molecular structure and natures are obviously different from the bulk state. It is generally believed that the state of the confined liquid has a different shape according to the working condition and phase transition occurs in the friction process.

Israelachvili *et al.* proposed that in sliding, the molecular film in the interface has three forms: solid-like, amorphous and liquid-like as shown in Figure 15.25 [16]. They used this viewpoint to explain the boundary lubrication properties and had satisfactory results.

As shown in Figure 15.25, the forms and their transformation are not the inherent characteristic of the liquid. They are only the dynamic natures emerging in the interface film forming and sliding process. The surface energy is the main factor in deciding the ordered arrangement of the molecules. It is influenced by

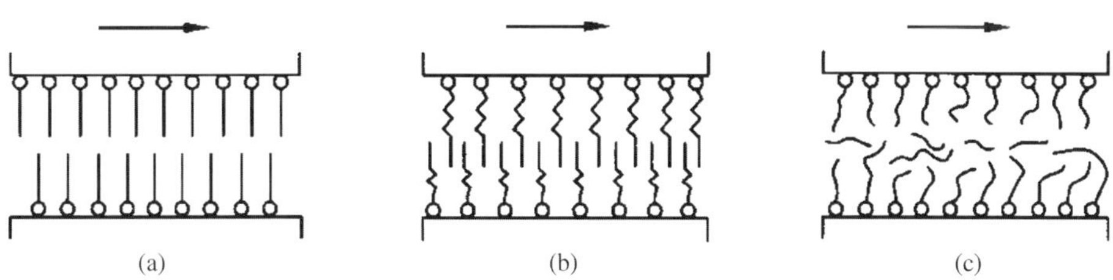

Figure 15.25 Three physical forms [16].

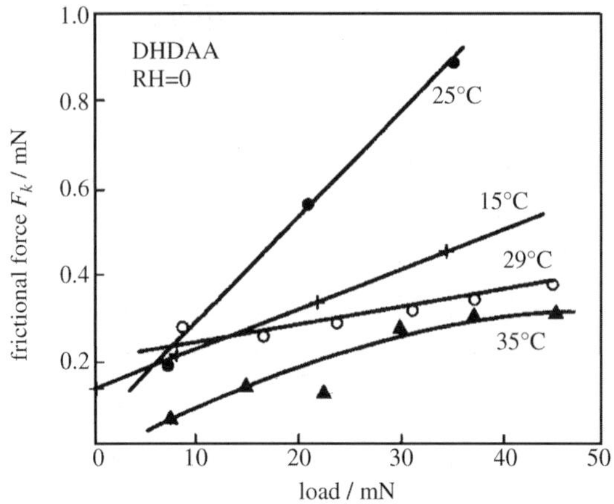

Figure 15.26 Temperature effect [16].

temperature, load, sliding velocity and shear rate. Usually, at a high sliding velocity and a lower temperature, the molecular film tends to be the solid-like form, while at a low sliding velocity, the molecular film tends to be the liquid-like form.

15.4.4 Temperature Effect and Friction Mechanism

The friction mechanism of the molecular lubricating film has been studied from the physical properties of the molecular film and the energy conversion. Yosmzawa *et al.* experimentally studied the temperature effect of the DHDAA monolayer in sliding [16]. Figure 15.26 shows their experimental results under the conditions of the dry friction, that is, the relative humidity $RH = 0$. From the figure we can see that the friction first increases and then decreases with increase of temperature. The maximum occurs at about 25 °C. This tendency is the same as the energy loss in the visco-elastic polymer. The polymer bond winding reaches the maximum at the temperature at which the energy loss is the maximum.

It can be thought that the friction behaviors of DHDAA are related to viscoelasticity. At the lower temperature, the molecular film is in a solid-like form, shown in Figure 15.25a. Because the winding does not easily occur between the molecules, the energy loss and the friction are small. At high temperatures, the film is in a liquid-like form, shown in Figure 15.25c. Although more chain-like molecules are wrapped around each other, the activity makes the large liquid molecules eliminate the winding much easier so that the energy loss and the friction are quite low. At the middle temperature, the molecular film is in the non-crystalline form. It is difficult to eliminate the winding of the molecules. Therefore, the frictional force is larger.

15.4.5 Rheological Property of Molecular Film

The dynamic shear property of the lubricant film, that is, the rheological property, is very important to engineering design. Granick *et al.* studied the relationship of equivalent viscosity η_e and film thickness of $CH_3CH_3(CH_2)_{10}CH_3$ (hexadecane) under the low shear rate condition at 28 °C [17]. The results are as shown in Figure 15.27. When the film thickness is 4.0 nm, $\eta_e = 10$ Pa·s. Because the hexadecane is a simple liquid, its bulk viscosity is only 0.0001 Pa·s, much lower than that of the molecular film. With decrease of the film thickness, the equivalent viscosity sharply increases. When the monolayer thickness is 2.6 nm $\pm$ 0.1 nm, the gap is very close to the diameter of the hexadecane molecule. This prevents the film from flowing so the equivalent viscosity suddenly diverges to infinity.

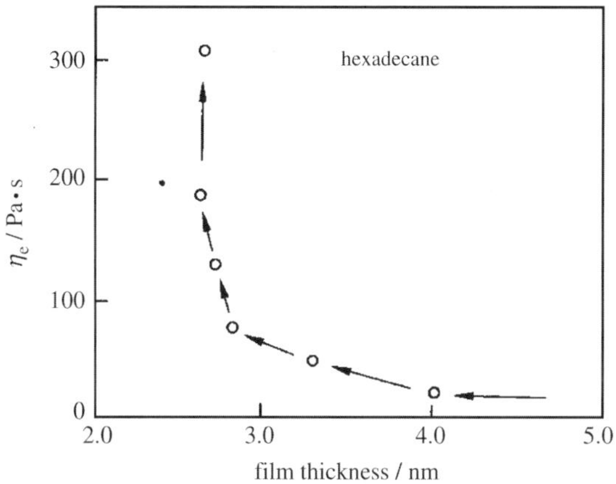

Figure 15.27 Relationship of equivalent viscosity and film thickness of hexadecane film [17].

Usually, the bulk hexadecane is the Newtonian fluid. However, if it is constrained by a small gap, its rheological behavior is quite complex. Figure 15.28 gives the relationship of the equivalent viscosity η_e of the hexadecane and the shear rate $\dot{\gamma}$ in the logarithmic coordinates. From the figure, we can see that at a very low shear rate, the equivalent viscosity is near a constant, that is, the hexadecane is of the Newtonian fluid property. As the shear rate increases, the equivalent viscosity decreases exponentially to appear as a nonlinear (shear-thinning) phenomenon.

Alsten and Granick studied the relationships of viscosity and film thickness of the dodecane molecular film, as shown in Figure 15.29 [18]. The numbers in the figure are the average pressure, MPa. The experiments show that the viscosity of the molecular film is not only related to film thickness, but also to pressure.

Figure 15.30 shows the relationship of the limit shear stress τ_s and the average pressure p of the dodecane film. In the figure, the limit shear stress increases linearly with the average pressure, and their relationship can be expressed as the following formulas:

$$\tau_s = \tau_{s0} + \alpha p, \tag{15.11}$$

where τ_{s0} is the limit shear stress at $p = 0$; and $\alpha \approx 20$.

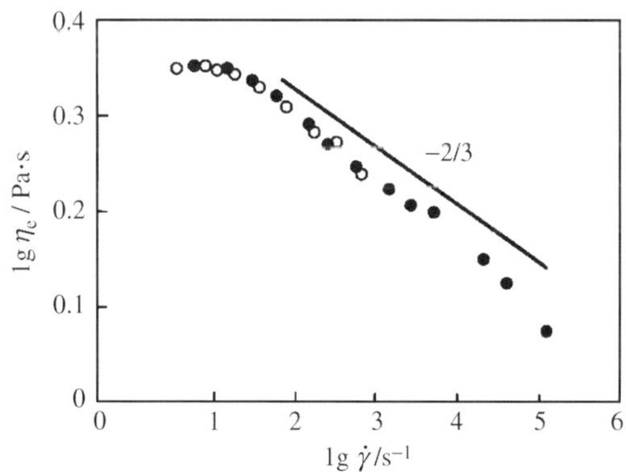

Figure 15.28 Relationship of viscosity and shear strain rate of hexadecane film [17].

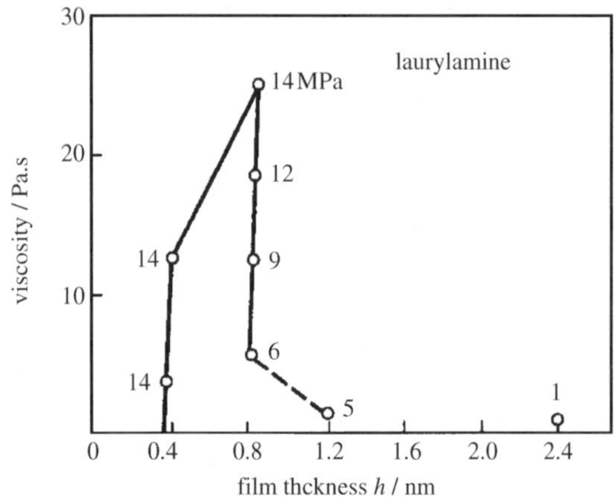

Figure 15.29 Relationship between film thickness and pressure of dodecane film [18].

In fact, the limit shear stress of the solid-like body is the starting yield shear stress. It is related to the static friction and the molecular film natures.

Alsten and Granick pointed out that yield stress of the molecular film is related to its experience process. After solidification, the limit shear stress of liquid increases with time. Figure 15.31 gives the curve that the limit shear stress of an amorphous polymer (poly methyl siloxane) film varies with time. The starting limit shear stress is measured to be 1 MPa; it increases 3 times after 4 hours. This shows that in the solidified film, the molecules rearrange. At the beginning, the structure may contain a number of cavities. After continuously adjusting over time, it becomes solid.

The above analysis helps us understand the mechanism of static friction. The static friction coefficient of a solid continuously increases with contact time. The traditional views believe that this is because the normal load causes the roughness of the two contact surfaces embedding deeper into each other and increasing the contact deformation with time, thus increasing the static friction. However, for molecular film lubrication, the reason for the variation of the static frictional force is because the confined liquid needs adequate time to complete the molecule re-arrangement.

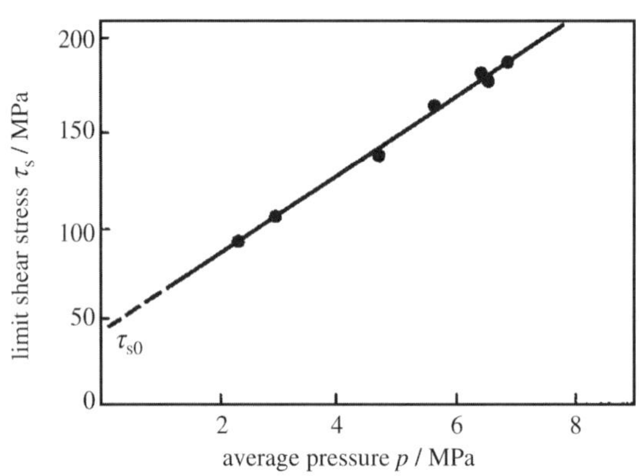

Figure 15.30 Relationship of limit shear stress and pressure of dodecane film [18].

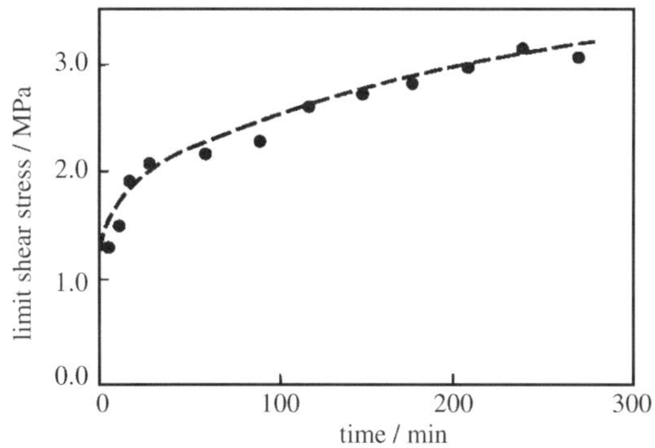

Figure 15.31 Variation of limit shear stress with time [18].

15.4.6 Ordered Molecular Film

Ordered molecular film is when one or more molecular layers cover the solid surface to arrange an orderly and compact film. With preparation, the composition of the ordered molecular film can be changed or we can introduce some special functional gene to meet the requirement. Therefore, it provides a way to control the tribological properties of the surface with the molecular engineering.

At present, ordered molecular films are mainly LB film (Langmuir-Blodgett film) and self-assembled membrane (self-assembled monolayer, SAM). Their structure, preparations and applications are introduced below.

15.4.6.1 LB Film

LB Film is a monolayer to range the organic amphiphilic molecules on the interface of water and air to generate a highly ordered film. Then, it is transferred onto the solid surface to form an orderly and ultra-thin layer. The forming molecules of the LB film are amphiphiles. It cannot be dissolved in water. One end of the molecule is hydrophilic and the other end is oleophylic and hydrophobic. When the amphiphilic molecule contacts water, the hydrophilic end sinks into water, but the hydrophobic end keeps in air. So, the molecule floats on water. A typical amphiphilic molecule is the fatty acid C_nH_{2n+1} COOH. Its hydrophilic end is the carboxyl group (−COOH). The hydrophobic end is the alkyl chain.

Water is the most common liquid as the sub-phase material. It serves as a carrier to form the membrane in the interface of water and air. In order to prepare a LB film, water must be deionized, and its PH value and surface tension must be controlled. The formed LB film can be transferred to the solid surfaces, such as glass, silicon, various metals or their oxides. According to the nature of the prepared LB film, the solid surface must be hydrophilicly or hydrophobicly treated.

Figure 15.32 gives the LB film preparation process. Figure 15.32a shows that the amphiphilic molecules are dissolved into the organic solvent. Then they are dropped onto water. After the solvent

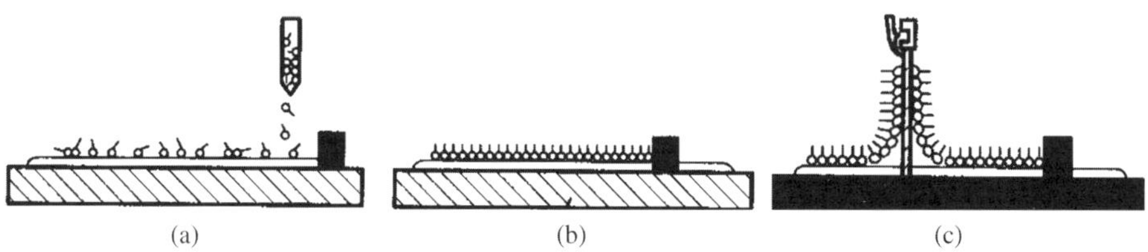

Figure 15.32 Preparation process of LB film [19].

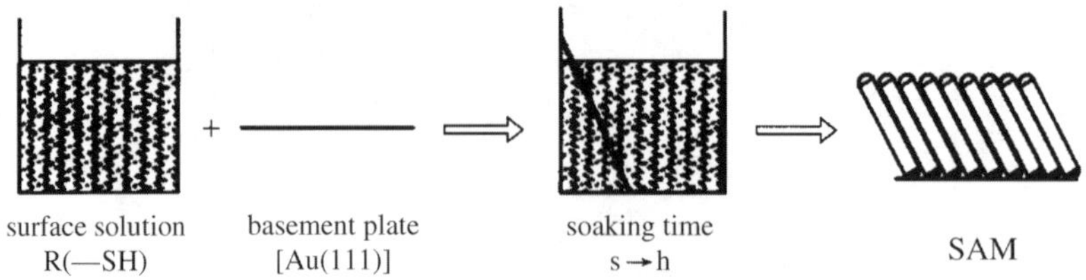

Figure 15.33 Preparation of self-assembled monolayer [1].

evaporates, a monolayer is left in the interface of water and air. As shown in Figure 15.32b, because the distance between molecules is large, the mobile barrier plate compresses the interface molecules gradually to form a close and orderly arranged layer. In Figure 15.32c, the layer is transferred to the substrate surface. The most common transformation method is the vertical lifting method.

One application of a LB film in tribology is as a lubricant. A perfluorinated polyether (PFPE) film with the thickness of 1 nm–10 nm is often prepared on the magnetic recording medium surface to improve the friction and wear properties of the disk and head.

There are many factors that influence the lubrication performances of a LB film. In addition to sliding velocity, the load, working temperature, film material, layer number, layer structure, substrate surface state, sub-phase liquid natures and interface chemical reaction are the important factors.

15.4.6.2 Self-Assembled Monolayer

The formation of the self-assembled monolayer mainly relies on the chemical action of the solid-liquid interface. The appropriate substrate is immersed in the organic solution with the surfactant. The reactive molecules (or head group) of the surfactant automatically adsorbs or reacts with the substrate to form a monolayer which is of the chemical bond on the substrate surface and is closely arranged. In the same layer, the connection of the molecules relies on the Van der Waal force. If the tail group of the surfactant molecule has the chemical reactivity, it can act with the other material to build a homogeneous or heterogeneous multi-layer. However, the surface chemical reaction should be selective so that it will react

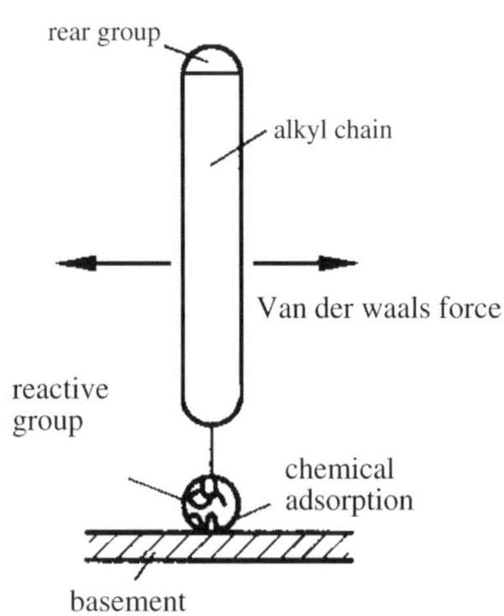

Figure 15.34 Molecular structure of surfactant.

with the mating substrate material according to different requirements to realize self-assembly. Figure 15.33 indicates the preparation process of the self-assembled monolayers. Figure 15.34 gives the structure of the surfactant molecule.

As shown in Figure 15.34, the structure of an active molecule of the self-assembled monolayer includes three parts: the head group with the chemical adsorption function to the substrate, the alkyl chain through the Van de Waals force connecting with the neighboring molecules, and the functional tail group. There are many factors affecting quality and formation capacity of the self-assembled monolayer. The primary ones are substrate material, surface roughness, activity of the molecular reactive group, molecular chain size, polarity of the activity of the tail group, polarity and concentration of the solvent in the film forming solution.

Because the structure of the self-assembled membrane is compact and stable, it can be used in lubrication and wear protection.

References

1. Wen, S.Z. (1998) *Nano-Tribology*, Tsinghua University Press, Beijing.
2. Bhushan, B. and Koinkar, V.N. (1993) Tribological studies of silicon for magnetic recording applications. *Journal of Applied Physics*, **75** (10), 5741–5746.
3. Ruan, J. and Bhushan, B. (1994) Atomic-scale and microscale friction studies of graphite and diamond using friction force micooscopy. *Journal of Applied Physics*, **76** (9), 5022–5035.
4. Ruan, J. and Bhushan, B. (1994) Frictional behavior of highly oriented pyrolytic graphite. *Journal of Applied Physics*, **76** (12), 8117–8120.
5. Guo, Q., Ross, J.D.J., and Pollock, H.M. (1988) What part do adhesion and deformation play in fine-scale static and sliding contact? *Proceedings of the Materials Research Society, Boston*, **140**, 51–66.
6. Pivin, J.C., Takadoum, J., Ross, J.D.J., and Pollock, H.M. (1987) *Tribology: 50 Years On*, Mech. Eng. Publications, London, pp. 179–181.
7. Chowdhury, S.K.R., and Pollock, H.M. (1981) Adhesion between metal surfaces: the effect of surface roughness. *Wear*, **66** (3), 307–321.
8. Landman, U. and Luedtke, W.D. (1993) Interfacial junctions and cavitations. *MRS Bulletin*, **18** (5), 36–43.
9. Jiang, Z.G., Lu, C., Bogy, D.B., and Miyamoto, T. (1995) An investigation of the experimental condition and characteristics of a nano-wear test. *Wear*, **181–183**, 777–783.
10. Bhushan, B., Miyamoto, T., and Koinkar, V.N. (1995) Microscopic friction between a sharp diamond tip and thin-film magnetic rigid disks by friction force mi- 28. 42 croscopy. *Advanced Information Storage System*, **6**, 151–161.
11. Qian, L.M., Luo, J.B., Wen, S.Z., and Xiao, Xudong (2000) Study on micro friction and adhesion force of OTS self-assembled membrane (I) experiment and analysis of friction. *Acta Physica Sinica*, **49** (11), 2240–2253.
12. Jiang, W.J., Luo, J.B., and Wen, S.Z. (2000) Friction characteristics of OST molecular film. *Chinese Science Bulletin*, **45** (17), 1900–1904.
13. Ko, J.S. and Gellman, A.J. (2000) Friction Anisotropy at Ni(100)/Ni(100) Interfaces. *Langmuir*, **16** (22), 8343–8351.
14. Vinet, P. (1986) PhD Dissertation. Ecole Centrale de Lyon. ECL Lyon.
15. Israelachvili, J.N. (1991) *Intermolecular and Surface Force*, Academic Press, New York.
16. Yoshizawa, H., Chen, Y.L., and Israelachvili, J. (1993) Recent advances in molecular level understanding of adhesion, friction and lubrication. *Wear*, **168** (1–2), 161–166.
17. Granick, S. (1991) Molecular tribology. *MRS Bulletin*, **16** (10), 1–6.
18. Alsten, J.V. and Granick, S. (1988) Molecular tribology: recent results and future prospects. *Proceedings of the Materials Research Society, Boston*, **140**, 125–130.
19. Xue, Q.J. and Zhang, J. (1996) *Ultra-Thin Film of Molecular Ordering System and its Application in Tribology*, Liaoning Science and Technology Press, Shenyang.

16

Metal Forming Tribology

Although natural metal has been forged since ancient time, study on the tribological problems of the forming process is very recent. The early lubricant in the metal forming process was natural oil or fat. Since the 1940s, study on metal forming lubricant had been rapidly developed and strict lubrication conditions have been proposed and a number of new metallic materials developed to meet the needs of aviation and aerospace technology and the electronics industry. Because the forming technology is quite different from that of common metals, some lubrication difficulties emerged so there were a lot of concerns about metal forming tribology [1–3].

In the metal forming process, friction exists between the die and workpiece, such as the friction between the forge piece and anvil, friction between the blank and extrusion die, friction of metal flowing from the impression and friction between the workpiece and roller. Metal forming friction is different from conventional friction because it is brought about under high pressure up to 2500 MPa. Furthermore, the forming is mostly carried out at high temperatures, usually 800–1200 °C. At such a temperature, the structure and properties of the metal are significantly changed and this cause great difficulties for lubrication. In addition, metal forming friction is often due to the new contact surface during the deformation and variation of the contact conditions between the tool and the workpiece. Because the displacement of the contact points of the metal surface is different, this also brings about some difficulties for lubrication.

Due to the action of friction, the die will be worn and some scratches may appear on the workpiece surface which shorten the die life and decrease the product surface quality. In addition, friction will cause deformation of the metal such that it is difficult for the workpiece to be demolded. Furthermore, this may also cause metal non-uniform deformation or even create severe cracks and adhesion.

However, friction sometimes has good functions, such as increasing the rolling frictional force between the roller and the blank that improve rolling ability.

16.1 Mechanics Basis of Metal Forming

16.1.1 Yield Criterion

According to plasticity theory, the stress components of metal plastic deformation satisfy some relationships. Generally, there are two common yield criteria. One is the Tresca yield criterion as follows:

$$\frac{\sigma_{\max} - \sigma_{\min}}{2} = \frac{\sigma_s}{2}. \tag{16.1}$$

Principles of Tribology, First Edition. Wen Shizhu and Huang Ping.

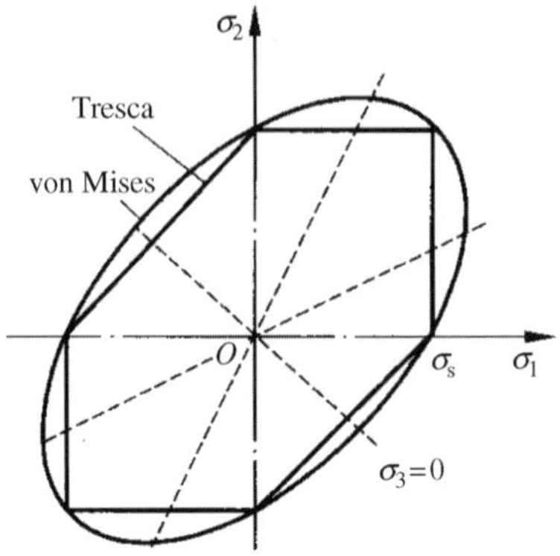

Figure 16.1 Two plane stress yield criteria.

The other is the von Mises yield criterion, which can be written as

$$(\sigma_1 - \sigma_2)^2 + (\sigma_2 - \sigma_3)^2 + (\sigma_3 - \sigma_1)^2 = 2\sigma_s^2, \tag{16.2}$$

where σ_1, σ_2 and σ_3 are the three principal stresses; and σ_s is the material yield stress.

The yield criteria can be easily expressed by the stress diagrams for the plane stress state ($\sigma_3 = 0$). As shown in Figure 16.1, the Tresca yield criterion is a hexagon, while the von Mises is an ellipse.

Some usual plastic deformations are as follow.

Tension: When the simple tension stress reaches the tensile yield stress σ_s, the material begins to be plastically deformed.

Compression: The material begins to be deformed under the yield compression stress. For ductile material, compression yield stress is generally equal to tensile yield stress.

Plate shaped: When two principal stresses of a plate sheet are equal and reach σ_s, plastic deformation appears.

Pure shear: The two principal stresses are equal to the yield shear stress τ_s and in the opposite directions, plastic deformation appears. Sometimes, the yield shear stress is represented by k.

Plane strain: When the deformation of the workpiece is limited only in one direction, the principal stress in this direction is equal to the average of the other two principal stresses. For the Tresca criterion, the yield stress is still equal to σ_s, while for the von Mises criteria, it is equal to $1.15\sigma_s$.

16.1.2 Friction Coefficient and Shear Factor

As mentioned above, the friction existing between the die and workpiece will influence the forming process. The conventional plasticity theory usually simplifies the force and friction on the surface.

16.1.2.1 Friction Coefficient and Interface Adhesion

The friction coefficient can be expressed as follows:

$$f = \frac{F}{N} = \frac{\tau}{p}, \tag{16.3}$$

where p is the pressure and τ is the shear stress. They are obtained by dividing the tangential force F and the normal force N by the apparent area A.

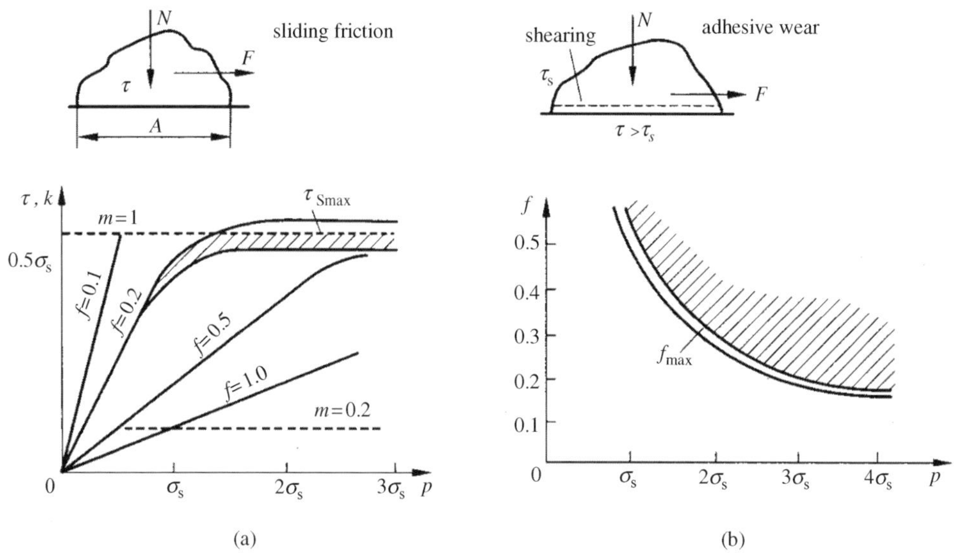

Figure 16.2 Variations of sliding shear stress and adhesive friction coefficient with surface pressure.

When τ is equal to k, sliding does not necessarily appear between the interface of the tool and the workpiece. Sliding may occur within the workpiece, but the interface remains still. In this situation, the interface adhesion will be formed, see Figure 16.2. The adhesion condition is:

$$\tau = fp > k. \tag{16.4}$$

16.1.2.2 Shear Factor

During the forming process, the friction coefficient obtained from Equation 16.3 changes significantly. If the interface is no relative sliding, the friction coefficient f may be very low. On the contrary, if the tensile stress is large, the interface pressure drops so f will be very high. Therefore, it is suggested that the shear factor m is used to express the interface friction, that is:

$$\tau = mk, \tag{16.5}$$

where m is the shear factor.

For the friction-free interface, $m = 0$, while in the adhesion, $m = 1$ (see Figure 16.2). However, m is not very convenient because the interface is usually closely related to the properties of the workpiece material. With Equations 16.4 and 16.5, we have

$$\tau = fp = mk \approx m\sigma_l/2, \tag{16.6}$$

where σ_l is the tensile stress.

It can be seen that the difference between the two equations will increase with increase of the interface pressure p.

16.1.3 *Influence of Friction on Metal Forming*

The influences of friction on the metal forming are reflected in many aspects and the influence factors affect each other.

16.1.3.1 Influence of Friction on Deformation Force

The forming pressure consists of three parts as follows:

$$p = f_1(\sigma_I)f_2(\tau)f_3, \tag{16.7}$$

where f_1 is the pure deformation force, expressing the influence of material properties. It usually fluctuates in ±5%; f_2 is the frictional force. In the good lubricated cold-rolled or drawing, it is less than 5% of the pure deformation force; and f_3 is the function of the geometric condition. It reflects the influence of non-uniform deformation, which may largely increase the deformation force or even completely covers the effect of friction.

When the tool and the workpiece move relatively, friction inevitably exists between them and affects the deformation force. For example, when a billet is extruded along the extrusion cylinder wall, the sliding friction is produced due to the forward movement (Figure 16.3a). The material deformation force is increased due to the friction between the die and workpiece as shown in Figure 16.3b.

In the upsetting, the friction caused by sliding is not very significant. As shown in Figure 16.4a, with decrease in height and increase in diameter, the extended surface of the cylinder slides outwards along the die surface. The interface will generate frictional resistance. In order to overcome the resistance, interface pressure must be increased. With increase in the distance of the edges, the end-surface friction

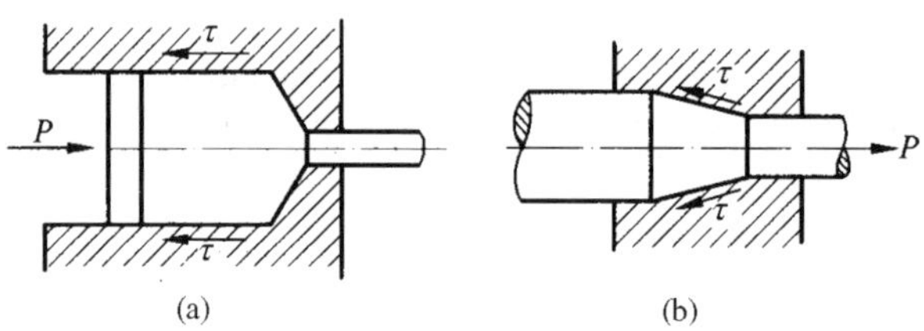

Figure 16.3 Deformation force increased by friction.

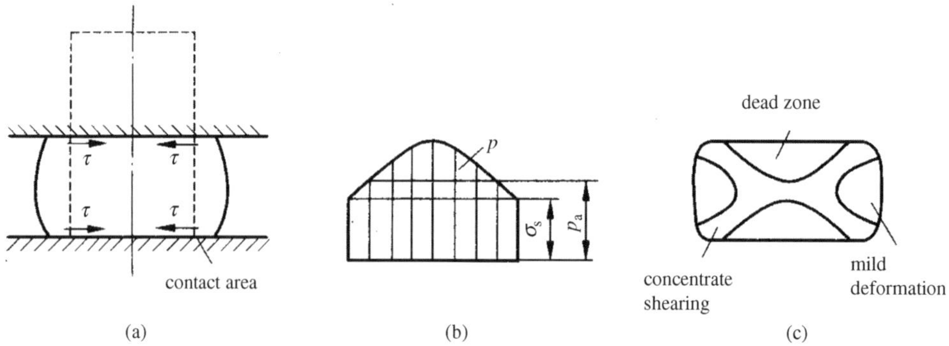

Figure 16.4 Influence of friction in upsetting. (a) Shear stress directions; (b) increase of interface pressure; (c) non-uniform deformation.

and the die pressure must be increased to lead to friction peak appearance (see Figure 16.4b). As a result, the die pressure may be much higher than the material yield stress σ_s.

16.1.3.2 Non-Uniform of Deformation

For convenience, in the metal forming analysis, the uniform deformation is often assumed. In fact, the assumption is not real.

First, the friction limits the deformation to have the workpiece surface under the action of shearing. Therefore, the workpiece deforms in the minimum energy. Under the constraint of the adhesive friction zone, the adjacent local deformation is brought about to produce some metal dead zones, as shown in Figure 16.4c. Also, because of heating or cooling, the internal deformation of the workpiece may be quite different from the external deformation.

Furthermore, the geometric conditions in the machining process may cause the heterogeneity of deformation. In a forming process, when the whole thickness of the workpiece deforms, the non-uniform deformation caused by the geometric conditions is not very obvious.

In many processes, the friction and geometric conditions may cause non-uniform deformations simultaneously. The non-uniform deformations sometimes reinforce and sometimes buck each other.

16.2 Forging Tribology

Forging is the earliest plastic processing technique, including hot forging, cold forging, free forging, die forging, open die forging and so on. These are all intermittent processes. Therefore, in forging, the remaining lubricant, wear or product surface changes continually. Because the forging process is rarely in a stable state, it is very difficult to carry out a systematic analysis. Here, we will introduce tribological analysis methods of upsetting, open and closed die forgings of the axisymmetric cylinder.

16.2.1 Upsetting Friction

16.2.1.1 Cylinder Upsetting

Because the sides of the forge piece in upsetting can freely deform, the required energy is minimal. The stress state is composed of yield stress σ_s and shear stress τ. The influences of the load on the stress, strain and friction peak are as shown in Figure 16.5.

Based on the ratio d/h_0 of the cylinder diameter and the height, the following situations occur:

1. Zero friction in Figure 16.5a. Because there is no friction, the ends can expand freely to maintain the workpiece as a cylinder, that is, $p_a = \sigma_s$.
2. Low friction in Figure 16.5a. The end extention is limited by the shear stress τ. When $\tau < k$, the sides of the cylinder significantly appear convex. The end faces slide outwards from the center and the cylindrical central line is the neutral line. The pressure has a low friction peak, as shown in the left picture of Figure 16.5d. The average pressure is only a little higher than the yield stress σ_s.
3. Adhesive friction in Figure 16.5a. When the ratio d/h_0 is large enough, it is in the adhesive state, that is, $k = fp_a$ If the friction is small, the deformation is not significant because the dead zone is near zero and the friction peak is arc-shaped, as shown in the right picture of Figure 16.5d. The pressure p_α increases sharply.
4. The large ratio d/h_0 in Figure 16.5b, e and f. When $\tau > k$ for all points of the contact surface, the dead zone increases and friction peak sharply increases, as shown in the right picture of Figure 16.5e and f. At this time, even in rough die upsetting without a lubricant, the total pressure is not large because the workpiece sides are compressed and bent, as shown in the right picture of Figure 16.5e. When $\tau < k$, the average pressure will increase dramatically, as shown in Figure 16.6b. The friction peak and shear deformation are shown in the left picture of Figure 16.5e and f.

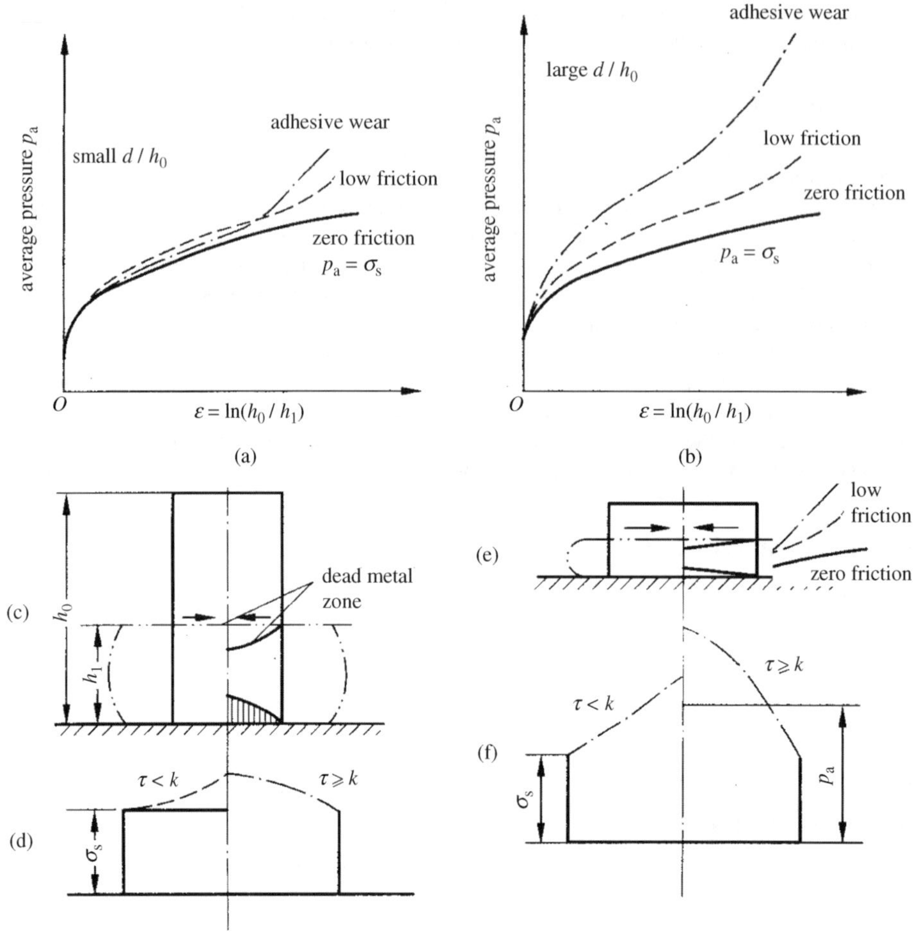

Figure 16.5 Deformation and stress of cylinder upsetting.

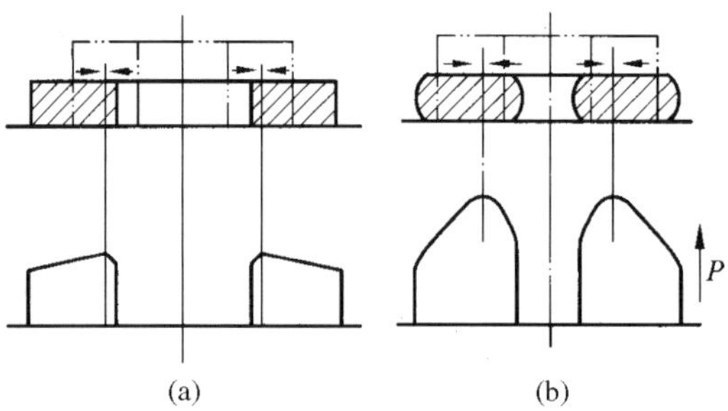

Figure 16.6 Deformation and pressure distribution of ring upsetting. (a) Low friction; (b) high friction.

16.2.1.2 Ring Upsetting

As shown in Figure 16.6, when the surface friction is zero or small, the circular workpiece is compressed to expand easily. The radial deformation is faster than the deformations in the other directions and friction increases slightly. By the minimum energy principle, it is known that the diameter of the central hole increases less than the diameter of the neutral circle. The deformation and pressure distribution are shown in Figure 16.6a. When friction is high, the diameter of the circular piece decreases, the inner and outer sides are drum-shaped and the pressure peak considerably increases, as shown in Figure 16.6b.

The ring upsetting can be used to evaluate the rolling performance of the lubricant. If the diameter is only reduced by a little amount, it means that shear resistance is small so the friction is low and the rolling performance of the lubricant is good.

16.2.2 Friction of Open Die Forging

In the open die forging, the metal forms a flash in the die parting face. Its function is to fill the liquid metal into the entire die chamber. When the blank begins to be deformed, the metal flows into the die chamber and also into the flash gutter. Subsequently, the extra metal of the blank is filled with the flash gutter to cause resistance to have the die filled with the metal. When the parting side forms a complete flash, the metal is full of the die chamber. Therefore, there should be high friction in the flash gutter, while low friction is required in the die chamber so as to be beneficial in shaping.

Usually, in the open-die forging, the lubricant can form a low-friction film in the die chamber and hardly forms a high friction film in the gutter at the same time. Therefore, the lubricant is of little use. When forging a larger cross-section piece, a lubricant is usually used to fill the metal fully with the die. It has been found that in open-die forging, the lubricant may prevent the metal from filling with the die because lubrication will cause the metal to overflow from the gutter easily.

According to measurement of the push force in forging, the adhesive frictional force between the die and workpiece can be estimated. For example, by using a ring die with a conical hole in Figure 16.7, a cylindrical billet is forged full of the hole. Then, turning the die upset to measure the push force P_E, we can determine the adhesive friction.

16.2.3 Friction of Closed-Die Forging

In the closed die forging process, the die chamber is closed and the gap of the parting face remains unchanged. The basic deformation procedure is the combination of forward and backward extrusions. Extrusion is always accompanied with friction to impede the metal's flow. Therefore, effective lubrication is necessary.

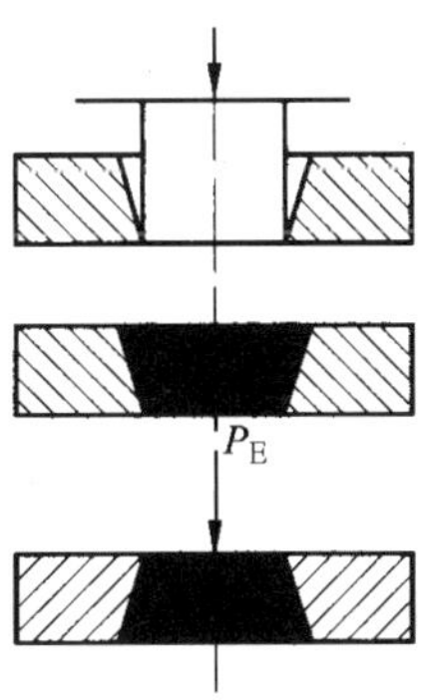

Figure 16.7 Measurement of frictional force by pushing experiment.

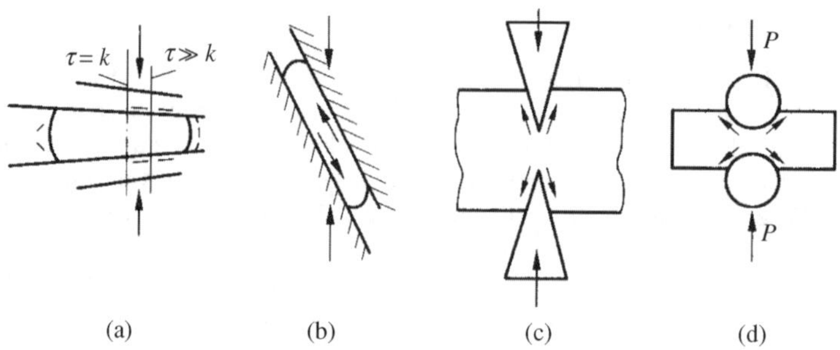

Figure 16.8 Evaluation methods of forging lubricant.

In forging, the function of the lubricant is to pretect the formed surface of the product during the relative sliding to the die wall because lubrication failure often cause adhesion during this period.

16.2.4 Lubrication and Wear

The lubricant is usually selected depending on technology, workpiece geometry, contact pressure, material ductility and relative sliding velocity.

The evaluation of the applicability of a lubricant is mostly determined by the experiment. Usually, the pressure variation or die adhesion is used as the evaluation standard. Four evaluation methods of a forging lubricant are shown in Figure 16.8. Figure 16.8a indicates that the two tilted surfaces force the central surface of the flat blank to move to one side. This method is widely used to evaluate lubricant performance because the location of the central surface not only depends on the tilt angle of the die surfaces, but also depends on frictional resistance.

If the lubricant is good, the metal flow in the width direction is always more. Therefore, we can evaluate the grade of the lubricant according to the material flow, determine the average surface friction coefficient f, or obtain the surface shear factor m with the appropriate way.

The other evaluation methods of the forging lubricant include:

1. Sandwich a metal sheet by the two inclined surfaces as shown in Figure 16.8b. According to the applied force, calculate the ratio of the horizontal and vertical forces to determine the friction coefficient.
2. Use the wedge with an angle of 30° to carry out the indentation experiment as shown in Figure 16.8c. When the force is applied, the penetration of the wedge can be calculated by the slip-line theory. This experimental method has a sufficient sensitivity for the high frictional force.
3. As shown in Figure 16.8d, use two cylindrical indenters to apply a force on a blank. The penetration ability of the lubricant is related to the force and shape of the indenters. First, determine the indentation force of the rough indenters and then the smooth. According to the indentation force, the friction coefficient can be calculated. This experimental method is simple and reliable, commonly used to evaluate the high-temperature property grade lubricant.

Although by using a liquid lubricant in forging, the partially plastic hydrodynamic (PHD) lubrication is possible in some area; the major area is in the boundary lubrication so it belongs to the mixed lubrication.

In the cold forging process, a lubricant is required for the following reasons: to withstand the pressure of more than 20MPa; to maintain a low friction coefficient; to possess a good thermal stability at 300–400 °C without reducing the lubricating effect; to be of a good coating ability and to be easily removed. Some examples would be light mineral oil, a mixture of molybdenum disulfide, graphite or oleic acid.

At high pressure and large deformation, a liquid lubricant may cause severe adhesive wear. Therefore, solid synthetic lubricant should be used. For example, thermoplastic material such as polyethylene or polyvinyl chloride can be used to form a solid lubrication film on the surface, but it is expensive.

For a forging process, the study of the sliding friction between the deformed metal and die wall has not yet been advanced enough to determine the friction coefficient accurately. The typical friction coefficient is low, about 0.05–0.1, with a good lubrication additive. If adhesion exists between the workpiece and die, the friction coefficient may increase to as high as 0.3 or above.

The surface roughness of the forging die significantly influences the friction resistance of the metal flow and the wear life of the die. A roughness with $R_a = 0.125$–2.50 mm is usually selected. For some cold punch dies, the surface roughness should be a little lower, $R_a = 0.125$–0.25 mm.

In general, in a forging process the contact surfaces will slide relatively under very high compressive stress at a very high temperature although the sliding velocity and displacement are not very large. In such a working condition, the wear mechanism of the die includes adhesive wear, oxidation wear, abrasive wear and even fatigue wear. The main wear form and locations of the die should be determined by the die material, die shape, forging technology and other factors.

The working life of a cold forging die is usually very long. When it is used for cold extrusion, a steel die can process about 100 000 workpieces. The tungsten carbide die can process 3–5 times more. The die material should have sufficient hardness and wear resistance. Furthermore, it cannot react with the extreme-pressure additive to prevent corrosive wear.

Practice has proved that lubrication can significantly reduce die wear. Figure 16.9a can be used to determine the wear according to the variation of the punch diameter. It shows that by using the light

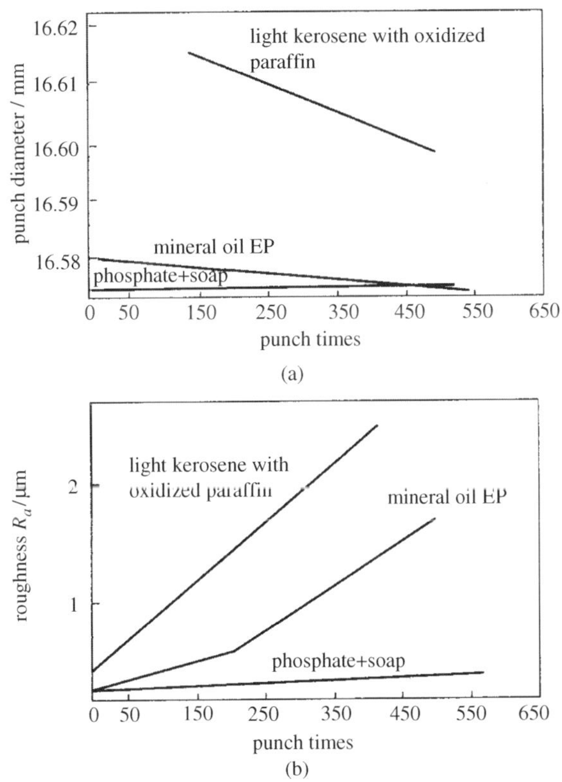

Figure 16.9 Wear curves of die.

kerosene with some oxidized paraffin, wear is very severe, while for the mineral oil with the extreme-pressure additive wear is less. Wear is quite minimal for phosphate and soap. In Figure 16.9b, the average peak-valley height is used to express the wear variation of the die surface roughness during punching.

16.3 Drawing Tribology

16.3.1 Friction and Temperature

As shown in Figure 16.10, in a drawing process, under the action of the drawing force P_0, the pressure of the metal on the die wall is p, the deformation of the metal is ε, the friction shear stress is τ, and the axial and radial normal stress are σ_t and σ_r respectively.

From the figure we know that the frictional force is equal to

$$F = A\tau, \tag{16.8}$$

where A is the contact area; and τ is the shear stress.

If the yield stress of the metal is σ_s, the friction coefficient f will be

$$f = \frac{F}{N} = \frac{A.\tau}{A.\sigma_s} = \frac{\tau}{\sigma_s}. \tag{16.9}$$

In drawing, the energy is mainly used for the effective deformation of the metal, non-uniform deformation, sliding friction inside the metal and external friction between the metal and die. Study has shown that the friction approximately consumes 10% of total energy. With the increase of the friction coefficient, the friction energy consumption increases proportionally to the total energy consumption. When the friction coefficient f changes between 0.02 and 0.1 and the reduced rate of the cross-sectional area is about 10–40%, the proportion of the friction power consumption is between 6% and 40%.

Almost all of the friction work is converted into heat. The heat generated by friction has a harmful influence on the drawing process. The friction work w_f can be expressed as follows:

$$\frac{dw_f}{dt} = fpv, \tag{16.10}$$

where v is the drawing velocity.

The wire temperature after drawing can be calculated below.

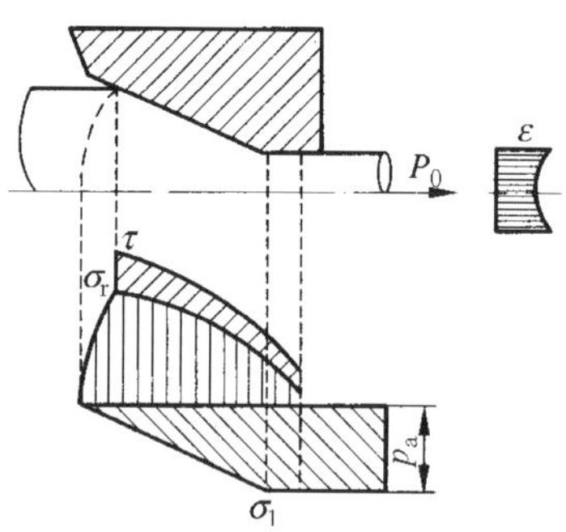

Figure 16.10 Stress distribution in drawing.

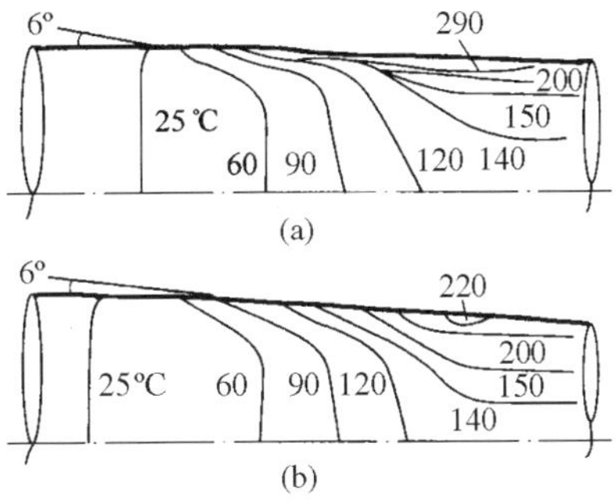

Figure 16.11 Temperature distribution in drawing wire.

First, the drawing stress σ is equal to:

$$\frac{\sigma}{\sigma_y} = \ln\frac{1}{1-\varphi} + \frac{2}{\sqrt{3}}\left(\frac{2}{\sin^2\alpha} - cot\,\alpha\right) + \frac{f}{\sin\alpha}\left(1 + \frac{1}{2}\ln\frac{1}{1-\varphi}\right)\ln\frac{1}{1-\varphi}, \tag{16.11}$$

where σ_y is the average yield stress; f is the friction coefficient; φ is the reduced area rate; and α is the half-cone angle of the die.

Equation 16.11 includes three parts: pure deformation work, shearing work and friction work. 80–90% of the heat generated in drawing is saved in the metal. If we assume pure deformation work and shear deformation work are both used to raise the metal temperature and only m of the friction work is used for heating, the temperature rise ΔT is equal to

$$\Delta T = \frac{\sigma_y}{J\rho c}\left\{\ln\frac{1}{1-\varphi} + \frac{2}{\sqrt{3}}\left(\frac{2}{\sin^2\alpha} - \cot\alpha\right) + \frac{mf}{\sin\alpha}\left(1 + \frac{1}{2}\ln\frac{1}{1-\varphi}\right) \times \ln\frac{1}{1-\varphi}\right\}, \tag{16.12}$$

where ρ is the metal density; c is the specific heat; and J is the mechanical equivalent of heat.

Figure 16.10 shows the temperature distribution in the deformed zone to draw the carbon steel wire with the sulfur content of 0.62% and the reduced area rate of 39.2% at the drawing velocity from 10 m/min to 100 m/min.

It is known from Figure 16.11 that at the exit of the die the temperature of the wire core is about 135 °C and is almost not affected by the drawing velocity. However, the temperature of the wire surface in the contact with the die wall increases significantly with increase of the drawing velocity. This is because the drawing velocity is high and surface heat is generated by the friction between the die walls. The heat cannot be conducted in time so friction work is all transformed into heat, which is absorbed by the wire surface and as a result, temperature increases quickly.

16.3.2 Lubrication

Lubrication is an important factor that affects the drawing process. A poor lubrication not only decreases the quality of products, but may also prevent the drawing from occuring.

The purposes of the drawing lubrication include: to reduce the friction so as to reduce the drawing energy consumption; to reduce the surface temperature, reduce non-uniformity of stress distribution so as to avoid breakage; reduce wear, prevent corrosion and prolong die life.

In order to reduce friction in drawing, the lubricant can be directly added on the wire surface before it enters the die. The lubricant can also play the function of cooling. The lubricant is required to be of high pressure resistance, under high temperature to maintain its lubrication performance and keep lubrication film effective. For example, in the high-carbon steel wire drawing, lubricant suffers the pressure in the die up to 2100 N/mm^2 and the temperature in the deformation zone up to 200 °C. If the lubrication condition is poor, the temperature may be even higher. In addition, the lubricant should be of good adsorption ability to the surface of the wire, otherwise the lubricant may easily be wiped off from the wire. However, after drawing, the lubricant needs to be easily removed.

In metal drawing, the lubrication mechanism is a combination of a variety of lubrication states. Hydrodynamic lubrication, boundary lubrication and mixed lubrication may exist at the same time in the deformation zone.

In drawing, because of the cone angle of the die, the lubricant in the deformation zone at the inlet has a strong hydrodynamic lubrication effect. The lubricant is dragged into the deformation zone to form a hydrodynamic lubricant film. In addition, because there are many lubricant pockets on the die and metal surface, the lubricant will be dragged into the deformation zone in drawing. The rougher the surface, the more the lubricant is dragged and the thicker is the lubricant film. In the boundary lubrication area, the lubricant is adsorbed by the metal surface. If we add sulfur, phosphorus, chlorine or other active additives into the lubricant, they react chemically with the metal to form the chemical reaction film with low friction under high temperature conditions because of the heat generated by the friction.

16.3.2.1 Establishment of Hydrodynamic Lubrication

In good condition, the lubricant can form a partial or full hydrodynamic lubrication film. The lubrication pressure in a drawing pipe is analyzed as follows:

The drawing pipe is as shown in Figure 16.12.

In the figure, r is the radius coordinate of the pressure pipe; r_0 is the inner radius of the pipe; r_D is the wire radius; h is the gap between the wire and the pressure pipe wall, $h = r_0 - r_D$; L is the pipe length; v is the velocity of the steel wire; p_E and p_A are the fluid pressures at the inlet and the outlet, respectively; and τ_0 is the fluid shear stress on the inner surface of the pipe.

When the pressures are difference, that is, $\Delta p = p_A - p_E \neq 0$, according to the balance condition, in the axial direction, we have

$$\tau_0 \pi (d + 2h) L + \tau_D \pi d L = p_A \pi (d + h) h$$

$$p_A = \left(\tau_0 \frac{d + 2h}{d + h} + \tau_D \frac{d}{d + h} \right) \frac{L}{h}, \tag{16.13}$$

where d is the wire diameter; τ_D is the shear stress on the wire surface.

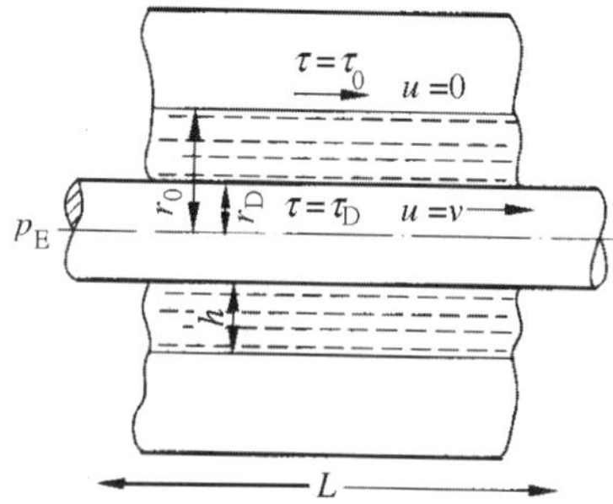

Figure 16.12 Lubrication pressure in pipe.

Because $h \ll d$, Equation 16.13 can be simplified to

$$p_A \approx (\tau_D + \tau_0)\frac{L}{h}. \tag{16.14}$$

We can see from the above equation, if τ_D and τ_0 are fixed, p_A varies with L/h accordingly. Therefore, we can reasonably select h and L to get a required lubricant pressure. If we increase lubricant viscosity so that τ_D and τ_0 are increased, the lubricant pressure can also be increased. However, because the lubricant viscosity is generally small, that is, τ_D and τ_0 are small, in order to realize a full film lubricated drawing, the lubricant pressure p should reach the value to have the drawing metal yielded so that the lubricant film can maintain a certain thickness to separate the steel surface and die wall completely. Therefore, in order to realize the hydrodynamic lubrication, and use a relatively small gap h, the length L of the pipe should be long enough.

16.3.2.2 Hydrodynamic Lubrication Calculation of Drawing

The soap powder is often used as a lubricant in drawing. It is a non-Newtonian fluid. Its constitutive equation can be approximately expressed as follows:

$$\tau = A + B\frac{du}{dz}, \tag{16.15}$$

where τ is the shear stress; du/dz is the shear rate; and A and B are the Reynolds factors of the lubricant. The relationships between A or B and the pressure index ϕ and p can be expressed as:

$$\begin{aligned} A &= A_0 \exp(\phi p) \\ B &= B_0 \exp(\phi p), \end{aligned} \tag{16.16}$$

where A_0 and B_0 are the Reynolds factors while pressure p is equal to zero.

From the one-dimensional Reynolds equation and Equation 16.15, we have

$$h^3\frac{dp}{dx} = -6Bv(h - h^*). \tag{16.17}$$

Substitute Equation 16.16 into Equation 16.17, the non-Newtonian lubrication equation of drawing is:

$$\exp(-\phi p)\frac{dp}{dx} = -6B_0 v\left(\frac{h - h^*}{h^3}\right), \tag{16.18}$$

where h^* is a constant.

By using equation (16.18), the hydrodynamic pressure distribution of the non-Newtonian fluid lubrication of drawing can be obtained.

16.3.3 Wear of Drawing Die

The drawing is a deformation process while the metal wire passes through the die hole. Because pressure is very high, the die hole may produce the adhesive wear. The wear causes the diameter of the die hole to increase so that the diameter of the product increases. In the production the wear value must be limited to keep the quality of the product.

In addition, because the wear of the die in drawing is non-uniform, the die geometry will be changed. This can not only damage the lubricant film, but also decrease the quality of the drawing product.

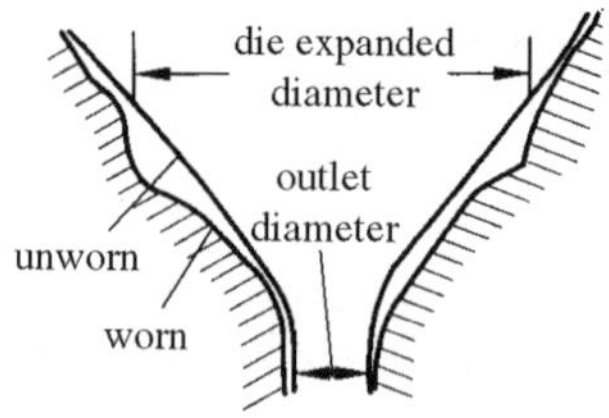

Figure 16.13 Cross-section shape after drawing worn.

16.3.3.1 Wear of Die Shape

After drawing, the die shape is worn as shown in Figure 16.13. The wear mainly occurs in the following three positions.

1. **The entrance.** It is severely worn. The worn shape is a ring. Because of wear, the shape and position of the entrance are changed and that cause the product surface and the lubricant film thickness to change as well.
2. **The cone angle.** The wear of the angle will change the die cone shape and thus directly affect lubricant film thickness and drawing force.
3. **The diameter.** The wear of the diameter enables the diameter of the product to be larger. Because there is a designed allowance, the die life will be reduced. If the die wear is non-uniform, the drawing product will form a noncircular cross section such that the roundness of the finished product does not meet the requirement.

16.3.3.2 Wear Mechanism

The main kinds of the drawing die wears are the adhesive wear, the abrasive wear and the fatigue wear. In many cases, the chemical erosion and physical damage are also included. When the severe adhesion exists or the significant vibration appears on the moving part, serious adhesive wear occurs on the friction surface. The wear curve I in Figure 16.14 shows this situation.

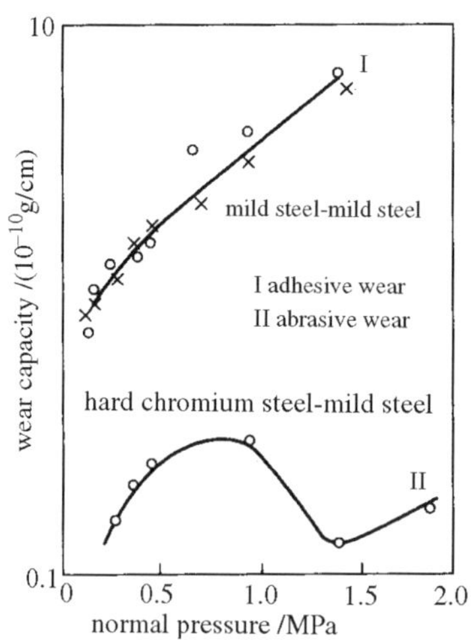

Figure 16.14 Relationship between wear and pressure without lubrication.

Abrasive wear is because most hard and brittle oxides on the metal surface are peeled off in the hot drawing to form the abrasive particles. In addition, the middle particles exist in the lubricant or coating. However, in the wet drawing, these particles cannot be effectively removed and formed abrasive wear. Furthermore, when the metal contacts with the die, the two body abrasive wear is likely to occur. The wear severity depends on the mating materials of the workpiece and the die.

Fatigue wear is due to the high stress gradient in the die hole under continuous loading. Vibration will worsen the wire fatigue wear. The high temperature will cause the thermal fatigue wear which may bring about some cracks in the steel die, but the fatigue wear is usually not the main wear form in drawing.

16.3.3.3 Measures to Reduce Wear

1. Improve the die material
   Use the high anti-wear material to make the die. The general drawing die is often made of WC-CO alloys; the fine wire drawing die is made of diamond graphite or ceramics. Particularly, a die with the ZrO_2 ceramic coating can be used on he drawing stainless steel wire, which is considered as the most promising die. On the die surface, a hard chromium coating can also be plated to improve the die life.
2. Cool die or cool wire directly
   The die can be heated during drawing. Because a lubricant is only valid for the specific temperature range, the lubricant film will fail after temperature exceeds the limit due to decomposition and coke so the wear of the die will also increase. During drawing, about 20% of the heat accumulates in the die. If the heat is not removed in time, the temperature will be raised much higher. If the temperature distribution is not uniform, the local high temperature will cause the die to wear seriously, or cause the die and the steel sleeve to separate, causing the mold core to break. Although by directly cooling the die the temperature can be reduced, the more important thing is to improve temperature distribution inside the die so as to reduce the sleeve temperature significantly. In addition, during drawing, cooling the drawing wire directly will not only improve the mechanical properties of the steel wire, but also reduce the die temperature and enhance the die life.
3. Improve lubricant and lubrication method
   If the drawing lubrication method is different, the die wear is also different. Figure 16.15 gives the wears of the different lubrication methods in drawing. The wear capacity is expressed by the diameter. Table 16.1 gives the influences of the lubrication methods on the die life. The die life of the forced lubricating drawing or the direct cooling drawing is usually two times higher than that of the

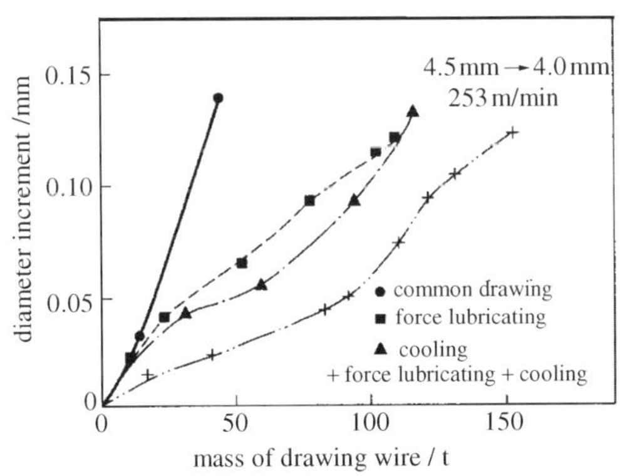

Figure 16.15 Die wear of different lubrication methods.

Table 16.1 Life ratio of die for different lubrication

	Dry friction	Lubrication
Common drawing	1.0	1.0
Forced lubricating drawing	2.0	—
Cool drawing	2.1	—
Forced lubricating and cooling drawing	3.0	—
Forced lubricating drawing	—	3.0–3.8

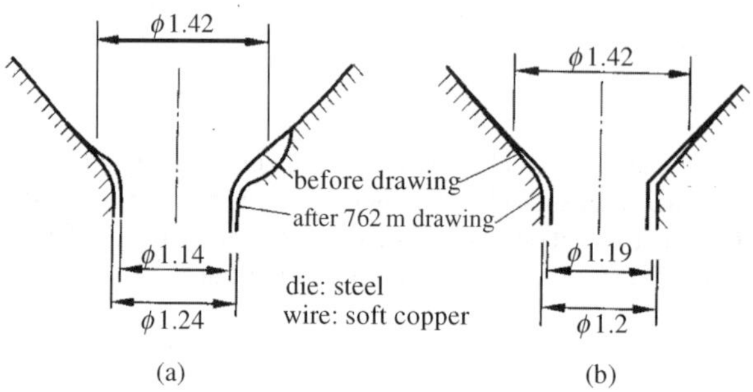

Figure 16.16 Die wear.

common drawing. If the forced lubricating drawing and direct cooling drawing are used together, the life can be increased by twofold. For the wet drawing, the forced lubricating drawing can be used to increase the die life 2–3 times.

Figure 16.16 shows the worn sections of the die in the common dry drawing and the forced lubricating drawing. It can be seen that under the same conditions, after drawing 762 m, the die wear amounts of the two methods are significantly different.

4. Use counter pull drawing
   The counter pull drawing is a method of applying force at the entrance in the opposite direction of the motion. Before the metal enters the die hole, it is pulled and thus elastically deformed such that the diameter becomes smaller. This reduces the friction and wear of the die hole.
5. Using rotary die drawing
   The rotary die drawing is to fix the die on a rotating cylinder. Then, drive the cylinder to rotate the die in drawing. Therefore, when the metal wire in the rotating die is deformed, its surface has a relative spiral motion to the die hole that changes the friction direction. This will decrease friction and thus reduce die wear. During drawing, because the die rotates at a high speed, the wear of the die-hole wall is quite uniform and this enhances die life. Study has shown that die life can be increased 10–100 times when the rotary die is used.

16.3.4 Anti-Friction of Ultrasound in Drawing

In 1955, Blaha and Lan-glueker found that ultrasound can significantly reduce the resistance of plastic deformation of the single-crystal zinc, so-called the Blaha effect. Subsequently, a variety of ultrasonic methods are explored to apply to the plastic forming process. Study has also shown that the ultrasound can reduce the wire drawing force and frictional force to improve the surface quality of the wire, reduce the number of the intermediate annealing and benefit the low plastic and hard processing materialing to produce the fine wire. In drawing, applying the ultrasound to the metal forming process can not only

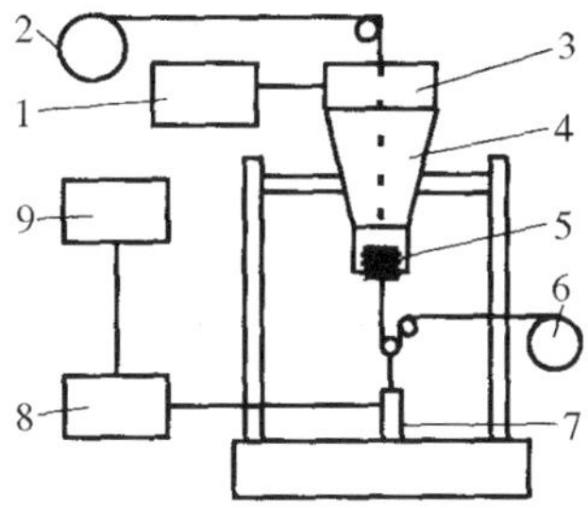

Figure 16.17 Ultrasonic drawing experimental tester: (1) generator; (2) supplement wheel; (3) transducer; (4) horn; (5) die; (6) reel; (7) strain gage; (8) strain gage; (9) computer.

reduce frictional force, but also improve the reduced rate of the cross-section. Thus, this is particularly suitable for the thin-walled pipe forming. In addition, when the ultrasound is applied in the deep drawing, the drawing force can be reduced and the deep punch ratio can be increased. Therefore, it can be said that the application of the ultrasonic plastic processing is a promising and special processing technique. In the following section, the work to reduce the frictional force with the ultrasound in drawing is discussed [4].

The ultrasonic wire drawing experimental device is shown in Figure 16.17, which includes the ultrasonic generator, the ultrasonic transducer, the ultrasonic horn, and the wire-drawing die. The ultrasound generator transmits the 50 Hz AC power into a high-frequency oscillation signal which provides the energy to excite the die.

In testing, the drawing force is measured by a strain gage. One end of the copper wire will be pre-grinded through the wire-drawing die and wound on the reel, which is driven by a motor controlled by a DC thyristor-supply to change the drawing speeds. Three kinds of lubricants, the saponification solution, the lubricating oil with the viscosity of 17 mPa·s and water, are used in drawing.

Figure 16.18 is the ultrasonic output current curves of the drawing force with a speed of 131.3 mm/s and saponification solution. The dashed line is the drawing power without applying ultrasound, while the solid line is the result of applying the ultrasonic vibration with output current of 0.7 A. The drawing force drops drastically when the ultrasound is applied. The average drawing force has decreased about 37% in the period of applying the ultrasound. It can also be seen from the figure that after applying ultrasound, the drawing force fluctuates significantly over time. The average drawing force F_{av} changes with the output current I of the ultrasonic generator, as shown in Figure 16.19. In these experimental conditions, the average drawing force increases approximately linearly with decrease of the output current.

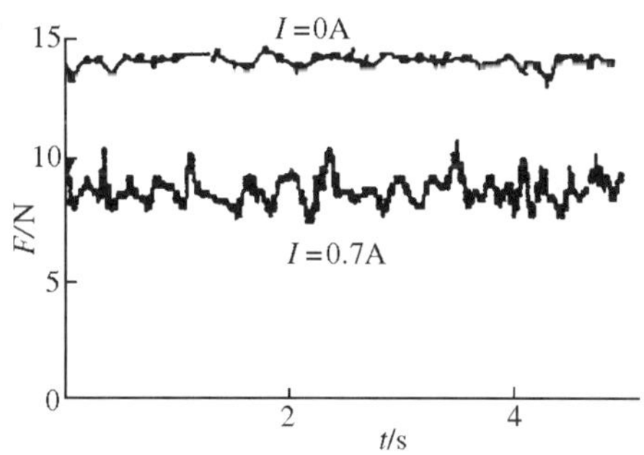

Figure 16.18 Drawing force varying with time for different output currents.

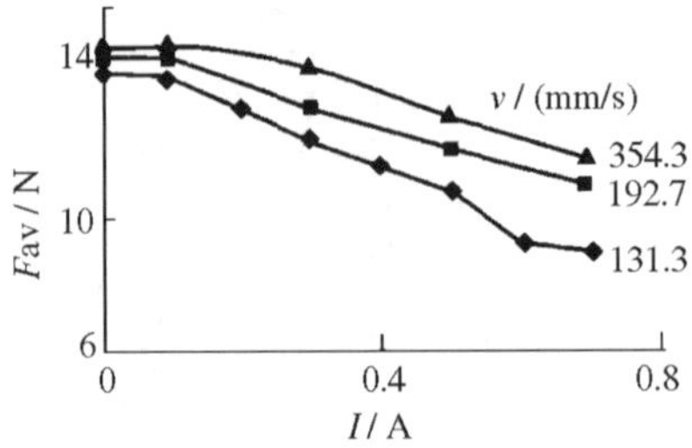

Figure 16.19 Average drawing force varying with different output current.

Figure 16.19 also gives the drawing forces under several drawing speed with the saponification solution. Under the same intensity of ultrasonic vibration, the greater the speed, the lesser is the force reduced by ultrasound.

Although the chemical composition and viscosities of the three kinds of experimental lubricants are quite different, the experimental results showed that if the other processing conditions are the same, the lubricating materials do not affect the drawing force. The ultrasound may improve the lubrication and reduce the friction, but it cannot significantly change the drawing force. If we only consider the experimental results of the drawing force, it is difficult to distinguish the influence of the ultrasound on the lubrication and friction. Therefore, the surface topographies with and without the ultrasonic drawing are also compared. Figure 16.20 gives the scanning electron topographies of the wire surface with the output current of 0.5 A and 0 A respectively. Without ultrasonic vibration, the wire surface has many potholes and tiny cracks, processing textures are not clear, and there are signs of adhesion. Under the action of the ultrasonic vibration, the surface textures are very clear, uniform, smooth, and without any small cracks. The results show that the ultrasonic vibration can improve lubrication between the die and wire, reduce surface adhesion and damage, and enhance surface quality.

According to experimental studies, Meng *et al.* proposed that the following factors may decrease friction and reduce the drawing force in the plastic processing by ultrasound: (1) the Blaha effect; (2) the high-speed impact of the die on the workpiece; (3) the vibration making the die move forward relatively to the workpiece so as to create a positive friction which reduces the original friction; (4) the ultrasound forcing the lubricant into the contact interface that easily enhance the lubricating performances; (5) the ultrasonic vibration causing the workpiece temperature to increase and thus decreasing deformation resistance.

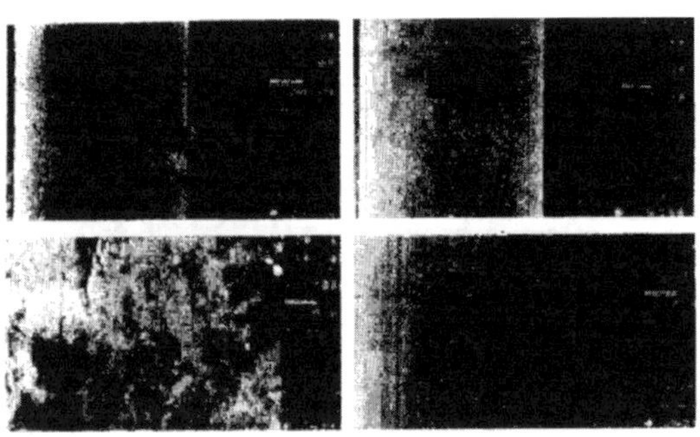

(a) (b)

Figure 16.20 Scanning electron topography of drawing wire surface. (a) $I = 0$ A; (b) $I = 0.5$ A.

16.4 Rolling Tribology

16.4.1 Friction in Rolling

16.4.1.1 Pressure Distribution and Frictional Force

The experimental results show that the friction varies along the contact arc in the rolling process. It is complicated, neither obeying the dry friction law nor the adhesive friction theory. It is closely related to the sliding of the metal and thus the pressure and frictional force are also related. In the rolling deformation zone, the friction points in the direction of the neutral face to limit the rolled part to move along the contact arc. This is the reason that the friction peak appears along the deformation zone. The characteristics of the frictional force and geometric shape of the deformation zone significantly influence the height of the peak. Figure 16.21 plots the pressure and frictional force under different conditions. The friction peak changes significantly.

Figure 16.21a: When $l/h > 5$, the part of the contact arc close to the inlet and outlet is the sliding zone so in the area the dry friction law should be obeyed. The unit pressure p gradually increases to the center of the contact arc. When the frictional force increases to the half shearing yield stress, that is, $\tau = fp = k/2$, the frictional force is constant. The central adhesion area is the plastic deformation stagnation zone, that is, no plastic deformation. For such a situation, the frictional force is approximately a straight line and is very steep.

Figure 16.21b: When $l/h = 2$–5, the constant friction area disappears. The friction distribution along the contact arc is triangular. This is because the contact arc is not long enough to achieve the maximum friction so the plastic deformation stagnation zone appears. For this situation, the frictional force in the pressure distribution zone is a little steep.

Figure 16.21c: When $l/h = 0.5$–2, the adhesion covers the whole deformation zone. Metal sliding tendency is small. Friction distribution can be expressed by a triangle. Frictional force varies gently.

Figure 16.21d: When $l/h < 0.5$, the metal sliding tendency along the contact arc is obvious so the influence of pressure on frictional force is small. Therefore, the frictional force is very flat.

16.4.1.2 Friction Coefficient of Rolling

Friction in rolling is complicated because there are many influencing factors. The following ways can usually be used to approximately calculate the friction coefficient.

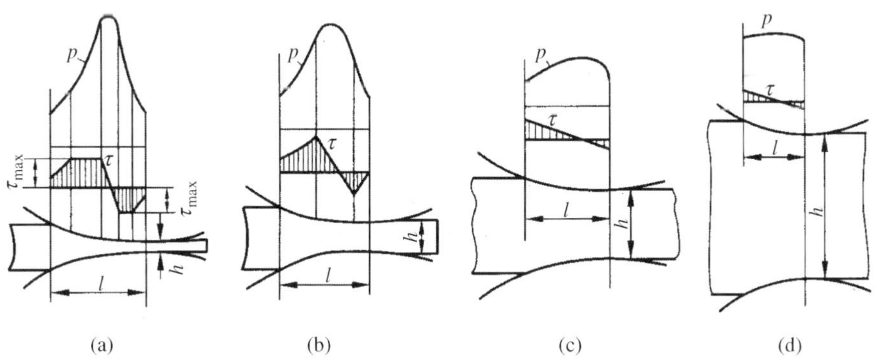

Figure 16.21 Pressure and friction along contact arc.

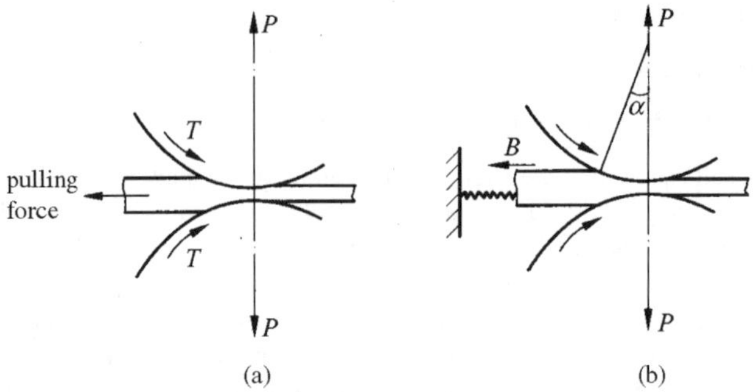

Figure 16.22 Determine frictional force through slip.

1. Calculate coefficient friction by torque
   In strip rolling, the tension grows continually until the plane moves to the neutral point of the outlet and the strip begins to slip (Figure 16.22a). According to the slip torque, the friction coefficient can be determined as follows:

$$f = \frac{T}{PR}, \tag{16.19}$$

where T is the torque of the two rollers; P is the rolling force; and R is the roller radius.

If there is no torque measuring device, a spring tension device can be used to be fixed on the strip, shown in Figure 16.22b. According the Pavlov theory, the friction coefficient is equal to

$$f = \frac{B}{4P} + \tan\frac{\alpha}{2}, \tag{16.20}$$

where B is the tension; and α is the entrainment angle.

2. Calculate friction coefficient by pre-slip
   In the neutral surface, the velocity of the strip is the same as that of the roller. Therefore, the velocity at the outlet is higher. As shown in Figure 16.23, the forward slip is defined as

$$S_f = \frac{v_1 - v_0}{v_0}, \tag{16.21}$$

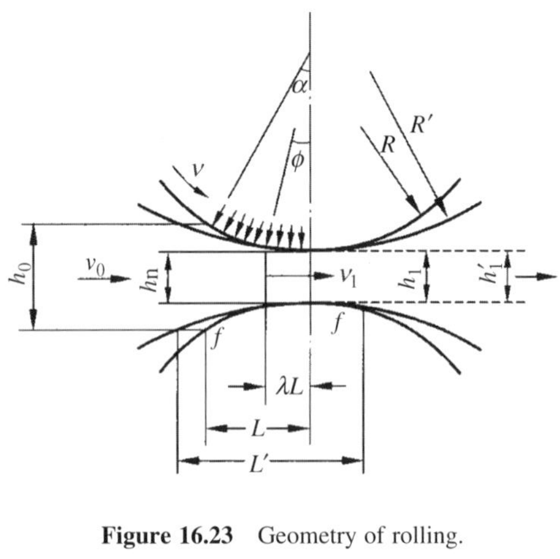

Figure 16.23 Geometry of rolling.

where v_0 is the velocity at the position of the entrainment angle α; v_1 is the velocity at the position of the neutral angle ϕ.

According to the geometric parameters of the rolling system, the forward slip is approximately equal to

$$S_f = \frac{1}{2}\phi^2\left[\frac{2R'}{h_1} - 1\right], \tag{16.22}$$

where h_1 is the rolling thickness.

From Figure 16.23, it can be known that the neutral angle is equal to

$$\phi = \frac{\alpha}{2} - \frac{1}{f}\left(\frac{\alpha}{2}\right)^2, \tag{16.23}$$

where

$$\sin\alpha = \frac{L}{R} = \left[\frac{h_0 - h_1}{R}\right]^{\frac{1}{2}}. \tag{16.24}$$

Substitute Equation 16.24 into Equation 16.23 and consider $\sin\alpha \approx \alpha$ because α is very small, we have

$$\phi = \left[\frac{h_0 - h_1}{4R}\right]^{\frac{1}{2}} - \frac{1}{f}\left[\frac{h_0 - h_1}{4R}\right]. \tag{16.25}$$

The forward slip can be measured by the indentation method on the roller surface. Therefore, after the neutral angle ϕ has been determined from Equation 16.22, the friction coefficient f can be obtained from Equation 16.25.

3. Calculation of sliding friction coefficient

With increase of the indentation, the neutral angle gradually moves toward the outlet. When the indentation reaches the critical value, the neutral surface moves at the outlet section, that is, $\phi = 0$. Then, slip occurs. If there is no tension and α is relatively small, Equations 16.24 and 16.25 can be used to obtained the slip entrainment angle

$$\alpha_s - \left(\frac{h_0 - h_1}{R}\right)^{\frac{1}{2}} = 2f. \tag{16.26}$$

After the entrainment angle has been measured, the friction coefficient can be obtained by the above equation.

16.4.2 Lubrication in Rolling

Lubrication in rolling is very important and many types of lubricants for rolling have been developed. The commonly used lubricants are mineral oil, animal and plant oils as well as fats, containing various additives. At the same time, water-based media and emulsions are also used as the lubricant in rolling.

In most rolling strips, a low entrainment angle and high rolling velocity are helpful in forming an oil film. However, the forward slip should be stable. Together with the limitations of the strip surface quality and the annealing color spot, the actual lubrication is often the mixed lubrication so that the workpiece will contact with the roller surface. Despite the mixed lubrication, some other lubrication states also play important roles.

16.4.2.1 Full Film Lubrication

Based on the elasto-hydrodynamic lubrication theory, the plastic hydrodynamic lubrication theory has been proposed for rolling, combining with fluid mechanics and elastic-plastic mechanics to carry out the numerical analysis. A number of lubrication film thickness formulas for rolling have also been proposed. Obviously, compared with the point or line contact EHL, it is very difficult to establish a complete mathematical model for a rolling problem to consider the plastic flow. Moreover, the full film lubrication cannot be realized in the actual rolling process. Therefore, based on some formula, an approximate analysis of the film thickness is usually used and the study focuses on the mixed lubrication condition.

A commonly used film thickness formula of the plastic hydrodynamic lubrication for rolling is as follows:

$$h = \frac{6\eta v}{2k \tan\theta}, \tag{16.27}$$

where η is the viscosity; v is the roller circumferential velocity or the average velocity of the strip and roller at the inlet; k is the yield stress. The plane strain yield shear stress is generally used, that is, $2k = \sigma_1 - \sigma_3 = \sigma_f$; and θ is the wedge angle at the inlet.

From Equation 16.27, it can be known that the film thickness increases as the yield stress or the rolling pressure decreases. Usually, the rolling pressure can be adjusted by changing the tension. Therefore, for a given metal, adjusting the tension can vary the film thickness.

It should be noted that the plastic hydrodynamic lubrication can also be applied to other metal forming problems. For example, Meng *et al.* carried out the finite element analysis on the plastic hydrodynamic lubrication for cold forging [5].

16.4.2.2 Mixed Lubrication

The actual rolling process is carried out under the mixed lubrication condition. The lubrication condition depends on the strip surface quality as well as the lubricant properties.

1. Film thickness

   In the mixed lubrication condition, the average film thickness reflects the effect of the hydrodynamic lubrication.

   The general influence factors on the film thickness are the inlet geometric parameters, lubricant viscosity and rolling velocity.

   a. Inlet geometric parameters

      While applying Equation 16.27, the entrainment rolling angle α is often used instead of $\tan\theta$. From the geometric relationships of Figure 16.23, it is known that with increase of the roll diameter and decrease of the indentation, α decreases so that the film thickness of Equation 16.27 increases.

   b. Viscosity

      With increase in lubricant viscosity, the average film thickness increases and the lubrication tends to be the hydrodynamic lubrication.

      While considering viscosity, the lubricant composition and its pressure-viscosity coefficient cannot be ignored. For example, the indentation of a low viscosity oil is sometimes deeper than that of a high viscosity oil. For another example, the viscosity of paraffin is the same as that of the naphthenic synthetic oil, but the lubrication ability of paraffin is much better. If we add 3% of the low density extreme pressure additive into oil, the equivalent viscosity increases so that a very low friction coefficient can be obtained.

   c. Rolling velocity

      Although a high rolling velocity is beneficial to the formation of a hydrodynamic lubrication film, it is limited by the technology.

2. Lubrication failure

   The lubricant film includes the fluid film and boundary film. When the film breaks down, surface adhesion and damage will appear. When rolling aluminum, under critical indentation, adhesion appears first on the serious injured surface. If indentation is shallow, the rolling can well operate, but if indentation is deep, adhesion appears non-uniformly on the roll surface and expands quickly. Finally, because the strip surface becomes very rough, has been cracked and even covered with the metal powder, the rolling has to stop. For the given film thickness, rolling velocity and indentation, the additive can be used to slow down the adhesion.

   For steel and stainless steel rolling, when the speed is high enough, large amount of heat will be produced to bring about some partial adhesion. Thus, the local adhesion or the elongated defects on the strip surface are known as thermal abrasion or friction adhesion. Thermal abrasion can be prevented by decreasing the rolling velocity below the critical point.
3. Water-based lubricant

   Water is a good cooling agent but a poor lubricant. It cannot prevent adhesion so water is occasionally used to roll the hardly adhesive metal, such as ordinary carbon steel or copper. However, in practical applications, water is often used together with a lubricant. For a cold-rolled steel strip, two popular application methods are as follows.

   a. Use water and lubricant separately

      The pure lubricant without water is pre-added onto the strip surface, and then some water is added to roll. This technique is mainly used in the series mill to roll the steel plate while palm oil is used as a lubricant. According to the surface quality of the rolled strip, it can be known that adding water to the oiled surface has little effect on improving lubrication, but it has a cooling effect.
   b. Emulsion

      When emulsion is used as a lubricant, the metal surface has good infiltration. The mineral oil-based emulsion has poor lubricity, but for non-ferrous metal, the lubricity can be improved by adding some oil compounds. The lipid-based emulsion has good lubricity in the steel rolling. The lubricity of the composite mineral oil-based emulsion is gradually improved with increase of the concentration before the concentration reaches 10%. Therefore, the lubricity of the low stable emulsion is better than that of the high stable emulsion. Pre-coating some emulsion on the strip can reduce the sensitivity of lubrication.

16.4.3 Roller Wear

In most cases, the normal wear is proportional to the sliding distance. In rolling, the relative sliding distance is only part of the whole rolling length. Thus, the wear capacity is very small in rolling. However, when the initial finish of the roll surface disappears in rolling, the product surface quality will be affected. Usually, the adhesive wear and abrasive wear are the common wear forms of the roller, but the roll discard is mainly due to peeling off the surface caused by fatigue.

For soft rolling material such as aluminum, the adhesive wear is mainly caused by the local adhesion, while for the harder materials such as titanium, the adhesive wear may occur directly. When surface oxide is very hard and is supported by the solid base, abrasive wear is very clear. When we roll the hard aluminum alloy, the stainless steel or nickel-based alloys, most wear is of the abrasive type. When we use the lubricant to prolong roller life, it might lead to corrosion wear, but the damage is only local.

The main reason to discard a roller is spalling. Spalling is sometimes very deep, and most of the cold-rolled hardened layer is worn. Because the local wear results in stress distribution being non-uniform, the cracks and residual stresses generated in grinding will make spalling worse. The spalling is formed by shallow crack or ring crack extending to the boundary between the hardened layer and roller center. The shallow debris peeled off can repair the roller surface by re-grinding.

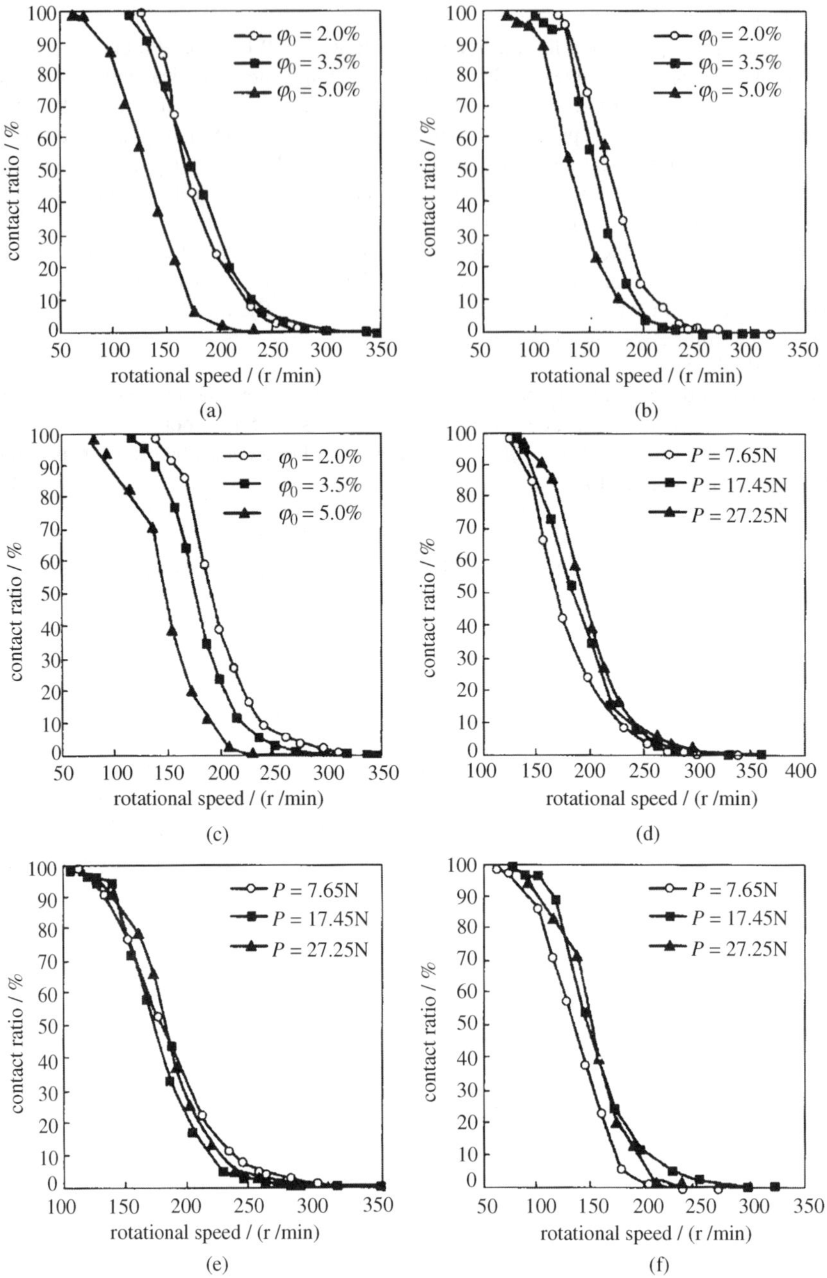

Figure 16.24 Lubrication performances of emulsion.

Sometimes the roller surface is occasionally damaged seriously. For example, at the welding point, object occasionally entering the deformation zone or the surface fold of the strip may damage the roller surface. Moreover, local thermal shock, or in cases where the high-speed continuous mill becomes out of control, may also destroy the roller. According to statistics, such damages are about 20–50% of the total roller damage.

16.4.4 Emulsion Lubricity in Rolling

In recent years, the emulsion is widely used for lubricating and cooling in rolling. Qian *et al.* experimentally studied the lubrication performances of the emulsion in rolling [6].

The experiment was carried out on a mixed lubrication tester. The block is made of rolled mild steel and the disc is made of No. 52100 steel with the hardness of HRC58-62. They form the line contact friction pair and their surface roughness are $R_a = 0.130\ \mu\text{m}$ and $R_a = 0.157\ \mu\text{m}$ respectively.

There are three types of experimental lubricants. Their volume fractions are respectively 2.0%, 3.5% and 5.0% of N54 oil with water as the rolling emulsion for the cold-rolled strip. The experiments were carried out under the same concentration with three kinds of loads: $P = 7.65$ N, 17.45 N and 27.25 N respectively, and the rotational speed is 0–400 r/min. The resistance method is used to measure the surface contact time rate in the mixed lubrication. The relationships of the oil volume fraction φ_0 (%) and the contact time rate with the rotational speed are as shown in Figure 16.24.

From Figure 16.24, it can be seen that according to the rotational speed, the curves under different working conditions can be divided into three zones: the low speed zone (0–150 r/min), medium speed zone (151–300 r/min) and high speed zone (301–500 r/min). In the low-speed zone, the contact time rate is nearly equal to 1, indicating that it is the continuous contact at the boundary lubrication condition. In the medium-speed zone, the contact time rate is between 0 and 1, showing that it is the intermittent contact in the mixed lubrication state. In this region, the contact time decreases sharply with increase in speed. In the high speed zone, the contact time rate is almost zero so there is a full lubricant film, indicating full-film hydrodynamic lubrication. If we increase the rotational speed further, the lubrication state has no significant change.

In summary, under sufficiently high speed and oil phase volume fraction conditions, the rolling emulsion shows good lubricating performances. It reduces the surface contact time of the mixed lubrication.

References

1. Ru, Z., Yu, W., Ruan, X.H., and Meng, X.T. (1992) *Plastic Working Tribology*, Science Press, Beijing.
2. Li, X.H. (1993) *Friction and Lubrication in Press Working Process*, Metallurgical Industry Press, Beijing.
3. Grudev, A.P., Zilberg, Yu.V., and Tilik, V.T. (1982) Friction and Lubricants during Metal Forming. Moscow: Metallurgia.
4. Meng, Y.G., Liu, X.Z., and Chen, J. (1998) Investigation on the effect of ultrasonic vibration on reduction in drawing force. *Journal of Tsinghua University*, **38** (4), 28–32.
5. Meng, Y.G. and Wen, S.Z. (1993) A finite element approach to PHD in cold forging. *Wear*, **160**, 163–170.
6. Qian, L.M., Meng, Y.G., Huang, P., and Shi, X.L. (1996) Monitoring the transition of lubrication status of O/W emulsion with electric resistance technique. *Tribology*, **16** (3), 239–246.

17

Bio-Tribology

There are a lot of friction phenomena in the human body, such as friction in the joints, the lumens (blood vessel, trachea, digestive tract, etc.), the muscle, and the tendon.

Since the 1980s, bio-mechanics, bio-chemistry, bio-rheology, bio-tribology and other disciplines within biology have been rapidly developing [1].

In bio-tribology, the study of lubrication mechanism of the human and animal joints as well as artificial joints has developed particularly rapidly. In recent years, research in joint lubrication and wear behavior have obtained many significant advances.

17.1 Mechanics Basis for Soft Biological Tissue

17.1.1 Rheological Property of Soft Tissue

In order to understand the physiological function of an organ, we should first understand the rheological behavior of soft tissue. Biological tissue is the composite material so its rheological property can be regarded as the mechanical property of the composite material. According to the stress range, the bearing elements are different. The constitutive equations during loading and unloading are different and have remarkable anisotropy.

Figure 17.1 shows the stress and strain curve of the muscle fiber after contraction. The solid line is the theoretical curve drawn by Equation 17.1. It meets with the experimentally measured results quite well:

$$\sigma = \frac{1}{3} G\sqrt{N}\left[L^{-1}\left(\frac{\gamma}{\sqrt{N}}\right) - \gamma^{-3/2}L^{-1}\left(\frac{1}{\sqrt{\lambda N}}\right)\right], \tag{17.1}$$

where σ is the stress before deformation; λ is the strain; γ is the constant; $G = nkT$; n is the chain number of the mesh per volume; k is the Boltzmann constant; T is and the absolute temperature; N is the random chain number of the mesh; L is the Langevin function: $L(x) = \cot\text{h}\, x - 1/x$; L^{-1} is the inverse function of L.

17.1.2 Stress-Strain Curve Analysis

Figure 17.2 is the measured stress-strain curve of an animal heart to be loaded and unloaded along the axial direction with the same strain rate.

Figure 17.3 shows the curve of T and $dT/d\lambda$ of an animal aortic muscle slice in the longitudinal direction, where T is the tension load across the sectional area and λ is the strain.

Principles of Tribology, First Edition. Wen Shizhu and Huang Ping.

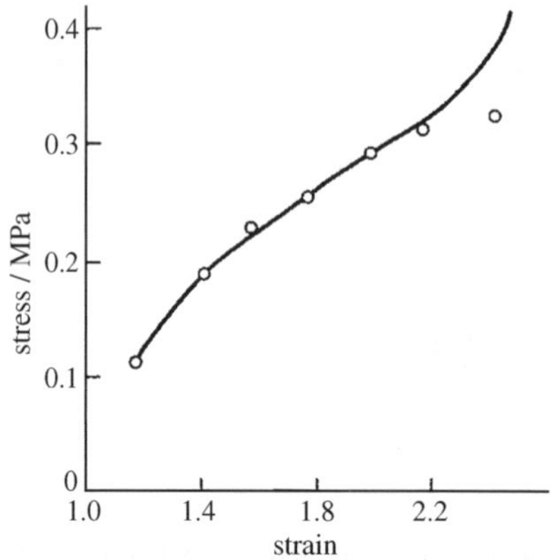

Figure 17.1 Relationship of stress and strain of muscle fibers after contraction [2].

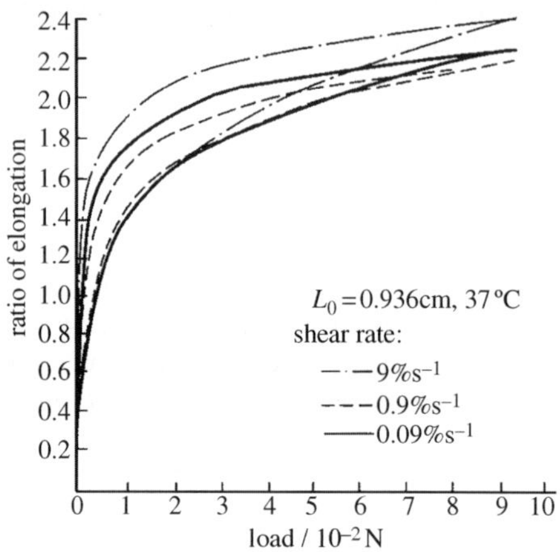

Figure 17.2 Load-ratio elongation curve of animal heart [2].

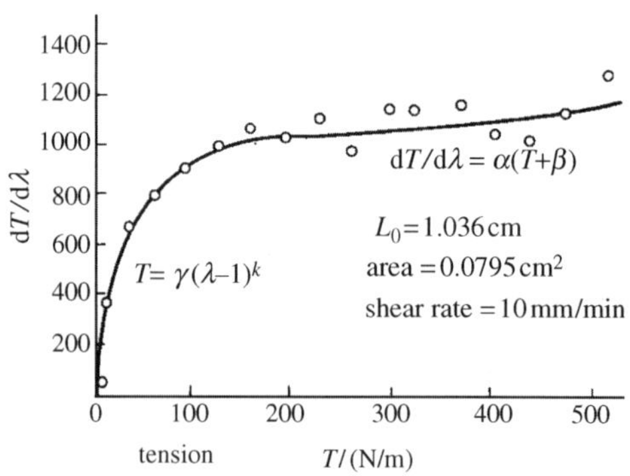

Figure 17.3 Curve of $dT/d\lambda - T$ of animal aortic muscle slice [2].

As shown in Figure 17.3, when the tension $T > 200$ g/cm, we can use the following equation to calculate the tension rate:

$$\frac{dT}{d\lambda} = \alpha(T+\beta). \tag{17.2}$$

Integrating the above equation, we have

$$T = (T^{*}+\beta)e^{\alpha(\lambda-\lambda^{*})} - \beta, \tag{17.3}$$

where T^{*} and λ^{*} are the initial tension and strain.

Similarly, from Figure 17.3, as $T < 200$ g/cm, we can obtain the relationship of the tension and the strain as follows:

$$T = \gamma(\lambda-1)^{k}. \tag{17.4}$$

The above relationship of the stress and the strain can be used for aortic as well as for mesentery, skin, catheter, heart and so on. An important feature of the relationship is that the curves of the loading and unloading are not the same. In other words, the parameters α, β, γ, and k are different in the loading or unloading processes.

The other feature of the stress and strain curve of the biological tissue is that the influence of the strain rate is very little whether during loading or unloading. In Figure 17.2, the variation of the strain rate of the different curves is about 100 times, but the curves vary little. Therefore, the specimen is stretched in the sinusoidal tension and the hysteresis loop does not change with the frequency. Fung *et al.* carried out the tensile measurements at period ranging from 1 s to 1000s. The results showed that the above conclusion for mesentery, arteries, skin, muscles, catheter, and so on are correct [3].

17.1.3 Anisotropy Relationship

Most biological tissues are anisotropic. At present, the data of the rheological properties of the soft tissues are mostly measured in one-dimensional condition, such as the tensile test with a slender round tubular specimen. In order to study the anisotropic characteristic of the tissue, two-dimensional measurement must be used.

Fung *et al.* carried out the two-dimensional tensile test on the animal stomach skin [3]. The results are shown in Figure 17.4. Their two-dimensional tensiles are applied to the two perpendicular directions respectively. Choose the longitudinal and the lateral directions of the abdominal skin as the axes of x and y. The corresponding stresses are σ_{xx}, τ_{xy} and σ_{yy} and the strains are ε_x and ε_y. In the figure, the curves are obtained while ε_x or ε_y are fixed:

$$\begin{aligned}
&\left(\partial\sigma_{xx}/\partial\varepsilon_y\right) \sim \sigma_{xx}\\
&\left(\partial\sigma_{yy}/\partial\varepsilon_x\right) \sim \sigma_{yy}\\
&\left(\partial\tau_{xy}/\partial\varepsilon_y\right) \sim \tau_{xy}.
\end{aligned} \tag{17.5}$$

Furthermore, their experiments on the animal mesenterium showed that the shear elastic modulus is not a constant but increases with increase of the stress. If the relationship between the stresses τ_{ij} and the shear strains γ_{ij} are written as follows,

$$\tau_{ij} = G\gamma_{ij}, \tag{17.6}$$

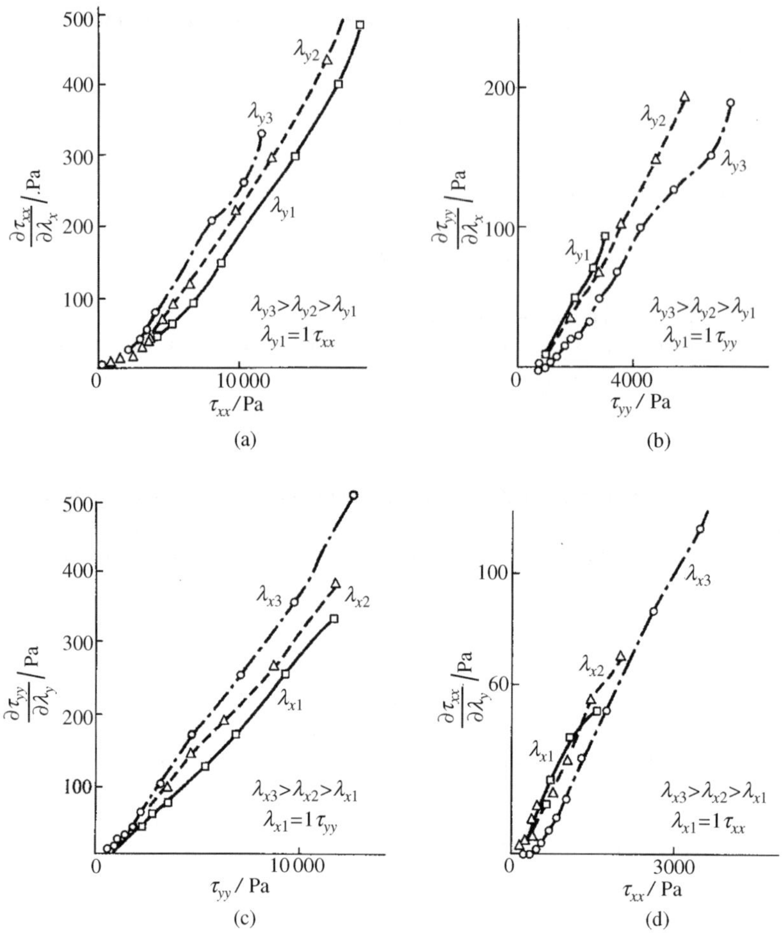

Figure 17.4 Two-dimensional tensile results of animal abdominal skin [3].

the experimental data can be expressed as:

$$G = G_0 + f_1(I_1) + c_2 I_2^{1/2} \tag{17.7}$$

where G_0 and c_2 are the constants; I_1 and I_2 are the first and second stress tensor invariants, respectively:

$$I_1 = \tau_{11} + \tau_{22} + \tau_{33}, \tag{17.8}$$

$$I_2 = \frac{1}{6}\left[(\tau_{11}-\tau_{22})^2 + (\tau_{22}-\tau_{33})^2 + (\tau_{33}-\tau_{11})^2\right] + \tau_{12}^2 + \tau_{23}^2 + \tau_{31}^2. \tag{17.9}$$

17.2 Characteristics of Joint Lubricating Fluid

17.2.1 Joint Lubricating Fluid

The joint lubricating fluid includes mucin or glycoprotein. They are not simple viscous liquid but visco-elastic liquid with drawing character. Arthritis will significantly influence the drawing character of the

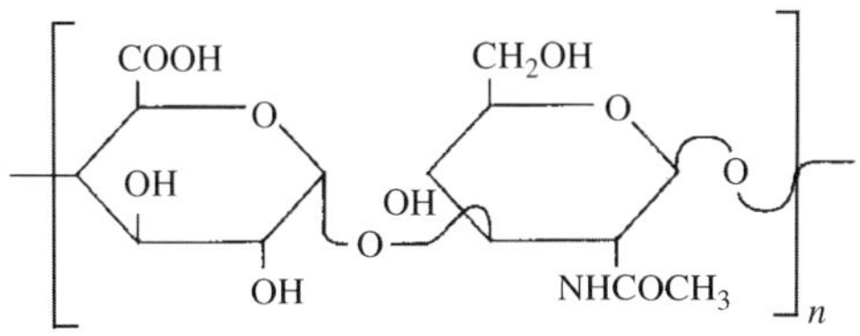

Figure 17.5 Molecular structure of hyaluronic acid [4].

joint lubricating fluid. For example, the protein or ground potato-like substance is also the viscoelastic fluid with drawing character. The so-called drawing character is like steamed moldy beans which can be extracted. In colloid chemistry, this fluid elasticity is called flow elasticity.

Joint lubricating fluid is dialyzed from the plasma. It does not contain fibrinogen so it cannot solidify. The main component of mucin is hyaluronic acid, and it mainly brings about the visco-elasticity of joint lubricating fluid. Hyaluronic acid is usually combined with protein in the complex form as shown in Figure 17.5.

In this complex, the protein is about 2%. If hyaluronic acid does not decompose or compose, it is very difficult to separate the protein. Therefore, the visco-elasticity of the hyaluronic acid is not obviously related to the formation extent of the complex. The hyaluronic acid molecular weight is about 10^6 and its molecule is composed of randomly curling spiral chains. Under the influence of external elements, such as chemicals, radiation, heat, solvents, and heavy metal ions, the molecular chain can be easily decomposed or composed. Because the shear stress can decompose the hyaluronic acid molecule, its viscosity measured under the static or the dynamic state is not the same.

17.2.2 Lubrication Characteristics of Joint Fluid

Below, we will introduce the research results of Myers *et al.* on joint lubricating fluid [4]. The experimental device is the torsional vibration concentric viscometer. It often works in the low frequency of 5–25 Hz. In the gap between the inner and outer cylinders the joint lubricating fluid is filled. The outer cylinder is acted by the sinusoidal torsional vibration with amplitude of 3°. From retarding the amplitude and phase angle of the inner cylinder, the dynamic viscosity η' and dynamic shear elastic modulus G' of the fluid can be obtained, which are the functions of the frequency. The joint fluid specimens are taken from patients and stored below 0 °C.

Generally, the relationship of the stress and strain of a dynamic visco-elastic fluid can be expressed by the following complex:

$$\frac{\tau}{\gamma} = G' + iG''. \tag{17.10}$$

If the complex viscosity is η, $\eta\dot{\gamma} = \tau$. If the angular frequency is ω, $\dot{\gamma} = i\omega\gamma$. Therefore,

$$\eta = \frac{G''}{\omega} - i\frac{G'}{\omega}. \tag{17.11}$$

The dynamic viscosity η' is equal to

$$\eta' = \frac{G''}{\omega}. \tag{17.12}$$

As the dynamic viscosity η' of the joint lubricating fluid varies with the angular frequency ω and the temperature T, the measured results should be converted to the value at 25 °C according to the time

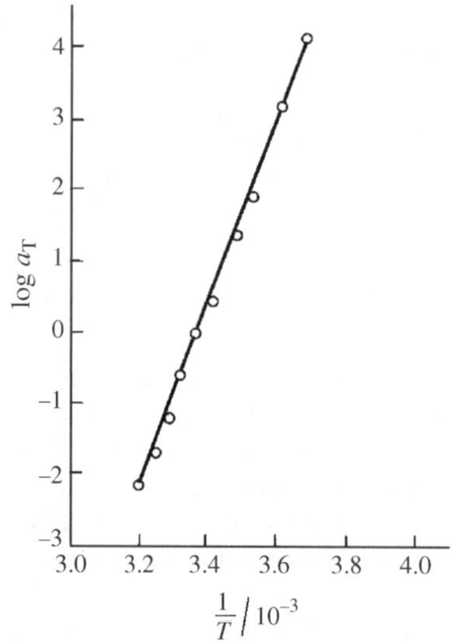

Figure 17.6 Relationship of log a_T and $1/T$ for joint lubricating fluid [4].

and the temperature conversion regulation. If the static viscosity at 25 °C is η_0, the reduced factor a_T of the static viscosity η at T is equal to

$$a_T = \frac{\eta}{\eta_0}. \tag{17.13}$$

Figure 17.6 shows the relationship of log a_T and $1/T$.

Figure 17.7 shows the relationship between the dynamic viscosity η' and the frequency. Figure 17.8 shows the relationship between the dynamic shear modulus G and the frequency. It can be seen that η' decreases as the frequency increases, while G increases when the frequency increases.

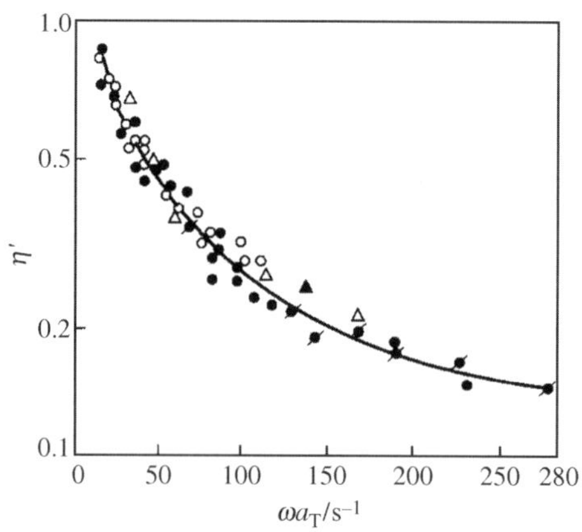

Figure 17.7 Dynamic viscosity of joint lubricating fluid [4].

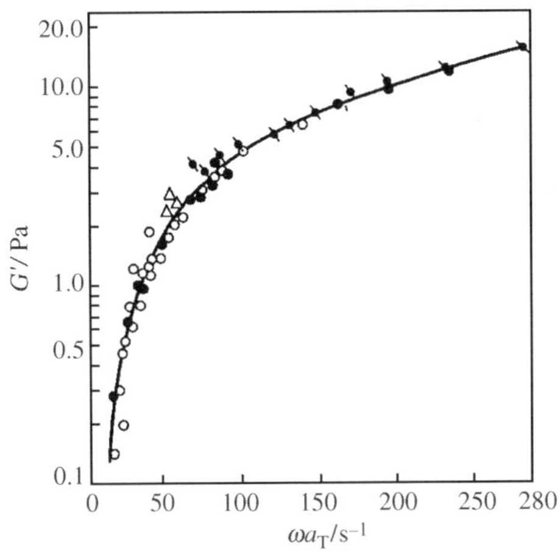

Figure 17.8 Dynamic shear modulus of joint lubricating fluid [4].

Since $\omega \to 0$, G decreases sharply, the joint lubricating fluid has no yield stress. We can see from Figure 17.9 that the static viscosity depends on the shear strain rate and therefore the fluid joint lubrication is non-Newtonian. When $\dot{\gamma} \to 0$ and $\omega \to 0$, the static viscosity limit and the dynamic viscosity limit are equal.

The above experimental results show that when the joint lubricating fluid works under low frequency, it is similar to a viscous liquid, while it tends to be an elastomer when the joint lubricating fluid works in the high-frequency region.

Although the hydrocortisone does not directly affect the static viscosity of the joint lubricating fluid, the dynamic viscosity η' and the elastic modulus G increase when the hydrocortisone is injected into the joint lubricating fluid.

The dynamic viscosity of the joint lubricating fluid depends on the concentration of the hyaluronic acid and partly depends on the formation of the complex of the hyaluronic acid and protein. On the other hand, the dynamic elastic modulus of the joint lubricating fluid is proportional to the formation of the complex.

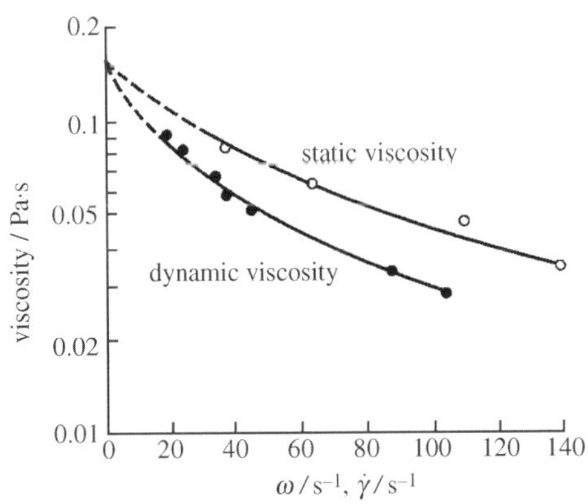

Figure 17.9 Static and dynamic viscosities of joint lubricating fluid [4].

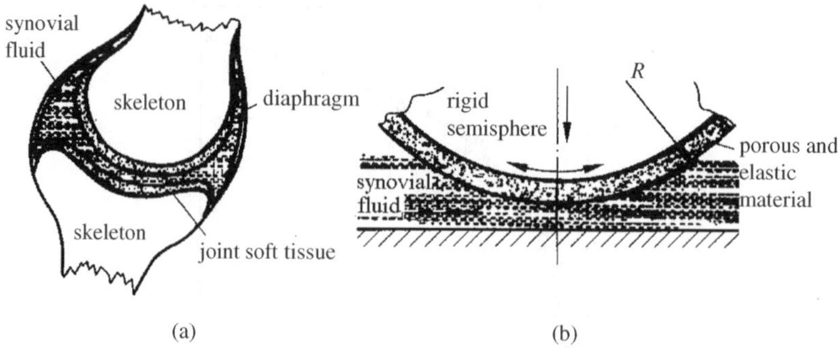

Figure 17.10 Lubrication model of human joint [4]: (a) human articular; (b) equivalent lubrication model.

17.3 Lubrication of Human and Animal Joints

Figure 17.10a gives a schematic of the human joint, which roughly draws the knee, the hip and spine. The bone of the joint is used to transfer the load. Its end can be expressed by a sphere or an oval which is used to provide the load-bearing area. In some other cases, a cylinder (knee) may be more appropriately used.

Figure 17.10b gives the equivalent joint lubrication model. The bone surface of the joint is covered by a layer of soft or porous articular cartilage tissue, which is the supporting material. The upper and lower cartilages are separated by the lubricating fluid so the lubricating fluid is contained in the diaphragm and supplies the necessary lubrication to the joint.

The normal human joint has a friction coefficient in the range of 0.001–0.03, which is far less than that of the hydrodynamic journal bearing. The early explanation of the joint lubrication was based on the hydrodynamic effect, but it was soon discovered the explanation is incorrect because the actual friction coefficient is extremely low and the relative velcosity between the bone surfaces is no more than a few centimeters per second such that it is impossible to generate a sufficient load-carrying capacity.

The boundary lubrication, the secretion or the elasto-hydrodynamic lubrication play important parts in lubrication of the joints. The squeezing lubrication film exists between the cartilage tissues and prevents the bone surfaces from contacting each other. The film thickness is especially thin, for example, 0.25 μm. The boundary lubrication mechanism of the human and animal joints can also be considered as the main part, in addition to the secretion and the EHL effects. The latter as the squeezing effect plays a complementary part.

17.3.1 Performance of Human Joint

The human articular cartilage is the smooth cartilage lining the surface of the grease of the joint bone. Its function is to absorb the wear debris caused by the joint movement, decrease the friction by lubricating, and transfer the load inside the body. The thickness of the articular cartilage is different for different joints. In addition, it may be different in the different positions on the surface of the same joint. In young people, the thickness of the large articular cartilage can reach up to 4–7 mm, the average thickness of the small articular cartilages is between 1 and 2 mm.

The structure of the cartilage consists of single cells distributing throughout the three-dimensional mesh of the collagen fiber in the bone tissue. The mesh is buried in chondroitin sulfate and is filled with a liquid, which is dispersed throughout the solid skeleton and is attached to the skeleton with the help of molecular attraction. Some liquid elements are firmly connected with the fiber structure, mostly in the clearance of the fiber. The molecules of the liquid pass through the cell base to transport under the

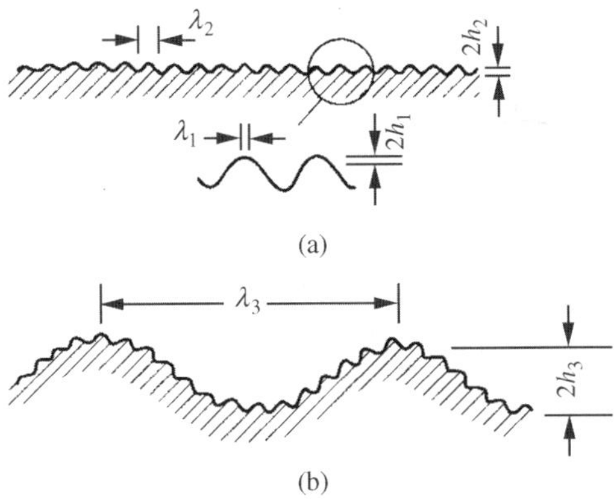

Figure 17.11 Typical surface roughnesses of joint soft tissues [4]: (a) healthy young; (b) elder.

combined action of the flow caused by the pressure gradient and the diffusion caused by the non-uniform chemical concentration. Consolidation and expansion are two main physical and chemical processes in the cartilage tissue. They are the results of the liquid flowing into and out of the cell tissue. The consolidation process occurs when the external compressive force is added to the cartilage causing the decrease of liquid. The expansion occurs when the liquid content increases. The variation rate of the liquid determines the time of the variation of the cartilage thickness. The most important structural feature in the articular cartilage is porosity. Usually, the average diameter of the pore is about 6 nm. The section of the pores in the cartilage surface plays an important part in lubrication. With a scanning probe apparatus, the cartilage surface roughness can be obtained. Usually, the surface texture is much rougher than that in engineering. Figure 17.11 shows two surface profiles of the cartilage. Figure 17.11a is the profile of a healthy young adult while Figure 17.11b is from an elderly person. The macro-texture of the former is obviously corrugated and the micro roughness is superposited on the macro roughness. The latter demonstrates larger macro fluctuation.

17.3.2 Joint Lubricating Fluid

The joint lubricating fluid is a transparent, yellow and sticky substance, existing in the cavity pores of the free moving joint. It can realize lubrication with interaction to the cartilage tissue. From the engineering viewpoint, we can take the lubricating fluid structure composed of the sticky acid protein to form a honeycomb mesh tissue wall which comprises the water-like component. Therefore, the lubricating fluid in the healthy state is of a sponge-like structure. The chemical composition analysis shows that the fluid is plasma dialysis with sticky protein acid and a tiny cell-like component. The most important property of the lubricating fluid is viscosity, which seems to be related to the components of the protein acid. As a boundary lubricant, the sticky protein acid can affect the friction property of the cartilage.

The joint lubricating fluid is non-Newtonian. It is obviously of the shear-thinning property, that is, the viscosity linearly declines with increase of the shear rate. If the film thickness of the cartilage surfaces is less than about 1 μm, the liquid molecule significantly affects the sliding properties. This is the characteristic of the boundary lubricant. It is because the lubricating fluid condenses under the action of the load so a gel forms on the surface of the cartilage. The viscosity of the gel is much higher than that of the bulk fluid. The formation of the gel is because the low viscosity fluid passes through the sponge-like structure of the cartilage under the normal load. The gel-like lubricating fluid is trapped in the recess of the cartilage surface. The recess serves as storage in order to maintain the necessary boundary lubrication.

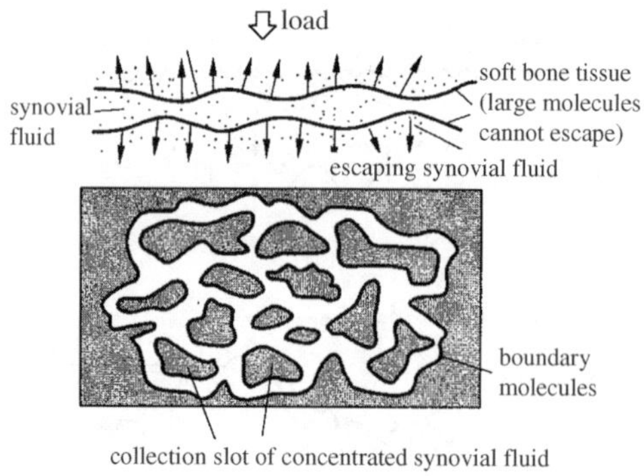

Figure 17.12 Contact zone of soft tissue containing lubricating fluid (synovial fluid) [4].

Figure 17.12 gives the secretion effect of the boundary lubrication. The lubricating fluid secretes from the pores of the cartilage, and is trapped and isolated in the pit by the gel.

17.3.3 Lubrication Mechanism of Joint

As mentioned above, the lubrication mechanism of the joint is very complex. In the joint, the hydrodynamic effect is weak, but under the light load condition (such as swing in walking), the two surfaces of the cartilage may be separated by the hydrodynamic lubrication film. This effect can be considered as a complement to the main part of the squeezing effect.

The squeezing effect of two parallel surfaces approaching each other can be expressed as the following formula:

$$t = \frac{K\eta L_T^4}{W}\left(\frac{1}{h^2} - \frac{1}{h_0^2}\right), \tag{17.14}$$

where t is time; K is the surface shape factor; η is the lubricant viscosity; L_T is the equivalent length; W is the load; h is the thickness; and h_0 the initial film thickness.

The above equation is the general form of the squeezing lubricating film, which can be applied to the joint situation. It can be seen that when the squeezing film thickness h is thin enough, the approaching time tends to infinity. It is amazing that the squeeze time observed in the human joint experiment is much larger that that predicted in Equation 17.14. This is because the viscosity of the joint lubricating fluid will substantially increase. For example, when $t = 40$ s and the minimum film thickness is equal to the diameter of the sticky protein acid molecule (about 0.5 μm), it can be found that the average viscosity is 20 Pa·s. It is much greater than that of bulk viscosity of the lubricating fluid, about 0.01 Pa·s. The result shows that in squeezing, a very thick material or gel forms on the surface of the cartilage, that is, the small molecules can leak out from the pores, but the large gel-like molecules are left behind, as mentioned earlier in Figure 17.13. Before swing in walking, the loading time is less than 1 s. It is certain that before the hydrodynamic effect works, the thinning of the squeeze film is very weak. If one stands for a long time, the squeezing film effect will produce a thick gel material to provide boundary lubrication such that the friction coefficient still remains low, approximately 0.15. Figure 17.13 gives the comparison of the squeezing effects of the sick and healthy cartilages. The ratio of the time reaching the boundary lubrication state to the squeezing time is used to show the effect, where the squeezing time is defined as the time that the lubricating fluid of the cartilage tissue drops from the initial fully saturated state to the dry friction state under the squeeze load.

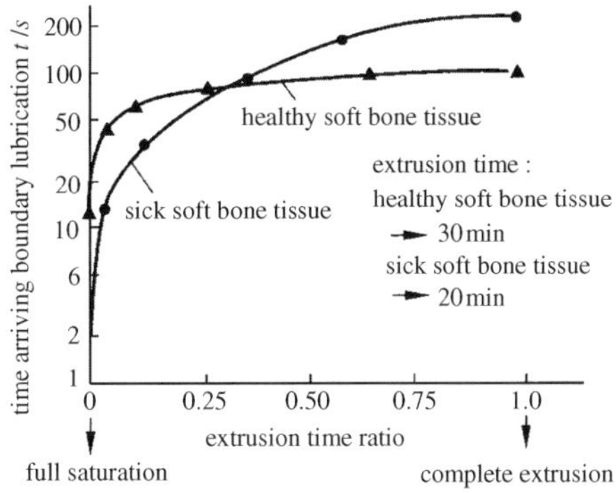

Figure 17.13 Comparison of compression performances of healthy and sick cartilages [4].

From Figure 17.13, it can be seen that the time from the full saturation to the boundary lubrication gradually increases for the healthy and sick cartilages, but the sick cartilage needs much more time. This may be partly because the low viscosity lubricating fluid infiltrates to the porous structure and gelates so the healthy and sick cartilages have different load capabilities.

Figure 17.12 shows that if one stands for a long time, the liquid left around the pit edge is in the boundary lubrication condition and the edge size increases in the squeezing process. At the same time, because the cartilage surface is rough (see Figure 17.11) the elastohydrodynamic effect will emerge and it is stronger in the squeezing process than in boundary lubrication. Figure 17.14 shows the comparisons of the EHL film thickness, the molecular length of sticky protein acid, squeeze film and surface roughness of the cartilage.

The research on human joint lubrication will help the elderly to eliminate the negative influence of joint disease. Osteoarthritis is one of the common diseases affecting the hip and knee operations. Although its causes are not yet entirely understood, the pain and stiffness are closely related to the wear and lubrication failure of the cartilage. In order to prevent further deterioration, the synthetic lubricating fluid may be injected into the joint. If the viscosity of the synthetic lubricating fluid is high enough, it will form a thick film to separate the surface and the bone so as to reduce pain and wear. High viscosity requires high shear stress, which needs strong muscles. However, the muscle of the joint suffering arthritis has no ability to overcome the high shear stress to allow the joint to move freely.

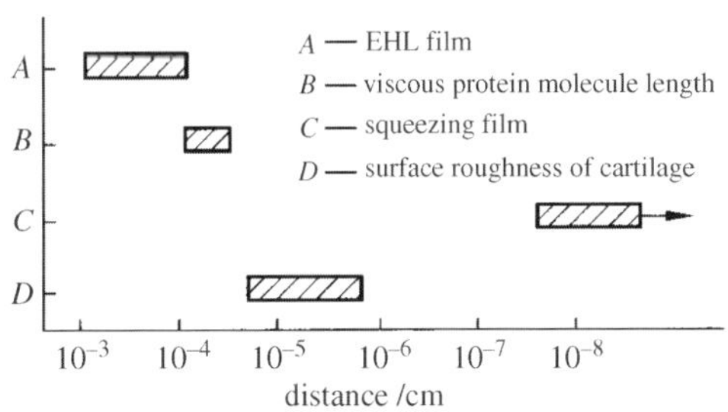

Figure 17.14 Relative sizes in human joint lubrication [4]. A: EHL film; B: sticky protein acid molecule length; C: squeeze film; D: surface roughness of cartilage.

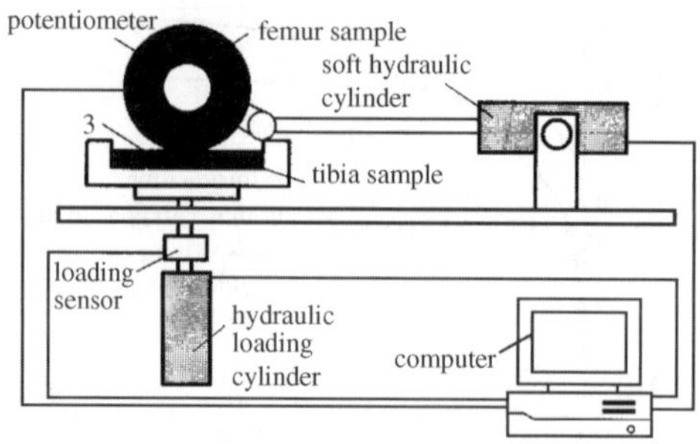

Figure 17.15 Knee simulating tester [5].

17.4 Friction and Wear of Artificial Joint

17.4.1 Friction and Wear Test

Teruo *et al.* carried out the artificial joint simulation test [5]. Their results are as follows.

1. Simulating tester
   The principles of the tester are as shown in Figure 17.15.
   The tester was designed for simulating the bending and stretching knee exercises and the walking movement of the tibia under the axial load. The load is applied by the two hydraulic vibration excitors at the side and the bottom. The frictional force between the femur and the tibia is obtained by measuring the torque of the femoral shaft. Figure 17.16 is the diagram of the structure and size of the sample.
   The cylindrical femur joint was made of SUS316 stainless steel, and the tibial component was made of an ultrahigh-molecular-weight polyethylene (UHMWPE) plate. During testing, PVA hydrogel and polyurethane materials are used for the artificial cartilages. The equilibrium water in the PVA hydrogel is 79%, the average polymerization degree of PVA is 2000 and the average saponification degree of PVA is 99%. The polyurethane resin is the medical material. The tibial component is made of polymethyl, an artificial cartilage material. The surface is made of polymethyl. Then, they are connected with cyanoacrylate. All tests were carried out at $14 \pm 1\,°C$.
2. Testing results
   Figure 17.17 shows the percentage curve of the torque of the UHMWPE tibial component and the stainless steel femoral component. The results show that the slurry protein can increase the friction in walking. The serum albumin raises the frictional force between the stainless steel femur and the UHMWPE tibia, while globulin has a more significant effect.

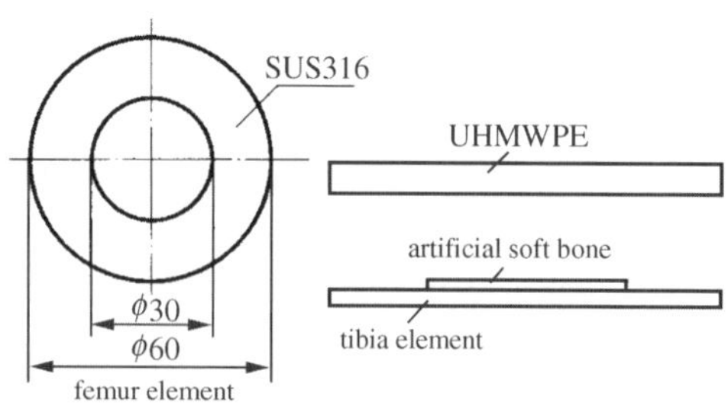

Figure 17.16 Structure and size of specimen [5].

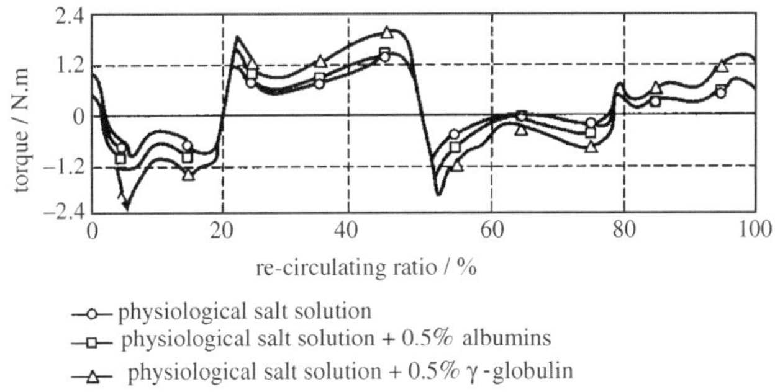

Figure 17.17 Torques in cycle [5].

The long-term experiments of the two artificial materials affected by the protein show that the frictional force of polyurethane is increased by the slurry protein, while the influence of globulin is more apparent. However, the slurry protein reduces the frictional force of PVA hydrogel.

17.4.2 Wear of Artificial Joint

When the human knee joint wears, it will produce abrasive particles. These particles are composed of the material of the articular surface. With the abrasive analysis, we can know the wear situation of the surface.

The ferrography analysis method can be used by magnetizing the abrasive particles in the synovial fluid. The results of ferrography analysis are introduced below [6].

For osteoarthritis and the rheumatoid arthritis patients, the wear debris can be extracted from their synovial fluid and analyzed by the ferrography method. By observing through the bi-color polarized light microscope and the scanning electron microscope, we can identify and distinguish the abrasive particles of osteoarthritis and rheumatoid bones or cartilage and fibrous tissue.

Combined with other synovial fluid analysis techniques, the above method can serve as a noninjury and repeated diagnosis method to study the abrasive wear mechanism and joint pathology.

17.4.2.1 Experimental Method and Apparatus

The testing process includes the magnetization and separation of abrasive particles in the joint synovial fluid. Then, use a ferrogram to observe and analyze under a bi-color microscope to gather information about the wear debris.

1. Preparation of pretreatment liquid

   First, extract 3 ml synovial fluid from the patient joint and dilute with an equal volume of saline. The synovial fluid of patients with rheumatoid arthritis should be treated with fungal hyaluronidase to avoid wear debris causing massive coagulation and sedimentation, but the synovial fluid of osteoarthritis patients does not need to be treated.

   After treatment, the sample is required to be maintained at 37 °C for about half an hour and then the diluted sample should be injected into a test tube to be centrifuged for 10 minutes at 6000 rpm. After the first centrifugation, remove the upper clear liquid and leave the bottom sediment. Then, add 1 ml saline solution and centrifuge again. Finally, the pretreatment liquid of the synovial fluid is obtained.
2. Preparation of bio abrasive ferrogram

   Add 1 ml Er^{+3} solution into the pre-processed solution for magnetization and fully oscillate it to suspend the abrasive. Send the sample liquid through the micro-pump (1), the magnetic field (2) and the

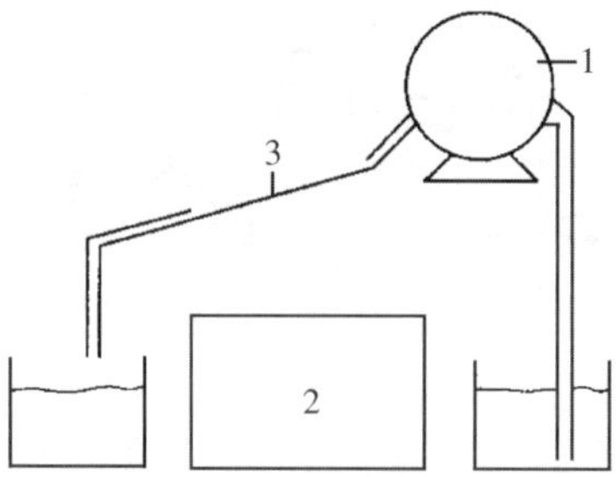

Figure 17.18 Ferrograph system [6].

glass substrate (3) of the ferrography as shown in Figure 17.18. After cleaning the residue liquid and fixing the particles, the ferrogram can finally be obtained.

Figure 17.19 shows the abrasive images under the bicolor microscope.

17.4.2.2 Testing Results

According to the ferrogram of the synovial fluid extracted from the human knee joint, it can be seen that the wear debris has a different shape and optical performances. The features can help us distinguish the abrasive debris type (such as bone, cartilage, and fibrous tissue). If we compare the sample with the normal bone, cartilage and fibrous tissue, we can distinguish the abrasive particles of the bone, cartilage or fibrous tissue by electron spectroscopy analysis to determine whether the osteoarthritis patient has the calcified cartilage abrasive layer in his synovial fluid or not.

According to the abrasive particle analysis of the human knee joint, the abrasive particles have the following characteristics:

1. The abrasive particles of the fibrous tissue are of high optical activity. They are thin strips 0–10 mm in length and usually deposited at the rear part of the ferrogram.
2. The abrasive particle of the cartilage is of mild optical activity and is yellow under polarized light. The osteoarthritis abrasive particles are often strips or blocks of a few microns to a few hundred microns in length. The rheumatoid arthritis abrasive particles are often blocks with a diameter of 10–25 μm. They often deposited at the rear part of the ferrogram.
3. The optical activity of the bone abrasive particles is high. They are dark under the polarized light and are generally granular or block. The abrasive particles of the osteoarthritis synovial fluid are fine. They deposit at the front of the ferrogram.

In the synovial fluid of the joint with osteoarthritis or rheumatoid arthritis, the cartilage abrasive particles are strips, blocks or balls. Among them, the ball abrasive particle often appears in rheumatoid arthritis. The strip and block abrasive particles of the cartilage are often in the synovial fluid of patients with osteoarthritis. Their surfaces are rough and the edges are sharp. Most of the particles are long and thin small flakes. They have typical characteristic of fatigue wear. This is because after the repeat friction between the articular cartilages, the contact surface is of high stress concentration. Under the periodical action of the load, fatigue peeling occurs on the surface.

The bone abrasive particles of the patient with osteoarthritis are significantly more than those of those patients with rheumatoid arthritis. The severer the osteoarthritis, the more bone abrasive particles there are. Abrasive particles are produced due to the following two reasons. One is that the articular cartilage is worn away to produce the crack which brings about bone cutting. The other is that one surface of the

(a)

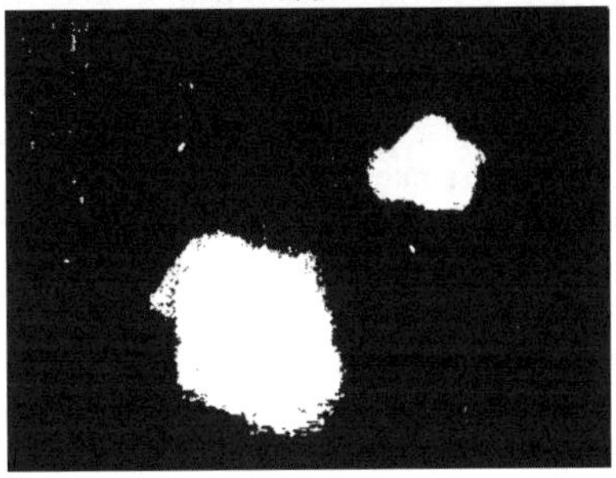

(b)

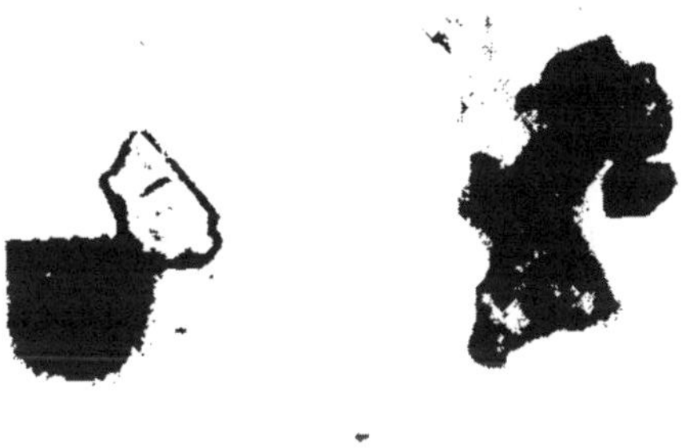

(c)

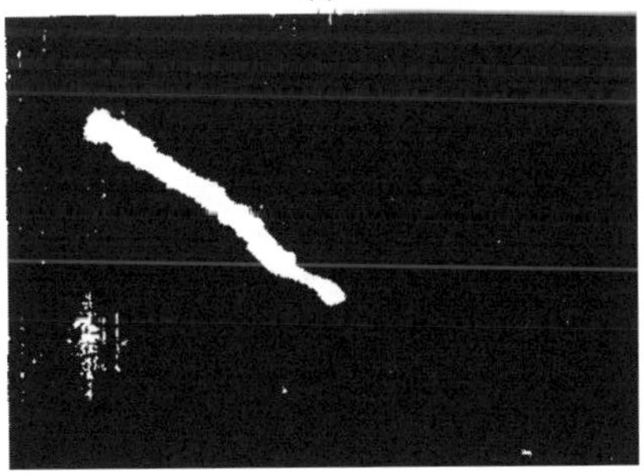

(d)

Figure 17.19 Images of bio-abrasive ferrograms [6]: (a) strip fiberous tissue of rheumatoid arthritis; (b) spherical abrasive particles of rheumatoid arthritis; (c) bone abrasive of osteoarthritis; (d) thin strip of osteoarthritis.

articular cartilage is worn away, and the protruding part embedded directly contacts with the other surface bone to plow a series of tiny abrasive particles. The abrasive particle surface is not smooth and presents flake or block, which are the typical characteristics of fatigue wear.

17.5 Other Bio-Tribological Studies

One important goal of bio-tribology is to develop low friction, small wear, and little pathological response artificial organs, mainly in artificial joints and the heart valve. These are sometimes referred to as bio-technology.

A large number of artificial joints have been used in late arthritis, trauma disabled and osteoma excised patients. According to the survey, there are 1–1.5 million patients needing artificial joint surgery in China [1].

Although the surgical short-term results after ten years are very satisfactory, the durability of the joint needs improvement. According to reports, 20 years after surgery for total knee and total marrow replacement patients experience a success rate of 93% and 86% [1].

Since the first artificial heart valve was used in 1960, artificial heart valves have been increasingly used to extend patients' life. However, heart valve material still needs improving because wear and fatigue caused by repeat friction may bring about severe results.

The human and animal joint lubrications have been discussed above. We only provide some basic principles of the knowledge. With these principles, other methods may be developed to improve prosthetic design, or the artificial joint design to restore the ill joint functions. Some other equally important examples include applications of tribological principles to blood vessels and blood flow in the capillary, body waste excretion as well as advanced membrane artificial heart valve and so on. Research in these areas will contribute to reduce blood coagulation, thrombosis, blood vessel expansion and heart disease.

Using a tribology system to measure skin disease is another applied biotechnology. Other practical techniques are at the developing stage. The friction coefficient of diseased skin is different from healthy skin. Therefore, it is very useful to design a portable instrument for skin friction testing, which can allow study of the friction properties of injured skin, such as burns, scars, and scratches.

References

1. Wen, S.Z. (2000) Central review and survey – the development tendency of tribology. *Chinese Journal of Mechanical Engineering*, **36** (6), 1–6.
2. Oka, S. (1980) *Bio Rheology*, Science Press, Beijing.
3. Fung, Y.C. (1973) Biorheology of soft tissues. *Biorheology*, **10** (2), 139–155.
4. Myers, R.R., Negami, S., and White, R.K. (1966) Dynamic mechanical properties of synovial fluid. *Biorheology*, **3**, 197–209.
5. Teruo, M. and Yoshinori, S. (1998) Effect of serum proteins on friction and wear of prosthetic joint material. Proc. of First Asia International Conf. On Tribology, Beijing, China, 2, pp. 828–833.
6. Gu, Z.Q. (1998) A preliminary ferrographic study of the wear particles in synovial fluid of Human Knee Joints. Proc. of First Asia International Conf. On Tribology, Beijing, China, 2, pp. 838–841.

18

Space Tribology

The progress of space technology has promoted the usage and development of tribology. The vehicle in space is exposed to an environment with low pressure, radiation, atomic oxygen and no gravitation. In the 1950s–1960s, the life of space equipment was only several tens of minutes or hours. Because electronic technology was still underdeveloped and most electronic devices were tubes and single electronic devices, these devices were of a high failure rate so the life of the mechanical part was not the main problem. However, with the development of large-scale integrated circuits and computers, the reliability of electronic devices has been greatly improved. The space working period has increased to more than ten years, some or even several decades long. Therefore, the life of mechanical components has become an urgent problem to be solved in space technology. Although tribological devices cost only a small portion of the space vehicle, their failure may cause the failure of the entire space vehicle and therefore they are comparatively more expensive.

Figure 18.1 is the growth of tribology requirements with advances in space given by Kannel and Dufrane [1]. The figure shows that although tribological technology has made some significant progresses in space, the development of tribology in space does not adequately meet the needs of space missions.

In this chapter, the characteristics of the problems of space tribology are first introduced. Then, the common phenomena in the space tribology, such as volatile, motility, and parched lubrication, are analyzed in detail. Finally, the characteristics of space lubricants and rolling contact bearing lubrication technology are discussed.

18.1 Features of Space Agency and Space Tribology

18.1.1 Working Conditions in Space

The lubrication instruments in a spacecraft include solar array, the torque wheel, reaction wheel, filter wheel, tracking antenna, scanning device and sensor. These instruments need separate lubrication. Table 18.1 shows the speed ranges of some space instruments and their working conditions.

A gyroscope usually works at the speed of 8000–20 000 rpm with high accuracy. The bearings are the most important parts of the gyroscope. The bearings present the fluctuation of the torque. The noise and heat will cause the bearings to fluctuate, thus increasing the gyroscope bias. The lubricant used in the gyroscope should be of high wear resistance, low friction and low evaporating rate. Furthermore, fixed and small amount of lubricant (3 mg) are often used in gyroscope lubrication. Because the universal joint of the gyroscope runs at low speed, it belongs to the boundary lubrication state.

Principles of Tribology, First Edition. Wen Shizhu and Huang Ping.

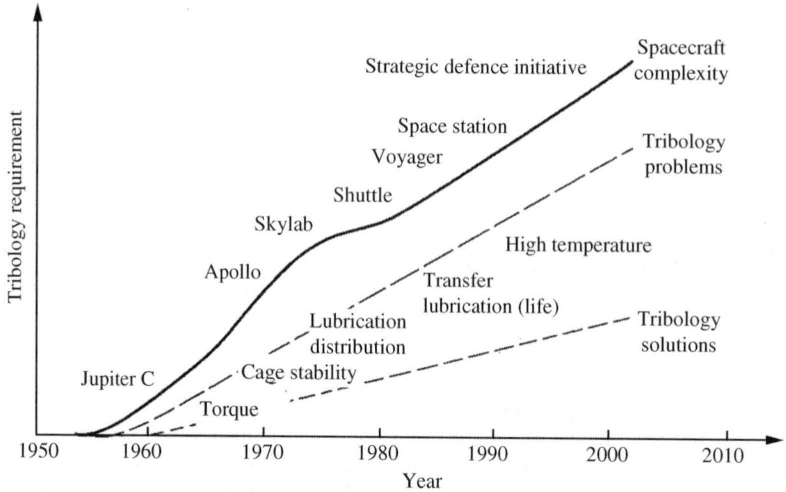

Figure 18.1 Growth of tribology requirements with advances in space [1].

The normal working speed of the torque wheel is 3000–10 000 r/min. The main problems in torque wheel lubrication are inadequate lubrication, lubricant loss, lubricant degeneration and so on. These are the main reasons for a torque wheel failure. The torque wheel works at high speed so the lubricant should be able to withstand a high operating temperature. The high temperature increases the rates of creep and degeneration. Methods to reduce the lubricant loss include using synthetic lubricant, labyrinth seal, plating leak-proof coating and so on.

The reactor is similar to the torque wheel but its working speed is slow. Therefore, most of its bearings work in mixed lubrication. So, the reactor lubricant should have good boundary lubrication properties.

A control moment gyroscope (CMG) is a combined device of the gyroscope and torque wheel, which can be used to control the attitude of a space vehicle. Therefore, when selecting a lubricant, both devices need to be taken into consideration.

The scanning and rotation sensors are another type of space agency which need lubricating. The horizontal scan sensor is such an example. It is used to measure the horizon of the Earth to let the space craft operate itself. Because its bearings work at the medium speed (400–1600 r/min) and under low load, the lubricant is easy to choose. On the other hand, the sensor in the swinging movement needs lubricating too. Because the swing angle is small, the external lubricant cannot be brought into the contact zone so the bearing works in the boundary lubrication condition.

Table 18.1 Speed ranges and working conditions of some space instruments

Agencies	Rotational speed r/min	Working conditions
Gyroscope	8000–30 000	High-speed, EHL of rolling contact bearing
Torque wheel	3000–10 000	Medium speed
Scanner device	400–1600	Medium and low speed
Tracking antenna	100	Low-speed, boundary lubrication
Reaction wheel	10	Low-speed, boundary lubrication
Filter wheel	$<\pm 10$	Low-speed, boundary lubrication
Sensors	minor swing	Low-speed, possible lubricating
Sliding ring	several to 20 000	Large speed range; difficulty in lubricating at low speed

The sliding ring is another application which also needs lubricating. It will normally works at the speed of 100 rpm. However, when the operation speed is low or high (several or 20 000 rpm) the speed and conductivity are the two important factors affecting lubricant choice. If the electrical noise is excessive, it may cause the failure of the sliding ring. This is usually due to surface contamination. Therefore, we should select the appropriate lubricant to reduce surface contamination.

There are many other space agencies needing lubricating, such as the solar array drive (SAD) which rotates the solar panels of the space vehicle, balls, rollers, trapezoidal screws and a variety of gear transmission equipment.

18.1.2 Features of Space Tribology Problems

Space tribology is the tribological branch to study the reliability of the satellite and space vehicle. It covers almost all the normal tribological conditions, such as hydrodynamic lubrication, elastohydrodynamic lubrication, parched lubrication, mixed lubrication and boundary lubrication. However, because of the uniqueness in space, many lubrication conditions bring about some different problems [2]. Table 18.2 lists several major space tribology phenomena.

Volatility refers to the process that a material from the solid or liquid state changes into the gas or steam state. The absolute pressure determines the return rate of the evaporation molecules. In the absolute vacuum, the molecules do not return so the lubricant continues to lose the molecules. In most space cases, the lubricant is sealed in a container so that a pressure balance can be set up. However, the higher the vapor pressure of the lubricant, the faster the escape rate is. Therefore, the vapor pressure of the lubricant should be chosen to be as low as possible.

Viscosity is extremely important in influencing tribological properties. With increase of the lubricant viscosity, the evaporation rate decreases. However, a low viscosity also meant low resistance and thus slipping occurs easily. High viscosity can reduce slip but increases the resistant torque.

The creep of a lubricant refers to the phenomenon that the lubricant freely expands on the contact surface without any action. Generally, the creep is caused by surface tension and viscosity. The greater the surface tension and the higher the viscosity, the slower the creep rate is. Generally, a lubricant cannot climb over the steep edge, the slag dam and the large thermal gradient region.

Reactants are lacking in the space environment. In the atmospheric environment metal is oxidized and forms the oxide film, which significantly reduces the friction coefficient. In the space environment, the lack of substances such as oxygen molecules means a new oxide film cannot be formed after the initial oxide film on the surface is worn away. This significantly increases the friction coefficient, leading to lubrication failure.

Table 18.2 Features of parts in space

Phenomenon	Result	Measures
Volatilization	Loss of lubricant by evaporation	Choose lubricant with low vapor pressure and evaporation rate; use seal, tight tolerance and protective gaskets
Low temperature	Relative sliding due to high viscosity, large torque, low stability	Choose lubricant with low viscosity and small temperature index
Creeping	Loss of lubricant by different liquid-gas and liquid-solid tension and wetting	Select lubricant with large surface tension, control temperature distribution uniform
Lack of reactants	No oxide layer, high friction coefficient, easy surface scuffing	Add oxidation and extreme pressure additives into lubricant
Radiation	Lubricant decomposition by ionization and replacement of organic lubricant	Remove lubricant from radiation as far as possible

The space radiation is strong, due to ultraviolet and X-rays which are very harmful to the organic lubricant. These may cause the lubricant to be ionized or excited to a high electronic energy state that increase the reaction ability of the material. In addition, absorption of the ultraviolet radiation may lead to cross-coupling, chain scission and random breakage of the molecular chain. Infrared radiation may cause the lubricant to thermally decompose. Usually, we can isolate the lubricant from radiation.

Atomic oxygen exists in the low orbit atmosphere of the Earth, which can rapidly react with the carbon molecule (one of the main components of the lubricant) to form unstable oxides. The polymeric materials such as epoxy, polyimide resin and polyimide can also react with atomic oxygen to bring about some negative impacts.

In addition, condensation, the non-gravity condition, heat conduction, cosmic dust, space impurities and other factors may also affect the normal work of lubrication equipment and cause the spacecraft to work improperly.

18.2 Analysis of Performances of Space Tribology

18.2.1 Starved Lubrication

Elastohydrodynamic lubrication, mixed lubrication and boundary lubrication can occur in space instruments. In space, because oil cannot be continuously supplied or reduces because of leakage, the space parts often work under "starved" lubrication conditions. The theory of starved lubrication was put forward many years ago and described the condition where there is a limited oil supplement. Because the hydrodynamic pressure cannot be generated far away from the inlet, the actual lubrication film thickness is thinner than theoretically predicted [3]. The author has carried out a systematical study with an optical interferometry to show the formation of elastohydrodynamic lubrication film under starved lubrication conditions. The thickness is in the order of 10 nm [4].

Reference [5] analyzed the starved lubrication problem of elastohydrodynamic lubrication on the point contact if the oil supplement is decreased. The results are shown in Figure 18.2. When the oil supplement is adequate enough, the oil leakage is also more severe. The film thicknesses h_{oil}, h_{cen} and h_{min} decrease rapidly when the operation time increases, where h_{oil}, h_{cen} and h_{min} are the initial, the center and minimum film thicknesses respectively. With increase of the lubrication number n, the supplement oil gradually decreases. Now, almost all the oil flowing into the inlet flows out of the outlet. Then, the film thickness tends to be stable. The effective film forms at the position close to the Hertzian contact zone. EHL gradually keeps the stable state which is the extreme starved lubrication.

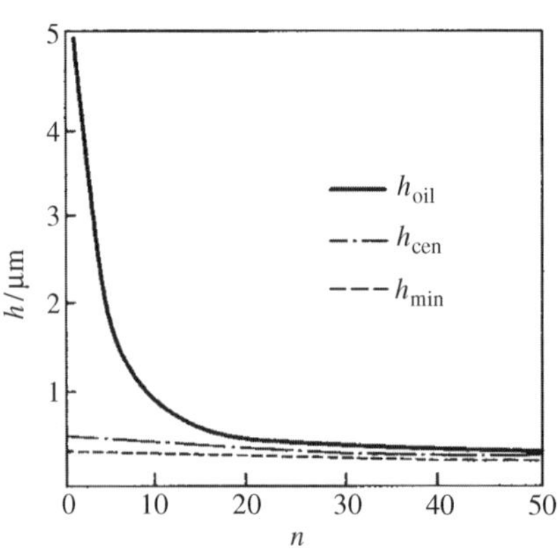

Figure 18.2 h_{oil}, h_{cen} and h_{min} via lubrication number n.

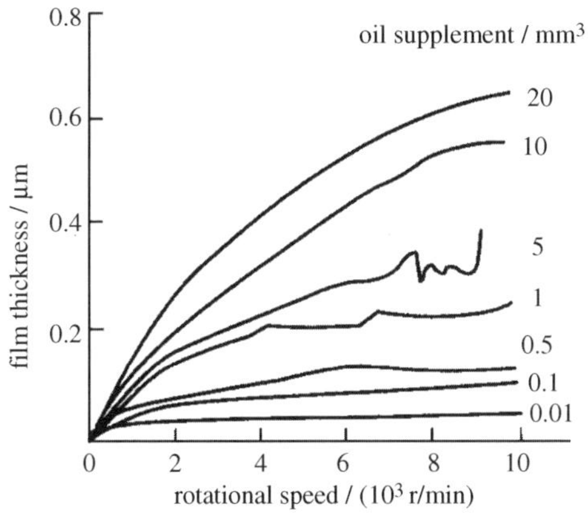

Figure 18.3 Film thicknesses of different oil supplements [6].

However, the starved EHL theory is sometimes difficult to describe some phenomena. For example, in the starved lubrication, the necking phenomenon cannot be observed. However, such a phenomenon can be explained by the parched EHL lubrication.

18.2.2 Parched Lubrication

The parched lubrication theory believes that under the extreme lack of oil, there is no free lubricant. The parched lubricant film is very thin and it is fixed in the narrow Hertzian contact zone. This state is particularly important to the space instruments because the parched lubricating bearing needs the minimum driving torque and it can rotate very precisely around the axis.

The parched lubrication is commonly found in the self-lubricating retainer and the grease lubrication. Akagami, with a very small amount of oil to lubricate a bearing, measured film thickness with the deformation method. He found that when the oil supplement decreases, oil film becomes thinner [6]. As shown in Figure 18.3, when the speed increases, the film does not increase. His further study also showed that such a thin film can be kept a long time, as shown in Figure 18.4.

Liu and the author carried out a series of analysis on parched lubrication [7]. The results showed that the lubrication behaviors are related to the rotational speed and the oil supplement (see Figure 18.5). Under high speed and lack of oil, parched lubrication appears when the oil supplement is extremely limited.

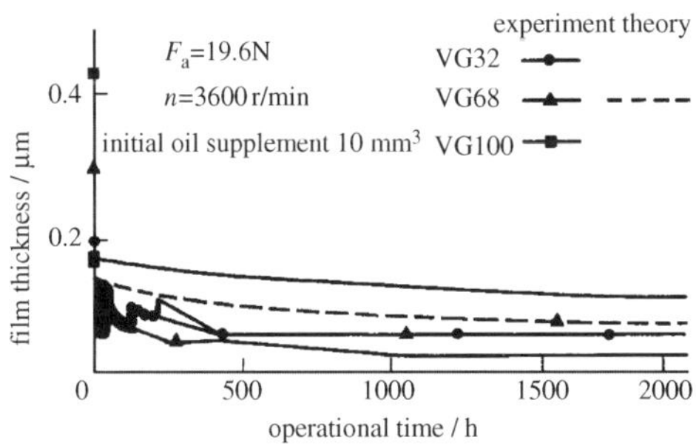

Figure 18.4 Film thicknesses during long period experiment [6].

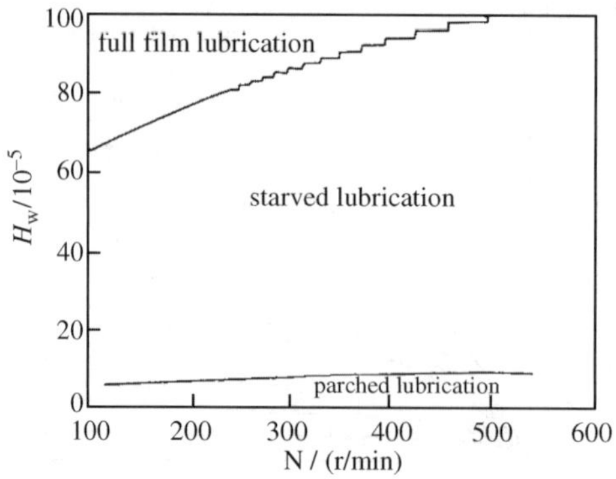

Figure 18.5 Lubrication state with initial thickness and speed [7].

Under this condition, the hydrodynamic film will be completely destroyed. The following formula can be used to determine the lubrication state [7]:

$$\frac{h}{h_w} = 1 - c_1 H_w^p H_D^q W^r k^s e^{-uT}, \tag{18.1}$$

where h is the thickness; h_w is the initial thickness far away from the inlet, H_w is the nondimensional form of h_w; h_D is the EHL film thickness by Dowson formula, H_D is the non-dimensional form; W is the dimensionless load; k is the ratio of the curvature radius; T is the dimensionless temperature; c_1, p, q, r, s and u are constants, and their values are referred to in reference [7].

In addition, the bearing cage can increase the necessary lubricating oil in the Hertzian contact zone. This plays an important role in maintaining parched lubrication.

EHL is important to space instruments but it is in the transition or unstable state. However, many space mechanical parts, such as bearings, gears, cams and traction drive components, work in such an unstable condition. The load, speed and contact geometry are not constants during working. For example, a stepper motor works in this state and is used in many parts of the space instruments.

18.2.3 Volatility Analysis

Although the labyrinth seal is widely used in space instruments, the loss of lubricant is still a problem for space instruments which are expected to work over a long period of time. If a certain temperature and the outlet shape, the lubricant loss is directly related to the evaporation pressure. The influence of the temperature is significant on the evaporation rate. Because most low-Earth orbit satellites operate at the temperature of 280–320 K, we should pay particular attention to the volatilization problem.

The reason that leads to volatilization can be analyzed by thermodynamics. Away from the atmosphere, the absolute pressure is about 10^{-11} Pa, which is a "near-vacuum" environment. In the first year in the orbit, a satellite usually has a water vapor pressure larger than 10^{-5} Pa. Therefore, the volatilization speed of the liquid lubricant is great. For a certain liquid film, we can use the Langmuir expression to estimate the volatilization rate [2]:

$$R = \frac{d_m}{d_t} = \frac{p}{17.14}\sqrt{\frac{m}{T}}, \tag{18.2}$$

where R is the volatilization rate; p is the saturation pressure; m is the molecular weight; and T is the temperature, K.

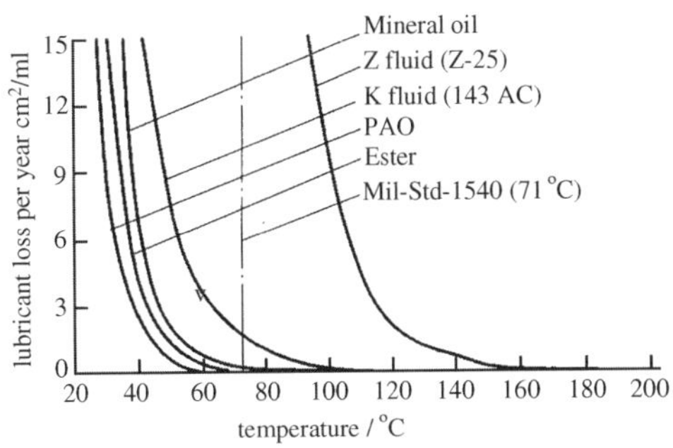

Figure 18.6 Relative loss rates of some space lubricants [8].

The balanced sub-pressure is a thermodynamic function of the specific molecule. When the velocity of the molecule returning to the surface is equal to the velocity of the molecule moving away from the surface, the sub-pressure is balanced. The absolute pressure does not affect the balanced sub-pressure of the solid or liquid, but the absolute pressure decides the returning velocity. This is because the evaporated molecule may collide with the other molecule and return to the surface. The molecule in the absolute vacuum does not return so all volatilized molecules are lost forever. Therefore, the balance cannot be reached. Strictly speaking, only in the immense vacuum with no container can such a situation appear. In the majority of the space vehicles, the lubricant is inclusively sealed in a container, though the high vapor pressure means that the volatilization speed will be high. Therefore, we should choose the low vapor pressure lubricant. The different types of mineral oils have different saturated pressures. A mineral oil can rapidly volatilize thoroughly in a few minutes or up to a day, but the volatility of fluorinated polyethylene is much lower, as shown in Figure 18.6.

The seal approach and the lubricant escaping amount from the oil lubrication system are as follows. Under normal operating conditions, there are two ways to consume oil. First, through the rotating parts of a shaft, the oil is pumped out in the steam form or directly transmitted from the hot area to the cold area in the vapor form. The other way is via surface transmission. For the first way, a reliable labyrinth seal must be designed at the rotational shaft end.

Figure 18.7 is the sealing structure of an annular pipe, where r_1 is the inner diameter; r_2 is the outer diameter. Because for gas molecules, the mean free path λ is much larger than the shaft radius, the annular pipe gas derivative $C(L/s)$ can be expressed as follows:

$$C = 30.48\sqrt{\frac{T}{M}\frac{(r_2^2 - r_1^2)(r_2 - r_1)}{l}}, \tag{18.3}$$

where T is the absolute temperature; M is the molecular weight of gas, g; l is the length of the pipeline.

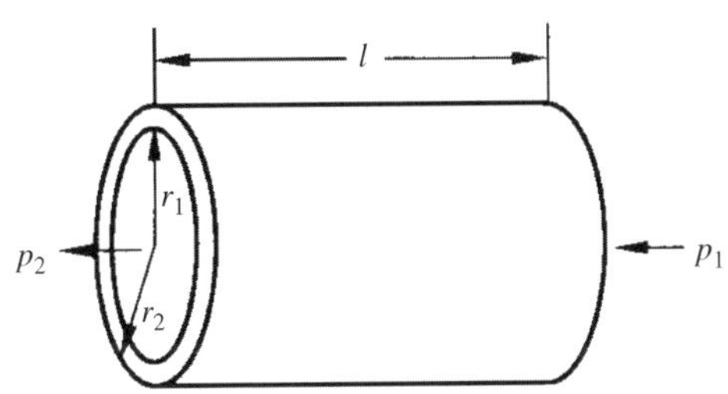

Figure 18.7 Seal of annular pipeline [9].

The gas flow Q of the annular pipe is equal to:

$$Q = C(p_1 - p_2), \quad (18.4)$$

where, $p_1 - p_2$ is the pressure difference on the two sides of the annular pipe.

From Equation 18.4, we can see that the escaping gas flow from the annular pipe is related to the pressure difference and the air conduction. In the outer space, $p_2 \approx 0$ so the pressure difference on both sides of the annular pipe depends primarily on the internal gas pressure of the rotating device. According to the above analysis, by simply choosing a low saturation vapor pressure lubricating oil, controlling the gap and the length of the rotational shaft, and filling with enough lubricating oil in the motor storage device and bearing cage, a suitable saturated vapor pressure can be kept for a long time to achieve bearing lubrication.

In order to increase the sealing effect, the ring circumference and the end labyrinths can be used together. Then, depending on the structure and with a different formula, we can calculate the total air conduction C of the labyrinth. If the saturated vapor pressure is p, the in-orbit time is t, and air conduction is C, during the period in the orbit, the total escaping lubricating oil rate is equal to $q = Cpt$. And, through the ideal gas equation, the total amount of escaping lubricant can be obtained. Finally, depending on the working life (usually several decades long), we can determine the dipping cage oil volume.

18.2.4 Creeping

The tendency for the liquid lubricant to creep on the bearing surface is inversely proportional to its surface tension. The surface tensions of various liquids are shown in Table 18.3. The surface tension of PFPE is very low so it creeps more easily than conventional fluids, such as hydrocarbon materials, esters, and silicone oils. These fluids can be left in the bearing raceway with the fluorinated carbon barrier film with low surface energy. However, the PFPE fluid will dissolve the barrier film during long-time contact. Therefore, they cannot effectively prevent the creeping of PFPE.

The Pennzane lubricant has a high surface tension so its creeping trend is weak. For example, the chlorophenyl silicone oil has the surface tension of 2.06×10^{-2} N/m. The surface tension of most metals can reach 1 N/m or above. Therefore, the adhesion work between the bearing raceway, steel ball surface and oil molecules is larger than the cohesive work of lubricant molecules such that the interaction between liquid and solid can make the liquid spread on the solid surface. This phenomenon is more significant in space where gravity is particularly small.

Therefore, the lubricant can quickly form a film to cover all the bearing raceway and steel ball surface. Also, the polar substances in the lubricant oil and the additives can be adsorbed on the bearing raceway and steel ball surface to form the physical adsorption film or react with the metal elements to form the chemical adsorption or reaction film to lubricate. However, the spreading can cause lubricant loss which is called the creeping loss. Hence, it is necessary to take some anti-creeping measures.

The lubricant creeping is also affected by the temperature gradient. As shown in Figure 18.8, super-refined mineral oil KG280 creeps at 2.2 °C thermal gradient and in a gravity-free environment [1]. This figure shows that the oil creeps from the hot area to the cold.

Table 18.3 Surface tension of some liquids (20 °C)

Liquid	Surface tension (mN/m)	Liquid	Surface tension (mN/m)
Pure water	72	PAO	28.5
Mechanical oil	29	Dioctyl sebacate	31
Pentaerythritol ester	30	Pentaerythritol tetraacetate	24
PFPE	20	Methyl silicone oil	21

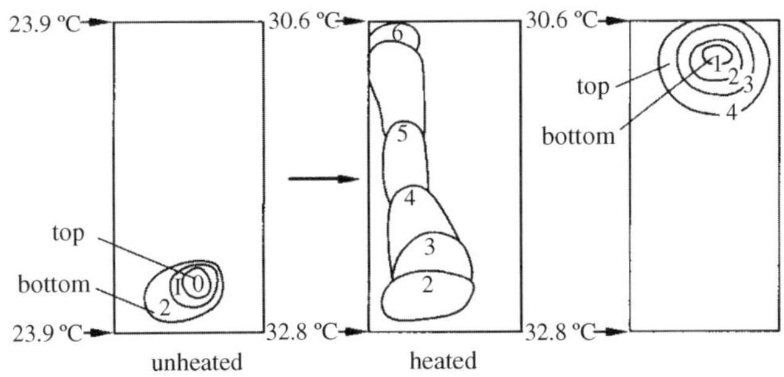

Figure 18.8 Creeping mode of KG-80 at 2.2 °C temperature gradient [1].

The film ratio, the minimum film thickness to the surface equivalent roughness, is an important parameter to express the characteristic of a bearing lubrication [3]:

$$\lambda = h_0/\sigma \tag{18.5}$$

where σ is the surface equivalent roughness, $\sigma = \sqrt{\sigma_1^2 + \sigma_2^2}$; h_0 is the minimum film thickness which can be estimated by the following formulas:

$$h_0 = 0.04(\phi GU)^{0.74} W^{-0.074} R_x \tag{18.6}$$

where ϕ, G, U and W are the nondimensionless parameters; R_x is the equivalent radius.

Figure 18.9 shows the relationship between the bearing fatigue life and the film ratio λ. The figure shows that when λ is equal to 1.5, it can be used as the design fatigue life. When λ is smaller than 0.5, the bearing will not work properly.

The loss caused by the oil creeping is because the surface tension of oil is much lower than the metal surface tension such that it is of benefit to lubrication because the lubricant can quickly spread over the raceway and the ball surface. However, this also makes the lubricant creep out of the labyrinth mouth. In order to prevent the loss, a layer of the low surface energy material should be plated on the labyrinth mouth and the bearing cover so that the surface tension is smaller than the surface tension of the lubricant. Therefore, the lubricant cannot lubricate the layer; or at least, spreading cannot be extended.

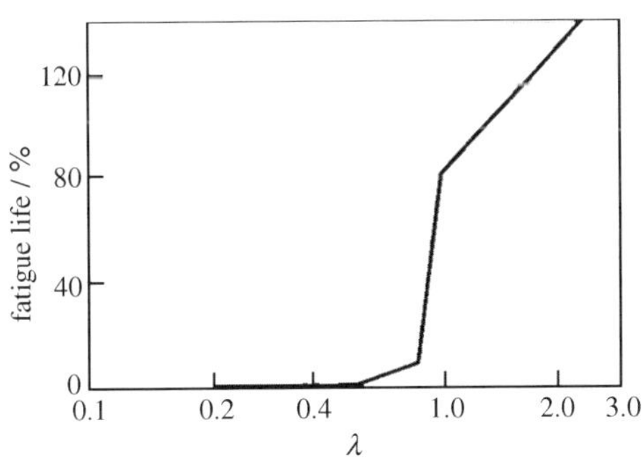

Figure 18.9 Relationship between fatigue life and λ [1].

Many materials can be used to prevent creeping, such as the alkyl silanes with the high fluorine functional group. When it forms the high-fluorine dodecanoic acid monolayer, its surface critical tension is only 6×10^{-3} N/m so it is able to effectively reduce the creeping loss of the lubricant.

18.3 Space Lubricating Properties

18.3.1 EHL Characteristics of Space Lubricant

In medium or high speed operation, the parts in the line and point contacts may form an elastic hydrodynamic lubricant film. For EHL, after the surface dimensions and topography of the parts have been determined, the lubricant film thickness is related to working conditions. The classical EHL film thickness calculation formulas in the line or point contacts have been given in Equation 18.7 or 18.8.

Line contact Dowson-Higginson formula:

$$H^*_{\min} = 1.6 \frac{G*^{0.6}\, U*^{0.7}}{W*^{0.13}}. \tag{18.7}$$

Point contact Hamrock-Dowson formula

$$H^*_{\min} = 3.63 \frac{G*^{0.49}\, U*^{0.68}}{W*^{0.073}} \left(1 - e^{-0.68k}\right). \tag{18.8}$$

The relationship of the lubricant viscosity, the working temperature and pressure are as follows:

Barus formula

$$\eta = \eta_0 \exp[\alpha p - \beta(T - T_0)]. \tag{18.9}$$

Reolands formula:

$$\eta = \eta_0 \exp\left\{ (\ln \eta_0 + 9.67) \left[\left(1 + 5.1 \times 10^{-9} p\right)^{0.68} \times \left(\frac{T - 138}{T_0 - 138}\right)^{-1.1} - 1 \right] \right\}. \tag{18.10}$$

According to the given geometric parameters, the lubricant properties, and the working conditions, the lubricant film thickness and the film ratio can be calculated. Therefore, we can determine the lubrication state of the part.

From the above it is known that the dynamic viscosity η and the pressure viscosity coefficient α are the two important physical properties to influence the formation of the EHL film. The molecular weight and the chemical structure of the lubricant will affect the viscosity. Except for low molecular weight fluid, α is also related to the structure of the lubricant. The pressure viscosity coefficient can be directly measured by using a traditional high-pressure viscometer or indirectly obtained from an optical EHL experiment. Table 18.4 gives the pressure–viscosity coefficient α of several space lubricants at temperature 38, 99 and 149 °C.

Table 18.4 Pressure viscosity coefficient α of several space lubricants at three temperatures ($\times 10^8\,\text{Pa}^{-1}$) [10]

Lubricant	38 °C	99 °C	149 °C
Ester	1.3	1.0	0.85
Synthetic paraffin	1.8	1.5	1.1
Z fluid (Z-25)	1.8	1.5	1.3
Naphthenic mineral oil	2.5	1.5	1.3
Traction oil	3.1	1.7	0.94
K fluid (143AB)	4.2	3.2	3.0

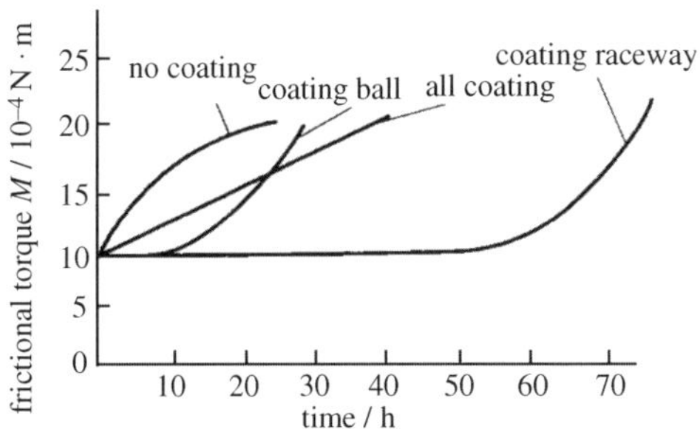

Figure 18.10 Bearing frictional torques of MoS_2 coating [11].

18.3.2 Space Lubrication of Rolling Contact Bearing

18.3.2.1 Bearing Coating

The mating of the coated bearing parts has a significant effect on the friction torque. Particular attention should be paid to the design and coating [11]. There are three kinds of coating on the bearing parts: (1) coating on the inner and outer raceways; (2) coating on the inner and outer raceway and rolling body; (3) coating on the rolling body. The measured frictional torques of the above three cases are shown in Figure 18.10. The corresponding working conditions are: the vacuum of 10^{-3}–10^{-4} Pa, the load of 25 N and the rotational speed of 800 r/min. It can be seen that only when the coating is on the inner and outer raceways, the bearing presents the best friction performance.

The common coating methods are the evaporation plating, the centrifugal plating and the RF sputtering. The ion plating is generally used for steel ball coating. The processing temperature is about 120 °C, the film thickness is about 0.3 μm, and the combined strength of the coating is strong. After coating, the steel ball still keeps a good precision. The coating materials are TiN, TiC or Ti (Al, V) N. The sputtering is generally used for plating the ring. The sputtering materials are the soft metals, such as Ag, Au and Pb, or nonmetals, such as MoS_2, PTFE and WS_2. The plating temperature is about 150 °C. In the short term, the surface temperature may reach up to 180 °C, and the coating thickness is about 1 μm. However, the sputtering may change the bearing dimensional accuracy and precision slightly, but the geometric accuracy, such as the roundness and ellipticity changes greatly. The hardness will also drops about 1–2 HRC. Because the coating changes the geometrical precision of the part, in order to maintain a high precision, the proposed heat treatment of the part is to temper it at 200 °C and keep it at 150 °C for the stability treatment. Because the load-carrying capability of the MoS_2 coating is large and the friction coefficient is small, it has a good lubricating performance. Especially in the vacuum, the friction coefficient and wear life are better than in the atmospheric environment so it is widely used in space. The commonly used form is the compound of MoS_2 + Au + rare earth elements.

18.3.2.2 Lubricant Film Transfer Technology

The method to improve the MoS_2 coating life of the ball bearing is to use the PTFE retainer at the same time. In the universal joint bearing life test, the life of the advanced MoS_2 coating with the synthesis PTFE retainer exceeds 45×10^6.

The bearing retainer made up of the compound lubricating material can deliver the lubricant to the rolling body and then to the raceway. Figure 18.11 has shown how the film is formed. Generally, this lubrication can only be used well under a light load condition. However, this technique has now been used in the turbo pump inside the space shuttle to induce rolling ball bearing lubrication, and has also been

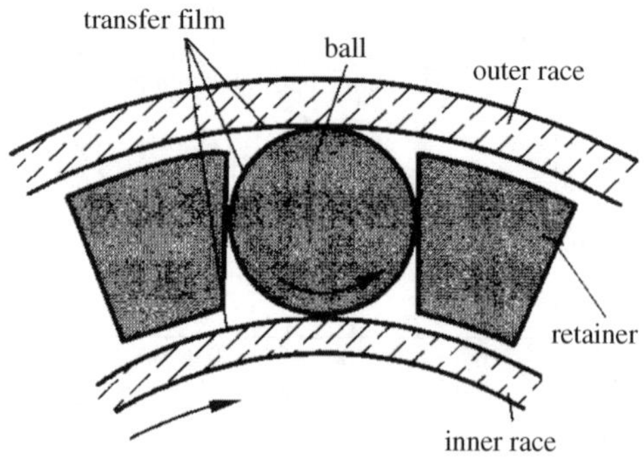

Figure 18.11 Ball bearing film transfer mechanism [12].

partially successful in the application of the liquid hydrogen pumps. Whether the technique works well in the liquid oxygen pump needs to be carefully studied.

18.3.2.3 Cage Instability

In the roller bearing, there is a cage instability problem, which is one of the most serious failure reasons. The expression of the instability is that the driving torque increases three times more than usual and the vibration is associated with a high noise. The instability may exist continuously or it may suddenly appear or disappear. Because there is noise at the same time, the instability is also known as the "howling". A space bearing with a small amount of howling can normally run for several years with no problem.

The typical mechanical frequency for the roller bearing is about several hundreds Hz. However, while howling occurs, the measured frequency is about several thousands Hz. Sometimes the higher frequencies are close to the first-order vibration model of the bending of the retainer.

Some people believe that the howling is caused by the whirl of the cage as a rigid body. The whirl model has become the basis for the stability analysis. It can explain many howling features. In order to effectively control the howling, the following measures can be used:

1. Use the porous polyimide as the cage material.
2. Divide the ball pocket holes into different intervals.
3. Use the mismatch steel balls (ball diameter difference $\leq 0.5\,\mu m$).

In order to control the cage howling, different cage pocket holes or different guide clearance fits can be used to find the correct solutions. These measures have obtained satisfying results to ensure completion of the corresponding mission.

References

1. Kannel, J.W. and Dufrane, K.F. (1986) Rolling Element Bearings in Space. The 20th Aerospace Mechanisms Symposium, NASA CP-2423, pp. 121–132.
2. Yao, Z.X., Huang, L.F., and Huang, J. (2005) The environmental factors to affect the performance of lubrication for space application. *Lubrication Engineering*, **169** (3), 155–157.
3. Wen, S.Z. and Huang, P. (2002) *Principles of Tribology*, 2nd edn,Tsinghua University Press, Beijing.
4. Huang, P., Luo, J.B., and Wen, S.Z. (1994) NGY-2 interferometer for nanometer lubrication film thickness measurement. *Tribology*, **14** (2), 175–179.

5. Tan, H.E., Yang, P.R., and Yi, C.L. (2007) Analysis of starvation in elastohydrodynamic lubrication point contacts with degradation of the oil-supply condition. *Lubrication Engineering*, **32** (4), 50–54.
6. Liu, C.H. and Li, J.D. (1999) Latest trends in research and development of NSK rolling contact bearing the. *Bearings*, **10**, 34–39, 8.
7. Liu, J.H. and Wen, S.Z. (1992) Fully flooded, starved and parched lubrication at point contact system. *Wear*, **159** (1), 135–140.
8. Conley, P. and Bohner, J.J. (1990) Experience with Synthetic Fluorinated Fluid Lubricants. Proc. of the 24th Aerospace Mech. Symp., NASA CP-3062, pp. 213–230.
9. Yuan, J. (2006) Lubrication of long duration medium to high speed ball bearing using in space effective. *Lubrication Engineering*, **175**, 156–158.
10. Jones, W.R. Jr., Johnson, R.L., Winer, W.O. *et al.* (1975) Pressure-viscosity measurements for several lubricants to 5.5 × 108 newtons per square meter (8 × 104 psi) and 149C (300F). *ASLE Transactions*, **18** (4), 249–262.
11. Liang, B., Ge, D., and Xi, Y.K. (2001) Aerospace solid lubricating bearing technology. *Bearings*, **5**, 8–12. 45.
12. Brewe, D.E., Scibbe, H.W., and Anderson, W.J. (1966) Film-Transfer Studies of Seven Ball-Bearing Retainer Materials in 60°R (33°K) Hydrogen Gas at 0.8 Million DN Value. NASA TN D-3730.

Index

abrasive wear 264, 327, 435
additives of lubricating oil 142, 177, 191, 321
adhesive wear 272, 328
adhesive wear theory 308
adsorption heat of boundary film 187
analysis of space tribological properties 442
analysis of wear failure 365
appearance and structure of surface coating 339
autocorrelation function of surface profile 225

basic characteristics of friction 233
bead weld technique 331
boundary condition of Reynolds equation 28
boundary element method 50
boundary lubrication model 186
boundary lubrication theory 177, 186, 385

cam lubrication 119
cavitation erosion 292
chemical absorption 178
chemical reaction film 183
choice of lubricants and additives 321
circle constitutive equation 12
common surface coating methods 331
constant delivery pump 94
contour map 219
corrosion wear 289
creeping 446
curves of wear processes 300

density of lubricating oil 6
detection of wear states 363
directional factor model 166
double electric layer effect 143
drawing tribology 406
dynamic bearing 100
dynamic journal bearing 101
dynamic viscosity 7

EHL inlet analysis 36
elastic deformation solution 36, 67
elasticity theory basic of contacts 30
elastohydrodynamic lubrication 36, 62, 155, 349
electro brush plated coating 331
energy wear theory 309
equivalent radius of curvature 30, 33
error of distribution curve 224
extreme pressure additive 191

fatigue wear 280, 311, 329
fatigue wear on subsurface 280
fatigue wear on surface 280
features of space lubrication 448
film thickness formula of EHL 67
finite difference method 44
forge tribology 401
friction and wear of joint 434
friction control 257
function of soften surface 183
function of thin polish film 183

gas bearing 107, 168
gear lubrication 113
grease lubrication 40

heat conduction equation 57
height distribution function 222
high stick thick film 185
hydrodynamic lubrication 23, 27, 154
hydrostatic bearing 92

Principles of Tribology, First Edition. Wen Shizhu and Huang Ping.

hydrostatic journal bearing 94
hydrostatic thrust disk 93

IBM wear calculation method 313
influence of fluid limiting shear stress on lubrication failure 202
influence of geometric quality on wear 304
influence of temperature on lubrication failure 207
initial condition of Reynolds equation 28

journal bearing 82, 94, 99, 101

kinematics viscosity 7

liquid crystal lubrication 135
liquid lubricant of joints 426
local thermal stable and unstable 208
lubricated wear model 199
lubrication failure 197, 202, 207
lubrication states 3, 153, 365

macro friction theory 236
magnetic fluid lubrication 123
materials of friction pair 298
mathematical model of thin film lubrication 166
measurement and exchange of viscosity 19
measurement of coating properties 339
measurement of wear 352
mechanical wear 263
mechanics basis of metal forming 397
mechanics on joints and soft tissue 423
men and animal joint lubrications 430
methods and equipment of tribological experiments 347
micro contact and adhesive phenomena 377
micro friction 246, 371
micro friction theory 246
micro wear 380
micro-polar fluid lubrication 129
milling tribology 415
mixed lubrication 177, 215
molecular film and boundary lubrication 385
molecular-mechanical wear 263
multi-grid integration method 76
multi-grid level method used in lubrication problems 70

nano film lubrication of gas 168
nano film lubrication of liquid 159

nonlinear visco-elastic constitutive equation 15
non-Newtonian properties 12
numerical analysis of TFL 165
numerical analysis of thin film lubrication 165
numerical method for solving ultra thin gas lubrication 168
numerical methods of Reynolds equation 44
numerical solutions of elastohydrodynamic lubrication 62
numerical solutions of energy equation 57

ordered phenomena of thin film lubrication 159

p_0U_s criterion 277
parameters of surface topography 219
parched lubrication 443
plowing effect 241, 375
pseudo plastic 12

Ree-Eyring constitutive equation 12
restriction throttle 94
Reynolds equation 24, 37, 101, 172
Reynolds equation derivation 24
rolling contact bearing 111, 449
rolling friction 254, 415
rough surface contacts 227

scuffing factor criteria 277
simple dynamic bearing 103
slider and thrust bearing 81
sliding friction 251
slurry coating technique 331
solid lubricant 325
Sommerfeld boundary condition 86
Sommerfeld integration 88
space tribology 439
spalling theory 311
squeeze bearing 96
squeezing disk 98
squeezing film journal bearing 99
starved lubrication 442
step bearing 82
strength of boundary film 190
superficial structure and surface properties 226
superposition method of carrying loads 104
surface absorption 186
surface coating 321, 331, 339
surface coating design 338
surface profile method 352

surface quality and wear 304
surface roughness 219
surface tension 15

test of surface coating performances 339
TFL and EHL testers 347
thin film lubrication 159, 165, 349
thixotropy 17
thixotropic feature of thin film lubrication 161
topographical parameters and statistics 222
topography analysis of friction surface 358
transformation of lubrication states 153
two-body abrasion wear 267
two-dimensional profile curves 219
types of boundary lubrication 177

V loop and *W* loop 70
variation of surface property 226
viscosity of fluids 7
viscosity-pressure effect 11
wear calculation 313
wear classification 263
wettability 17
WV_n criterion 277